Springer Handbook of Auditory Research

For further volumes:
http://www.springer.com/series/2506

Fan-Gang Zeng • Arthur N. Popper
Richard R. Fay
Editors

Auditory Prostheses

New Horizons

Springer

Editors
Fan-Gang Zeng
University of California–Irvine
Department of Otolaryngology - Head & Neck Surgery
Hearing & Speech Laboratory
Irvine, CA 92697
USA
fzeng@uci.edu

Arthur N. Popper
Department of Biology
University of Maryland
College Park, MD 20742
USA
apopper@umd.edu

Richard R. Fay
Marine Biological Laboratory
Woods Hole, MA 02543
USA
rfay@luc.edu

ISBN 978-1-4419-9433-2 e-ISBN 978-1-4419-9434-9
DOI 10.1007/978-1-4419-9434-9
Springer New York Dordrecht Heidelberg London

Library of Congress Control Number: 2011934480

Printed on acid-free paper

Springer is part of Springer Science+Business Media (www.springer.com)

We take pleasure in dedicating this volume to Dr. Robert V. Shannon, Director of Auditory Implant Research at the House Research Institute, Los Angeles, CA, in honor of his contributions to and leadership in the field of auditory prostheses for over three decades. In addition, Bob has been a wonderful mentor, colleague, and friend. Finally, we note that the publication of this volume coincides with Bob's Award of Merit from the Association for Research in Otolaryngology in 2011.

Fan-Gang Zeng, Arthur N. Popper, and Richard R. Fay

Series Preface

The Springer Handbook of Auditory Research presents a series of comprehensive and synthetic reviews of the fundamental topics in modern auditory research. The volumes are aimed at all individuals with interests in hearing research including advanced graduate students, post-doctoral researchers, and clinical investigators. The volumes are intended to introduce new investigators to important aspects of hearing science and to help established investigators to better understand the fundamental theories and data in fields of hearing that they may not normally follow closely.

Each volume presents a particular topic comprehensively, and each serves as a synthetic overview and guide to the literature. As such, the chapters present neither exhaustive data reviews nor original research that has not yet appeared in peer-reviewed journals. The volumes focus on topics that have developed a solid data and conceptual foundation rather than on those for which a literature is only beginning to develop. New research areas will be covered on a timely basis in the series as they begin to mature.

Each volume in the series consists of a few substantial chapters on a particular topic. In some cases, the topics will be ones of traditional interest for which there is a substantial body of data and theory, such as auditory neuroanatomy (Vol. 1) and neurophysiology (Vol. 2). Other volumes in the series deal with topics that have begun to mature more recently, such as development, plasticity, and computational models of neural processing. In many cases, the series editors are joined by a co-editor having special expertise in the topic of the volume.

Richard R. Fay, Falmouth, MA
Arthur N. Popper, College Park, MD

Volume Preface

There have been marked advances in the development and application of auditory prostheses since the first book on cochlear implants in this series, *Cochlear Implants: Auditory Prostheses and Electric Hearing* (SHAR, Zeng, Popper, and Fay, 2004). These advances include not only new approaches to cochlear implants themselves but also new advances in implants that stimulate other parts of the auditory pathway, including the middle ear and the central nervous system. This volume, then, provides insight into the advances over the past 7 years and also examines a range of other current issues that concern complex processing of sounds by prosthetic device users.

Chapter 1 (Zeng) provides an overview of the volume, insights into the history of development of prostheses, and thoughts about the future of this burgeoning field. In Chapter 2, van Hoesel examines the natural extension from single to bilateral cochlear implants. This is followed by Chapter 3 in which Turner and Gantz focus on the improved performance of combined electro-acoustic stimulation over electric stimulation alone.

In the near term, implantable middle ear devices have satisfactorily filled a gap between hearing aids and cochlear implants. Snik (Chap. 4) clearly delineates the complex technological and medical scenarios under which implantable middle ear devices can be used.

Dizziness and balance disorders are other major ear-related diseases that may also be treated by electric stimulation but have received little attention until recently. Golub, Phillips, and Rubinstein (Chap. 5) provide a thorough overview of the pathology and dysfunction of the vestibular system as well as recent efforts and progress in animal and engineering studies of vestibular implants.

New technologies are also being developed to advance significant problems associated with current cochlear implants that use electrodes inserted in the scala tympani to stimulate the auditory nerve. Taking one approach, Richter and Matic (Chap. 6) advocate an optical stimulation approach that should significantly improve spatial selectivity over the electric stimulation approach. This is followed by Chapter 7 by Middlebrooks and Snyder, which considers an alternative approach that uses traditional electric stimulation but places the electrodes in direct contact with the neural tissue to achieve selective stimulation.

In patients lacking a functional cochlea or auditory nerve, higher auditory structures have to be stimulated to restore hearing. McCreery and Otto (Chap. 8) present an account of research and development of cochlear nucleus auditory prostheses or the auditory brainstem implants. This is followed by Chapter 9 by Lim, M. Lenarz, and T. Lenarz, which discusses the scientific basis, engineering design, and preliminary human clinical trial data of auditory midbrain implants.

While it is important to continue to develop innovative devices, it is equally important to evaluate their outcomes properly and to understand why and how they work. Sharma and Dorman (Chap. 10) review both deprivation-induced and experience-dependent cortical plasticity as a result of deafness and restoration of hearing via cochlear implants, while Fu and Galvin (Chap. 11) document both the importance and effectiveness of auditory training for cochlear implant users. The significance is considered further for understanding the development of language in children following pediatric cochlear implantation in Chapter 12 by Ambrose, Hammes-Ganguly, and Eisenberg. Still, music perception remains challenging to cochlear implant users. McDermott (Chap. 13) reviews extensive research and recent progress in this area and identifies both design and psychophysical deficiencies that contribute to poor implant musical performance. Similarly, Xu and Zhou (Chap. 14) not only summarize acoustic cues in normal tonal language processing but also identify the design and perceptual issues in implant tonal language processing. Finally, in Chapter 15, Barone and Deguine examine multisensory processing in cochlear implants and present future research and rehabilitation needs in this new direction.

The material in this volume very much relates to material in a large number of previous SHAR volumes. Most notably, the aforementioned volume 20 has much material that complements this volume. But, in addition, issues related to music perception in patients with cochlear implants are considered in a number of chapters in volume 26, *Music Perception* (Jones, Fay, and Popper, 2010) while computational issues related to implants are discussed in chapters in volume 35 on *Computational Models of the Auditory System* (Meddis, Lopez-Poveda, Popper, and Fay, 2010). Finally, hearing impairment and intervention strategies in aging humans is considered at length in volume 34, *The Aging Auditory System* (Gordon-Salant, Frisina, Popper, and Fay, 2010).

Fan-Gang Zeng, Irvine, CA
Arthur N. Popper, College Park, MD
Richard R. Fay, Falmouth, MA

Contents

Chapter 1
Advances in Auditory Prostheses

Fan-Gang Zeng

1 Introduction

Advances in auditory prostheses were accompanied by competing ideas and bold experiments in the 1960s and 1970s, an interesting and exciting time that was reminiscent of the Era of Warring States in ancient China (for a detailed review see Zeng et al. 2008). The most contested technological issue was between a single-electrode (House 1974) and a multi-electrode (Clark et al. 1977) cochlear implant, with the former winning the battle as the first commercially available auditory prosthesis in 1984, but the latter winning the war because it has become the most successful neural prosthesis: it has restored partial hearing to more than 200,000 deaf people worldwide today. For cochlear implants to achieve this remarkable level of success, not only did they have to compete against other devices such as tactile aids and hearing aids, but they also had to overcome doubt from both the mainstream and deaf communities (for a detailed review see Levitt 2008). Many technological advances, particularly innovative signal processing, were made in the 1980s and 1990s to contribute to the progress in cochlear implant performance (Loizou 2006; Wilson and Dorman 2007).

Figure 1.1 shows sentence recognition scores with different generations of the cochlear implant from three major manufacturers. At present, all contemporary cochlear implants use similar signal processing that extracts temporal envelope information from a limited number of spectral bands and delivers these band-limited temporal envelopes non-simultaneously to 12 to 22 electrodes implanted in the cochlea. As a result, these implants produced similarly good speech performance (70–80% sentence recognition in quiet), which allows an average cochlear implant user to carry on a conversation over the telephone.

F.-G. Zeng (✉)
Departments of Otolaryngology–Head and Neck Surgery,
Anatomy and Neurobiology Biomedical Engineering, and Cognitive Science,
University of California–Irvine, 110 Med Sci E, Irvine, CA 92697, USA
e-mail: fzeng@uci.edu

F.-G. Zeng et al. (eds.), *Auditory Prostheses: New Horizons*, Springer Handbook of Auditory Research 39, DOI 10.1007/978-1-4419-9434-9_1,

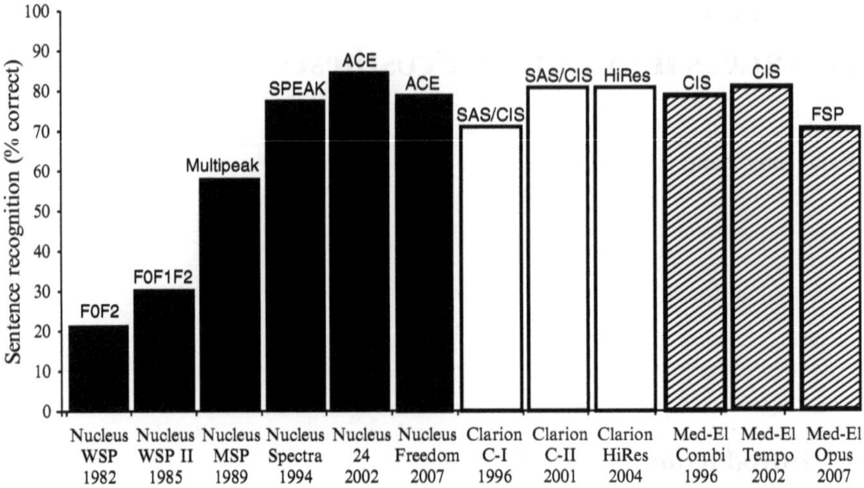

Fig. 1.1 Progressive sentence recognition with different generations of cochlear implants from the three major manufacturers, including the Nucleus device from Cochlear Corporation, the Clarion device from Advanced Bionics Corporation, and the devices from Med El (Adapted from Fig. 3 in Zeng et al. 2008)

Despite the good performance in quiet, there are still significant gaps in performance between normal-hearing and cochlear-implant listeners (Fig. 1.2). For example, the implant performance is extremely poor in noise, producing a 15-dB loss in functional signal-to-noise ratio with a steady-state noise background, and an even greater 30-dB loss with a competing voice (Zeng et al. 2005). Music perception is also extremely limited in implant users who can access some rhythmic information but little melody and timbre information (McDermott 2004). Finally, both tone perception and production are severely compromised in implant users who speak tonal languages such as Mandarin, Thai, and Vietnamese (Peng et al. 2008).

To close the performance gap between implant and normal listeners, new ideas and tools are needed and indeed have been developed intensely in recent years. Compared with the first 5 years of the new millennium, the number of publications related to cochlear implants has increased from 1196 to 1792 in the past 5 years (Fig. 1.3).

Where did the growth in publications come from? Bilateral cochlear implants were one area of such growth, with the number of related publications almost doubling, while the combined hearing aids and cochlear implants were another area of publication growth, with publications increasing fourfold in the same period. New tools such as midbrain stimulation and optical cochlear implants have also emerged. In contrast with a previous *Springer Handbook of Auditory Research* volume on cochlear implants (Zeng et al. 2004), which focused on the basic science and technology of electric stimulation, the present volume goes beyond traditional cochlear implants and presents new technological approaches, from bilateral cochlear implantation to midbrain prostheses, as well as new evaluation tools from auditory training to cross-modality processing.

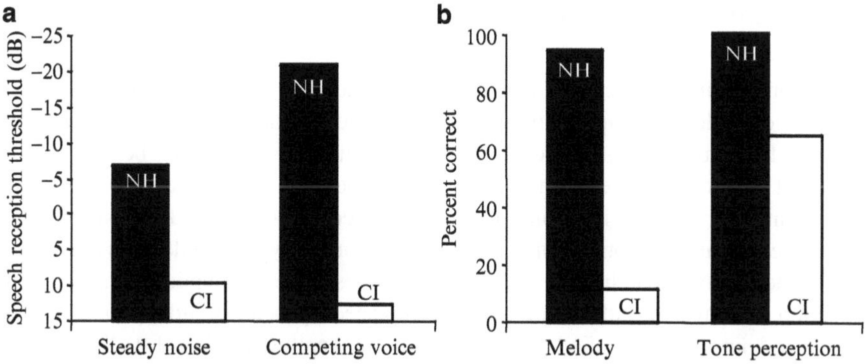

Fig. 1.2 Speech perception in noise (**a**) and music and tone perception (**b**) between normal-hearing (NH) and cochlear-implant (CI) listeners. Speech perception in noise is represented by signal-to-noise ratio in dB, at which 50% of speech is recognized. Music perception is percentage of melodies correctly recognized, while tone perception is percentage of Mandarin tones correctly recognized (Adapted from Fig. 21 in Zeng et al. 2008)

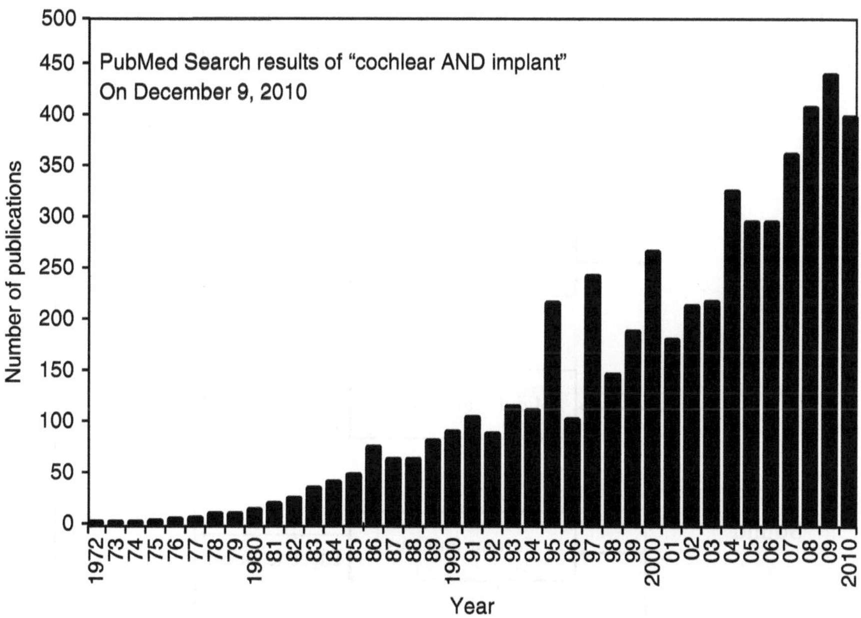

Fig. 1.3 Annual number of publications since 1972 on cochlear implants retrieved from PubMed (http://www.ncbi.nlm.nih.gov) on December 9, 2010

2 Advances in Technological Approaches

Cochlear implants have greatly expanded their function and utility through improvement in technology and application to a broad range of hearing related disorders. One aspect of the advances is the realization that auditory sensation can be induced by different physical energies (Fig. 1.4). In normal hearing, acoustic energies are converted into mechanical vibrations and then into electric potentials. In impaired hearing, different interventions are needed depending on the types and degrees of hearing impairment. For most listeners with cochlear loss, the mechanical amplification function is damaged and can be partially replaced by hearing aids, which take in sound and output its amplified acoustic version (first pathway in Fig. 1.4). To increase amplification and avoid acoustic feedback, sound can be converted into mechanical vibration to stimulate the middle ear (second pathway). In cases of profound deafness, sound has to be converted into electric pulses in a conventional cochlear implant, bypassing the damaged microphone function and directly stimulating the residual auditory nerve (third pathway). Recently, optic stimulation has also been found to be able to activate the nerve tissue directly (fourth pathway), providing potentially advantageous alternative to traditional electric stimulation.

The other aspect of advances is stimulation at different places of the auditory system, which can be used to treat different types of hearing impairment. The eardrum

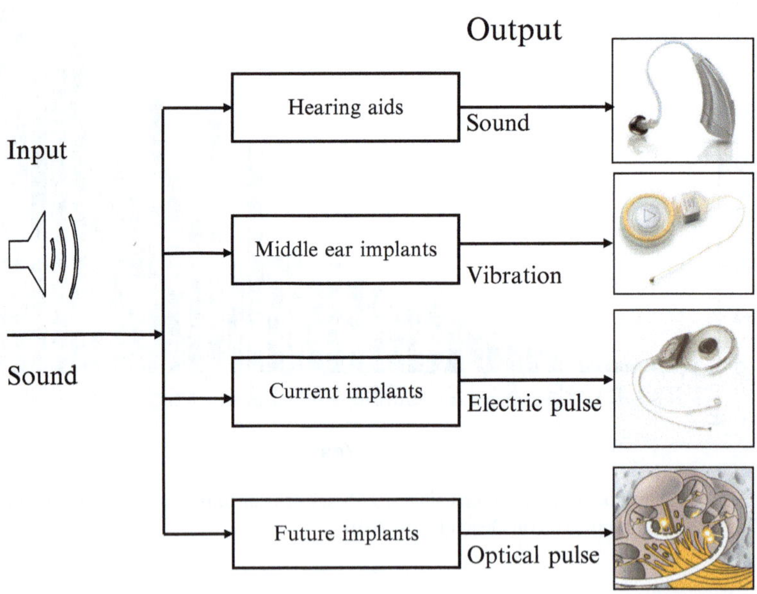

Fig. 1.4 Different approaches to stimulation of the auditory system. Hearing aid image is from www.starkey.com, middle ear implant image from www.medel.com, cochlear implant image from www.cochlear.com, and optical stimulation from www.optoiq.com

is stimulated acoustically in normal hearing and by amplified sound in a hearing aid to treat cochlear loss. The entire middle ear chain from incus to stapes can be stimulated mechanically to provide higher amplification and to treat persons with conductive loss related to collapsed ear canal and chronic ear diseases. The auditory nerve can be stimulated electrically, or optically, to provide functional hearing to persons with damaged inner hair cells. The entire central system from cochlear nucleus to cortex can also be stimulated to treat persons with acoustic tumors and other neurological diseases. Although not covered by the present volume, electric stimulation has been applied to treat auditory neuropathy, tinnitus, and multiple disabilities (Trimble et al. 2008; Van de Heyning et al. 2008; Teagle et al. 2010).

As the most natural extension to a single cochlear implant, bilateral cochlear implantation has experienced significant progress in terms of both clinical uptake and scientific understanding in the last decade. Van Hoesel (Chap. 2), who conducted the first study on bilateral cochlear implantation (van Hoesel et al. 1993), systematically reviews the rationale, progress, and remaining issues in this rapidly growing area. Compared with single cochlear implantation, bilateral implantation guarantees that the better ear is implanted. Although bilateral speech perception in noise and sound localization are improved by bilateral implants, the improvement is still modest and mostly comes from the acoustic head shadow effect that utilizes interaural level differences. There is little evidence that bilateral implant users take advantage of the interaural time difference to improve their functional binaural hearing, partially because of deprivation of binaural experience in typical users (Hancock et al. 2010) and partially because of the lack of encoding of low frequency fine structure information in current cochlear implants. One means of providing such low frequency fine structure information is to complement the cochlear implant with a contralateral hearing aid in subjects who have residual acoustic hearing.

Turner and Gantz (Chap. 3) focus on the improved performance of combined electro-acoustic stimulation (EAS) over electric stimulation alone. Compared with the typical 1 to 2 dB improvement in speech perception in noise with bilateral implants over unilateral implants, EAS can improve speech perception in noise by as much as 10 to 15 dB, depending on noise type and quality of residual hearing. The mechanisms underlying the improvement are also totally different between bilateral implantation and EAS, with the former relying on loudness summation, whereas the latter utilizes voice pitch to separate signals from noise or glimpsing signals at time intervals with favorable signal-to-noise ratios (Li and Loizou 2008). EAS, with its promising initial outcomes, improved surgical techniques, and signal processing, will likely continue to expand its candidacy criteria to include those who have significant residual hearing and possibly become the choice of treatment for presbycusis in the future.

In the near term, implantable middle ear devices have satisfactorily filled the gap between hearing aids and cochlear implants. Snik (Chap. 4) clearly delineates the complex technological and medical scenarios under which implantable middle ear devices can be used. Technologically, the middle ear implants avoid several pitfalls associated with the use of ear molds in most conventional hearing aids. These include the so-called occlusion effect where the hearing aid wearers' own voice sounds louder than normal, feedback squeal because of acoustic leakage between microphone and speaker, and undesirable blockage of residual hearing at low frequencies. Medically, for persons with conductive or mixed conductive and sensorineural loss, such as collapsed or lacking ear canals, chronic ear infection, and severe to profound hearing loss, hearing aids cannot be applied, and cochlear implants are not likely as effective as the implantable middle ear devices.

Dizziness and balance disorders are other major ear-related diseases that may also be treated by electric stimulation, but they have received little attention until recently. Golub, Phillips, and Rubinstein (Chap. 5) provide a thorough overview of the pathology and dysfunction of the vestibular system, as well as recent progress in the animal and engineering studies of vestibular implants. Especially interesting is their novel concept and design of a vestibular pacemaker that can be relatively easily fabricated and used to control dizziness. In October of 2010, the University of Washington group successfully implanted such a device in the first human volunteer. Compared with cochlear implantation, the enterprise of vestibular implantation is small but ready to take off, owing to the clinical need, encouraging animal studies, and the borrowing of similar cochlear implant technologies. Sophisticated sensor-based vestibular implants, a totally implantable device, and even vestibular brainstem implants, are likely to be developed and trialed by persons with severe balance disorders in the near future.

New technologies are also being developed to advance significant problems associated with current cochlear implants that use electrodes inserted in the scala tympani to stimulate the auditory nerve. With a bony wall separating the electrode and the nerve, the current implant not only requires high currents to activate the nerve, but also is severely limited by broad spatial selectivity and lack of access to apical neurons. Taking one approach, Richter and Matic (Chap. 6) advocate optical stimulation that should significantly improve spatial selectivity over the electric stimulation approach. The authors probe the mechanisms underlying optical stimulation and present promising preliminary animal data to demonstrate the feasibility of an optical cochlear implant. Middlebrooks and Snyder (Chap. 7) investigate an alternative approach that uses traditional electric stimulation but places the electrodes in direct contact with the neural tissue to achieve selective stimulation. In a cat model, this "intraneural stimulation" approach has produced not only low stimulation thresholds and sharp spatial selectivity, as expected, but more surprisingly and importantly, access to apical neurons that are more capable of transmitting temporal information than basal neurons. Both optical and intraneural stimulation approaches have the potential to improve current cochlear implant performance by quantum steps but are likely years away from human clinical trials: they have to overcome challenging technical issues such as size (for optical stimulation) and stability (for both).

In patients lacking a functional cochlea or auditory nerve, higher auditory structures have to be stimulated to restore hearing. Along with pioneers such as Robert Shannon, Derald Brackmann, and William Hitselberger, McCreery and Otto (Chap. 8) present a uniquely personal as well as masterfully professional account of research and development of cochlear nucleus auditory prostheses or auditory brainstem implants (ABI). ABIs have evolved from a simple single surface electrode device to sophisticated devices with multiple surface and penetrating electrodes. Their utilities have also been expanded from initial treatment of patients with bilateral acoustic tumors to current inclusion of non-tumor patients with ossified cochleae and damaged auditory nerves. The unexpected yet surprisingly good performance with the non-tumor patients is especially encouraging, because it not only allows many more suitable patients but also presents unique opportunities for improved understanding of the basic auditory structures and functions.

Because of its well defined laminated structure and easy access in humans, the inferior colliculus has also been targeted as a potential site of stimulation. As the inventors of the auditory midbrain implant (AMI) stimulating the inferior colliculus to restore hearing, Lim, M. Lenarz, and T. Lenarz (Chap. 9) discuss the scientific basis, engineering design, and preliminary human clinical trial data of the AMI. Although still in its infancy, AMI continues to push the technological and surgical envelope and to expand the horizon for wide acceptance and high efficiency of central auditory prostheses. For example, it may build a bridge between auditory prostheses and other well established neural prostheses, e.g., deep brain stimulation that have been used to treat a wide range of neurological disorders from Parkinson's disease to seizures. It is possible that future central prostheses will be integrated to treat not only one disability but also a host of disorders including hearing loss and its associated symptoms, such as tinnitus and depression.

3 Advances in Functional Rehabilitation and Assessment

While it is important to continue to develop innovative devices, it is equally important to evaluate their outcomes properly and to understand why and how they work. Rehabilitation and assessment of auditory prostheses can be challenging, due to the complexity and diversity at the input and output of the auditory system (Fig. 1.5). The input can be based solely in the hearing modality via either acoustic or electric stimulation or both; the auditory input can be combined with visual cues (e.g., lip-reading) and tactile cues. The output can be measured by speech perception, music perception, language development, or cross-modality integration. The deprivation of auditory input and its restoration by various auditory prostheses provide opportunities to study the physiological processes underlying brain maturity, plasticity, and functionality. Functionally, research has taken advantage of brain plasticity to improve cochlear implant performance by perceptual learning and training. In recent years, significant advances have been made in understanding these input–output relationships, the feedback loop, and their underlying physiological processes. Quantitatively, the number of publications in the last 5 years has doubled that of the previous 5 years in essentially every category, including cochlear implant plasticity (37 vs. 67), training (102 vs. 223), language development (151 vs. 254), music (27 vs. 112), tonal language (137 vs. 264), and cross-modality (62 vs. 126) research. Chapters 10 through 15 qualitatively present advances in these areas.

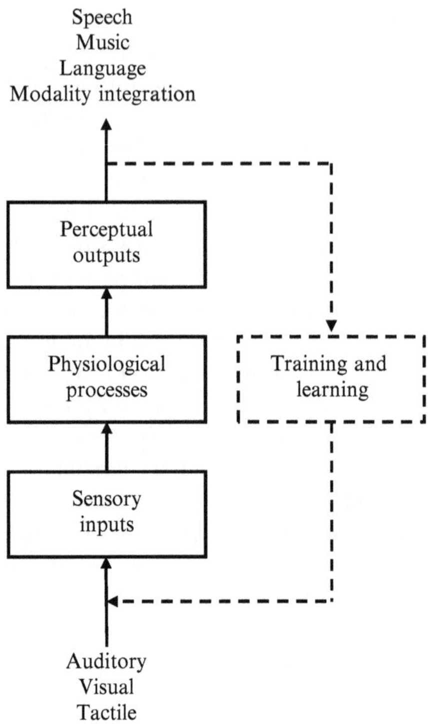

Fig. 1.5 A system approach to understanding of cochlear implant performance and function

Sharma and Dorman (Chap. 10) review both deprivation-induced and experience-dependent cortical plasticity as a result of deafness and restoration of hearing via cochlear implants. Coupled with language outcome measures and assisted by innovative non-invasive technologies from cortical potentials to brain imaging, central development research has identified a sensitive period up to 7 years, with an optimal time of the first 4 years of life, for good cochlear implant performance in prelingually deafened children. In postlingually deafened adults, central plasticity studies have identified non-specific cortical responses to electric stimulation due to cross-modal reorganization as one cause for poor cochlear implant performance. These central studies will continue to reveal neural mechanisms underlying cochlear implant performance, and more importantly, will guide development of effective rehabilitation for cochlear implant users.

Fu and Galvin (Chap. 11) document both the importance and effectiveness of auditory training for cochlear implant users. Because electric stimulation is significantly different from acoustic stimulation and usually provides limited and distorted sound information, auditory learning, sometimes referred to as adaptation, is needed to achieve a high level of cochlear implant performance. Compared with costly updates in hardware and software, structured auditory training can be much cheaper but equally effective if adequate information is provided. Auditory training will continue to grow in both basic and clinical areas, but research questions about the limit, optimization, and generalization of learning need to be answered.

One example of human learning, language development, particularly spoken language development, seems to be so effortless for a normal-hearing child but so challenging, if not impossible, for a deaf child. Can normal language develop following pediatric cochlear implantation? This has been a classic question facing researchers in the auditory prosthesis field. By reviewing normal language development, its negative impact by hearing impairment, and remarkable progress made by cochlear implantation, Ambrose, Hammes-Ganguly, and Eisenberg (Chap. 12) convincingly answer this question: despite great individual differences, many pediatric implant users have developed language capabilities on par with their hearing peers. This is a remarkable triumph not only by cochlear implant researchers and educators, but more importantly, for half of the pediatric users of the total 200,000 cochlear implants worldwide. It is expected that language development performance will increase while individual variability will decrease as technology continues to advance and more children receive the cochlear implant in the first 3 to 4 years of life, the optimal time within the sensitive period (see Chap. 10).

However, music perception remains challenging to cochlear implant users. Except for rhythmic perception that is similar to normal hearing persons, cochlear implant users perform much poorer in melody and timbre perception. McDermott (Chap. 13) reviews extensive research and recent progress in this area and identifies both design and psychophysical deficiencies that contribute to poor implant music performance. The key to improving cochlear implant music perception seems to lie in the encoding of pitch and related temporal fine structure, which not only form the basis of melody and timbre perception but also are critical to separating multiple sound sources, including different musical instruments in an orchestra.

Similarly, tone production and perception are a challenge to cochlear implant users who speak a tonal language. Xu and Zhou (Chap. 14) summarize acoustic cues in normal tonal language processing and, not surprisingly, isolate the lack of temporal fine structure in current devices as the culprit for their users' poor tone production and perception. They also identify age of implantation and duration of device usage as two demographic factors that influence tone production and perception in pediatric cochlear implant users. It is important to note that poor tone representation in cochlear implants not only affects tonal language processing, as expected, but it also disrupts or delays other important tasks such as vocal singing and even generative language development in non-tonal languages (Nittrouer and Chapman 2009).

In a natural environment, communication is usually multi-modal, involving auditory, visual, and other senses. In fact, cochlear implants were used mostly as an aid to lip-read in early days. Recently, multisensory processing in cochlear implants has become a hot topic, providing a unique and interesting model to study brain plasticity and integration in humans. Barone and Deguine (Chap. 15) review the latest advances in this new direction of research and present a unifying compensation model to account for the observed greater than normal cross-modality activation before and after cochlear implantation. Despite rapid progress in neuroscience of multisensory processing in cochlear implants, cross-modal applications to rehabilitation are still lagging but have great potential to improve overall cochlear implant performance in the future.

4 Summary

After steady progress in cochlear implant performance, mostly because of improved signal processing with multi-electrode stimulation in the 1990s, auditory prostheses entered a new era in the first decade of the twenty-first century. Three distinctive features mark the new ear. The first feature is "multiple stimulation in different places." The multiple stimulation includes bilateral electric stimulation, combined acoustic and electric stimulation, mechanical and optical stimulation, and visual and tactile stimulation. The different places include not only traditional acoustic and electric pathways, namely, the ear canal for hearing aids and scala tympani for cochlear implants, but also new stimulation sites from the auditory nerve to brain stem and midbrain structures that form direct contact with surface or penetrating electrodes. The second feature is the improvement of cochlear implant outcomes beyond speech perception, including language development, music perception, and tonal language processing. The means to improve cochlear implant performance has also been expanded to include identification of optimal cochlear implant time and candidacy as well as applications of auditory training and multisensory integration. The third feature is to apply the principles and successes of cochlear implants to the treatment of other neurological disorders such as auditory neuropathy, tinnitus, and dizziness.

The present volume is intended not only to capture these advances in auditory prostheses but also to extend the new horizon for future research and development. There is no question that current technological trends will continue, including fine timing control and sharp spatial selectivity in the device and the electronics-neuron interface, more and better use of residual low frequency acoustic hearing, structured learning and multisensory training, and biological means of preserving or even increasing nerve survival. There are also several new development efforts that will either significantly improve cochlear implant performance or change the face of auditory prostheses altogether. First, the rapid progress in bioengineering and regenerative medicine will produce more natural, highly efficient and effective electronics-neuron interfaces, including possibly a fifth pathway, chemical stimulation via reconstructed synapses, to evoke auditory sensation (Fig. 1.4). Second, auditory prostheses will be integrated with other peripheral and central prostheses (e.g., vestibular and deep brain implants) to treat not just one symptom but to address its whole spectrum (for example, hearing loss and its associated problems in tinnitus, dizziness, and depression). Finally, progress in neuroscience, particularly non-invasive brain monitoring will allow a full account of individual variability in cochlear implant performance, monitoring presurgical prediction of postsurgical performance, and more importantly, closed-loop fitting, operation and optimization of cochlear implants.

Acknowledgements The author would like to thank Grace Hunter, Tom Lu, and Dustin Zhang for technical assistance. The author's work was supported by NIH grants RO1-DC-008858 and P30-DC-008369.

References

Clark, G. M., Tong, Y. C., Black, R., Forster, I. C., Patrick, J. F., & Dewhurst, D. J. (1977). A multiple electrode cochlear implant. *Journal of Laryngology and Otology*, 91(11), 935–945.

Hancock, K. E., Noel, V., Ryugo, D. K., & Delgutte, B. (2010). Neural coding of interaural time differences with bilateral cochlear implants: effects of congenital deafness. *Journal of Neuroscience*, 30(42), 14068–14079.

House, W. F. (1974). Goals of the cochlear implant. *Laryngoscope*, 84(11), 1883–1887.

Levitt, H. (2008). Cochlear prostheses: L'enfant terrible of auditory rehabilitation. *Journal of Rehabilitation Research and Development*, 45(5), ix–xvi.

Li, N., & Loizou, P. C. (2008). A glimpsing account for the benefit of simulated combined acoustic and electric hearing. *Journal of the Acoustical Society of America*, 123(4), 2287–2294.

Loizou, P. C. (2006). Speech processing in vocoder-centric cochlear implants. *Advances in Oto-Rhino-Laryngology*, 64, 109–143.

McDermott, H. J. (2004). Music perception with cochlear implants: a review. *Trends in Amplification*, 8(2), 49–82.

Nittrouer, S., & Chapman, C. (2009). The effects of bilateral electric and bimodal electric-acoustic stimulation on language development. *Trends in Amplification*, 13(3), 190–205.

Peng, S. C., Tomblin, J. B., & Turner, C. W. (2008). Production and perception of speech intonation in pediatric cochlear implant recipients and individuals with normal hearing. *Ear and Hearing*, 29(3), 336–351.

Teagle, H. F., Roush, P. A., Woodard, J. S., Hatch, D. R., Zdanski, C. J., Buss, E., & Buchman, C. A. (2010). Cochlear implantation in children with auditory neuropathy spectrum disorder. *Ear and Hearing*, 31(3), 325–335.

Trimble, K., Rosella, L. C., Propst, E., Gordon, K. A., Papaioannou, V., & Papsin, B. C. (2008). Speech perception outcome in multiply disabled children following cochlear implantation: investigating a predictive score. *Journal of the American Academy of Audiology*, 19(8), 602–611.

Van de Heyning, P., Vermeire, K., Diebl, M., Nopp, P., Anderson, I., & De Ridder, D. (2008). Incapacitating unilateral tinnitus in single-sided deafness treated by cochlear implantation. *Annals of Otology, Rhinology, and Laryngology*, 117(9), 645–652.

van Hoesel, R. J., Tong, Y. C., Hollow, R. D., & Clark, G. M. (1993). Psychophysical and speech perception studies: a case report on a binaural cochlear implant subject. *Journal of the Acoustical Society of America*, 94(6), 3178–3189.

Wilson, B. S., & Dorman, M. F. (2007). The surprising performance of present-day cochlear implants. *IEEE Transactions on Biomedical Engineering*, 54(6, pt. 1), 969–972.

Zeng, F. G., Popper, A. N., & Fay, R. R. (2004). *Cochlear implants: Auditory prostheses and electric hearing* (Vol. 20). New York: Springer.

Zeng, F. G., Rebscher, S., Harrison, W., Sun, X., & Feng, H. H. (2008). Cochlear implants: System design, integration and evaluation. *IEEE Reviews in Biomedical Engineering*, 1(1), 115–142.

Zeng, F. G., Nie, K., Stickney, G. S., Kong, Y. Y., Vongphoe, M., Bhargave, A., Wei, C., & Cao, K. (2005). Speech recognition with amplitude and frequency modulations. *Proceedings of the National Academy of Sciences of the United States of America*, 102(7), 2293–2298.

Chapter 2
Bilateral Cochlear Implants

Richard van Hoesel

1 Introduction

Over the last decade bilateral cochlear implantation has seen increasing clinical uptake. Particularly in the case of young children, a growing number of recipients are being fitted bilaterally, with the expectation that providing input from both ears early in life will lead to improved outcomes compared to implanting only one ear. However, a wide range of factors is likely to influence the extent to which binaural hearing advantages can be imparted to bilateral implant users. Some of those relate to the availability of binaural cues in sound processing methods for implant users, some to the nature of the neural responses to electrical stimulation, and others to pathology, developmental considerations, and plasticity in relation to listening experience with two, one, or no ears. As implant outcomes continue to improve over time and candidacy criteria moderate, more unilateral implant recipients with useful residual hearing in the contralateral ear are combining an implant with a contralateral hearing aid. Outcomes for those listeners are also reviewed and compared with those seen in bilateral cochlear implant (BiCI) users.

2 Listening with Two Ears (Normal Hearing)

2.1 Spatial Hearing

Spatial hearing cues are the signal characteristics at the ears that provide listeners with information about the location of sound sources, and through reflections also

R. van Hoesel (✉)
The Hearing CRC, University of Melbourne,
550 Swanston St., Parkville, VIC 3010, Australia
e-mail: richardv@unimelb.edu.au

F.-G. Zeng et al. (eds.), *Auditory Prostheses: New Horizons*, Springer Handbook of Auditory Research 39, DOI 10.1007/978-1-4419-9434-9_2,

about the surrounding environment. A comprehensive review of spatial hearing and localization in listeners with normal hearing can be found in Blauert (1997). The orientation of the ears in the horizontal plane, with the head as an intervening acoustic barrier, causes signals originating from outside the vertical median plane to be higher in intensity and arrive first at the ear closest to the sound source. The interaural level differences (ILDs) and interaural time delays (ITDs) are essential to sound localization. The physical dimensions of the head produce ILDs that are minimal up to several hundred Hz, but can exceed 15 dB at frequencies beyond 2 kHz (Shaw 1974; Duda and Martens 1998). The ITD cue arises from the additional time it takes for sound to travel to the farther ear. For a sound source at 90° the ITD is almost 700 μs at frequencies above 2 kHz, and because of diffraction effects, about 800 μs for low frequencies below 600 Hz (Kuhn 1977). The smallest change in azimuth a listener can detect is referred to as the minimum audible angle (MAA), and is about 1° or 2° for pure tones arriving from around 0° when they are in the range 500 to 1000 Hz, or 3 to 6 kHz (Mills 1958). That result is in good agreement with the duplex theory of sound localization in which low frequencies are localized largely using ITDs and high frequencies using ILDs (Rayleigh 1907). For a fixed change in azimuth, the changes in ILD and ITD cues are smaller near 90° than 0°, leading to discrimination that can be up to ten times worse than at 0°. Whereas the MAA describes the ability to detect relative cues when comparing two source locations, the ability to use absolute cues to determine spatial position is usually measured using pointer methods or sound-direction identification paradigms in which listeners select from a finite number of response locations (Stevens and Newman 1936). Listeners can benefit from head turns to minimize ambiguity or combine information from multiple orientations. The ability to track dynamic interaural cues is, however, restricted to a fairly slow rate (Grantham 1995; Blauert 1997). While the availability of multiple cues can sometimes be used to decrease ambiguity, disregarding some of those cues when they are in conflict with one another can be equally beneficial (Zurek 1980; Hartmann 1997; Wightman and Kistler 1997).

2.2 *Binaural Sensitivity*

ILD thresholds in listeners with normal hearing for many signals are in the range 0.5 to 1 dB, and often lateralization is nearly completely towards the ear with the higher signal level when ILDs exceed about 10 dB (Yost and Hafter 1987). Ongoing ITD sensitivity is highly frequency dependent, and for pure tones thresholds decrease from around 75 μs at 90 Hz to as little as 11 μs as frequency increase to 1 kHz but increase again at higher frequencies to become immeasurable beyond 1500 Hz (Klumpp and Eady 1956). With broadband noise and click trains, thresholds can also approach 10 μs. The very poor sensitivity to ongoing ITDs at frequencies above 1500 Hz can be much improved using low frequency amplitude modulation (AM) of the envelope. Envelopes comprising half-wave rectified low frequency sinusoids ("transposed tones," van de Par and Kohlrausch 1997) have been shown to produce thresholds on the order of 100 μs as long as envelope fluctuation rates remain below a few hundred Hz (Bernstein and Trahiotis 2002; Oxenham et al. 2004). The influence of stimulus onset cues compared to later arriving cues has most often been studied under the heading "precedence," and results often implicate higher level processing (for a review, see Litovsky et al. 1999). Hafter and colleagues (1983, 1990) also proposed a low level binaural adaptation mechanism that gradually reduces both ILD and ITD cue effectiveness over time for high-rate click trains, but can be "restarted" through signal anomalies. However, such gradually decreasing effectiveness has not been demonstrated in observer weighting experiments that directly measure the contributions from the cues applied to each click (Saberi 1996; Stecker and Hafter 2002). When the same signal and noise are presented at each ear, the listening condition is referred to as diotic (as opposed to dichotic). The improvement in binaural detection thresholds in noise, when binaural cues instead differ for concurrent target and interfering signals, is referred to as the binaural masking level difference (BMLD, Hirsch 1948). For diotic broadband masking noise the pure-tone BMLD obtained by inverting the phase of the tone at one ear (SπN0) is about 10 to 15 dB for target frequencies in the range 200 to 500 Hz and gradually reduces to a few dB as frequency increases to beyond 1.5 kHz (Durlach and Colburn 1978). The smaller benefit at high frequencies occurs predominantly as a result of the loss of sensitivity to fine-timing ITD cues, but also because of critical bandwidth and spectral interference considerations (Zurek and Durlach 1987). Narrowband BMLDs on the other hand can be larger than 20 dB at low frequencies, and as much as 10 dB even for unmodulated high frequencies (van de Par and Kohlrausch 1997). An influential model accounting for many but not all of the broadband BMLD data is the Equalize and Cancel (EC) model (Durlach 1963). Various alternative models have also been proposed and are typically based on interaural correlation or directly on binaural cues. In all of these approaches, the strong role of ITDs in eliciting BMLDs is acknowledged. In addition to providing interaural difference cues, listeners with normal hearing experience an increase in loudness when using both ears, which for many signals in the range 40 to 80 dB is equivalent to a monaural level increases of about 4 to 8 dB (Scharf 1978). Interaural frequency differences appear to exert little effect on the amount of binaural loudness summation when care is taken to ensure listeners estimate overall loudness rather than that of readily discriminated monaural components.

2.3 *Speech Intelligibility Gains*

When target speech and interfering noise are both in front of a listener (S0N0) the signals are to a first order approximation diotic, and the benefit of listening with both ears, rather than only one, is small. In terms of speech reception thresholds (SRTs) that describe the signal-to-noise ratio (SNR) required to achieve 50%-correct performance, the improvement is on the order of 1 dB for listeners with normal hearing (Bronkhorst and Plomp 1988). The change in speech intelligibility when target speech and interfering signals are spatially separated is referred to as spatial release from masking. Speech studies mainly concerned with energetic aspects of spatial unmasking show intelligibility gains of up to about 10 dB (Bronkhorst 2000). That result arises from the combined effects of the monaural headshadow, which improves the SNR at one ear, and binaural unmasking gains due to low frequency ITDs. Note that when using only one ear, spatial separation is only beneficial when it improves the SNR at that ear. The benefit of using both ears rather than just the one with the better SNR is sometimes referred to as binaural squelch (Koenig 1950; Carhart 1965) because it acts to "squelch" the effect of the noise. In studies with normal-hearing listeners, squelch is largely attributed to binaural unmasking that arises particularly from differences in low frequency ITDs for the target and interfering signals (Bronkhorst 2000). The speech intelligibility benefit obtained through binaural unmasking is also referred to as the binaural intelligibility level difference (BILD). The application of ITDs to speech in diotic noise has been shown to result in a BILD on the order of 5 dB (Carhart et al. 1967; Bronkhorst and Plomp 1988) and is well predicted from intelligibility weighted pure-tone BMLDs (Levitt and Rabiner 1967). Because a large BILD can be obtained by inverting the speech signal in one ear, ITDs need not be consistent across frequency bands and therefore need not correspond to a single spatial position to produce the benefit. The relative contributions of headshadow and binaural unmasking to spatial unmasking depend on the frequency content of the speech material, and the two do not sum independently. For example, while head-derived ITDs alone can produce a BILD of up to 5 dB in the absence of headshadow effects, their maximal contribution when headshadow is introduced is reduced to about 2 to 3 dB (Bronkhorst and Plomp 1988). When there is perceptual uncertainty regarding which parts of the total signal relate to the target, or when interferers contain competing information, additional difficulties in understanding speech may arise due to "informational masking" even in the absence of energetic overlap. Under those conditions, spatial cues may be used to improve separation between target and interferer streams, and the amount of spatial unmasking can be considerably larger than is accounted for by purely energetic considerations (see Kidd et al. 2008, for an overview).

3 Sound Coding for Bilateral CI

Interaural cues in BiCI processors are first modified by the microphone response characteristics, which modify the signal's amplitude and phase for each frequency and direction of incident sound. Spatial selectivity can be improved by using directional microphones that combine acoustic signals arriving via different physical travel paths, and are most frequently used to decrease levels for signals arriving from behind the listener. Placement of the microphone on the head alters its spatial response in accordance with the acoustic headshadow. The left panel of Fig. 2.1 shows the broadband spatial response plot as a function of azimuth for a typical ear-mounted directional microphone presented with speech weighted noise. For a microphone at the contralateral ear, the response is approximately inverted from left

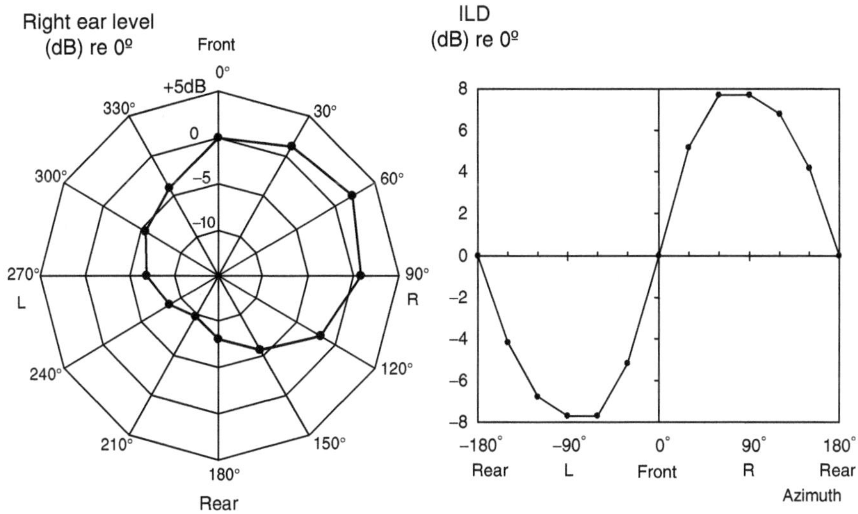

Fig. 2.1 (*Left*) Horizontal plane response as a function of presentation azimuth (dB re: 0°) for a directional microphone placed behind the right ear of a KEMAR manikin, presented with speech weighted noise. (*Right*) Estimated broadband ILD as a function of azimuth for the same signal, calculated as the difference between the response shown in the left plot and a left-right inverted version of that plot (corresponding to an ideal matched microphone at the left ear)

to right, and subtraction of the two responses provides an estimate of the broadband spatial ILD function shown in the right panel. At higher frequencies, narrowband ILD curves as a function of azimuth are steeper, but become non-monotonic at larger azimuths and therefore introduce greater ambiguity in those regions, whereas at lower frequencies the cue is smaller but less ambiguous. The shape of the broadband ILD function therefore depends on the frequency content of the signal and the microphones used, but the flattening of cues at larger azimuths will generally be present for broadband signals.

The ILD cue is preserved reasonably well in CI processors due to the emphasis on salient, frequency-specific level information in sound coding strategies, although factors such as asymmetric activation of automatic gain control circuits or mismatched acoustic-to-electrical mapping can lead to misrepresentation of ILDs. In contrast, the low frequency fine-timing ITD cues that contribute strongly to binaural speech unmasking in normal hearing are discarded in the clinical CI strategies used in most BiCI studies to date. In those strategies the microphone signal is separated into multiple frequency bands using a bank of bandpass analysis filters. The number of filters used is typically equal to the number of available electrodes. The output of each analysis filter is subjected to envelope extraction. Regardless of the method used to extract the envelope, the maximal rate at which the envelope can fluctuate is limited by the bandwidth of the analysis filter. To provide sufficient spectral detail to code speech well, filters employed are often no more than a few hundred Hz wide, which means binaural envelope cues are also limited to that rate, regardless of the envelope sampling rate (update rate) or electrical stimulation rates used. Envelope information from each selected filter is used to determine the stimulation current applied to a corresponding electrode, usually at a fixed stimulation rate. Because that stimulation rate is unrelated to the signal properties, the fine timing contained in the electrical pulse rate provides no useful information. In fact, if implant users can hear ITDs associated with those stimulation rates, a concern sometimes expressed is that the use of two independent processors results in disruptive timing cues. Note however that synchronous electrical stimulation (ITD = 0) in both ears does not offer any benefit in that regard because the fine-timing cue remains incorrect for signals other than those arriving from 0° azimuth, and the envelope ITD cue is unaffected as long as each pulse accurately represents the envelope amplitude at the time of stimulation. The degree to which fine-timing cues conflict with those contained in the envelope is also not improved (on average) by using synchronous stimulation. A predetermined outcome with these clinical strategies is that the benefits that depend on fine-timing ITDs in normal hearing will be absent in CI users, because that information is discarded, and any ITD-based benefits must be derived from low-rate envelope timing cues. An experimental strategy designed to preserve ITDs better uses positive peaks in the fine timing of each band-pass filter output to determine when electrical pulses are applied to the associated electrodes (Peak Derived Timing, PDT, van Hoesel 2002). As a result both fine-timing and envelope cues are consistent with the signal at each ear and interaural cues are correctly represented despite the use of independent processors. While such a strategy can present fine-timing information that is absent in envelope based strategies, the associated benefits seen in normal hearing need not ensue in BiCI users if electrical ITD sensitivity is relatively poor. With the exception of low rates below a few hundred Hz, psychophysical and physiological data indicate that is indeed the case (see Sects. 5 and 6 in this chapter).

4 Outcomes with Bilateral CI Users

4.1 A Priori Considerations

According to multiple-looks considerations, or integration of information across the two ears (Green and Swets 1966; Viemeister and Wakefield 1991), a binaural benefit may be derived to the extent that the *effective* noise in each ear is independent. Such noise may be imparted by various non-linear asymmetries between the ears with BiCIs and could lead to larger diotic benefits than seen in normal hearing. If those asymmetries are substantial, performance may also favor the binaural condition in experienced BiCI users because the speech representation with either ear alone is unfamiliar. It is also possible, however, that increased asymmetries in BiCI users could lead to worse binaural performance than with either ear alone because of binaural interference. For low level speech or speech components in quiet, intelligibility may be governed largely by audibility, so that binaural loudness summation may provide larger binaural benefits than at higher levels. For spatially separated speech and noise, the squelch benefit derived from adding an ear with poorer SNR in listeners with normal hearing can be several times larger than the diotic redundancy benefit and is therefore largely attributed to binaural unmasking. However, when the squelch benefit is more comparable in magnitude to the diotic benefit, it needs not involve binaural unmasking and particularly in listeners with asymmetric performance in the two ears, can arise in the absence of any binaural processing. Indeed, if binaural speech unmasking in normal hearing occurs largely as a result of low frequency fine-timing ITDs, its role in most BiCI studies to date would be expected to be minimal because the clinical sound processors discard those cues. In the same way that the term "squelch" has been used to express the benefit derived from adding an ear with a poorer SNR, the benefit of adding an ear with a *better* SNR has in some BiCI studies been referred to as a "headshadow benefit." However, that measure does not result from the monaural effect of the headshadow alone. It also includes contributions from binaural listening and performance asymmetry between the ears. In other studies the same term has been used to describe the performance difference between the two ears when noise is on one side, which avoids binaural contributions but can retain those associated with performance asymmetry. In the present chapter, the effect of the headshadow is described strictly in terms of monaural performance at a fixed ear when spatial positions of target and/or interferers are varied.

When listeners with normal symmetrical hearing are assessed under conditions for which the SNR differs at the two ears, benefits from the use of both ears are calculated relative to the ear with the better SNR, which offers better performance. It is a conservative measure of binaural benefit that describes the advantage that cannot be attributed to better-ear listening. A consistent approach in listeners with asymmetric performance is to calculate binaural benefits relative to the ear with the better monaural result for each spatial configuration. While other measures are possible, they are potentially confounded by better-ear listening contributions. To illustrate, consider a BiCI user with an SRT of 2 dB in the better ear, and 4 dB in the poorer ear. If the listener attends only the ear with better performance and ignores the other ear, the binaural SRT will also be 2 dB. Calculation of the "binaural benefit" relative to the better ear provides the correct value of 0 dB. In contrast, if the benefit is calculated relative to the poorer ear, the estimated benefit is 2 dB, or when calculated relative to the average of monaural left and right conditions, it is 1 dB. More generally, when performance differs between ears, calculation of the binaural benefit relative to a fixed ear (e.g., left ear, or first implanted ear) will inflate the benefit that must be attributed to the use of both ears. The same applies if results for a fixed ear are averaged across subjects before calculating the benefit. The inflationary effect is likely to be largest when the SNR is comparable in both ears. When noise arrives from one side, the ear with the better SNR is more likely to provide better performance. However, in listeners with larger asymmetries, that may not be the case, so squelch measures can also overestimate true binaural benefit. Listening experience in bilateral CI users favors the binaural condition and therefore potentially also imparts a positive bias to estimated binaural benefits. While comparison of binaural performance with the better monaural result avoids inflation because of better-ear listening, it potentially underestimates the binaural benefit for statistical reasons. That arises because better-ear performance in each subject is based on selecting the better of two sample means (the left ear test average, and the right ear average), whereas the binaural performance is sampled half as often (Byrne and Dillon 1979; Day et al. 1988). An initial evaluation of that effect in BiCI users, however, suggests it is likely to be small (van Hoesel and Litovsky, in preparation).

4.2 Speech Intelligibility Results for Adult BiCI Users

Initial case studies (Green et al. 1992; van Hoesel et al. 1993) established that BiCI users were able to listen to electrically coded speech from both ears without performance degradation resulting from mismatched percepts. A considerable number of subsequent studies have reported a small diotic benefit for speech in quiet, or for speech and noise presented to the front, and larger gains when speech and a single interferer are spatially separated. The benefit in the latter case occurs mainly as a result of the physical effect of the headshadow, which improves the SNR at the ear that is contralateral to the noise. To minimize the variability across studies that results from different measures of benefit, the reported benefits have been recalculated in this section where needed (and possible) to determine monaural headshadow effects and binaural advantages relative to the better monaural result for each listener and condition. A small number of recent studies have investigated speech intelligibility in the presence of multiple independent noise sources.

4.2.1 Speech in Quiet (Adults)

Many of the BiCI studies in which speech performance has been assessed in quiet conclude that there is a small benefit from using both ears. Comparison across studies with more than 5 subjects, and for which better ear performance can be determined, shows outcomes range from no benefit to slightly over 10 percentage points (pp) (e.g., Gantz et al. 2002, 6 pp; Müller et al. 2002, 11 pp; Laszig et al. 2004, 0 pp and 4 pp; Tyler et al. 2007, 4 pp; Wackym et al. 2007, 12 pp; Buss et al. 2008, 12 pp; Laske et al. 2009, 0 pp; Mosnier et al. 2009, 10 pp). Mosnier et al. (2009) found greater benefit in 10 listeners with symmetric performance in the two ears than in 9 listeners with larger asymmetries, and the average benefit for the former group approached 20 pp. Evidence of greater benefit in more symmetric listeners can also be found in the individual subject data reported by Gantz et al. (2002) and Müller et al. (2002). Clearly it must be the case that when asymmetry is large enough, the contribution from adding an ear with poorer performance diminishes.

4.2.2 Speech in Noise (Adults)

Single-Interferer Studies

Speech intelligibility in the presence of a single interfering noise source is perhaps the most frequently reported performance metric in BiCI users. Figure 2.2 shows results for BiCI studies that included around 10 or more adult subjects, and in which noise was presented from a single loudspeaker. Results in the left panel show (predominantly) the monaural benefit because of headshadow. Those in the middle panel show the (diotic) binaural benefit relative to the better monaural result for spatially coincident speech and noise. Results in the right panel show the binaural benefit relative to the better monaural result for spatially separated speech and noise. Despite the considerable range of methods and materials, outcomes across studies are in good agreement. Large increases in monaural performance result from the headshadow. For the studies in which noise was presented at 90° (S0N90), the average increase in monaural performance is around 5 dB when noise is moved from the ipsilateral to contralateral side relative to the implant. The effect is larger for the studies with signal and noise on opposite sides of the head at 45° (+ and * symbols) because both signal and noise levels are affected by the headshadow. For three studies (open circles, triangles, and diamonds) monaural headshadow measures were unavailable, and those data instead show the total benefit of adding a second ear with a better SNR. While that measure combines monaural headshadow, ear asymmetry, and binaural contributions, results are only about 1 dB higher than for the studies describing monaural results, which suggests only a small binaural benefit in agreement with results shown in the middle and right panels. Diotic benefits (S0N0) in the middle panel are on average a little less than 1 dB when referred to the better monaural result. Binaural benefits for spatially separated speech and noise in the right panel are also close to 1 dB when calculated relative to the better monaural result (closed symbols). The similarity of these last two outcomes suggests that, on average, contributions from binaural unmasking are minimal or absent.

Systematic Influences on Measured Benefits

Figure 2.3 shows the influence of high SNRs needed for poor performers when using SRT methods targeting a fixed performance criterion and indicates reduced estimates of headshadow benefit and increased variability in binaural benefit for spatially separated speech and noise. As for the speech outcomes in quiet, there are suggestions from some studies that benefits in noise are greater for more symmetrical performers, but data from more subjects are needed to verify that conjecture. Variation in loudness mapping procedures can also impact on reported benefits. In assessing the effects of loudness mapping and processor gain settings when switching between monaural and binaural listening conditions, van Hoesel et al. (2005) found that lowering electrical stimulation levels by the amount required to compensate for binaural

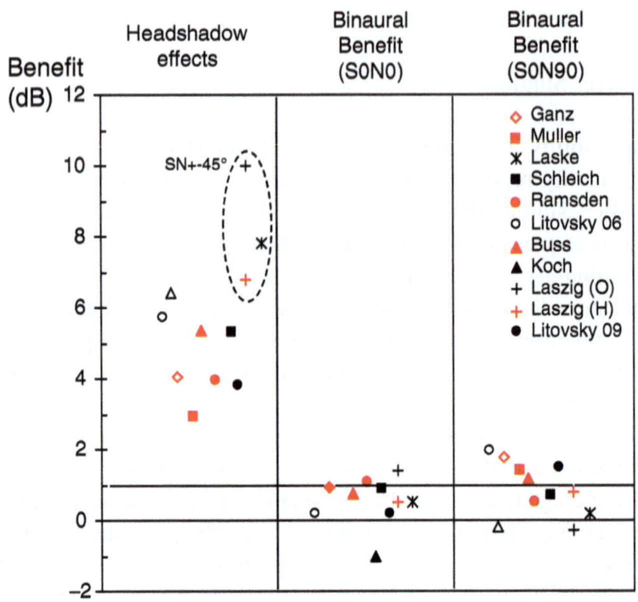

Fig. 2.2 Speech-in-noise results from 11 studies with BiCI users. (*Left*) Benefit because of monaural headshadow, preferably calculated by comparing performance for noise ipsilateral to the implant with that for noise contralateral to the implant. Where that value could not be determined, the benefit of adding an ear with better SNR is shown instead (*open symbols*). (*Middle*) Diotic benefit for speech and noise presented to the front (S0N0), calculated by comparing the better monaural result with binaural performance. (*Right*) Binaural benefit for spatially separated speech and noise, preferably calculated by comparing binaural performance with the better monaural result. Where that value could not be determined, the squelch benefit of adding an ear with poorer SNR is shown (*open symbols*). Black symbols are from studies that used SRT methods. *Red symbols* are estimated equivalent SRT benefits (reductions) for studies in which fixed SNRs were used. In the absence of performance intensity (PI) functions relating percent correct to SNR for the subjects in each of those studies, an average PI gradient of 7%/dB has been assumed. While that assumption is clearly an oversimplification, it may be reasonably used for group data, and the derived estimates are well matched in magnitude to those from studies using SRT tests. For the spatially separated speech and noise condition, most studies report results for speech at 0° and noise at 90° to the left or right (S0N90). In two studies speech and noise were presented on opposite sides of the head at 45° (+ and * symbols). Data are from Gantz et al. (2002), Müller et al. (2002), Laszig et al. (2004), Schleich et al. (2004), Ramsden et al. (2005), Litovsky et al. (2006c), Buss et al. (2008), Koch et al. (2009), Laske et al. (2009) (O = Oldenburg Sentence SRTs, H = HSM Sentences at fixed SNR), and Litovsky et al. (2009)

loudness summation reduced performance by about 1 dB, which would be sufficient to eliminate the small binaural benefits shown in the middle and right panels in Fig. 2.2.

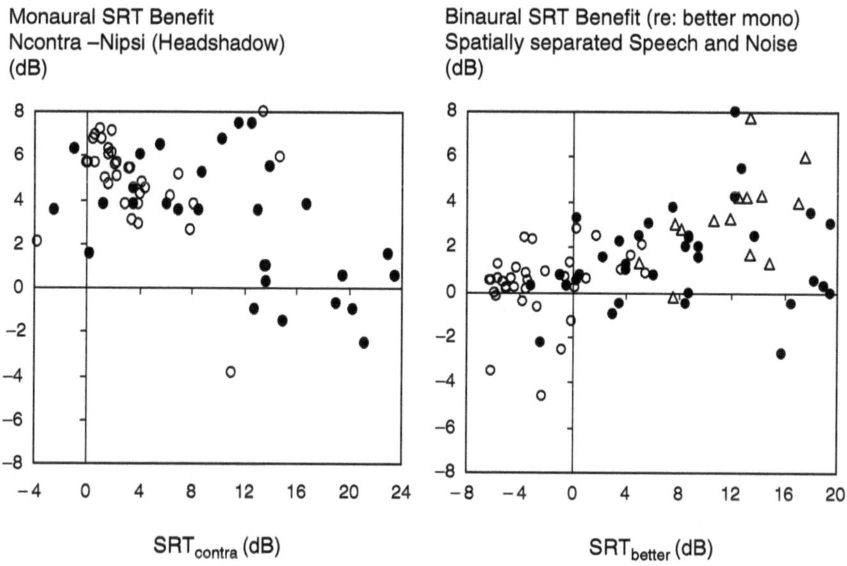

Fig. 2.3 Individual subject speech benefit measures as a function of SNR from three studies using SRT methods. *Open circles* display data from Schleich et al. (2004); *filled circles* from Litovsky et al. (2009); and *open triangles* from the multi-interferer study by Ricketts et al. (2006). The *left panel* describes the monaural benefit as a result of the headshadow (contralateral versus ipsilateral noise) for the first two of those studies. The right panel describes the binaural benefit relative to better-ear performance for spatially separated speech and noise. It can be seen that poor performers who required high SNRs to achieve the fixed 50% performance criterion used in these studies show reduced estimates of headshadow benefit and increased variability in binaural benefit, suggesting the assumptions underlying the SRT method may not be valid in those cases

Multiple Interferer Studies

Ricketts et al. (2006) tested BiCI users with independent cafeteria noise presented from 5 loudspeakers at 30°, 105°, 180°, 255°, and 330°, and target speech at 0°. Because the noise configuration is symmetrical, headshadow effects were assumed to be minimal. While the authors reported a relatively large binaural benefit of 3.3 dB relative to the better monaural result when using an adaptive SRT method, many of the subjects required high SNRs to reach the target criterion (Fig. 2.3, right panel). Additional tests were conducted with 10 of the 16 subjects in that study at a fixed SNR of 10 dB, and in that case a benefit of only 10 pp was found. Assuming a mean PI slope of at least 7%/dB (Cox et al. 1988) that corresponds to less than 1.5 dB benefit, which is more similar to that seen in the single-interferer studies. Loizou et al. (2009) tested BiCI users' abilities to understand speech from a male talker presented at 0° in the presence of 1 to 3 independent interferers placed either symmetrically or asymmetrically around listener. The interferer was speech modulated noise, or in a separate condition, a female talker. Consistent with the results from the single-interferer studies, the largest effects resulted from the monaural headshadow, and addition of an ear with a poorer SNR (squelch) provided smaller benefits of 2 dB or less. Binaural spatial unmasking was not found to be dependent on whether the interferer was the female talker or modulated noise. That outcome is in contrast to listeners with normal hearing who obtain substantially greater benefit from spatial separation because of informational unmasking with interferering speech than noise.

Binaural Unmasking (Binaural Intelligibility Level Difference)

Because binaural squelch is a poor indicator of binaural unmasking in BiCI users (see Sect. 4.1), van Hoesel et al. (2008) used a more direct method to assess the BILD. Binaural SRTs were measured in diotic noise for speech that was either also diotic (S0N0), or else contained an ITD of 700 μs (S700N0), which is close to the maximal ITD delay imparted by the human head. An additional advantage of that method is that the amount of unmasking in normal hearing listeners is maximized because of the lack of interaural level differences (Durlach and Colburn 1978). Results using both clinical strategies and PDT, however, showed no evidence that the 700 μs ITD elicited binaural unmasking. That outcome is in strong contrast to results for listeners with normal hearing, who under similar conditions show a BILD of about 5 dB (Bronkhorst and Plomp 1988). While the result is largely as expected with clinical processors that discard fine-timing cues, the same result using PDT indicates that the provision of additional fine-timing cues did not invoke binaural unmasking. That outcome is in accord with the much poorer ITD sensitivity seen in BiCI users than in normal hearing, particularly as rates increase beyond a few hundred Hz (see Sect. 5). Additional contributions to the lack of unmasking in this study may include inter-subject variations (see Sect. 5.5) and disruptive timing effects from multiple electrodes with out-of-phase temporal cues and broad current spread (Jones et al. 2008).

4.2.3 Time Course Considerations (Speech)

While both monaural and binaural performance generally improves over time following implantation, changes in binaural benefits remain unclear. Several studies have shown little or no change in binaural benefits over time courses ranging from 6 to 17 months (Laszig et al. 2004; Schleich et al. 2004; Ricketts et al. 2006). In contrast, Buss et al. (2008) reported increasing squelch but not diotic benefits during the first year, and again at 4 years (Eapen et al. 2009). Litovsky et al. (2009) also reported a larger benefit at 6 months than at 3 months when adding an ear with a better SNR and possibly also changes in squelch, but again not for diotic signal presentation, and attributed those outcomes to improved spatial hearing over time. Koch et al. (2009) reported increased binaural benefit in quiet at 6 to 8 months compared to 3, and in noise when adding an ear with a better SNR, but not when adding an ear with a poorer SNR (squelch). The reasons for these different outcomes in relation to time course effects are not clear but may include differences in subject groups and test methods. Further assessment is needed, and the use of more direct measures of binaural abilities (such as binaural unmasking) may provide greater insight.

4.3 *Localization in Adult BiCI Users*

4.3.1 Localization in Quiet

The term "localization" is used in the rest of this chapter and much of the CI literature in a somewhat inaccurate sense to describe the ability to relate percepts to external event locations, without considering actual perceived locations, which may for example be intracranial. A few early studies (e.g., Gantz et al. 2002) showed that BiCI users were better able to discriminate sounds originating from the left or right side of a listener with both ears than with either alone. More detailed evaluations in subsequent work have typically used the sound-source direction identification paradigm with an array of loudspeakers placed in an arc around the listener. In that paradigm, a stimulus is presented on each trial from a randomly selected loudspeaker, often at a randomized level to reduce monaural level cues, and the listener is required to identify the activated loudspeaker. Analysis of errors is based on the differences between the source and response azimuths over numerous presentations from each loudspeaker. The left panel in Fig. 2.4 shows an example bubble plot describing localization responses for a BiCI user tested with pink noise bursts presented from an 8-loudspeaker array spanning a 180° arc in the forward direction. The leading diagonal in such plots corresponds to correct identification, and increased deviation from the diagonal reflects larger errors. When using only one CI, performance is usually poor and responses show large variation and/or bias towards the side of the ear being used. When using both implants, results are generally much better aligned with the correct response diagonal, particularly for loudspeakers that are not too far to the left or right. Strong overall bias shifts to the left or right are much less common than for unilateral listening. While details differ for individual listeners, response patterns for loudspeakers nearer the ends of the array are often compressed to varying degrees, as is evident in Fig. 2.4.

Overall performance in localization experiments is frequently reported using either the root-mean-squared metric (RMS, Rakerd and Hartmann 1986) or the mean-absolute error (MAE). The MAE is generally smaller than the RMS error by a factor related to the variance in responses. Summary error values calculated over the entire loudspeaker array are shown for various studies in Table 2.1. Values in italics show MAE measures for those studies that report that metric instead of the RMS error. The results from Seeber et al. (2004), and Agrawal (2008), who used pointer-methods that avoid the response quantization inherent in the source-direction identification task (Hartmann et al. 1998), are in good agreement with those from the identification studies.

Figure 2.5 shows those results from Table 2.1 that describe RMS errors. Solid symbols show errors when using both ears, and unfilled symbols are for unilateral

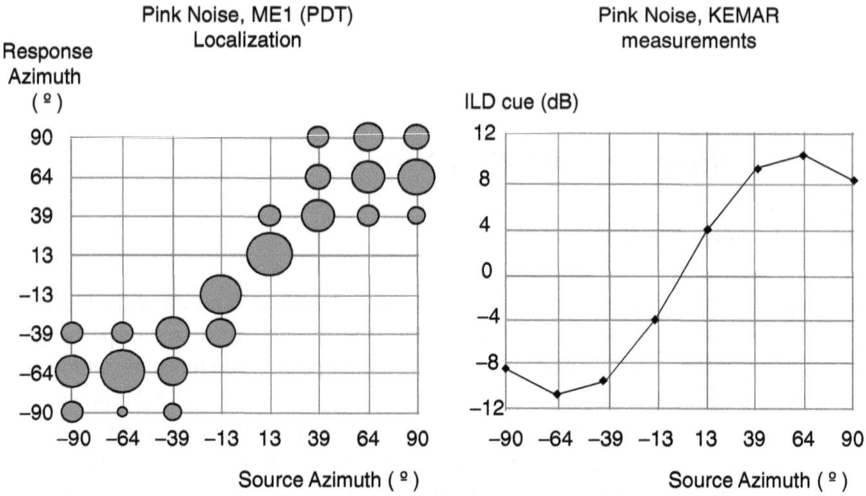

Fig. 2.4 (*Left*) Example bubble plot describing localization responses from a BiCI user tested by van Hoesel (2004) in a sound-direction identification task using 8 loudspeakers spanning 180°. The abscissa describes the source position for each loudspeaker, and the ordinate the response positions. Relative bubble diameters indicate the fraction of total responses at each position. Perfect performance corresponds to bubbles only along the leading diagonal. The signal was pink noise, roved in level over an 8 dB range. (*Right*) Broadband ILD cues measured for the same signal recorded using a KEMAR manikin in the same position as for the BiCI user data shown in the left *panel*

device use. Also shown are errors that would be obtained by a subject responding randomly (upper dashed line) and for fixed responding at 0° (lower dashed line). The results in Fig. 2.5 (and Table 2.1) show that unilateral performance is often near chance, although occasional subjects in some studies show reasonably good results even when levels are roved, suggesting an ability to use spectral cues. Localization errors when using both ears are about 2 to 4 times smaller than with either ear alone. The actual improvement in localization may be underestimated by that factor because the monaural error is limited by chance performance. The RMS error when using both ears is about 10° to 15° for small loudspeaker spans of about 100° and increases to about 30° for spans approaching 180°.

4.3.2 Available Cues

The increase in binaural localization error with span is well predicted by the spatial ILD function (van Hoesel 2004; van Hoesel et al. 2008), which describes interaural level cues available as a function of source azimuth. The right panel in Fig. 2.4 shows broadband ILDs measured using ear-level microphones for the same signal that was used to obtain the subjective results in the left panel. For narrow spans the ILD function is steeper and unambiguous across the entire array, although even then loudspeakers at the ends of the array are more often confused because of the

Table 2.1 BiCI localization results for various studies using multiple loudspeaker arrays

Study	Span° (spkrs)	Signal (rove dB)	Left (SD)	Bin (SD)	Right (SD)
van Hoesel and Tyler (2003) N=5	108° (8)	Pink noise (8 dB)			
		PDT	33° (10)	9° (2)	33° (6)
		Clinical	36° (18)	12° (4)	35° (8)
Laszig et al. (2004) N=16	360° (12)	Speech (5 dB)	87° (9)	50° (16)	89° (10)
Nopp et al. (2004) N=18	180° (9)	Speech-noise (20 dB)	*53° (15)*	*19° (10)*	*51° (17)*
Seeber et al. (2004) N=4	100° (11)	BB noise bursts (12 dB)	*30° (9) (1st CI)*	*15° (6)*	*30° (6) (2nd CI)*
Verschuur et al. (2005) N=20	180° (11)	Various (10 dB)	*67° (9)*	*24° (5)*	*67° (10)*
Grantham et al. (2007) N=18	160° (17)		Better ear		
		Noise	76° (13)	31° (10)	
		Speech	69° (15)	29° (13)	
Neuman et al. (2007) N=8	180° (9)	Pink noise (6 dB)	58° (17)	30° (13)	46° (7)
		Speech (6 dB)	60° (22)	32° (13)	53° (17)
Tyler et al. (2007) N=7	108° (8)	"Everyday sounds"		29° (11)	
Agrawal (2008) N=9	100° (11)	Speech (12 dB)	Better ear 60°	15°	
Laske et al. (2009) N=29	360° (9)	Speech		*57°*	
Litovsky et al. (2009) N=17	140° (8)	Pink noise (12 dB)	57° (15)	28° (13)	60° (15)

Column 1 shows the total error using either RMS or MAE (italicized) metrics for all loudspeakers, averaged across subjects in each study (and standard deviations). Column 2 lists the loudspeaker span used in each study, and column 3 describes signals and the amount of level roving employed. Columns 4–6 show results for left, binaural, and right-ear CI use, respectively. Two studies reported the better monaural result rather than left and right ear performance

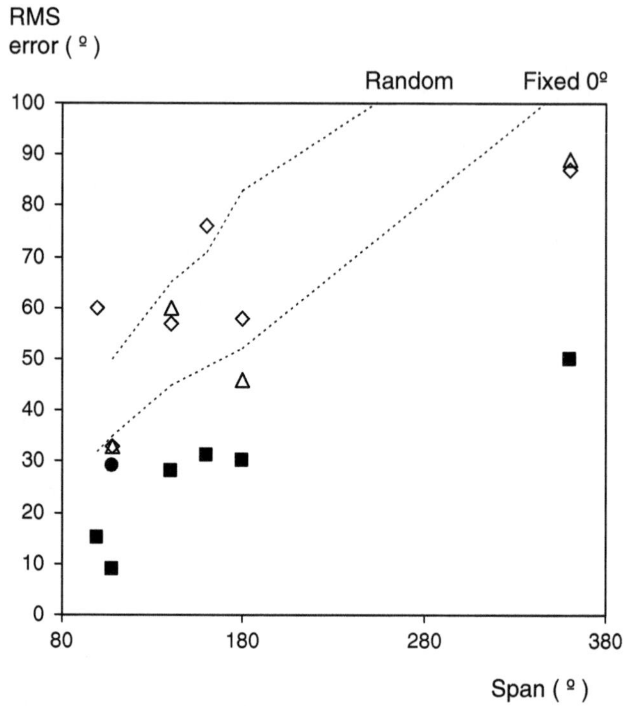

Fig. 2.5 Total RMS errors as a function of loudspeaker array span, for those sound-direction identification experiments listed in Table 2.1 for which the RMS error metric was reported. *Open symbols* describe monaural performance and *filled symbols* binaural performance. Data shown from left to right in order of increasing span are from Agrawal (2008), van Hoesel and Tyler (2003), Tyler et al. (2007), Litovsky et al. (2009), Grantham et al. (2007), Neuman et al. (2007), and Laszig et al. (2004). Additional details are described in Table 2.1. Chance performance values for random responding are shown by the *upper dashed line* and for fixed responding at 0° by the *lower dashed line*, and have been estimated here from the experimental conditions described in each study. The binaural datum from Tyler et al. (2007) (*filled circle*), which shows a relatively large error for the span used, is for a subject group who were nearly all implanted with a second device after more than 10 years of unilateral CI use, and were tested with less predictable signals than those used in most other studies. In the study by Agrawal (*left*) the active loudspeakers spanning 100° were concealed, and listeners were allowed to respond over a full range from −90° to +90°. Monaural chance errors therefore are higher than if responses were limited to the 100° presentation range (the upper dashed chance curve does not apply to those data)

shallower ILD slope. As the span increases beyond about 100°, cue ambiguity increases substantially because multiple loudspeakers nearer to the ends of the array produce similar ILDs, and sometimes even produce decreasing ILDs with increasing azimuth. When arrays span a full 360° circle (Laszig et al. 2004; Laske et al. 2009), much larger errors result from the large number of front-back confusions because ILDs are similar for sources to the front or rear of the listener (see Fig. 2.1). Although ILDs are smaller at lower frequencies, they are also less ambiguous (see Seeber and Fastl 2008, for a frequency dependent spatial response plot), and subjects who can selectively attend ILDs at different frequencies may obtain better performance when the broadband ILD is ambiguous. The ability of a listener to localize on the basis of ILDs will also be dependent on the translation of acoustic to electrical levels in the sound processors, front-end processing asymmetries such as independent activation of automatic gain control circuits (van Hoesel et al. 2002), electrical stimulation interactions among multiple electrode sites, the listener's sensitivity to electrical ILD cues, and the (cognitive) ability to relate those cues to sound-source direction. The evidence from several studies suggests that envelope ITDs are ineffective for signals containing discernable ILDs (van Hoesel 2004; Verschuur et al. 2005; Grantham et al. 2007; Neuman et al. 2007; Seeber and Fastl 2008). Only when ILDs are unavailable or ambiguous, and envelope fluctuation rates are sufficiently low, is there some indication that envelope ITDs can contribute. Similarly, inclusion of fine-timing ITD cues in the PDT strategy has so far not resulted in substantial reduction of localization errors (van Hoesel and Tyler 2003; van Hoesel et al. 2008) when compared to clinical strategies.

4.3.3 Minimum Audible Angle (MAA)

The ability to discriminate between sounds arriving from two different locations was reported for five BiCI users by Senn et al. (2005). For pairs of loudspeakers placed symmetrically in front of the listener, MAAs measured for white noise and click trains were between 4° and 8° for bilateral CI use, compared to 12–35° with either ear alone. When loudspeakers were placed symmetrically around 45°, binaural MAAs increased to between 5° and 20°, and at 90° (or –90°), BiCI MAAs were greater than 45° for all but one subject. Results were similar for signals presented from the front and the rear. The moderate increase in MAA when comparing results at 0° and 45° and the much larger increase at 90° are in good agreement with ILD cue considerations discussed above.

4.3.4 Localization in Noise

Localization in noise has been assessed in a smaller number of studies. Mosnier et al. (2009) reported percent correct scores for disyllabic words presented from a 5-loudspeaker array spanning 180°, while identical cocktail party noise was presented from all 5 loudspeakers. Van Hoesel et al. (2008) tested BiCI users' abilities to localize click trains in spectrally matched noise using PDT and two clinical strategies. Noise was presented at a fixed location, at either 0° or 90°, and at 0 dB SNR. Results were similar with PDT and the most familiar clinical strategy, and binaural RMS errors for the 180° span ranged from about 25° to 35° depending on click rate and noise position. Analysis of response patterns showed that 99% of the variance in the subjects' responses was accounted for by the combined broadband target-plus-noise ILD cues, but listeners adjusted the response range for the different noise positions to compensate changes in absolute ILDs. Agrawal (2008) tested localization of a target word spoken by a male speaker using 11 (concealed) loudspeakers spanning 100°, both in quiet and in the presence of 2 competing sentences spoken by a female and presented randomly from 2 of 4 locations at +–35° and +–55°. Results showed RMS errors of about 15° in quiet, which is in good agreement with results from other studies using a comparable span (Table 2.1). At SNRs of 10, 0, and –5 dB, the RMS errors increased to about 20°, 25°, and 30° respectively. Results with normal-hearing control subjects showed considerably smaller errors (RMS errors of 6° and 9° at SNRs of +10 and −5 dB respectively). In a further experiment, the target word was varied on each presentation, and the listener was instructed to identify the speech token in addition to determining its location. Comparison with results for each task performed separately showed that neither speech intelligibility nor localization was affected by the requirement to perform the two tasks simultaneously over the range of SNRs tested.

4.3.5 Time Course Considerations (Localization)

Grantham et al. (2007) reported no significant change in localization abilities for 12 subjects tested at 5 or 10 months after receiving bilateral implants. Litovsky et al. (2009) reported on localization performance in simultaneously implanted BiCI users after 3 months' experience with both devices. More listeners showed a bilateral benefit relative to unilateral results for a left-right discrimination analysis than for a within-hemifield analysis. That result was attributed to the use of a simple left-right discrimination mechanism in those inexperienced listeners, rather than a more fine-tuned localization mechanism that may develop with prolonged bilateral device use. However, the RMS error for that subject group appears in good agreement with other studies involving subjects with more bilateral listening experience, and smaller binaural benefit for the within-hemisphere analysis than left-right discrimination is also expected on the basis of the spatial ILD function because it is steepest at 0° (Figs. 2.1 and 2.4). Long term BiCI use in the study by Chang et al. (2010) showed little change in RMS errors when results at 12 months were compared with subsequent outcomes in 10 subjects who were assessed at multiple times up to at least 4 years after bilateral implantation.

4.4 Subjective Measures and Cost Effectiveness

To assess the effectiveness of bilateral cochlear implantation in ways that may not be easily or efficiently determined in laboratory settings, subjective self-rating questionnaires have been used. Results demonstrate various benefits from bilateral implant use (e.g., Summerfield et al. 2006; Laske et al. 2009; Litovsky et al. 2006c; Tyler et al. 2009; Veekmans et al. 2009), particularly in relation to spatial hearing, although less so for elderly listeners (Noble et al. 2009). Cost-effectiveness of bilateral implantation has been assessed by comparing the total cost of receiving and maintaining two implants (over the anticipated lifetime of a patient) against the total improvement in quality of life, measured as the increase in number of quality adjusted life years. Summerfield and colleagues over the course of several studies (e.g., Summerfield and Barton 2003; Summerfield et al. 2006) concluded that bilateral implantation was unlikely to be cost effective (in the UK) unless further performance gains were achieved using sound processing or the cost of bilateral implantation was reduced. In contrast, Bichey and Miyamoto (2008) showed a favorable outcome that was considered well below the figure suggested to represent a reasonably effective treatment in the United States. Bond et al. (2009) concluded from a probabilistic modeling study that implantation with a second device leads to considerably higher overall cost utility than unilateral implantation, but also noted that the bilateral predictions are more error prone than are the unilateral ones, especially for children.

4.5 *Pediatric BiCI Outcomes*

While the outcomes in adult BiCI users show clear benefits of bilateral implantation, that outcome may be dependent on having had hearing in both ears early in life. An important question, therefore, is whether similar benefits are available to congenitally deaf children or to those who lose their hearing very early in life.

4.5.1 Speech Intelligibility and Detection

The left panel in Fig. 2.6 shows results for speech intelligibility in quiet from those pediatric BiCI studies with larger numbers of subjects that used fixed level tests. The benefit of adding the second ear (CI-2) is shown as a percentage point increase, usually relative to the result with the first implanted ear (CI-1) alone. Word recognition scores from several studies shown in the figure (Peters et al. 2007; Gordon and Papsin 2009; Scherf et al. 2009a) are averaged over various subgroups of subjects and are discussed in more detail in Sect. 4.5.2. The measured benefit ranges from about 2 pp to 12 pp, which is comparable to that seen in adults. While benefits described relative to CI-1 may be inflated by better ear contributions (see Sect. 4.1), the first implanted ear is often the ear with better performance, at least for sequentially implanted children with longer intervals between implantations. SRTs in quiet were measured by Litovsky et al. (2006b), and showed a larger binaural improvement of about 5 dB. That larger benefit presumably at least partly arises from increased audibility of low level signals because of loudness summation, and perhaps also the increased likelihood of presenting complementary information when audibility in each frequency region differs between ears. Results from several studies for speech and noise both presented at 0° (or for noise at 180°) are shown in the right panel of Fig. 2.6. Benefits are slightly larger than in quiet, ranging from 7 to 17 pp. While that seems slightly larger than the 1 dB diotic benefit seen in adults,

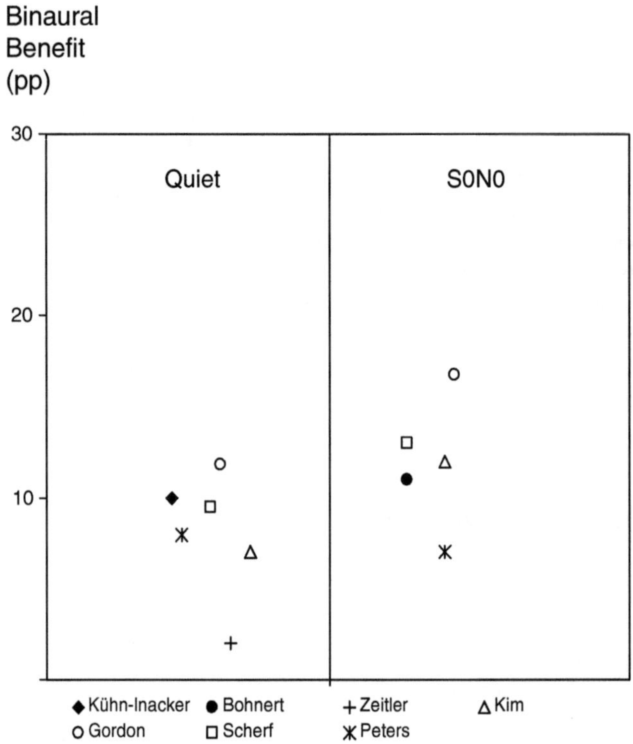

Fig. 2.6 Pediatric speech benefits in quiet and spatially coincident noise (fixed level test results) usually measured relative to the first implanted ear (except for the data from Scherf et al. which indicate the difference between binaural and group-averaged better monaural results). Results are from Kühn-Innacker et al. (2004), Bohnert et al. (2006), Peters et al. (2007), Zeitler et al. (2008), Kim et al. (2009), Gordon and Papsin (2009), and Scherf et al. (2009a)

the difference may be the result of better ear contributions. Two additional studies using SRT methods (Litovsky et al. 2006b; Wolfe et al. 2007) show considerably larger than expected binaural benefits given the absence of headshadow and binaural difference cues in this configuration. High variability for some subjects may account for the 5 dB benefit seen in the former of those studies. The 6 dB benefit reported in the latter study may be the result of presenting modest-level noise from behind listeners wearing directional microphones. That combination may have led to low level signals for which binaural loudness summation and increased complementary signal presentation may have played a larger role. A few studies have assessed speech outcomes with a single spatially separated interferer (Litovsky et al. 2006b; Peters et al. 2007; Galvin et al. 2007, 2008; Steffens et al. 2008). While the monaural benefit due to the headshadow approaches that found in adults, the binaural benefit obtained when adding CI-2 at a favorable SNR is often smaller than predicted by the headshadow because performance with the later implanted ear is poorer. Kühn-Innacker et al. (2004) reported a fairly large binaural benefit of 18 pp at a high SNR of 15 dB, but interpretation of the underlying contributions is complicated by the symmetric presentation of both target speech and noise from multiple locations. Schafer and Thibodeau (2006) presented sentences from 0° and two uncorrelated multi-classroom recordings at 135° and 225°, and found a bilateral benefit of 2 dB relative to the first implanted ear. Allowing for better ear-contributions that result appears in approximate agreement with the adult outcomes, which show only small binaural benefits on the order of 1 dB.

4.5.2 Time Course Considerations in Children (Speech)

Several studies have described longitudinal speech outcomes in children delineated according to age at implantation in one or both ears (e.g., Peters et al. 2007; Wolfe et al. 2007; Gordon and Papsin 2009; Scherf et al. 2009a). Monaural performance is generally more similar between ears when children are implanted early and the delay between implantations is small or absent. After about 1 year of bilateral listening experience, results are often comparable if both ears have been implanted before the age of 4 or 5. Larger monaural differences are seen if the first ear is implanted early and the second is implanted late, although with prolonged experience the difference is eliminated. In contrast, for listeners who are implanted late in the first ear and experience large delays before the second surgery, performance in the second ear remains considerably poorer than the first, even after several years. Most studies report that the binaural benefit relative to monaural performance in noise increases with ongoing listening experience. While that result is sometimes discussed in terms of developing binaural functionality, examination of the data shows that, at least in sequentially implanted children, it is also the result of reduction in monaural CI-1 performance over time. For example, in the study by Peters et al. (2007), comparison of results at 3 and 9 months shows considerably smaller increases in bilateral scores than decreases in monaural CI-1 scores. The magnitude of the bilateral advantage as a function of age at implantation in each ear shows more variable outcomes across studies. Gordon and Papsin (2009) showed that early implantation in both ears leads to a larger diotic binaural benefit relative to the first implanted ear. However, it is not clear from that outcome that binaural benefit *per se* increases with shorter delays. The reason for that is that shortening the delay also increases the likelihood of the second ear being the one with better performance, which leads to greater inflation of the estimated binaural benefit (see Sect. 4.1). The data from Scherf et al. (2009a) show that after sufficient experience diotic binaural benefits in children receiving CI-2 after the age of 6 can be as large as in those implanted bilaterally before that age.

4.5.3 Localization, Left-Right Discrimination, and MAA

Localization abilities using multiple-loudspeaker arrays in children implanted in the second ear at a relatively late age have been shown to be poor (Litovsky et al. 2004; Bohnert et al. 2006; Galvin et al. 2007; Steffens et al. 2008). Van Deun et al. (2010b) tested binaural localization abilities using 9 loudspeakers spanning 120° for children aged 4 to 15. Although the reported mean RMS error of 38° is fairly large for that span, best performers showed results that were comparable to those from adult BiCI users. The largest predictor of localization ability in that study was the availability of auditory stimulation early in life, either acoustically or electrically, in at least one ear. Age at second implantation was a significant predictor only when children who used a hearing aid in the second ear prior to implantation were excluded from the analysis. Using a simple left-right discrimination task, performance in young children under the age of 4 has been found to be much better when using both implants than either alone (Beijen et al. 2007; Galvin et al. 2008). Litovsky et al. (2006a) measured MAAs in children ranging between 3 and 16 years of age and found large variation in results, ranging from an inability to discriminate left from right at any loudspeaker separation, to inter-speaker thresholds as low as 10°. Unilateral thresholds were found to be better with CI-1 than CI-2, and MAAs were uncorrelated with speech results (Litovsky et al. 2006b) for the 6 subjects tested on both measures. Grieco-Calub et al. (2008) compared MAAs in children implanted bilaterally before the age of 2.5 and tested before the age of 3 with those of 8 age-matched unilateral CI users, as well as 8 children with normal hearing. None of the unilateral CI users could perform the task, whereas about half of the BiCI users could. Among those that could, MAA thresholds varied widely, ranging from as little as 10° in one case to about 110°. The two best performers showed comparable performance to the normal hearing children. Although those two children received their second implant earlier and experienced shorter delays between implantations than most children in the BiCI group, a third child with similar experience could not complete the task. While the best performers in some of these studies show results that are similar to the performance shown by normal-hearing children below the age of about 6, the considerably better performance seen at an older age in normal hearing remains to be demonstrated in BiCI users.

4.5.4 Subjective Measures in Children

Several studies (e.g., Beijen et al. 2007; Galvin et al. 2007, Van Deun et al. 2010b) have used the SSQ questionnaire (Gatehouse and Noble 2004) or a modified version (Galvin et al. 2007) to assess pediatric BiCI users. As in adults, significantly higher SSQ scores are seen with bilateral device use, particularly for the questions relating to the spatial hearing domain, although correlation with localization measures in some of those studies is moderate. Scherf et al. (2009b) used the Würzburg questionnaire (Winkler et al. 2002) and found more sustained benefit for children who received a second implant before the age of 6 than those implanted later in the second ear.

Those authors also noted that, in terms of auditory awareness and speech production, bilaterally implanted children achieved matched outcomes to unilateral CI users over shorter time periods.

5 Psychophysical Studies with Adult BiCI Users

Early case studies with BiCI recipients (Pelizzone et al. 1990; van Hoesel et al. 1993; van Hoesel and Clark 1997; Lawson et al. 1998) demonstrated that direct electrical activation of place-matched electrodes in each ear could lead to fused percepts, that loudness increased compared to stimulation in either ear alone, and that while lateralization could be readily affected by electrical ILDs, outcomes with ITDs were more variable and rate dependent.

5.1 ILDs and ITDs at Low Pulse Rates

Lateralization of electrically induced sound percepts has been shown to be affected by both ILDs and ITDs in low-rate unmodulated pulse trains (van Hoesel and Tyler 2003; Poon et al. 2009; Litovsky et al. 2010). Electrical ILD thresholds typically range from less than 0.17 dB to at most about 1 dB, and in many listeners changes in ILD of a few dB can produce lateralization shifts of more than 50% of the full range. The effect of ITDs on lateralization is more variable across listeners. For those with adult onset of deafness, ITD thresholds typically range from below 100 μs to about 500 μs, and in some cases changes in ITDs from –800 μs to +800 μs can also produce lateral shifts in excess of 50%. In contrast, none of a small number of subjects with early onset of deafness included in the study by Litovsky et al. (2010) showed any consistent lateral position shifts even with large ITDs. ITD thresholds have not been found to be systematically related to the matched place of stimulation in both ears (van Hoesel et al. 2009; Litovsky et al. 2010) but do increase when place is mismatched between ears (Poon et al. 2009) in a manner that is consistent with estimates of spread of excitation along the cochlea (Cohen et al. 2001).

5.2 Rate Effects

Figure 2.7 shows whole-waveform ITD thresholds from three studies as a function of pulse rate for 14 subjects who all displayed good ITD sensitivity at 100 pps. Overall, ITD sensitivity deteriorates as pulse rate increases. At pulse rates below about 200 pps the average threshold in these subjects is about 150 μs, whereas at 800 pps or higher most subjects are unable to detect ITDs within the range available from natural head-width considerations despite the presumed availability of the onset cue. However, in

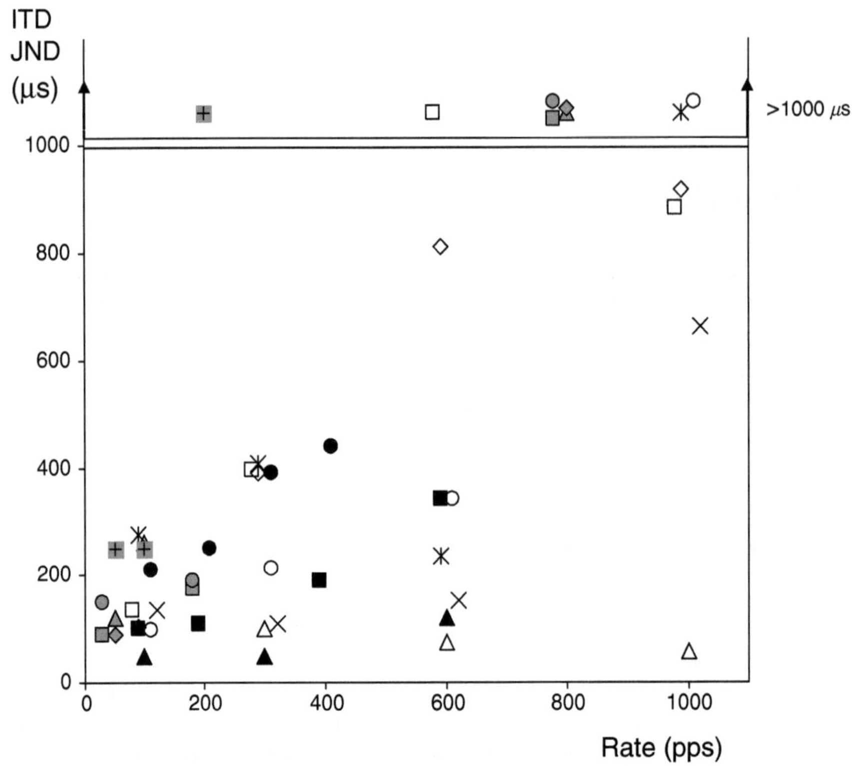

Fig. 2.7 Whole-waveform ITD thresholds as a function of pulse rate using unmodulated electrical pulse trains. Stimuli were 300 ms in duration with 0 ms rise times and were applied to place-matched electrodes at stimulation levels in the upper portion of the dynamic range. Each symbol shape represents 1 of 14 subjects in three different studies: van Hoesel and Tyler (2003); van Hoesel (2007); van Hoesel et al. (2009). For the last of those studies, the data from the most apical electrodes tested are shown. Symbols plotted above the graph at threshold values beyond 1000 μs indicate JNDs could not be determined for ITDs of up to 1 ms

some subjects measurable ITD sensitivity remains at high rates even when slow rise times are applied to the envelope to reduce the effectiveness of the onset (Majdak et al. 2006; van Hoesel et al. 2009). In agreement with decreased ongoing ITD-cue salience at high rates, the improvement in ITD threshold when adding more pulses by increasing stimulus duration is largest at low rates and, at 1000 pps, ITD sensitivity can in fact be poorer for 300-ms bursts than for a single pulse (van Hoesel et al. 2009). That result may be attributed to ITD ambiguity in relation to the interpulse interval and refractory effects that occur in the pulse-train but not the single-pulse stimulus. Poor whole-waveform ITD sensitivity at high pulse rates can be largely restored to that seen at low rates by applying deep low-rate amplitude modulation (AM) (van Hoesel and Tyler 2003; Majdak et al. 2006; van Hoesel et al. 2009). Laback and Majdak (2008) showed that high-rate ITD sensitivity can also be improved by applying diotic rate jitter to the pulse rate and attributed that result to binaural restarting because of the ongoing rate variations. An alternative explanation of the beneficial effect of jitter relates to its reduction of high-rate cue uncertainty and introduction of low-rate cues, both in the form of lengthened interpulse intervals and in the modulation of the responses of integrating neural circuits in the auditory system (van Hoesel 2008; van Hoesel et al. 2009). Although the introduction of AM or rate jitter can improve electrical ITD sensitivity at high rates, at best it restores sensitivity to that seen at lower electrical pulse rates. While that low-rate ITD sensitivity in BiCI users approaches the relatively poor ITD sensitivity seen in listeners with normal hearing attending 100-Hz pure tones, performance in normal hearing improves by about an order of magnitude as the pure-tone frequency increases to about 1 kHz. The effect of rate on electrical ITD sensitivity better resembles that seen with envelope ITD cues for high frequency signals in normal hearing (Hafter and Dye 1983; Bernstein and Trahiotis 2002). That outcome is unlikely to be the result of insufficient insertion depth of electrodes into the cochlea (Stakhovskaya et al. 2007). It may be largely the result of much more synchronous activation of nerves over a fairly broad region along the cochlea than occurs in normal hearing. In agreement with that conjecture, Colburn et al. (2009) showed that substantially rate-limited ITD sensitivity can arise from highly synchronous inputs using a simple model for electrical ITD sensitivity in the brainstem.

5.3 Onset Dominance and Precedence

If it is assumed that the onset response is unaffected by later arriving pulses, the decrease in ongoing ITD sensitivity at high rates means that in relative terms the onset becomes more effective. Several BiCI studies have confirmed that result by applying different ITDs to onset and later arriving pulses. Using a binaural beat paradigm, the salience of slowly varying ITD cues following a diotic onset degrades rapidly for pulse rates above about 200 pps (van Hoesel and Clark 1997; van Hoesel 2007; Carlyon et al. 2008). In agreement with precedence studies in normal hearing listeners, the ability to discriminate ITDs applied to different electrical pulses is consistently better for the first pulse than later arriving pulses when they are separated by only 1- or 2-ms intervals, but is more similar when they are separated by larger intervals (Laback et al. 2007; van Hoesel 2007; Agrawal 2008). Free-field precedence results from BiCI users (Agrawal 2008) have shown more moderate onset dominance and rate effects, which is in agreement with the greater role of ILDs (Sect. 4.3.2) and poor coding of ITDs in sound processors (Sect. 3). The greater effect of rate on ITD than ILD sensitivity is also evident in the data from van Hoesel (2008), who used an observer weighting paradigm to assess the relative strengths of cues applied to each pulse in the stimulus directly. Results from that study showed that post-onset ITDs and ILDs contributed strongly to perceived lateral positions at 100 pps. At 300 and 600 pps, those contributions remained substantial for ILD cues but were much reduced for ITDs, particularly at 600 pps. Further results from that study provided no evidence of a gradual adaptation process or binaural restarting following an irregular shortened interpulse interval with brief electrical pulse trains.

5.4 Through-the-Processor (TTP) Measures

Several studies have reported on the abilities of BiCI users to hear binaural cues contained in audio signals when processed by sound processors (Laback et al. 2004; van Hoesel 2004; Senn et al. 2005; Grantham et al. 2008). It should be noted that "thresholds" measured in that manner describe the combined effects of sound processor modifications and listeners' perceptual abilities. Accordingly, TTP thresholds for ongoing fine-timing ITDs will be unmeasurable with clinical strategies (Senn et al. 2005) because the cue is not coded electrically. TTP thresholds for whole-waveform ITDs or envelope ITDs in signals that have relatively fast or shallow modulations have also been found to be poor but can be comparable to results for low-rate direct stimulation when using low-rate click trains that have deep, slow modulations. TTP thresholds for acoustic ILDs, using typical acoustic-to-electrical mapping procedures, have been reported to be on the order of one to a few dB.

5.5 Binaural Masking Level Differences (BMLDs)

Van Hoesel (2004) compared TTP detection thresholds for diotic and phase-inverted 500 Hz pure tones in diotic noise using the PDT strategy and found BMLDs of 1.5 to 2 dB in two BiCI users, which is much less than is seen for those conditions in listeners with normal hearing. Long et al. (2006) found larger BMLDs of 9 dB for phase-inverted 125-Hz sinusoidal envelopes in narrow band (50 Hz) diotic noise, most of which was the result of envelope fluctuations below 50 Hz. Van Deun et al. (2010a) used the same stimuli with pediatric BiCI users and also showed a mean threshold reduction of about 6.4 dB. Note that phase inversion at 125 Hz corresponds to an ITD of about 4 ms, which is more than 5 times larger than can be imparted by the human head. Lu et al. (2010) extended the parameters used by Long et al. (2006) and found substantial inter-subject variations. Whereas 3 of 5 subjects showed only small BMLDs on the order of 1 dB, the other two showed BMLDs near 10 dB. The latter two showed large BMLDs even at the high modulation rate of 500 Hz, for which ITD sensitivity is expected to be relatively poor and for which phase inversions corresponds to a smaller ITD of 1.25 ms.

5.6 Additional Measures

Several psychophysical outcomes of interest have been assessed in only a few subjects and require further validation in larger numbers of subjects. Binaural loudness summation was assessed in 2 early BiCI recipients (van Hoesel and Clark 1997) and showed an approximate doubling of loudness when monaural loudness of single electrode components in each ear was matched. The same was observed in a different subject (van Hoesel 2004) using multi-electrode stimulation. Diotic rate discrimination in the study by van Hoesel and Clark (1997) was comparable to monaural rate discrimination, and central masking was clearly observed, but in contrast to normal hearing outcomes (Zwislocki et al. 1968) was not strongly place-match dependent. Two of the 3 subjects tested by van Hoesel (2007) showed better sensitivity to ITD cues at stimulation levels near 85% of the dynamic range, than at lower levels near 60%, suggesting better ITD sensitivity at higher levels.

6 Physiological Studies

6.1 Unit Responses

Smith and Delgutte (2007a) measured single unit responses to ITDs in the inferior colliculus (IC) of acutely deafened cats. Responses to unmodulated electrical pulse trains displayed peak firing rates for most neurons at preferred ITDs (ITD_{best}) within the natural head-width range of cats, although peaking behavior was evident over only a small dynamic range of at most a few dB. In contrast to studies with acoustic stimulation, a systematic relation between ITD_{best} and the tonotopic axis in the IC was not found. At low rates of 40 pps, individual neurons produced spikes on almost every stimulus pulse at favorable ITDs whereas at higher rates up to 320 pps responses became increasingly limited to the onset. That rate limitation is also seen in IC recordings for unilateral electric stimulation but not for stimulation with pure tones in normal hearing animals, and also to a much lesser extent in auditory nerve data for electric stimulation. The effect of rate, the difference with acoustic stimulation, and robust coding of onset cues are all consistent with the human psychophysical outcomes described in the previous section. Smith and Delgutte (2008) further reported IC responses to sinusoidally amplitude-modulated (AM) pulse trains at carrier rates of 1000 or 5000 pps. About half of the neurons responded only to ITDs applied to the envelope (ITD_{env}), whereas the other half also showed sensitivity to fine-timing ITDs (ITD_{fs}) when the carrier pulse rate was 1000 pps, but not when it was 5000 pps. At 1000 pps and 40-Hz AM, ITD_{fs} tuning was considerably sharper than ITD_{env}, and estimated ITD_{fs} thresholds were comparable to those for low rate unmodulated pulse trains. The improvement in ITD_{fs} resulting from AM in the physiological data was however not seen at pulse rates of 5000 pps, which appears to be in contrast to human behavioral data that show that whole-waveform ITD sensitivity for 100-Hz AM applied to 6000-Hz pulse trains approximately matches that seen with unmodulated pulse trains at 100 pps (van Hoesel et al. 2009). Kral et al. (2009) compared the propagation of local field patterns recorded at the surface of the auditory cortex in congenitally deaf cats and acutely deafened hearing cats in response to bilateral electrical stimulation, and found that while a fast cortical wave was generated by cochlear implant stimulation, it was modified in congenitally deaf cats that lacked hearing experience. Tillein et al. (2010) compared intracortical multi-unit responses from the primary auditory cortex in congenitally deaf cats and acutely deafened hearing cats using three-pulse electrical stimuli at 500 pps. Results showed that while some aspects of subcortical ITD coding were preserved in the congenitally deaf animals, fewer units responded to ITDs, maximum evoked firing rates were lower, fits to ITD response templates were poorer, and response patterns potentially reflecting precedence related processing were largely absent.

6.2 Evoked Responses

The binaural interaction component (BIC) is calculated as the difference between the binaural response and the sum of monaural responses, and is assumed to arise as result of binaural interaction. Pelizzone et al. (1990) measured electrical evoked auditory brainstem responses (EABRs) in an early BiCI recipient and reported a BIC that resembled that seen in normal hearing. Smith and Delgutte (2007b) confirmed Pelizzone's hypothesis that the BIC would be largest for optimally matched places of electrical stimulation in each ear by comparing responses with multi-unit recordings in the IC using a cat model. Thai-Van et al. (2002) recorded EABRs in two BiCI users and found larger wave V latencies in the ear with longer duration of deafness. Gordon et al. (2007, 2008) found the same result in children implanted in each ear at various ages and also reported that latency decreased over time with bilateral implant use in children with short delays between implantations. Children with long delays showed less evidence of change when the first implant was at a young age, and least change was found for children who received both implants at a later age and had a long delay between implants. Similarly, for sequentially implanted children implanted early in the first ear, BICs recorded shortly after receiving a second implant showed prolonged latencies that were largely eliminated with ongoing CI use in children with short delays, but not those with long delays between implantations. Sharma et al. (2007) examined P1 response latencies in cortical auditory evoked potentials measured in children implanted in both ears before the age of 3.5, either simultaneously or with delays of 0.25 to 1.7 years (mean ~1 year), to test the hypothesis that simultaneous implantation may offer more rapid maturation of the central auditory pathways. Results showed that the P1 response did not differ significantly and both groups reached normal hearing limits within about 3 months. Iwaki et al. (2004) and Sasaki et al. (2009) measured cortical P3 response latencies in both BiCI users and unilateral implant users with a contralateral hearing aid (CIHA) when presented with occasional 2-kHz target tones amongst frequent 1-kHz non-target tones. Results showed significantly reduced P3 latencies for the bilateral listening conditions, which was interpreted as evidence that the task required less effort with two ears than with one for both BiCI and CIHA users.

7 Combining a Cochlear Implant with a Contralateral Hearing Aid (CIHA)

Whereas signals are processed in a similar manner for both ears in bilateral implant users, a CI user with a hearing aid (HA) in the opposite ear receives substantially different information at the two ears. The HA provides low frequency temporal information that is absent or poorly presented by the CI, and the CI provides high frequency spectral information that is often largely inaudible with the HA. Accordingly, benefits in CIHA users are expected to be more complementary in nature than for listeners with similar hearing in both ears. Psychophysical studies by Francart et al. (2008, 2009) have shown that it is possible under well controlled conditions to elicit ILD and ITD sensitivity in CIHA users. However, ITD sensitivity in those studies was measurable only for carefully selected delay and level combinations, making practical application difficult, and ILDs are small at the low frequencies often available to HA users. For CIHA studies to date reporting on speech intelligibility and localization performance with clinical devices, binaural cues will have been much more poorly controlled, if not entirely unavailable. These considerations warrant caution in drawing parallels with mechanisms leading to binaural benefits in listeners with normal hearing (or BiCI users). In many CIHA users, speech intelligibility with the CI alone is substantially better than with the contralateral HA alone. Accordingly, the main emphasis in this section is on the incremental benefit provided by adding the HA in such listeners. In CI users with high performance approaching normal hearing in the acoustically stimulated ear (Vermeire and van de Heyning 2009; Cullington and Zeng 2010) outcomes and mechanisms may differ from those discussed here.

7.1 CIHA Benefits in Adults

7.1.1 Speech Outcomes

Most speech studies with adult CIHA users have used fixed level testing, with words or sentences in quiet, and in noise at SNRs of 5 to 10 dB. Figure 2.8 shows summary results from more recent studies (with presumably better quality hearing aids than earlier studies) that included larger numbers of subjects. The plotted values describe mean percentage point increases for the CIHA listening condition compared to CI alone in each study. For tests in quiet, mean speech scores with the CI alone are often 2 or 3 times better than with the HA alone. Despite that large difference, when the HA is added to the CI, mean scores improved in these studies by 9 to 17 pp for various word or sentence materials. Assuming typical PI gradients in the range 5 to 10%/dB, the benefit of adding the HA in quiet is at least as large as the performance gain

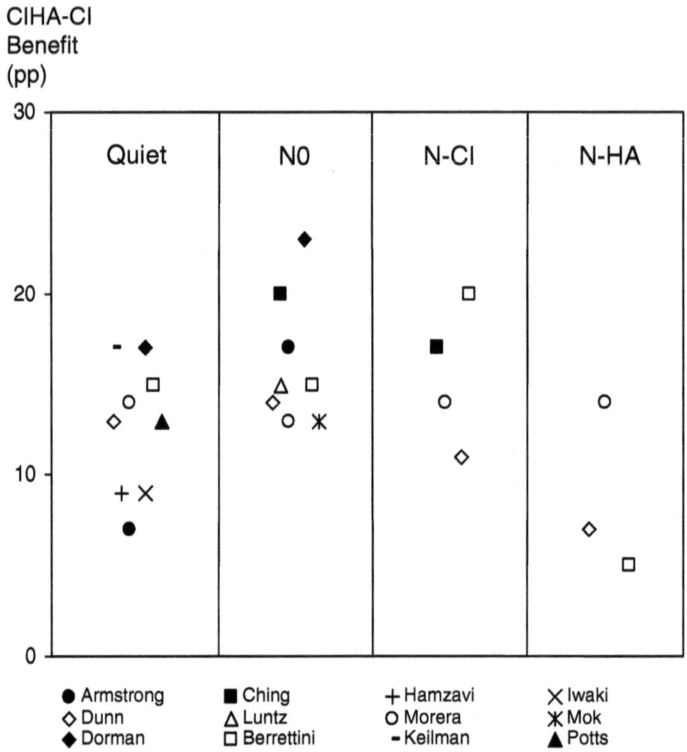

Fig. 2.8 Speech perception benefits derived from adding a HA contralateral to a CI from various studies, shown as percentage point increases in scores for combined CIHA use relative to CI only. Results in the four panels, from left to right, are for speech in quiet, in spatially coincident noise (S0N0), with noise on the implant side (S0N-CI), and noise on the hearing aid side (S0N-HA). Data for tests in spatially separated noise are for S0N90 conditions, except for those from Ching et al. (2004) who presented speech and noise on opposite sides of the head at 60°, with noise ipsilateral to the implant. *Filled symbols* are for the studies with the largest numbers of subjects (N = 15 to 21). Data are from Armstrong et al. (1997), sentences (average of the 2 groups of subjects); Ching et al. (2004), sentences (experienced HA users); Hamzavi et al. (2004), sentences; Iwaki et al. (2004), words in quiet, sentences in noise; Dunn et al. (2005), words in quiet, sentences in noise (0 to 10 dB SNR); Luntz et al. (2005), sentences (7 to 12 month data); Morera et al. (2005), words (+10 dB); Mok et al. (2006), sentences (+5 or +10 dB SNR); Dorman et al. (2008), sentences (+10 dB/+5 dB); Berrettini et al. (2009) words; Keilmann et al. (2009), (Optimized case); and Potts et al. (2009), words in quiet (roved location)

in BiCI users when adding the poorer performing ear, despite the fact that in BiCI users performance is usually more similar in the two ears. Similarly, for the S0N0 condition the addition of the HA to the CI improves scores between 13 and 23 pp despite the fact that the mean HA-alone score can be close to 0%. Again, this is more than is seen in BiCI users when bilateral performance is compared with the better ear alone. Iwaki et al. (2004) and Mok et al. (2006) reported adaptive SRT benefits of 4 dB and 2 dB respectively, when adding the HA in the S0N0 condition, whereas the benefit in adult BiCI users is about 1 dB. A small number of studies in which

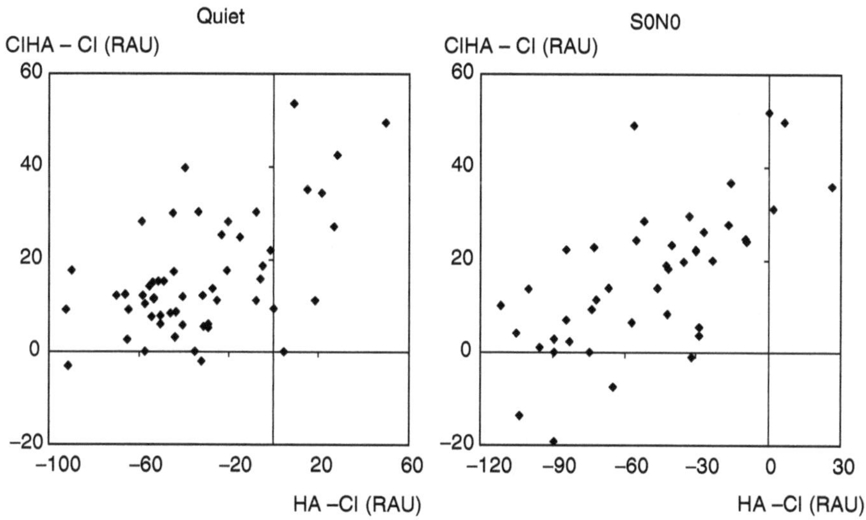

Fig. 2.9 Individual subject CIHA benefits, expressed as the increase in CIHA relative to CI only performance and plotted as a function of the difference between HA and CI alone scores (HA – CI) for each subject. To reduce the impact of floor and ceiling effects, results are shown in rationalized arcsine units (RAU) (Studebaker 1985). For speech in quiet, data are from Hamzavi et al. (2004); Dunn et al. (2005); Mok et al. (2006); Gifford et al. (2007) (words); Keilmann et al. (2009); and Potts et al. (2009). For speech in noise (S0N0), data are from Dunn et al. (2005); Luntz et al. (2005); Mok et al. (2006); and Gifford et al. (2007)

performance was assessed for spatially separated speech and noise (Fig. 2.8, right panel) suggest that the benefit of adding the HA at a favorable SNR provides no more benefit than when the HA is added in the N0 condition, which may be partly attributable to the fact that the headshadow is fairly ineffective at low frequencies. SRT data for that condition from Iwaki et al. (2004) show a large benefit of more than 6 dB but also show high variance, and the SRT benefit reported by Mok et al. (2006) is a more moderate 3 dB. When the noise is ipsilateral to the HA, benefits may be somewhat smaller than for the N0 condition, but that conclusion is limited by the small number of available data.

Several CIHA studies have stressed the importance of carefully adjusting both the CI and HA to obtain optimal absolute outcomes (e.g., Ching et al. 2004; Keilmann et al. 2009; Potts et al. 2009). Nevertheless, the results from the studies in Fig. 2.8 show fairly consistent mean benefits from adding a HA irrespective of whether that has been a key consideration or not in each study. Within-study inter-subject variation in the benefit derived from adding the HA to the CI, in contrast, is very large. Attempts to predict that benefit from hearing abilities with the HA ear alone show mixed outcomes. A potential contribution to those varied outcomes is that the measurement of benefit is relative to CI performance, which in itself is highly variable across subjects. A subject with good CI performance may reasonably be expected to gain less from a moderate HA result than would a subject with poor CI performance. To assess that conjecture, Fig. 2.9 shows benefits for individual subjects when adding the HA, as a function of how well the HA performs relative to the CI. Results are from six studies in which individual subject data were reported for words in quiet, and four studies for sentences in spatially coincident noise (S0N0). Both plots show increasing benefit as the HA performance improves relative to that with the CI alone. Regression analysis shows that the difference between HA and CI performance accounts for about 42% ($p<0.001$) of the variance for speech in noise and 28% ($p<0.001$) for speech in quiet. Similar evaluations for absolute performance scores with the HA alone account for about 21% ($p<0.002$) and 22% ($p<0.001$) respectively. A linear fit to the data for sentences in noise suggests a fairly shallow slope with an increase in benefit of about 1 pp for every 3 pp increase in HA-CI difference score. That shallow slope supports the conjecture that the CIHA benefit (relative to CI alone) is not strongly affected by the position of the noise.

Kong et al. (2005) found that the addition of the HA to a contralateral CI in 4 CIHA users offered improved ability to understand a male target speaker in the presence of another single interfering speaker, particularly when the interferer differed in gender from the target. However, an earlier study by Blamey et al. (2000) showed no benefit of adding the HA in three CIHA users when listening to a target female talker in the presence of a single competing male talker, so further data are needed from a larger number of subjects. The CIHA benefit found by Kong et al. was attributed to listeners abilities to correlate low frequency pitch in the HA signal with weak temporal pitch cues in the CI envelope. An even simpler interpretation, which would not require listeners to correlate electric and acoustic signals with substantially different properties, is that improved ability to detect the temporal target-speech boundaries in the HA signal may be used to better identify which changes in the spectrally rich CI signal are in response to the target signal. In accordance with the large asymmetries in signal representation in each ear and lack of fine-timing ITDs in the CI processors, Ching et al. (2005) found no evidence of binaural speech unmasking in CIHA users when applying target ITDs of 700 μs.

7.1.2 Localization

Figure 2.10 shows mean RMS errors for adult CIHA users participating in three sound-direction identification studies, plotted as a function of the loudspeaker span. The result on the left is from Dunn et al. (2005), who measured bilateral performance with a relatively small span of 108°. The best performer in that study showed an RMS error of about 28°, which is 2 to 3 times larger than typically seen in BiCI users for comparable spans. However, that result does match the performance of a group of 7 BiCI users who experienced very large delays between implantations and showed comparatively poor localization performance when tested with the same materials and methods (Tyler et al. 2007, Table 2.1). The data in the middle and on the right show results from Potts et al. (2009) and Ching et al. (2004) respectively. Both studies show that, while CIHA errors were smaller than for unilateral device use, the improvement is modest. The results from Ching et al. (2004), for a span of 180°, show that the average RMS error when using both ears reduced from about 80°

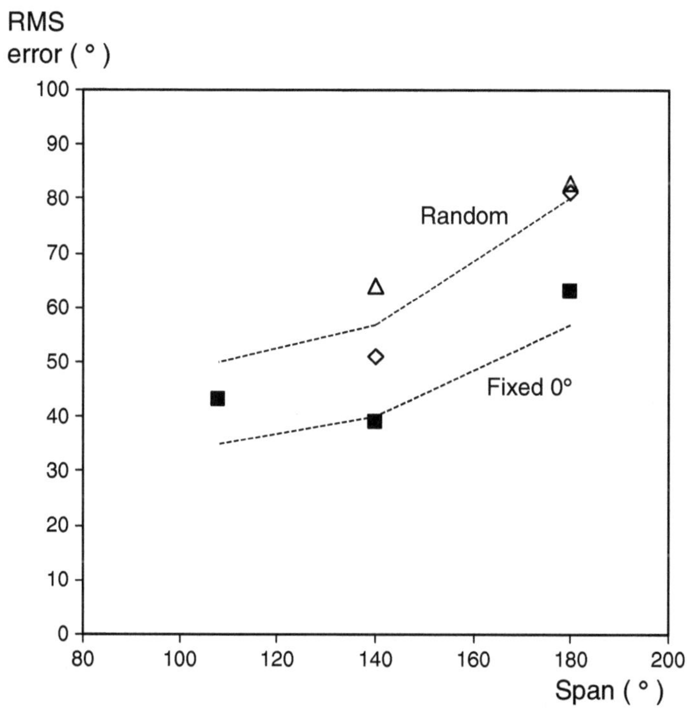

Fig. 2.10 RMS errors for localization in adult CIHA users, as a function of loudspeaker span in three different studies. *Solid squares* show errors for bimodal listening, *open diamonds* for CI only, and *open triangles* for HA only. Estimated chance performance is shown for random responding by the upper *dashed line* and for constant responding at 0° by the *lower dashed line*. Data from left to right are from Dunn et al. (2005), Potts et al. (2009), and Ching et al. (2004)

(for either HA or CI) to 64°. Only the best performer among 18 tested listeners showed a bilateral RMS error below 30°, which is typical for BiCI users with that span. The data from Potts et al. (2009), for a span of 140°, show an average RMS error of about 61° for the HA alone, 54° for the CI alone, and 39° for the CIHA condition. Although in this case 5 of 19 subjects showed errors under 30° performance was likely improved by the allowance of head turns in the direction of a preceding cueing phrase that was presented from the same loudspeaker as the target on each trial. Monaural RMS errors in that study were reported to be almost twice as small when signals were presented ipsilateral rather than contralateral to the ear being tested. That may however largely be the result of the use of the RMS metric, which produces smaller errors for ipsilateral signal presentations if the monaural response bias shifts towards the ear being tested. On a related note, it is interesting to consider that when signals were on the CI or HA side, average CIHA performance was almost identical to monaural performance with the device ipsilateral to the signal direction. Seeber et al. (2004) assessed localization abilities in CIHA users with a pointer method. Nine of the 11 subjects showed poor bilateral performance, half of whom displayed no systematic variation with source location at all. However, the two best performers showed good bilateral benefits with a MAE that is comparable to that seen in good performers among BiCI users.

7.2 *Pediatric CIHA Measures*

7.2.1 Speech Studies

Figure 2.11 shows benefits from more recent studies using fixed level tests in children, described as percentage point improvements when the HA was added to the CI. Overall benefits range from about 5 to 20 pp, and around 10 pp on average. Benefits are, as expected, more variable than in adult CIHA users. In addition to the general difficulties associated with testing children accurately, that variation is likely to be also the result of the large range of test materials used, which included words, sentences, and lexical tones. Benefits in noise are described in relatively few studies, and it is not clear from the limited data available whether the benefit differs from that in quiet, or varies with noise position. While the data from 5 subjects tested by Holt et al. (2005) suggest the CIHA benefit in children may improve with listening experience or age, more data are needed to confirm that conjecture. Nittrouer and Chapmann (2009) found no overall significant difference using several language measures when comparing different groups of children listening with one CI, BiCI, or CIHA. However, those authors did show a generative language advantage for children exposed to CIHA use early in life, which may be attributable to improved perception of prosody using the HA.

7.2.2 Localization

Sound-direction identification was tested in pediatric CIHA users by Ching et al. (2001) using 11 loudspeakers spanning 180°. Results were reported in terms of the sum of the 11 MAEs for each loudspeaker, RMS-averaged over multiple presentation blocks. That error was reduced from about 38 intervals for the CI-only condition, to 31 intervals when using CIHA. That reduction, by a factor of 1.23, corresponds well with an RMS error reduction from 80° to 64° (a factor of 1.25) for adult CIHA users tested using the same configuration (Ching et al. 2004). Assessment of the pediatric CIHA performance prior to optimization of the fitting of both devices showed a larger error of 35, rather than 31 loudspeaker intervals. Comparison of the MAAs reported by Litovsky et al. (2006b) for both pediatric CIHA and BiCI users shows that when a second device was added, the MAA reduced by a factor of about 2.5 for the BiCI users tested, and a more moderate factor of 1.4 for CIHA users. Beijen et al. (2010) generally found very large MAAs in most children using CIHA, with only 7 of the 20 subjects tested showing MAAs below 100°, even without level roving. The greatest decline in CIHA performance in that study was seen when spectra rather than amplitudes were roved.

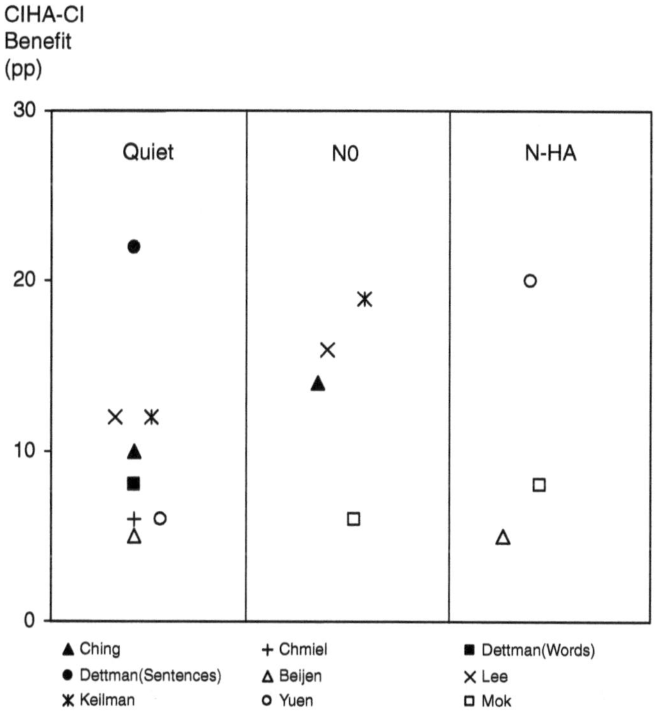

Fig. 2.11 Pediatric CIHA speech benefits, as percentage point increases for CIHA relative to CI only scores. Data are from Chmiel et al. (1995); Ching et al. (2001) sentences; Dettman et al. (2004), average for words and (live) sentences; Holt et al. (2005), words, 1-year data; Beijen et al. (2007), phonemes; Lee et al. (2008), monosyllables and bisyllables; Keilmann et al. (2009), mixed speech materials; Yuen et al. (2009), average for words and lexical tones; Mok et al. (2010), CNC words

7.3 Subjective Measures

As in BiCI studies, CIHA benefits have been found using self-report questionnaires. Ching et al. (2001) found benefits in relation to listening in quiet and in noise, as well as to environmental awareness. Potts et al. (2009) found high CIHA ratings for questions relating to qualities of sound, particularly with regard to the clarity and naturalness of speech. In contrast, spatial hearing ratings were relatively low, in good agreement with the finding that localization is only slightly improved in most CIHA users. Noble et al. (2008) compared subjective ratings of self-rated handicap for relatively large groups of BiCI, CIHA, and monaural CI users. Somewhat surprisingly, CIHA users showed significantly higher perceived handicap in relation to Emotional Distress than BiCI or monaural CI users. That outcome was considered to be the possible result of a change from purely acoustic to electro-acoustic stimulation. BiCI users showed significantly lower handicap with respect to both Social Restriction and Difficulty in Hearing, when compared to either CIHA or monaural CI users.

8 Summary

Compared with unilateral CI performance, substantial but different benefits can be derived through addition of either a second CI or HA in the contralateral ear. In the case of bilateral implants:

- The largest benefit is derived from the acoustic headshadow, which produces robust interaural level differences that greatly improve localization and allow listeners to attend the ear with the better SNR when speech and noise are spatially separated.
- Binaural speech benefits in quiet are modest, typically ranging up to about 10 pp, unless levels are low, in which case binaural loudness summation can contribute to improved audibility.
- Binaural speech intelligibility in noise *relative to the ear with better performance* shows binaural benefits that are on the order of 1 dB, irrespective of the noise position. The minimal effect of noise position implies that binaural unmasking generally does not play a significant role, unlike in normal hearing.
- Localization performance appears well accounted for by ILDs.
- Contributions from ITDs to speech and localization outcomes are largely absent because (a) low frequency fine-timing cues are discarded in clinical processors, (b) ITDs are poorly perceived when electrical stimulation or modulation rates increase beyond a few hundred Hz, and (c) ITD salience is likely to be disrupted in multi-electrode activation because of broad current spread if stimulation is substantially out-of-phase and/or contains conflicting timing cues on nearby electrodes.
- Large individual variability exists for reported binaural benefits, particularly in pediatric BiCI users. While the reasons for that variability are presently unclear, likely influences include listening experience and preference, performance and mapping asymmetries, and test methodologies.

In the case of adding a contralateral HA:

- Benefits appear related less to spatial hearing and more to the provision of complementary information, as expected from the substantial mismatch in signal representation in each ear and smaller effect of the headshadow at low frequencies.
- Speech intelligibility benefits at typical conversational levels in quiet, or for S0N0 presentation, are comparable, if not somewhat larger than in BiCI users.
- Speech intelligibility benefits available from adding the HA are not strongly dependent on noise position.
- Inter-subject variation in speech benefits derived from adding the HA to a CI is better predicted by the relative performance of the HA compared to the CI, rather than absolute HA performance.
- While localization is on average considerably poorer than in BiCI users, further work is needed to clarify why a few exceptional CIHA subjects show outcomes that match BiCI users with good performance.

Acknowledgements The author gratefully acknowledges the provision of additional data, re-calculations of benefits, or clarification of experimental procedures by Ruth Litovsky, Emily Buss, Theresa Ching, Karyn Galvin, and Lisa Potts. The author is further indebted to Andrew Vandali for helpful discussions in developing some of the ideas presented in this chapter, as well as to Fan-Gang Zeng, Art Popper, and Julie Murphy for helpful editorial suggestions.

References

Agrawal, S. S. (2008). Spatial hearing abilities in adults with bilateral cochlear implants (Unpublished doctoral dissertation). University of Wisconsin, Madison.

Armstrong, M., Pegg, P., James, C., & Blamey, P. (1997). Speech perception in noise with implant and hearing aid. *American Journal of Otology*, 8, S140–S141.

Beijen, J. W., Snik, A. F., & Mylanus, E. A. (2007). Sound localization ability of young children with bilateral cochlear implants. *Otology & Neurotology*, 28, 479–485.

Beijen, J., Snik, A. F. M., Straatman, L. V., Mylanus, E. A. M., & Mens L. H. M. (2010). Sound localization and binaural hearing in children with a hearing aid and a cochlear implant. *Audiology & Neurotology*, 15, 36–43.

Bernstein, L. R., & Trahiotis, C. (2002). Enhancing sensitivity to interaural delays at high frequencies by using transposed stimuli. *Journal of the Acoustical Society of America*, 112(3), 1026–1036.

Berrettini S., Passetti S., Giannarelli M., & Forli F. (2009). Benefit from bimodal hearing in a group of prelingually deafened adult cochlear implant users. *American Journal of Otology*, doi:10.1016/j.amjoto.2009.04.002.

Bichey, B. G., & Miyamoto, R. T. (2008). Outcomes in bilateral cochlear implantation. *Archives of Otolaryngology–Head & Neck Surgery*, 138, 655–661.

Blamey, P. J., James, C. J., & Martin, L. F. A. (2000). Sound separation with a cochlear implant and hearing aid in opposite ears. *Proceedings of the 8th Australian International Speech Science and Technology Conference*, Canberra, Australia, 4–7 December 2000, pp. 452–457. Canberra: Australian Speech Science and Technology Association.

Blauert, J. (1997). *Spatial Hearing*. MIT Press, Cambridge, MA.

Bohnert A., Spitzlei, V., Lippert, K. L., & Keilmann, A. (2006). Bilateral cochlear implantation in children: experiences and considerations. *Volta Review*, 106, 343–364.

Bond, M., Mealing, S., Anderson, R., Elston, J., Weiner, G., Taylor, R. S., Hoyle, M., Liu, Z., Price A., & Stein, K. (2009). The effectiveness and cost-effectiveness of cochlear implants for severe to profound deafness in children and adults: a systematic review and economic model. *Health Technology Assessment*, 13(44), 1–330.

Bronkhorst, A. W. (2000). The cocktail party phenomenon: a review of research on speech intelligibility in multiple-talker conditions. *Acustica*, 86, 117–128.

Bronkhorst, A. W., & Plomp, R. (1988). The effect of head-induced interaural time and level differences on speech intelligibility in noise. *Journal of the Acoustical Society of America*, 83, 1508–1516.

Buss, E., Pillsbury, H. C., Buchman, C. A., Pillsbury, C. H., Clark, M. S., Haynes, D. S., Labadie, R. F., Amberg, S., Roland, P. S., Kruger, P., Novak, M. A., Wirth, J. A., Black, J. M., Peters, R., Lake, J., Wackym, P. A., Firszt, J. B., Wilson, B. S., Lawson, D. T., Schatzer, R., D'Haese P. S., & Barco, A. L. (2008). Multicenter US bilateral MED-EL cochlear implantation study: speech perception over the first year of use. *Ear and Hearing*, 29, 20–32.

Byrne, D., & Dillon, H. (1979). Bias in assessing binaural advantage. *Australian Journal of Audiology*, 1, 83–88.

Carhart, R. (1965). Monaural and binaural discrimination against competing sentences. *International Audiology*, 4, 5–10.

Carhart, R., Tillman, T. W., & Johnson, K. R. (1967). Release of masking for speech through interaural time delay. *Journal of the Acoustical Society of America*, 42, 124–138.

Carlyon, R. P., Long, C. J., & Deeks, J. M. (2008). Pulse-rate discrimination by cochlear-implant and normal-hearing listeners with and without binaural cues. *Journal of the Acoustical Society of America*, 123, 2276–2286.

Chang, S.-A., Tyler, R. S., Dunn, C. C., Ji, H., Witt, S. A., Gantz, B., & Hansen, M. (2010). Performance over time on adults with simultaneous bilateral cochlear implants. *Journal of the American Academy of Audiology*, 21, 35–43.

Ching, T. Y., Psarros, C., Hill, M., Dillon, H., & Incerti, P. (2001). Should children who use cochlear implants wear hearing aids in the opposite ear? *Ear and Hearing*, 22, 365–380.

Ching, T. Y., Incerti, P., & Hill, M. (2004). Binaural benefits for adults who use hearing aids and cochlear implants in opposite ears. *Ear and Hearing*, 25, 9–21.

Ching, T. Y. C., van Wanrooy, E., Hill, M., & Dillon, H. (2005). Binaural redundancy and inter-aural time difference cues for patients wearing a cochlear implant and a hearing aid in opposite ears. *International Journal of Audiology*, 44, 513–521.

Chmiel, R., Clark, J., Jerger, J., Jenkins, H., & Freeman, R. (1995). Speech perception and production in children wearing a cochlear implant in one ear and a hearing aid in the opposite ear. *Annals of Otology, Rhinology, and Laryngology*, 166, 314–316.

Cohen, L. T., Saunders, E., & Clark, G. M. (2001). Psychophysics of a prototype peri-modiolar cochlear implant electrode array. *Hearing Research*, 155, 63–81.

Colburn, H. S., Chung, Y., Zhou, Y., & Brughera, A. (2009). Models of brainstem responses to bilateral electrical stimulation. *Journal of the Association for Research in Otolaryngology*, 10, 91–110.

Cox, R. M., Alexander, G. C., Gilmore, C., & Pusakulich, K. M. (1988). Use of the Connected Speech Test (CST) with hearing-impaired listeners. *Ear and Hearing*, 9, 198–207.

Cullington, H. E., & Zeng, F.-G. (2010). Bimodal hearing benefit for speech recognition with competing voice in cochlear implant subject with normal hearing in contralateral ear. *Ear and Hearing*, 31, 70–73.

Day, G. A., Browning, G. G., & Gatehouse, S. (1988). Benefit from binaural hearing aids in individuals with a severe hearing impairment. *British Journal of Audiology*, 22, 273–277.

Dettman, S. J., D'Costa, W. A., Dowell, R. C., Winton, E. J., Hill, K. L., Williams, S. S. (2004). Cochlear implants for children with significant residual hearing. *Archives of Otolaryngology–Head & Neck Surgery*, 130, 612–618.

Dorman, M. F., Gifford, R. H., Spahr, A. J., & McKarns, S. A. (2008). The benefits of combining acoustic and electric stimulation for the recognition of speech, voice and melodies. *Audiology & Neurotology*, 13(2),105–112.

Duda, R. O, & Martens, W. L. (1998). Range dependence of the response of a spherical head model. *Journal of the Acoustical Society of America*, 104(5), 3048–3058.

Dunn, C. C., Tyler, R. S., & Witt, S. A. (2005). Benefit of wearing a hearing aid on the unimplanted ear in adult users of a cochlear implant. *Journal of Speech, Language, and Hearing Research*, 48, 668–680.

Durlach, N. I. (1963). Equalization and cancellation theory of binaural masking-level differences. *Journal of the Acoustical Society of America*, 35, 1206–1218.

Durlach, N. I., & Colburn, H. S. (1978). Binaural phenomena. In E. C. Cartrette & M. P. Friedman (Eds.), *Handbook of perception, Vol. IV* (365–466). New York: Academic Press.

Eapen, R. J., Buss, E., Adunka, M. C., Pillsbury, H. C., & Buchman, C. A. (2009). Hearing-in-noise benefits after bilateral simultaneous cochlear implantation continue to improve 4 years after implantation. *Otology & Neurotology*, 30, 153–159.

Francart, T., Brokx, J., & Wouters, J. (2008). Sensitivity to interaural level difference and loudness growth with bilateral bimodal stimulation. *Audiology & Neurotology*, 13, 309–319.

Francart, T., Brokx, J., & Wouters, J. (2009). Sensitivity to interaural time differences with combined cochlear implant and acoustic stimulation. *Journal of the Association for Research in Otolaryngology*, 10, 131–141.

Galvin, K. L., Mok, M., & Dowell, R. C. (2007). Perceptual benefit and functional outcomes for children using sequential bilateral cochlear implants. *Ear and Hearing*, 28, 470–482.

Galvin, K. L., Mok, M., Dowell, R. C., & Briggs, R. J. (2008). Speech detection and localization results and clinical outcomes for children receiving sequential bilateral cochlear implants before four years of age. *International Journal of Audiology*, 47, 636–646.

Gantz B. J., Tyler R. S., Rubinstein J. T., Wolaver, A., Lowder, M., Abbas, P., Brown, C., Hughes, M., & Preece, J. (2002). Binaural cochlear implants placed during the same operation. *Otology & Neurotology*, 23, 169–180.

Gatehouse, S., & Noble, W. (2004). The Speech, Spatial and Qualities of Hearing Scale (SSQ). *International Journal of Audiology*, 43(2), 85–99.

Gifford, R. H., Dorman, M. F., McKarns, S. A., & Spahr, A. J. (2007). Combined electric and contralateral acoustic hearing: word and sentence recognition with bimodal hearing. *Journal of Speech, Language, and Hearing Research*, 50, 835–843.

Gordon, K. A., & Papsin, B. C. (2009). Benefits of short interimplant delays in children receiving bilateral cochlear implants. *Otology & Neurotology*, 30, 319–331.

Gordon, K. A., Valero, J., & Papsin, B. C. (2007). Auditory brainstem activity in children with 9–30 months of bilateral cochlear implant use. *Hearing Research*, 233, 97–107.

Gordon, K. A., Valero, J., van Hoesel, R., & Papsin, B. C. (2008). Abnormal timing delays in auditory brainstem responses evoked by bilateral cochlear implant use in children. *Otology & Neurotology*, 29, 193–198.

Grantham, D. W. (1995). Spatial hearing and related phenomena. In B. C. J. Moore (Ed.), *Hearing: Handbook of perception, and cognition, 2nd ed.* (297–345). London: Academic Press Limited.

Grantham, W., Ashmead, D. H., Ricketts, T. A., Labadie, R. F., & Haynes, D. S. (2007). Horizontal-plane localization in noise and speech signals by postlingually deafened adults fitted with bilateral cochlear implants. *Ear and Hearing*, 28, 524–541.

Grantham, D. W., Ashmead, D. H., Ricketts, T. A., Haynes, D. S., & Labadie, R. F. (2008). Interaural time and level difference thresholds for acoustically presented signals in post-lingually deafened adults fitted with bilateral cochlear implants using CIS processing. *Ear and Hearing*, 29, 33–44.

Green, J. D., Mills, D.M., Bell, B. A., Luxford, W. M., & Tonokawa, L. L. (1992). Binaural cochlear implants. *Otology & Neurotology*, 13, 495–616.

Green, D. M., & Swets, J. A. (1966). *Signal detection theory and psychophysics*. New York: John Wiley and Sons.

Grieco-Calub, T. M., Litovsky, R. Y., & Werner, L. A. (2008). Using the observer-based psychophysical procedure to assess localization acuity in toddlers who use bilateral cochlear implants. *Otology & Neurotology*, 29, 235–239.

Hafter, E. R., & Buell, T. N. (1990). Restarting the adapted binaural system. *Journal of the Acoustical Society of America*, 88, 806–812.

Hafter, E. R., & Dye, Jr., R. H. (1983). Detection of interaural differences of time in trains of high-frequency clicks as a function of interclick interval and number. *Journal of the Acoustical Society of America*, 73(5), 1708–1713.

Hamzavi, J., Pok, S. M., Gstoettner, W., & Baumgartner, W.-D. (2004). Speech perception with a cochlear implant used in conjunction with a hearing aid in the opposite ear. *International Journal of Audiology*, 43(2), 61–65.

Hartmann, W. M. (1997). Listening in a room and the precedence effect. In R. H. Gilkey & T. R. Anderson (Eds.), *Binaural and spatial hearing in real and virtual environments* (pp. 1–23). Mahwah, NJ: Lawrence Erlbaum.

Hartmann, W., Rakerd, B., & Gaalaas, J. (1998). On the source identification method. *Journal of the Acoustical Society of America*, 104, 3546–3557.

Hirsch, I. J. (1948). The influence of interaural phase on summation and inhibition. *Journal of the Acoustical Society of America*, 20, 536–544.

Holt, R. F., Kirk, K. I., Eisenberg, L. S., Martinez, A. S., & Campbell, W. (2005). Spoken word recognition development in children with residual hearing using cochlear implants and hearing aids in opposite ears. *Ear and Hearing*, 26, 82 S–91 S.

Iwaki T., Mathushiro N., Mah S.-R., Sato T., Yasuoka E., Yamamoto K., & Kubo, T. (2004). Comparison of speech perception between monaural and binaural hearing in cochlear implant subjects. *Acta Oto-Laryngologica*, 124, 358–62.

Jones, G., van Hoesel, R., Litovsky, R. (2008). Effect of channel interactions on sensitivity to binaural timing cues in electrical hearing. *Journal of the Acoustical Society of America*, 123(5, Pt. 2), 3055.

Keilmann, A. M., Bohnert, A. M., Gosepath, J., & Mann, W. J. (2009). Cochlear implant and hearing aid: a new approach to optimizing the fitting in this bimodal situation. *European Archives of Oto-Rhino-Laryngology*, 266, 1879–1884.

Kidd, G., Mason, C., Richards, V., Gallun, F., & Durlach, N. (2008). Informational masking. In W. A. Yost, A. N. Popper, & R. R. Fay (Eds.), *Auditory perception of sound sources: Springer handbook of auditory research, Vol.29* (143–190). New York: Springer.

Kim, L-S, Jang, Y.S., Choi, A-H., Ahn, S-Y., Park, J-S., Lee, Y-M., Jeong, S-W. (2009). Bilateral cochlear implants in children. *Cochlear Implants International*, 10(S1), 74–77.

Klumpp, R. G., & Eady, H. R. (1956). Some measurements of interaural time difference thresholds. *Journal of the Acoustical Society of America*, 28, 859–860.

Koch, D. B., Soli, S. D., Downing, M., & Osberger, M. J. (2009). Simultaneous bilateral cochlear implantation: prospective study in adults. *Cochlear Implants International*, 11(2), 84–99.

Koenig, W. (1950). Subjective effects in binaural hearing. *Journal of the Acoustical Society of America*, 22, 61–62.

Kong, Y. Y., Stickney, G. S., & Zeng, F. G. (2005). Speech and melody recognition in binaurally combined acoustic and electric hearing. *Journal of the Acoustical Society of America*, 117, 1351–1361.

Kral, A., Tillein, J., Hubka, P., Schiemann, D., Heid, S., Hartmann, R., & Engel, A. K. (2009). Spatiotemporal patterns of cortical activity with bilateral cochlear implants in congenital deafness. *Journal of Neuroscience*, 29(3), 811– 827.

Kuhn, G. (1977). Model for the interaural time differences in the azimuthal plane. *Journal of the Acoustical Society of America*, 62(1), 157–167.

Kühn-Innacker, H., Shehata-Dieler W., Muller J., & Helms J. (2004). Bilateral cochlear implants: a way to optimize auditory perception abilities in deaf children? *International Journal of Pediatric Oto-Rhino-Laryngology*, 68, 1257–1266.

Laback, B., & Majdak, P. (2008). Binaural jitter improves interaural time difference sensitivity of cochlear implantees at high pulse rates. *Proceedings of the National Academy of Sciences*, 105, 814–817.

Laback, B., Pok, S. M., Baumgartner, W. D., Deutsch, W. A., & Schmid, K. (2004). Sensitivity to interaural level and envelope time differences of two bilateral cochlear implant listeners using clinical sound processors. *Ear and Hearing*, 25, 488–500.

Laback, B., Majdak, P., & Baumgartner, W.-D. (2007). Lateralization discrimination of interaural time delays in four-pulse sequences in electric and acoustic hearing. *Journal of the Acoustical Society of America*, 121, 2182–2191.

Laske, R. D., Veraguth, D., Dillier, N., Binkert, A., Holzmann, D., & Huber, A. M. (2009). Subjective and objective results after bilateral cochlear implantation in adults. *Otology & Neurotology*, 30, 313–318.

Laszig, R., Aschendorff, A., Stecker, M., Müller-Deile, J., Maune, S., Dillier, N., Weber, B., Hey, M., Begall, K., Lenarz, T., Battmer, R. D., Böhm, M., Steffens, T., Strutz, J., Linder, T., Probst, R., Allum, J., Westhofen, M., & Doering, W. (2004). Benefits of bilateral electrical stimulation with the nucleus cochlear implant in adults: 6-month postoperative results. *Otology & Neurotology*, 25, 958–968.

Lawson, D. T., Wilson, B. S., Zerbi, M., Honert, C., Finley, C. C., Farmer, J. C., McElveen, J. T., & Roush, P. A. (1998). Bilateral cochlear implants controlled by a single speech processor. *American Journal of Otology*, 19, 758–761.

Lee, S.-H.., Lee, K.-Y., Huhm M.-J., & Jang, H-S. (2008). Effect of bimodal hearing in Korean children with profound hearing loss. *Acta Oto-Laryngologica*, 128, 1227–1232.

Levitt, H., & Rabiner, L. R. (1967). Binaural release from masking for speech and gain in intelligibility. *Journal of the Acoustical Society of America*, 42, 601–608.
Litovsky, R. Y., Colburn, H. S., Yost, W. A., & Guzman, S. J. (1999). The precedence effect. *Journal of the Acoustical Society of America*, 106, 1633–1654.
Litovsky, RY., Parkinson, A., Arcaroli, J (2004). Bilateral cochlear implants in adults and children. *Archives of Otolaryngology–Head & Neck Surgery*, 130, 648–655.
Litovsky, R. Y., Johnstone, P. M., Godar, S., & Agrawal, S. (2006a). Bilateral cochlear implants in children: localization acuity measured with minimum audible angle. *Ear and Hearing*, 27, 43–59.
Litovsky, R. Y., Johnstone, P. M., & Godar, S. P. (2006b). Benefits of bilateral cochlear implants and/or hearing aids in children. *International Journal of Audiology*, 45, 78 S–91 S.
Litovsky, R., Parkinson, A., Arcaroli, J., & Sammeth, C. (2006c). Simultaneous bilateral cochlear implantation in adults: a multicenter clinical study. *Ear and Hearing*, 27, 714–731.
Litovsky, R. Y., Parkinson, A., & Arcaroli., J. (2009). Spatial hearing and speech intelligibility in bilateral cochlear implant users. *Ear and Hearing* 30, 419–431.
Litovsky, R. Y., Jones, G. L., Agrawal, S., & van Hoesel, R. (2010). Effect of age at onset of deafness on binaural sensitivity in electric hearing in humans. *Journal of the Acoustical Society of America*, 127, 400–414.
Loizou, P., Hu, Y., Litovsky, R. Y., Yu, G., Peters, R., Lake, J., & Roland, P. (2009). Speech recognition by bilateral cochlear implant users in a cocktail party setting. *Journal of the Acoustical Society of America*, 125, 372–383.
Long, C. J., Carlyon, R. P., Litovsky, R. Y., & Downs, D. H. (2006). Binaural unmasking with bilateral Cochlear Implants. *Journal of the Association for Research in Otolaryngology*, 7, 352–360.
Lu, T., Litovsky, R., & Zeng, F.G. (2010). Binaural masking level differences in actual and simulated bilateral cochlear implant listeners. *Journal of the Acoustical Society of America*, 127, 1479–1490.
Luntz, M., Shpak, T., & Weiss, H. (2005). Binaural-bimodal hearing: concomitant use of a unilateral cochlear implant and a contralateral hearing aid. *Acta Oto-Laryngologica* 125, 863–869.
Majdak, P., Laback, B., & Baumgartner, W. D. (2006). Effects of interaural time differences in fine structure and envelope on lateral discrimination in electric hearing. *Journal of the Acoustical Society of America*, 120, 2190–2201.
Mills, A. W. (1958). On the minimum audible angle. *Journal of the Acoustical Society of America*, 30, 237–246.
Mok, M., Grayden, D., Dowell, R., & Lawrence, D. (2006). Speech perception for adults who use hearing aids in conjunction with cochlear implants in opposite ears. *Journal of Speech, Language, and Hearing Research*, 49, 338–351.
Mok, M., Galvin, K. L., Dowell, R. C., & McKay, C. M. (2010). Speech perception benefit for children with a cochlear implant and a hearing aid in opposite ears and children with bilateral cochlear implants. *Audiology & Neurotology*, 15, 44–56.
Morera, C., Manrique, M., Ramos, A., Garcia-Ibanez, L., Cavalle, L., Huarte, A., Castillo, C. , & Estrada, E. (2005). Advantages of binaural hearing provided through bimodal stimulation via a cochlear implant and a conventional hearing aid: a 6-month comparative study. *Acta Oto-Laryngologica*, 125, 596–606.
Mosnier, I., Sterkers, O., Bebear, J.P., Godey, B., Robier, A., Deguine, O. Fraysse, B., Bordure, P., Mondain, M., Bouccara, D., Bozorg-Grayeli, A., Borel, S., Ambert-Dahan, E., & Ferrary, E. (2009). Speech performance and sound localization in a complex noisy environment in bilaterally implanted adult patients. *Audiology & Neurotology*, 14, 106–114.
Müller, J., Schön, F., & Helms, J. (2002). Speech understanding in quiet and noise in bilateral users of the MED-EL COMBI 40/40+ cochlear implant system. *Ear and Hearing*, 23, 198–206.
Neuman, A. C., Haravon, A., Sislian, N., & Waltzman, S. B. (2007). Sound-direction identification with bilateral cochlear implants. *Ear and Hearing*, 28, 73–82.
Nittrouer, S., & Chapman, C. (2009). The effects of bilateral electric and bimodal electric–acoustic stimulation on language development. *Trends in Amplification*, 13, 190–205.

Noble, W., Tyler, R., Dunn, C., & Bhullar, N. (2008). Hearing handicap ratings among different profiles of adult cochlear implant users. *Ear and Hearing*, 29, 112–120.

Noble, W., Tyler, R., Dunn, C., & Bhullar, N. (2009). Younger- and older age adults with unilateral and bilateral cochlear implants: speech and spatial hearing self-ratings and performance. *Otology & Neurotology*, 30, 921–929.

Nopp, P., Schleich, P., & D'Haese, P. (2004). Sound localization in bilateral users of MED-EL COMBI 40/40+ cochlear implants. *Ear and Hearing*, 25, 205–214.

Oxenham, A. J., Bernstein, J. G. W., & Penagos, H. (2004). Correct tonotopic representation is necessary for complex pitch perception. *Proceedings of the National Academy of Science*, 101, 1421–1425.

Pelizzone, M., Kasper, A., & Montandon, P. (1990). Binaural interaction in a cochlear implant patient. *Hearing Research*, 48, 287–290.

Peters, B. R., Litovsky, R., Parkinson, A., & Lake, J. (2007). Importance of age and postimplantation experience on speech perception measures in children with sequential bilateral cochlear implants. *Audiology & Neurotology*, 28, 649–57.

Poon, B. B., Eddington, D. K., Noel, V., & Colburn, H. S. (2009). Sensitivity to interaural time difference with bilateral cochlear implants: development over time and effect of interaural electrode spacing. *Journal of the Acoustical Society of America*, 126, 806–815.

Potts, L. G., Skinner, M. W., Litovsky, R. Y., Strube, M. J., & Kukl, F. (2009). Recognition and localization of speech by adult cochlear implant recipients wearing a digital hearing aid in the nonimplanted ear (bimodal hearing). *Journal of the American Academy of Audiology*, 20, 353–373.

Ramsden, R., Greenham, P., O'Driscoll, M., Mawman, D., Proops, D., Craddock, L. Fielden, C., Graham, J., Meerton, L., Verschuur, C., Toner, J., McAnallen, T., Osborne, J., Doran, M., Gray, R., & Pickerill, M. (2005). Evaluation of bilaterally implanted adult subjects with the nucleus 24 cochlear implant system. *Otology & Neurotology*, 26, 988–998.

Rakerd, B., & Hartmann, W. M. (1986). Localization of sound in rooms III: onset and duration effects. *Journal of the Acoustical Society of America*, 80, 1695–1706.

Rayleigh, L. (1907). On our perception of sound direction. *Philosophical Magazine*, 13, 214–232.

Ricketts, T. A., Grantham, D. W., Ashmead, D. H., Haynes, D. S., & Labadie, R. F. (2006). Speech recognition for unilateral and bilateral cochlear implant modes in the presence of uncorrelated noise sources. *Ear and Hearing*, 27, 763–773.

Saberi, K. (1996). Observer weighting of interaural delays in filtered impulses. *Perception and Psychophysics*, 58, 1037–1046.

Sasaki, T., Yamamoto, K., Iwaki, T., & Kubo, T. (2009). Assessing binaural/bimodal advantages using auditory event-related potentials in subjects with cochlear implants. *Auris Nasus Larynx*, 36, 541–546.

Schafer, E. C., & Thibodeau, L. M. (2006). Speech recognition in noise in children with cochlear implants while listening in bilateral, bimodal, and FM-system arrangements. *American Journal of Audiology*, 15, 114–126.

Scharf, B. (1978). Loudness. In E. C. Cartrette & M. P. Friedman (Eds.), *Handbook of perception, Vol. IV.* New York: Academic Press.

Scherf, F., van Deun, L., van Wieringen, A., Wouters, J., Desloovere, C., Dhooge, I. J., Offeciers, F. E., Deggouj, N., De Raeve, L., Wuyts, F., & Van de Heyning, P. H. (2009a). Three year post-implantation auditory outcomes in children with sequential bilateral cochlear implantation. *Annals of Otology, Rhinology, and Laryngology*, 118, 336–344.

Scherf, F., van Deun, L., van Wieringen, A., Wouters, J., Desloovere, C., Dhooge, I. J., Offeciers, F. E., Deggouj, N., De Raeve, De Bodt, M., & Van de Heyning, P. (2009b). Functional outcome of sequential bilateral cochlear implantation in young children: 36 months postoperative results. *International Journal of Pediatric Oto-Rhino-Laryngology*, 73, 723–730.

Schleich, P., Nopp, P., & D'Haese, P. (2004). Head shadow, squelch, and summation effects in bilateral users of the MED-EL COMBI 40/40+ cochlear implant. *Ear and Hearing*, 25, 197–204.

Seeber, B. U., & Fastl, H. (2008). Localization cues with bilateral cochlear implants. *Journal of the Acoustical Society of America*, 123, 1030–1042.

Seeber, B.U., Baumann, U., & Fastl, H. (2004). Localization ability with bimodal hearing aids and bilateral cochlear implants. *Journal of the Acoustical Society of America*, 116, 1698–1709.

Senn, P., Kompis, M., Vischer, M., & Häusler, R. (2005). Minimum audible angle, just noticeable interaural differences and speech intelligibility with bilateral cochlear implants using clinical speech processors. *Audiology & Neurotology*, 10, 342–352.

Sharma, A., Gilley, P. M., Martin, K., Roland, P., Bauer, P., Dorman, M. (2007). Simultaneous versus sequential bilateral implantation in young children: effects on central auditory system development and plasticity. *Audiological Medicine*, 5(4), 218–223.

Shaw, E. A. G. (1974). Transformation of sound pressure level from the free field to the eardrum in the horizontal plane. *Journal of the Acoustical Society of America*, 56, 1848–1861.

Smith, Z. M., & Delgutte, B. (2007a). Sensitivity to interaural time differences in the inferior colliculus with bilateral cochlear implants. *Journal of Neuroscience*, 27, 6740–6750.

Smith, Z. M., & Delgutte, B. (2007b). Using evoked potentials to match interaural electrode pairs with bilateral cochlear implants. *Journal of the Association for Research in Otolaryngology*, 8, 134–151.

Smith Z. M. & Delgutte, B. (2008). Sensitivity of inferior colliculus neurons to interaural time differences in the envelope versus the fine structure with bilateral cochlear implants. *Journal of Neurophysiology*, 99, 2390–2407.

Stakhovskaya, O., Sridar, D., Bonham, B. H., & Leake, P. A. (2007). Frequency map for the human cochlear spiral ganglion: implications for cochlear implants. *Journal of the Association for Research in Otolaryngology*, 8, 220–233.

Stecker, G. C., & Hafter, E. R. (2002). Temporal weighting in sound localization. *Journal of the Acoustical Society of America*, 112, 1046–1057.

Steffens, T., Lesinski-Schiedat, A., Strutz, J., Aschendorff, A., Klenzner, T., Ru, S., Voss, B., Wesarg, T., Laszig, R., & Lenarz, T. (2008). The benefits of sequential bilateral cochlear implantation for hearing-impaired children. *Acta Oto-Laryngologica*, 128, 164–176.

Stevens, S. S. & Newman, E. B. (1936). Localization of actual sources of sound. *American Journal of Psychology*, 48, 297–306.

Studebaker, G.A. (1985). A rationalized arcsine transform. *Journal of Speech and Hearing Research*, 28, 455–462.

Summerfield, A. Q., & Barton, G. R. (2003). Getting acceptable value for money from bilateral cochlear implantation. *Cochlear Implants International*, 4(Suppl. 1), 66–67.

Summerfield, A. Q., Barton, G. R., Toner, J., McAnallen, C., Proops, P., Harries, C., Cooper, H., Court, I., Gray, R., Osborne, J., Doran, M., Ramdsen, R., Mawman, D., O'Driscoll, M., Graham, J., Aleksy, W., Meerton, L., Verschure, C., Ashcroft, P., & Pringle, M. (2006). Self-reported benefits from successive bilateral cochlear implantation in post-lingually deafened adults: randomized controlled trial. *International Journal of Audiology*, 45, S99–S107.

Thai-Van, H., Gallego, S., Veuillet, E., Truy, E., & Collet, L. (2002). Electrophysiological findings in two bilateral cochlear implant cases: does the duration of deafness affect electrically evoked auditory brainstem responses? *Annals of Otology, Rhinology, and Laryngology*, 111, 1008–114.

Tillein, J., Hubka, P., Syed1, E., Hartmann, R., Engel, A. K., & Kral, A. (2010). Cortical representation of interaural time difference in congenital deafness. *Cerebral Cortex*, 20, 492–506.

Tyler, R. S., Dunn, C. C., Witt, S. A., & Noble, W. G. (2007). Speech perception and localization with adults with bilateral sequential cochlear implants. *Ear and Hearing*, 28, 86 S–90 S.

Tyler, R. S., Perreau, A. E., & Ji, H. (2009). Validation of the Spatial Hearing Questionnaire. *Ear and Hearing*, 30, 466–474.

van de Par, S., & Kohlrausch, A. (1997). A new approach to comparing binaural masking level differences at low and high frequencies. *Journal of the Acoustical Society of America*, 101, 1671–1680.

van Deun, L., van Wieringen, A., Francart, T., Scherf, F., Dhooge, I. J., Deggouj, N., Desloovere, C., van de Heyning, P. H., Offeciers, F. E., de Raeve, L., & Wouters, J. (2010a). Bilateral cochlear implants in children: binaural unmasking. Audiology & Neurotology, 14, 240–247.

van Deun, L., van Wieringen, A., Scherf, F., Deggouj, N., Desloovere, C., Offeciers, F. E., Van de Heyning, P. H., Dhooge, I. J., & Wouters, J. (2010b). Earlier intervention leads to better sound localization in children with bilateral cochlear implants. *Audiology & Neurotology*, 15, 7–17.

van Hoesel, R. J. M. (2002). A peak-derived timing stimulation strategy for a multichannel cochlear implant. International Patent No. PCT/AU2002/000660.
van Hoesel, R. J. M. (2004). Exploring the benefits of bilateral cochlear implants. *Audiology & Neurotology*, 9, 234–246.
van Hoesel, R. J. M. (2007). Sensitivity to binaural timing in bilateral cochlear implant users. *Journal of the Acoustical Society of America*, 121, 2192–2206.
van Hoesel, R. J. M. (2008). Observer weighting of level and timing cues in bilateral cochlear implant users. *Journal of the Acoustical Society of America*, 124, 3861–3872.
van Hoesel, R. J. M., & Clark, G. M. (1997). Psychophysical studies with two binaural cochlear implant subjects, *Journal of the Acoustical Society of America*, 102, 495–507.
van Hoesel, R. J. M., & Tyler, R. S. (2003). Speech perception, localization, - and lateralization with bilateral cochlear implants. *Journal of the Acoustical Society of America*, 113, 1617–1630.
van Hoesel, R. J. M., Tong, Y. C., Hollow, R. D., & Clark, G. M. (1993). Psychophysical and speech perception studies: a case report on a binaural cochlear implant subject. *Journal of the Acoustical Society of America*, 94, 3178–3189.
van Hoesel, R., Ramsden, R., & O'Drisoll, M. (2002). Sound-direction identification, interaural time delay discrimination, and speech intelligibility advantages in noise for a bilateral cochlear implant user. *Ear and Hearing*, 23, 137–149.
van Hoesel, R., Bohm, M., Battmer, R. D., Beckschebe, J., & Lenarz, T. (2005). Amplitude-mapping effects on speech intelligibility with unilateral and bilateral cochlear implants. *Ear and Hearing*, 26, 381–388.
van Hoesel, R., Böhm, M., Pesch, J., Vandali, A., Battmer, R. D., & Lenarz, T. (2008). Binaural speech unmasking and localization in noise with bilateral cochlear implants using envelope and fine-timing based strategies. *Journal of the Acoustical Society of America*, 123, 2249–2263.
van Hoesel, R. J. M., Jones, G. L., & Litovky, R. Y. (2009). Interaural time-delay sensitivity in bilateral cochlear implant users: effects of pulse rate, modulation rate, and place of stimulation. *Journal of the Association for Research in Otolaryngology*, 10, 557–567.
Veekmans, K., Ressel, L., Mueller, J., Vischer, M., & Brockmeier, S. J. (2009). Comparison of music perception in bilateral and unilateral cochlear implant users and normal-hearing Subjects. *Audiology & Neurotology*, 14, 315–326.
Vermeire, K., & van de Heyning, P. (2009) Binaural hearing after cochlear implantation in subjects with unilateral sensorineural deafness and tinnitus. *Audiology & Neurotology*, 14, 163–171.
Verschuur, C. A., Lutman, M. E., Ramsden, R., Greenham, P., & O'Driscoll, M. (2005). Auditory localization abilities in bilateral cochlear implant recipients. *Otology & Neurotology*, 26, 965–971.
Viemeister, N. F., & Wakefield, G. H. (1991). Temporal integration and multiple looks, *Journal of the Acoustical Society of America*, 90, 858–865.
Wackym, P. A., Runge-Samuelson, C. L., Firszt, J. B., Alkaf, F. M., & Burg, L. S. (2007). More challenging speech perception tasks demonstrate binaural benefit in bilateral cochlear implant users. *Ear and Hearing*, 28, 805–855.
Wightman, F. L., & Kistler, D. J. (1997). Factors affecting the relative salience of sound localization cues. In R. H. Gilkey & T. R. Anderson (Eds.), *Binaural and spatial hearing in real and virtual environments* (pp. 1–23). Mahwah, NJ: Lawrence Erlbaum.
Winkler, F., Schön, F., Peklo, L., Müller, J., Feinen, C., & Helms, J. (2002). The Würzburg Questionnaire for assessing the quality of hearing in CI-children. *Laryngorhinootologie*, 81, 211–216.
Wolfe, J., Baker, S., Caraway, T., Kasulis, H., Mears, A., Smith, J., Swim, L., & Wood, M. (2007). 1-year post-activation results for sequentially implanted bilateral cochlear implant users. *Otology & Neurotology*, 28, 589–596.
Yost, W. A., & Hafter, E. R. (1987). Lateralization. In W.A. Yost & G. Gourevitch (Eds.), *Directional Hearing* (49–84). New York: Springer.
Yuen, K. C. P., Cao, K.-L, Wei, C.-G., Luan, L., Li, H., & Zhang, Z.-Y. (2009). Lexical tone and word recognition in noise of Mandarin-speaking children who use cochlear implants and hearing aids in opposite ears. *Cochlear Implants International*, 10(Suppl. 1), 120–129.

Zeitler, D. M., Kessler, M. A., Terushkin, V., Roland, T. J., Svirsky, M. A., Lalwani, A. K., & Waltzman, S. B. (2008). Speech perception benefits of sequential bilateral cochlear implantation in children and adults: a retrospective analysis. *Otology & Neurotology*, 29, 314–325.

Zurek, P. M. (1980). The precedence effect and its possible role in the avoidance of interaural ambiguities. *Journal of the Acoustical Society of America*, 67, 952–964.

Zurek, P. M., & Durlach, N. I. (1987). Masker-bandwidth dependence in homophasic and antiphasic tone detection. *Journal of the Acoustical Society of America* 81, 459–464.

Zwislocki, J. J., Buining, E., & Glantz, J. (1968). Frequency distribution of central masking. *Journal of the Acoustical Society of America*, 43, 1267–1271.

Chapter 3
Combining Acoustic and Electric Hearing

Christopher W. Turner and Bruce J. Gantz

1 Introduction

There are numerous advantages for patients who can add residual acoustic hearing to the electric stimulation of a cochlear implant. When cochlear implantation was first becoming accepted, patients were typically profoundly deaf in both ears, so any possible advantages of also using acoustic hearing were non-existent. However, as time progressed and the range of patients considered for implantation was expanded to include those with more and more residual hearing, this remaining acoustic hearing became a factor to consider. The residual acoustic hearing was at first most often located in the non-implanted ear, especially since the trend has been to place the implant in the poorer ear if there was any aidable hearing. In recent years, residual acoustic hearing has been preserved even in the implanted ear, whereby acoustic and electric hearing are combined in the same ear. It turns out that the acoustic hearing that remains in either ear after cochlear implantation can still contribute to overall performance in some very significant ways, including cases where, by itself, the acoustic hearing produces only minimal or no word recognition.

C.W. Turner (✉)
Department of Communication Sciences and Disorders,
University of Iowa, 121B SHC, Iowa City, IA 52242, USA
e-mail: christopher-turner@uiowa.edu

F.-G. Zeng et al. (eds.), *Auditory Prostheses: New Horizons*, Springer Handbook of Auditory Research 39, DOI 10.1007/978-1-4419-9434-9_3,

2 Limitations and Benefits of Residual Acoustic Hearing Versus Electric Stimulation

2.1 Acoustic Hearing

In the normally functioning ear, the cochlea operates to produce exquisite sensitivity for low level sounds and also very fine-grained frequency resolution that help to "separate" various sounds in everyday life. The normal ear can detect sounds at levels near or below the noise thresholds of most measuring instruments and can also separate and distinguish different frequencies to the point where even slight mistunings of musical instruments will be noticed. This is accomplished by sharply tuned vibration patterns of the cochlea and perhaps also by very accurate timing mechanisms that sample the firing patterns of individual nerve fibers and across groups of fibers. Most cases of sensorineural hearing loss are accompanied by a loss of hair cells. The inner hair cells are the primary receptor cells that send information to the brain. Outer hair cells, in contrast, are, in part, responsible for the finely tuned and sensitive cochlear vibration patterns. The degree of hearing loss is correlated with the type and pattern of hair cell loss (Liberman and Dodds 1984). As the outer hair cells are more susceptible to damage, they are the first to suffer. As the degree of hearing loss increases toward approximately 60 dB, more and more of the outer hair cells are missing and some damage to inner hair cells is also noted. This loss of outer hair cells will decrease the sensitivity of the cochlear vibration (resulting in a moderate to severe hearing loss as measured by pure-tone thresholds) and also destroys the sharp frequency tuning of the cochlea. Without the outer hair cells, the basilar membrane vibrates in its passive mode, which is less sensitive and more broadly tuned than normal. Thus in a cochlea with absent outer hair cells but a remaining population of inner hair cells, detection of sound will still be possible, but at elevated levels. And the ability of such a cochlea to separate and distinguish various frequencies will be reduced as compared to normal (Glasberg and Moore 1986; Trees and Turner 1986; Dubno and Dirks 1989). Of course, when even inner hair cells are missing in large numbers, a cochlear implant providing direct stimulation of the auditory nerve is necessary in order to transmit speech information to the brain.

2.2 Electric Hearing

The frequency selectivity of electric hearing is not dependent upon the basilar membrane vibration patterns; instead such factors as current spread from electrodes and the distance from each electrode to the stimulated neural elements presumably determine the independence of individual channels. Shannon et al. (1995) demonstrated that for understanding speech in quiet backgrounds, only a very limited degree of frequency selectivity is necessary. Even when speech with only 4 independent channels across a frequency range of 200 to 8000 Hz was presented to normal hearing listeners, recognition was nearly perfect. Thus each channel could be as wide as 2 octaves. Fishman et al. (1997) showed that this held true for the very best performing cochlear implant patients as well. Thus for speech in quiet, the limitations of the implant device itself, in terms of frequency resolution, do not appear to be significant. The poorer performing implant listeners (average implant patient performance is certainly not equal to normal-hearing listeners) presumably do not have a full complement of neural elements or suffer from more central deficits, caused by such factors as long durations of deafness or prelingual onset of hearing loss. That the very best implant patients do not seem to suffer too much when trying to understand speech in quiet suggests that in these optimal conditions, existing methods of electric stimulation provide adequate frequency selectivity.

If a noise background is present during speech recognition, the demands upon frequency resolution become greater. Then the physical limitations of electric hearing in terms of frequency selectivity become more visible (Fu et al. 1998a). When speech in a background of steady noise was presented to normal-hearing listeners, performance increased as the number of independent channels was increased to as many as 16 to 20 channels. However, even in the best performing cochlear implant users, performance was poorer than in the normal-hearing listeners and reached an asymptote at 6 to 8 channels. Adding more channels to electric hearing, which should, at first glance, seem to lead to better frequency resolution, does not actually increase the realized frequency resolution. As noted above, the independence of implant channels seems to have an upper bound well below that of the normal ear, presumably limited by current spread or possible central factors. When the background noise consists of more complex signals, such as modulated noises (Nelson et al. 2003) or other talkers (Qin and Oxenham 2003), then the frequency resolution requirements can become even more demanding. Kwon and Turner (2001), using modulated noise maskers to interfere with speech recognition, suggested that this is the result of the listener confusing the modulations of the target speech signal with the modulations of the background signals. This might occur because the background and target signals are not perceptually separated because of the limited frequency resolution and a type of "informational masking" is added to the normal "energetic masking." Current cochlear implants do not appear to be able to provide the necessary frequency resolution to recognize speech at high levels of performance when these types of complex background noise are also present (Stickney et al. 2004).

2.3 Comparing Electric and Acoustic Hearing

Of particular interest for the possible benefits of residual acoustic hearing for electric stimulation would be a comparison between two frequency resolutions: (a) that associated with the acoustic hearing of normal and impaired ears with various

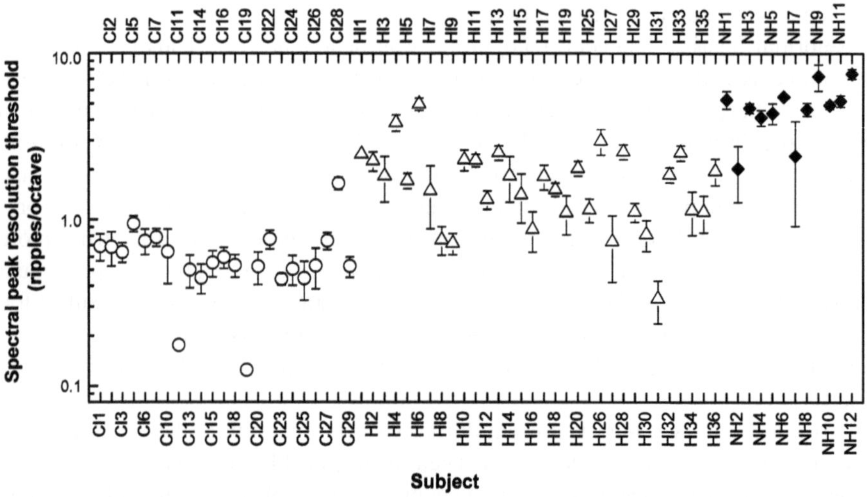

Fig. 3.1 The spectral resolution of normal-hearing, hearing impaired, and cochlear implant patients as measured by the spectral ripple test. From left to right individual subjects are ordered from cochlear implants (CI), sensorineural hearing loss (HI), and normal-hearing (NH). A higher measure of ripples per octave represents better performance. (Reprinted with permission from Henry et al. 2005)

degrees of sensorineural hearing loss and (b) that provided by electric stimulation via cochlear implants. Performing this comparison requires a measure that can be used in all types of listeners. Henry et al. (2005) directly compared these three types of listeners using a task that required them to discriminate between loudspeaker-presented spectral shapes. The implant subjects wore current design cochlear implants and listened through their everyday speech processor. These shapes were broadband stimuli shaped to have a rippled spectrum, with peaks and valleys occurring at log-spaced intervals in the frequency domain. The listeners had to discriminate between stimuli where the spectral valleys and peaks were interchanged between the members of a pair. The frequency spacing of the peaks and valleys was made smaller and smaller until the peaks and valleys were so close to one another that the members of the pairs of spectral shapes were indistinguishable. Thus the ability to distinguish closely spaced spectral peaks (one measure of frequency resolution) could be expressed in terms of the maximum number of spectral ripples per octave that could be distinguished. Figure 3.1 shows that the finest frequency resolution (higher number of ripples/octave) was found consistently in the normal-hearing listeners, followed by the listeners with sensorineural hearing loss, and the poorest performance was seen in the implant listeners. The subjects with sensorineural hearing loss ranged from approximately 30 dB of loss to as much as 90 dB of loss in this study. Thus for nearly all cases, even impaired acoustic hearing appears to provide better frequency resolution than electric stimulation.

As suggested by the earlier work of Fu et al. (1998a), frequency resolution abilities are related to the ability to recognize speech in background noises. In Fig. 3.2,

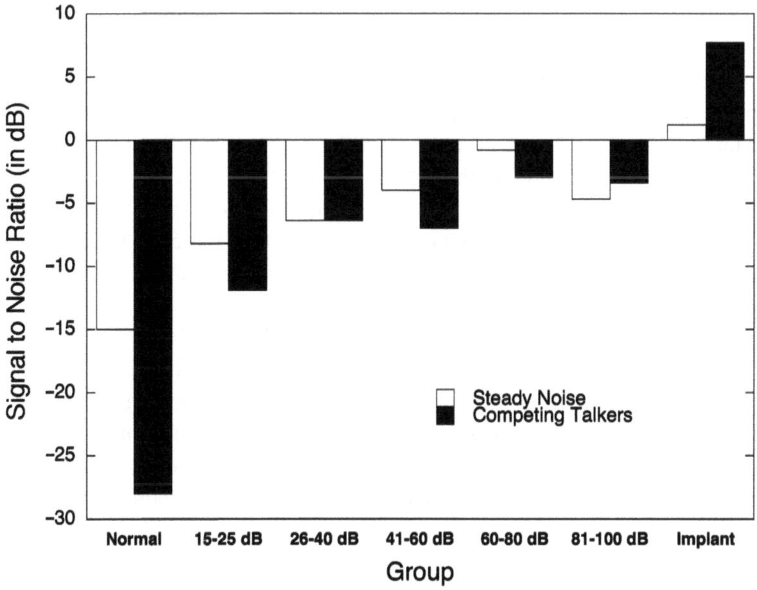

Fig. 3.2 Speech Reception Threshold (SRT) in dB signal-to-noise ratio, for various subject groups (pure-tone average hearing loss at 0.5, 1.0, and 2.0 kHz is listed for the non-implant subjects). Lower SRT's indicate better performance. (From Turner 2006)

the range of frequency resolution abilities across acoustic and electric hearing listeners is depicted along with their ability to recognize easy spondee words in two types of background noise, a steady state noise, and a competing-talkers background. The signal to noise ratio (SRT in dB) required for 50%-correct recognition of easy spondee words is plotted for various groups of listeners, separated into electric hearing and acoustic hearing and categorized by their degree of hearing loss. The target words here (spondees) are easily recognized in quiet listening conditions even without high frequency information; all subjects could score at high levels (>80%-correct) when listening to these words in quiet. Although this task is technically a speech recognition test, it can also be viewed as a measure of the subject's ability to resist the effects of noise. The ability to recognize speech in poorer (lower number) signal-to-noise ratio represents better performance.

In Fig. 3.2 the best performance was observed for those listeners with normal hearing. For these normal-hearing listeners, the competing-talkers background yielded the best performance, approximately 13 dB better than in the steady noise. This occurs because the fine frequency resolution of these ears allows the listeners to separate the various speakers, and the listener can take advantage of quieter portions of the fluctuating background to perceive additional information about the target speech (Duqeusnoy 1983). For listeners with sensorineural hearing loss the performance is poorer than normal for both types of background noises, and there is little or no advantage for the competing-talkers background. The limited frequency resolution of these listeners, because of the absence of very sharp cochlear tuning accompanying a loss of outer hair cells, probably leads to this poorer performance. The poorest performance of all occurs for the cochlear implant users, requiring signal-to-noise ratios 25 to 30 dB more favorable than is required for the normal-hearing listeners. In addition, the very poor frequency resolution of the implant listeners leads to even poorer performance for the competing-talkers background than in the steady noise. Presumably this occurs, as mentioned above, because the informational masking effect of the time-varying nature of the background is being confused with the target speech. It is noteworthy here that the even the listeners with severe sensorineural hearing loss tend to have better resistance to noise than do the implant listeners.

The limited frequency resolution of current cochlear implants is also seen when the measures involve stimuli where perceiving musical pitch is tested. Dorman et al. (1996) demonstrated that cochlear implant patients have poor frequency discrimination. Gfeller et al. (2002) and Kong et al. (2004) reported that implant patients have much poorer than normal music perception abilities, especially when the recognition of melodies is tested. Pitch cues can also be important for some tonal languages such as Cantonese and Mandarin, and implant listeners often have difficulty perceiving those cues. For Mandarin, temporal cues can sometimes compensate for the poorer pitch perception (Fu et al. 1998b; Xu et al. 2002), but in Cantonese, where temporal cues are not as strong, implant patients do poorly (Ciocca et al. 2002). Peng et al. (2008) demonstrated that even pitch cues in English speech, such as the rising intonation of a question, known as suprasegmental pitch cues, are also not well perceived by implant patients.

In summary, even the impaired frequency resolution of most listeners with sensorineural hearing loss appears to be superior to that provided by purely electric stimulation by today's implants. Thus the implant listeners are seen to perform particularly poorly in tasks involving resistance to competing backgrounds or the perception of musical pitches. This provides a strong rationale for using existing residual acoustic hearing, when possible, to supplement electrical stimulation for patients requiring a cochlear implant. In the simplest case, this might involve encouraging a patient to listen with the contralateral, non-implanted ear (which usually requires a hearing aid) along with their cochlear implant ear. In a more recent development, residual low frequency hearing in the implanted ear can be preserved even after surgery.

3 Preservation of Hearing Following Implantation

Using residual hearing was the impetus for the development of the Hybrid Short (S) Electrode. In the mid 1990s a few children with significant residual hearing received cochlear implants at several centers (Gantz et al. 2000). These results suggested that if enough residual hearing was present, the combination of acoustic and electric hearing could greatly improve the speech understanding in this group. Some of these children had significant low frequency acoustic hearing but had poor speech perception scores even with appropriately fit hearing aids. The concept of placing

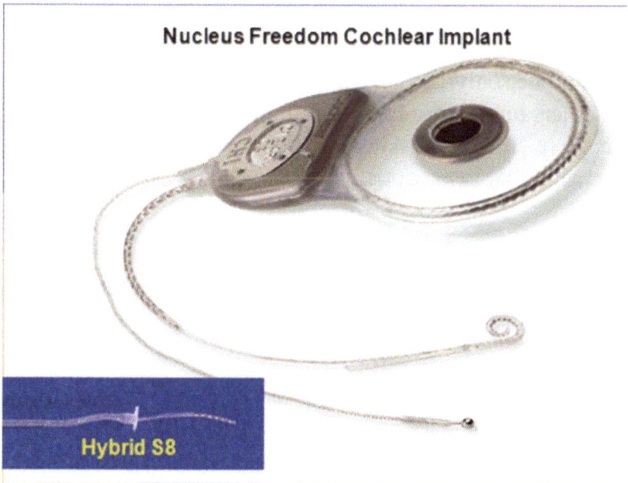

Fig. 3.3 A comparison of a standard-length cochlear implant (Nucleus Freedom) with a short (Hybrid S8) implant specifically designed for acoustic plus electric hearing. Picture is courtesy of Cochlear Americas Corp

an implant electrode limited to the basal turn was considered. Fortunately, in several studies using normal-hearing animals, hair cells have been observed to survive apical to a chronically implanted implant array (e.g., Shepherd et al. 1983; Ni et al. 1992; Xu et al. 1997). Even when the animals have a severe sensorineural hearing loss prior to implantation, remaining hair cells can survive (Coco et al. 2007). These studies are in agreement with occasional clinical reports over the years that human patients sometimes retained some sensitivity to acoustically presented sounds after being implanted with a traditional long-electrode array (Hodges et al. 1997). Thus the idea that residual hearing could be intentionally preserved during implantation is not as far-fetched as it might appear at first glance. Von Ilberg et al. (1999) reported preservation of residual hearing in one patient following implantation of a standard cochlear implant, and that this residual hearing was capable of some speech recognition postoperatively. When electric stimulation was added to the acoustic hearing, speech recognition for this patient increased. Gantz and Turner (2003) described a specifically designed electrode that was limited to the basal turn of the cochlea in order to preserve residual low frequency hearing. Six subjects had received two different versions of the device as part of an FDA feasibility trial that began in May of 1999. The initial results with a 6-mm, 6-channel electrode demonstrated that acoustic low frequency acoustic hearing could be preserved within a few decibels of the pre-implant hearing; however, the electrical processing was perceived as too high frequency. The design was modified to lengthen the electrode to 10 mm and place the six channels at the apical end of the electrode. The results of acoustic plus electric hearing in 3 subjects demonstrated that the acoustic hearing was preserved with the 10-mm electrode, and there was substantial improvement in speech perception combining acoustic and electric speech processing. Figure 3.3 displays this shortened electrode (inset) in comparison with a standard length cochlear implant.

Following these initial reports, the preservation of residual hearing in the implanted ear was pursued in a number of clinical trials. Some groups used a standard length electrode, only partially inserted, combined with "soft surgery" techniques, while others turned to smaller and shorter electrodes to attempt to minimize cochlear damage (e.g., Skarzynski et al. 2003; Gstoettner et al. 2004; Kiefer et al. 2005; James et al. 2005; Lenarz et al. 2009).

4 Candidacy for Hearing Preservation in Cochlear Implantation

While the general group of patients for whom hearing preservation surgery is designed continues to be those with severe to profound high frequency hearing loss, the specific characteristics of candidacy continue to evolve. The Hybrid S clinical trial began with recruitment of adults with normal to severe loss low frequency hearing at 125 to 500 Hz and then rapidly dropping to 80 dB at 1500 Hz as shown in Fig. 3.4. The ear to be implanted could be 50% CNC words while the contralateral ear could have 60% CNC word understanding with an appropriately fit hearing aid. Recent clinical trials for the Hybrid S12 and Hybrid L (16-mm electrode) include subjects with normal hearing to 1500 Hz and then a rapid decrease to 80 dB at 2000 Hz and beyond. CNC word understanding in the implant ear cannot be better

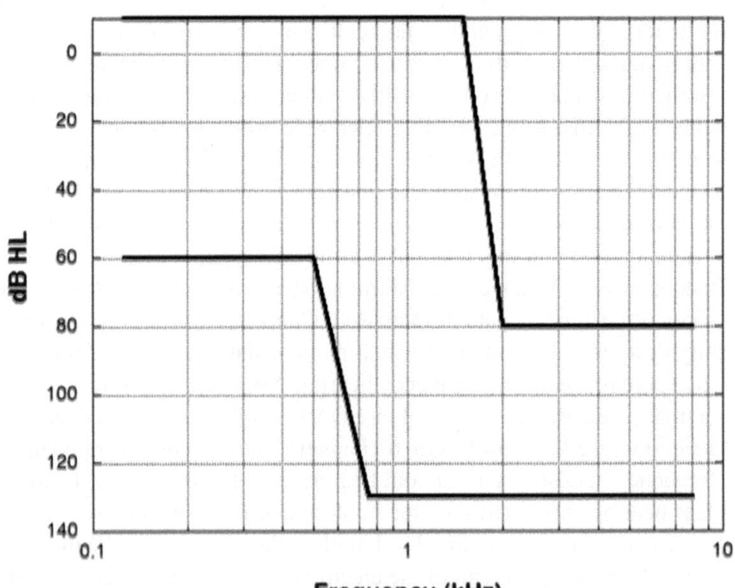

Fig. 3.4 Audiometric candidacy for implantation with the Hybrid short-electrode device. For low audiometric frequencies, hearing thresholds can range from normal to moderate-severe hearing loss

than 60%, and the contralateral ear can have CNC word understanding of less than 80%. The ongoing Med El 20-mm electrode trial is using similar criteria as the Hybrid trials.

The goals of these trials involving electrodes of varying length (10 to 20 mm) are to determine the optimal combination of hearing preservation and enhancement of speech perception. As individuals with more residual hearing are implanted, the risks to the remaining hearing increase. The risk-to-benefit ratio has not been defined, but it is likely that different lengths of electrodes will be necessary for different degrees of hearing impairment along with other predictive variables for the specific patient, such as duration of hearing loss, age of the individual, and the audiometric pattern of hearing loss.

Gantz et al. (2009) analyzed combined acoustic and electric speech recognition in a large sample ($n=68$) of subjects using the Nucleus Hybrid S8 implant (10 mm/6-channel electrode array). They identified that the strongest predictors of postoperative performance were preoperative speech recognition score along with the duration of high frequency deafness, presumably reflecting the respective contributions of the low frequency acoustic hearing and the ability of the implant to effectively provide meaningful speech information to the brain. Predictive factors such as these are just beginning to emerge as clinical trials are completing.

Trials with a standard length Nucleus advance-off stylet electrode have been performed in Europe and in the United States with mixed results (James et al. 2005; Balkany et al. 2006). This implant device uses an insertion tool designed to minimize perforations of the basilar membrane during deeper electrode insertions. Balkany et al. (2006) described some preservation of low frequency pure tone thresholds in 28 severely hearing impaired adults but could not preserve any acoustic word understanding using a stylet insertion tool with the Nucleus electrode. It is interesting to note that the subjects with preserved hearing did not demonstrate improved word understanding performance compared to those with no residual acoustic hearing.

5 Surgical Strategies for Preserving Residual Hearing

The surgical techniques for placing electrodes for preservation of hearing are different from those described for placement of standard cochlear implants. The electrode must be limited to the scala tympani. Past teaching was to create a cochleostomy anterior to the round window over the promontory. This type of cochleostomy usually results in placement of the electrode into the scala media, which would destroy cochlear structures necessary for acoustic hearing. Several modifications of the standard implantation procedure for preserving residual hearing are recommended. As described in Gantz et al. (2005), they all involve some special steps for minimizing cochlear damage. For example, Lehnhardt (1993) first described a soft surgical technique for placing electrodes. Several modifications of this strategy have been described for placement of acoustic plus electric speech processing (Adunka et al. 2004;

Briggs et al. 2005; Gantz et al. 2005). Yet others believe placement through the round window is the method of choice (Skarzynski et al. 2007; Lenarz et al. 2009).

6 Physiological Interactions Between Electric and Acoustic Hearing

What are the effects in the cochlea of electrical stimulation on the function of remaining acoustic hearing? And what are the effects of remaining hair cells on the electrical stimulation provided by a cochlear implant? In order for these interactions to occur within the cochlea, some overlap in the spatial excitation of the two modes of stimulation is necessary. For human patients with a severe to profound high frequency hearing loss receiving a shortened electrode such as the 10-mm Hybrid, no strong evidence of interactions has been noted. This is not surprising because the short electrode only reaches to approximately the 4000-Hz region of the normal cochlear map, and the residual acoustic hearing is limited to the lowest frequencies. However, for patients receiving a longer electrode, or for those with residual acoustic hearing at mid-frequencies, these interactions might be expected to be more common. To date, very few, if any, peripheral interaction in the form of one mode interacting with the other have been noted in humans (von Ilberg 1999; Wilson et al. 2002). Again, perhaps the spatial overlap in these patients is not sufficient to produce the necessary interactions, especially because the implants are usually intentionally inserted only to a depth shallow enough not to interfere with residual hearing.

In contrast, in animal experiments, where the same considerations of preserving the audiogram are not as important, and also where the animals do not necessarily have a severe hearing loss prior to implantation, it is quite possible to show peripheral interactions between electric and acoustic hearing. Moxon (1971) stimulated the cochlea with combined acoustic and electric signals. Low level electric current could produce responses in the surviving outer hair cells and thereby mechanically modulate the vibration of the basilar membrane to affect the responses of inner hair cells to fire the auditory nerve, whereas higher level electrical stimulation presumably stimulated the nerve fibers directly. Electrical stimulation can therefore produce responses in fibers that are connected to surviving hair cells as well as those that are not (van den Honert and Stypulkowski 1984; Stypulkowski and van den Honert 1984; von Ilberg et al. 1999).

Surviving hair cells, because they have their own responses to any acoustic activity and also have spontaneous activity, can also influence the responses to electrical stimulation. Miller and colleagues (Miller et al. 2006; Nourski et al. 2007; Miller et al. 2009) present evidence from animal experiments that surviving hair cells stimulated acoustically can influence the time course of recovery of fibers following electrical stimulation. In addition, the spontaneous activity and/or stochastic nature of acoustically sensitive fibers can influence the timing of nerve spikes to electrical stimulation.

In some fibers that Miller et al. (2009) recorded from, the acoustic stimulation could influence the spike timing in response to electric stimulation. Figure 3.5 shows

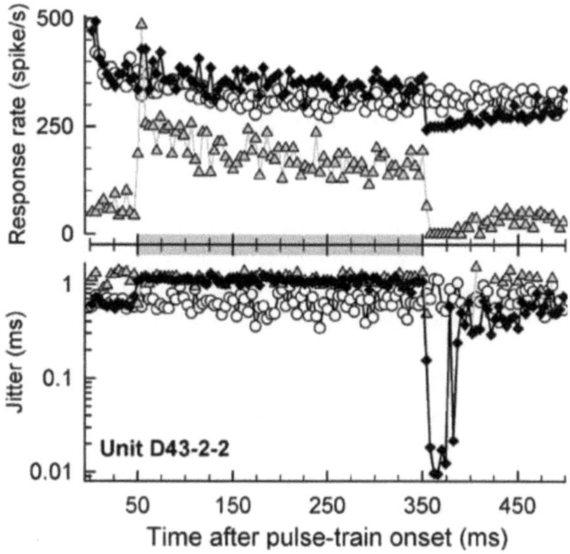

Fig. 3.5 The auditory nerve response to acoustic alone stimulation (*open triangles*), electric alone stimulation (*open circles*), acoustic plus electric stimulation (*filled diamonds*). The *upper panel* depicts spike rate, while the *lower panel* depicts spike-timing variability (jitter). (From Figure 11 in Miller et al. (2009) with permission from the Association for Research in Otolaryngology)

the auditory nerve spike rates as function of time (upper panel) and in the lower panel, the corresponding degree of nerve fiber "jitter" (stochastic nature of spikes) for acoustic-alone, electric-alone, and acoustic+electric stimulation. The electric pulse train stimulation was on continuously throughout the time frame, and the acoustic stimulation was added between 50 and 350 ms (shaded bar on x-axis). Even though the acoustic stimulation added very little to the driven spike rate, the amount of jitter in the spike timing was greater with acoustic stimulation added than in the electric-alone condition. Also note that following the offset of the acoustic stimulus, the electrically driven rate of the fiber declines to about 2/3 of its previous rate before slowly recovering. Although this behavior was noted in only some fibers, it does raise the possibility of potentially beneficial effects that acoustic stimulation may have for cochlear implant patients, because the lack of stochastic timing of electrically stimulated nerve responses has often been cited as one potential shortcoming of a cochlear implant compared to natural acoustic stimulation (Rubinstein et al. 1999b). These interactions will be dependent upon specific phases and magnitudes of the various stimuli, as well as the proximity of the stimulating electrode to the site of acoustic stimulation, and could be helpful or detrimental to an eventual goal of improving the auditory perception of patients using cochlear implants.

Higher-level interactions between acoustic and electric hearing are certainly common. For example, Snyder et al. (2004) found that acoustic stimulation to one ear and electric stimulation to the other ear both can stimulate the same regions of the Inferior Colliculus. And the numerous behavioral studies reviewed in the remainder of this article provide strong evidence that acoustic and electric stimulation can combine either within or across ears to produce potentially beneficial results for cochlear implantation.

7 Behaviorally Measured Interactions

The addition of the acoustic information (usually low frequency) from the either the implanted ear or the contralateral ear can provide supplemental information about a target speech signal both in quiet listening conditions (Shallop et al. 1992; Gantz and Turner 2003; Dorman et al. 2008) and in background signals (Turner et al. 2004; Ching 2005; Dorman et al. 2008). Use of acoustic information to supplement the electric stimulation can improve music perception as well (Gantz et al. 2006; Kong et al. 2005). For children, Nittrouer and Chapman (2009) demonstrated that having a period of development where stimulation is provided by an implant in one ear and a hearing aid in the other ear can result in stronger language abilities, as compared children with simultaneous implantation of bilateral devices. These advantages would presumably be a result of the better frequency resolution provided by thc acoustic hearing in the contralateral ear (as compared to what would be available from a second cochlear implant in that ear). This better frequency resolution would certainly be helpful in transmitting some of the intonational aspects of speech that are important in language development. The acoustic signal can also provide cues to the target speech signal that are complementary to the electrically presented information from the implanted ear. For example, the acoustic hearing can often provide more detailed information about the fundamental frequency and lower formant spectral regions of the target speech than can the implant. Zhang et al. (2010) showed that the majority of information provided by the acoustic hearing was located at 125 Hz and below, suggesting that more precise information on fundamental frequency and voicing is particularly important for this effect. At the same time, the higher frequency regions of the speech signal would be best transmitted by the cochlear implant ear, because the acoustic ear usually has a severe to profound hearing loss (and few, if any sensory hair cells) in the basal end of the cochlea, making the hearing aid ineffective for these higher frequency regions. This combination of two sources of information can sometimes combine in a "synergistic" manner to produce speech recognition scores that are even greater than the score that results from adding the two individual source scores.

How does the auditory system learn to use this "combined" acoustic plus electric information? Von Ilberg et al. (1999) presented evidence that 4 days of training and practice with A+E hearing (in the same ear) could improve speech recognition in quiet listening conditions. Longer term improvement over time was detailed in Gantz and Turner (2003) and Reiss et al. (2008), showing that performance continued to improve over the time course of 9 to 24 months. Figure 3.6 shows the time course of improved speech recognition in one patient using the 10-mm short electrode over a 1 year period using A+E hearing in the same ear.

This long time course of improvement (for ipsilateral A+E) seems to be a bit unusual in the literature on cochlear implants in general, with most traditional long-electrode patients reaching asymptotic performance after approximately 3 months following hookup. It may be that for same-ear stimulation, the process of adapting

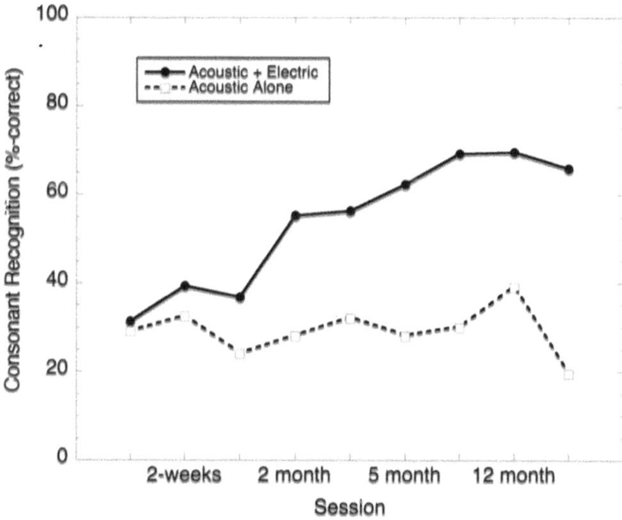

Fig. 3.6 The recognition of consonants by a patient implanted with a short-electrode implant that preserved hearing in the implanted ear as a function of the duration of experience with the device. Speech recognition scores are presented for both Acoustic plus Electric- and Acoustic-alone listening

to two different place-frequency maps (one for the implant and one for the acoustic hearing) in the same cochlea is difficult for patients to accomplish.

In addition to providing direct information on the target speech, acoustic hearing can also provide an advantage when the task is recognizing speech in the presence of background noises. In a simulation study using the vocoder method of producing cochlear implant processed speech (Shannon et al. 1995) presented to normal-hearing subjects, Turner et al. (2004) demonstrated that replacing the lowest 500 Hz of the vocoder speech with the original unprocessed acoustic stimuli yielded an advantage in terms of signal-to-noise ratio when the task was recognizing speech in a background of competing talkers. Chang et al. (2006), in another simulation study, demonstrated that providing only the acoustic spectrum below 300 Hz to either ear of the subject could produce a 10 to 15 dB advantage for recognizing speech in noise. In that example, the low-pass acoustic signal by itself provided no speech recognition. Thus supplementing the implant signal with a low frequency acoustic signal can provide improvements in listening situations that have been traditionally difficult for implant listeners. In these examples, the acoustic signal (and the cochlear implant simulations) was presented to ears with normal hearing. Can a similar advantage be obtained when the acoustic signal is presented to an ear with some degree of sensorineural hearing loss?

The answer appears to be yes. Turner et al. (2004) tested speech recognition in noise in three patients who were implanted with a short electrode cochlear implant and received low frequency acoustic stimulation in the same ear. These patients had mild to moderate low frequency hearing losses, accompanied by severe to profound

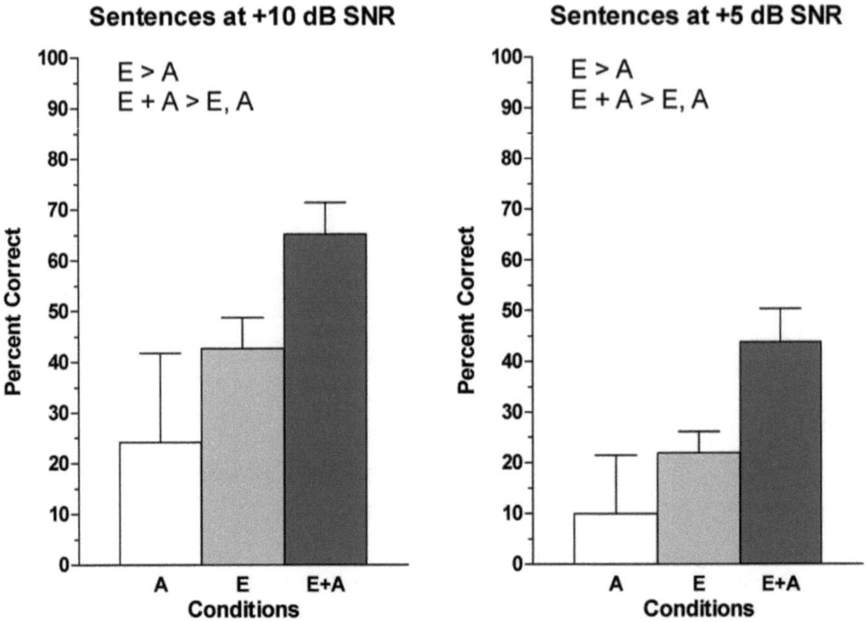

Fig. 3.7 The recognition of sentences presented in a multi-talker babble background (*left panel* +10 dB S/N; *right panel* +5 dB S/N). (From Dorman et al. 2008)

high frequency losses. When compared to long-electrode patients (matched for speech recognition in quiet) a substantial advantage for the A + E patients was noted. Figure 3.7 shows that adding low frequency acoustic information to the contralateral ear for sensorineural hearing loss patients implanted with a long electrode improved not only speech recognition in quiet and in noise, but also for the recognition of melodies and voices (Dorman et al. 2008). In these studies, the "acoustic-alone" speech recognition, either from the ipsilateral, implanted ear, as in Turner et al. (2004), or from the contralateral, non-implanted ear (Dorman et al. 2008), was at a low, but non-zero level.

An even more surprising demonstration of this effect was presented by Kong et al. (2005). In that study, unilaterally implanted patients were tested for speech recognition in background noise with (1) the implant alone, (2) the contralateral acoustic ear alone, and (3) the "bimodal" condition, where both ears, the electric and the acoustic, were used. Figure 3.8 displays the results of that study. What was striking in this demonstration was that the acoustic-alone contralateral ears (which had severe to profound hearing losses) had essentially zero speech recognition by themselves, but when added to the implant in the bimodal condition, could improve speech recognition in noise by 20 to 30 percentage points.

How does the contribution of the acoustic hearing depend upon the degree of hearing loss associated with this acoustic hearing? The Kong et al. study (2005) suggests that not much residual hearing is necessary in order to demonstrate an

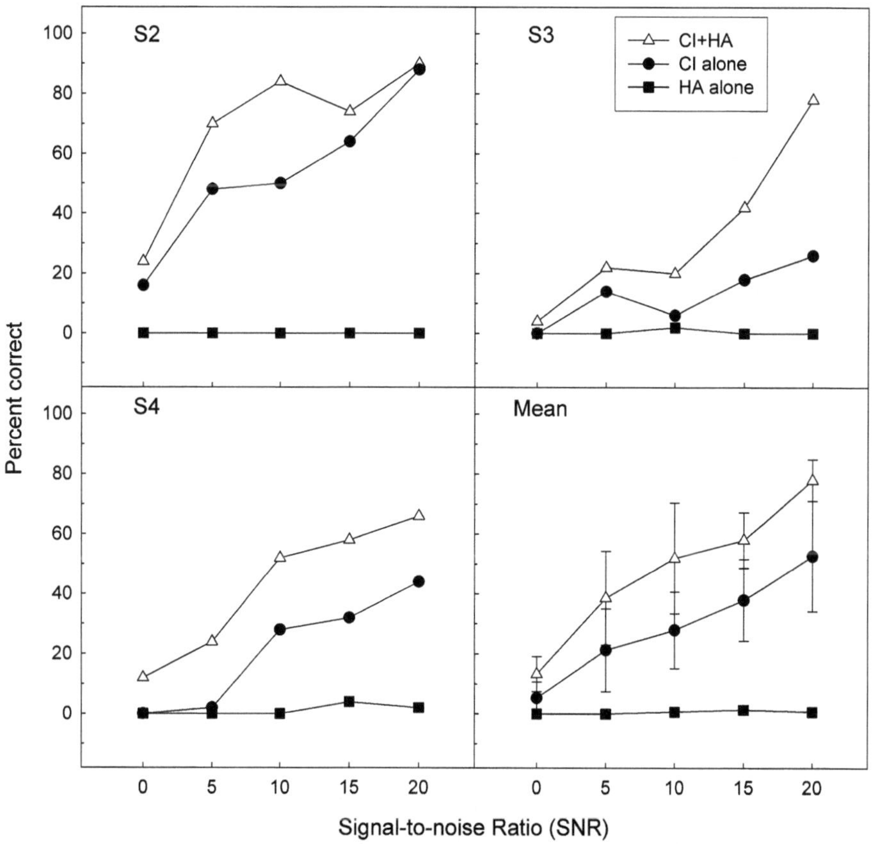

Fig. 3.8 The recognition of speech by three cochlear implant listeners in various levels of background noise. Displayed are the speech recognition scores for implant alone, contralateral (hearing aid) ear alone, and the combined (CI+HA) condition. (Reprinted with permission from Kong et al. 2005)

advantage over electric-alone stimulation at least in those particular patients who use the acoustic hearing of the non-implanted ear. Thus patients should be encouraged to try to use the acoustic hearing in the non-implanted ear to see if they can gain an advantage. Another way to look at this issue is to think in terms of clinical choices for patient treatment if preserving hearing in the implanted ear is a possibility. In other words, is a patient better off solely with a long-electrode implant (which presumably can provide a wider range of stimulation across the cochlea than a shallowly inserted implant designed to preserve hearing) or with the combination of acoustic plus electric hearing using a short electrode designed to preserve hearing? The question of whether adding acoustic hearing is beneficial for real patients, either from the contralateral ear, or from preserved hearing in the implanted ear, is not a simple one. Unfortunately, as is seen in most across-subject comparisons of real

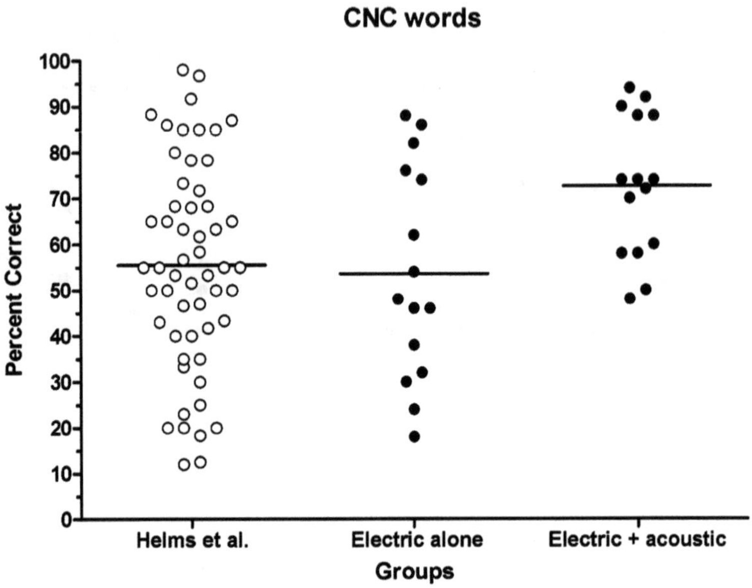

Fig. 3.9 Comparing the performance of implant alone subjects to subjects who add acoustic hearing to their implant (EAS). (Figure from Dorman et al. 2008)

world cochlear implant patients, the range of performance across groups is typically so large that a decisive conclusion is not obvious. Figure 3.9 shows that while there is an advantage in the group average scores for A+E hearing over electric-alone hearing, the large range of performance for all groups indicates that predicting whether an individual patient will do better with one approach or the other is risky (Dorman et al. 2008). Patient selection for assembling these types of comparisons would seem to be a critical factor in determining the outcome.

Turner et al. (2008) measured the signal-to-noise ratio required for ipsilateral A+E short-electrode patients to recognize spondee words in a background of other talkers at a 50%-correct level of performance as a function of the degree of their low frequency hearing loss (Fig. 3.10). These patients received acoustic information below 750 Hz (because of their audiogram shape) and electrically stimulated information on the speech signal from 700 to 8000 Hz through the short-electrode implant. Low frequency hearing loss was expressed as the average pure-tone thresholds at 125, 250, and 500 Hz. Only a mild correlation is noted ($r=0.56$). The two open circles represent patients who lost significant amounts of residual hearing some months following surgery; without the data from those patients, the correlation is not significant.

Figure 3.10 also shows the average results from two control groups of long-electrode patients. The upper dashed horizontal line represents the mean value of SRT

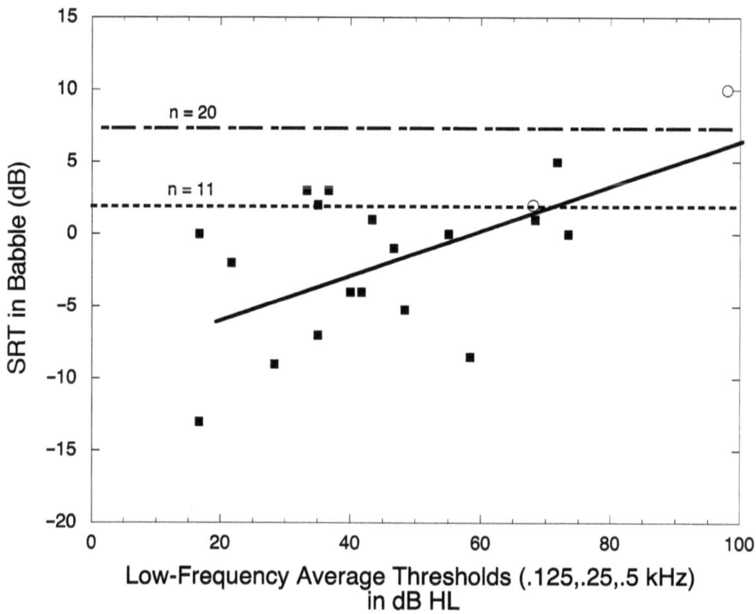

Fig. 3.10 The signal-to-noise ratio (SRT) in individual patients required to correctly recognize 50% of spondee words in a background of multi-talker babble by A+E patients (ipsilateral preserved hearing) plotted as a function of the degree of low frequency hearing loss in the preserved ear. (Figure from Turner et al. 2008)

from a random selection of long-electrode patients on the same speech-in-noise task. Comparing this group to the A+E patients, A+E offers an advantage of E-alone in nearly every case. However, not every implant candidate is an appropriate candidate for ipsilateral A+E stimulation; many of the members of this long-electrode control group had little or no residual hearing. In addition, some members of this control group may have suffered from other negative factors such as long duration of deafness or very poor preoperative speech recognition (both factors that can be related to poor performance with an implant as described by Rubinstein et al. 1999a). The lower dotted horizontal line is a subset of the original larger control group, chosen by taking only the top performers of the long-electrode patients based on their speech recognition in quiet scores, so that the A+E and long-electrode patient groups were equivalent in mean performance for speech in quiet. When this more select group is used for comparison, it appears that there is some limit on the amount of residual hearing that can provide an advantage over electric-only stimulation. The regression line suggests that severe to profound hearing loss (90 to 100 dB HL) for low frequency hearing is an approximate limit for how much residual acoustic hearing is worth preserving in order to realize a potential A+E advantage for speech in background noises.

8 Mechanisms for the Advantage of Residual Acoustic Hearing

These speech-in-noise studies imply that the acoustic signal provides important information that allows the patient to listen selectively to the target speech as opposed to the competing background noise. This ability to "stream" different signals is an important factor in being able to understand speech in a background of competing talkers (i.e., the "cocktail party effect"). Cues for streaming can include fundamental frequency, timing information, and spatial location. Acoustic hearing added to cochlear implant stimulation could potentially help with some of these cues.

One hypothesis to explain this effect is that by using the acoustic ear to identify the fundamental frequencies of the various talkers, and by associating each talker with a corresponding speech envelope, the listener can then better separate the various competing envelope speech streams presented by the cochlear implant ear. As discussed earlier, it is this confusion of competing envelope streams in speech processed through a cochlear implant that is suspected to be a major contributor to the problems that implant listeners have in understanding speech in a background of other talkers (Kwon and Turner 2001; Nelson et al. 2003; Qin and Oxenham 2003). Separating the various fluctuating streams also allows the listener to take advantage of the quieter portions of the background in order to better perceive the target speech (Miller and Licklider 1950). This hypothesis receives support in a study by Brown and Bacon (2009), who used normal-hearing listeners and presented only a sine wave at the fundamental frequency that was modulated by the target speech, in addition to the cochlear implant processed speech. They found that performance in a background of other talkers was significantly improved. Kong and Carlyon (2007) suggest that the acoustic portion of the speech signal assists in the perception of timing cues when listening to speech in these complex backgrounds. Of course, the information helpful in separating the target speech from the background would be in addition to any speech information provided by the acoustic portion of the signal that directly helps the recognition of the target signal, as would be observed in a no-background condition. It is clear that replacing at least some of the electric signal (with its necessarily poor frequency resolution) with acoustic hearing (and its usually better resolution) appears to be a key factor in this "speech in competing talkers" advantage.

Yet another cue for the "cocktail party effect" is spatial location. Preserved acoustic hearing might provide an advantage for determining spatial location because traditional cochlear implants tend to be notoriously poor in tests of localization (Litovsky et al. 2009). A number of studies have looked at localization abilities of patients who wear a cochlear implant on one side and a hearing aid on the other. For example, Ching et al. (2004) tested a large number of adult users on speech recognition in background noise where the speech and noise came from different locations. The noise was presented from the implant side and the speech was from directly in front. These patients had hearing losses of 100 dB HL in the non-implanted ear. When the subjects listened with both devices, their average scores were higher than the situation using the CI alone, and also considerably higher than the hearing aid alone. Clearly the head shadow effect, which could allow the listener to obtain a better signal-to-noise ratio on the hearing aid side offered some potential advantage. However, the hearing aid alone scores for these patients were near zero percent, suggesting that other factors were also involved. That the ability to use spatial separation cues was at least partly responsible for this advantage was supported by the additional finding that performance in a localization task, in which subjects identified the location of sounds presented from an 11-loudspeaker array, was best for the two-eared (CI plus contralateral hearing aid) condition. In general, however, the cues available for localization by the hearing aid ear (generally low frequency, because of the severe to profound hearing loss) and the cochlear implant (usually high frequency interaural loudness cues, since CIs are generally not good at providing the proper interaural timing cues) are inconsistent with each other. We might expect that even better perception of spatial location would occur if both ears had similar localization cues available, i.e., some acoustic hearing in both ears.

Indeed, Long et al. (2009) used vocoder simulation studies and demonstrated that in a speech in background noise task where spatial location is a beneficial cue, providing bilateral low frequency acoustic information, as opposed to simulated low frequency cochlear implant information, yielded an advantage for this task. Presumably the acoustic low frequency hearing allows the listeners to take advantage of precise interaural timing differences between the two ears to improve the perception of spatial location. Dorman and Gifford (2010), using real implant patients, demonstrated a similar advantage for patients who were allowed to use bilateral low frequency acoustic hearing as compared to patients using only electrical stimulation from one ear and a hearing aid in the other (Fig. 3.11). Patients had preserved hearing in the implanted ear and in the contralateral, non-implanted ear. The task was recognition of speech presented from a single loudspeaker in front, and a competing multi-talker background signal presented from a multi-speaker array surround the patient. The signal-to-noise ratio required for 50%-correct performance was the dependent measure. In the condition labeled Bimodal, the patient's implanted ear was plugged to permit only electric stimulation. This condition is equivalent to the case where hearing is not preserved in the implanted ear, and the A + E stimulation comes from combining two ears. In the Combined condition, the earplug was removed, allowing the patients to use the acoustic hearing from the implanted ear, thereby providing binaural (primarily low frequency) acoustic cues from both ears. As noted in the accompanying audiograms, the acoustic hearing remaining in the implanted ear was poorer than that in the contralateral ear. So not only do these results suggest that providing some type of combined acoustic and electric input to both ears is beneficial for many patients in real world situations, but they also imply that attempting to preserve acoustic hearing in the implanted ear can be quite beneficial, because this may allow the patient access to "true binaural" cues for spatial location when combined with the acoustic hearing in the contralateral ear.

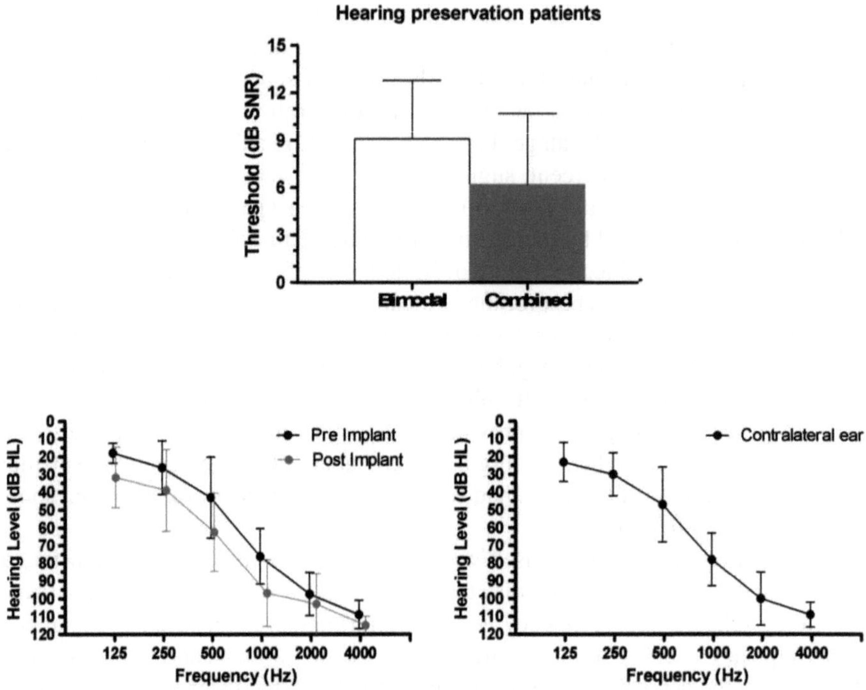

Fig. 3.11 *Upper panel* displays the signal-to-noise ratio required to recognize speech in the Bimodal condition (electric-only implant stimulation in one ear and a hearing aid in the other) compared to the Combined condition (same as Bimodal with the addition of acoustic hearing in the implanted ear). The *lower panels* display the average audiograms of the patient groups. (From Dorman and Gifford 2010)

9 Tonotopic Frequency Mapping in Implants

The combination of acoustic and electric hearing has also provided insights into several theoretical issues related to tonotopic mapping in the cochlea and how well the auditory system can adapt to distortions of place-frequency mapping. Fu and Shannon (1999) have previously demonstrated that patients with traditional long-electrode implants can adapt over time to a changed place-frequency map after the patients' speech processor mapping function was changed. With the recent experiences of A+E patients, in which there is usable hearing in either the contralateral, non-implanted ear, or even the ipsilateral, implanted ear, the patients' auditory stimulation comes from both the abnormal place-frequency mapping of electrical stimulation and also a "relatively normal" acoustic map (at least to the first approximation, given that sensorineural hearing loss can produce some distortions in the place-frequency map). In these patients, providing a new "electric map" will also change the relation between the acoustic mapping and electric mapping.

In the case where the acoustic hearing is in the other ear, the place of stimulation for this ear will be relatively normal, perhaps amplified by a hearing aid. For electric stimulation, the place of stimulation will be wherever the speech processor assigns the frequency of the stimulus. In most of these cases, these two places of stimulation will be different across ears. An interesting question arises from this mismatch. In the case of short-electrode patients with a 10-mm electrode, where even low and mid-frequency sounds are programmed to stimulate extremely basal locations in the cochlea, the places of stimulation for single-frequency stimulus in the real world will be very different between the two ears. In these cases, does a patient hear two different pitches in the two ears, or would the patient just hear one unitary percept? Reiss et al. (2007, 2008) addressed this question by asking patients to match the pitch sensation of an electrical pulse train presented to the basally located electrodes of the short electrode implant to a pure tone presented to the acoustic hearing in the contralateral ear. Many of the patient's pitch sensations from electrical stimulation appeared to change over time and quite often reflected a change from an initially high pitch sensation when they were first implanted, corresponding to the place of electric stimulation, to a pitch sensation that over time shifted to correspond to the frequency assigned to that electrode by their speech processor. In this manner, any discrepant pitch sensations between the two ears (because of a place of stimulation mismatch) will disappear with experience and the patient will eventually hear a unitary pitch between the two ears. Figure 3.12 summarizes these findings (Reiss et al. 2008).

In the case of A+E stimulation in an ipsilateral ear, the patient will have two potentially discrepant place-frequency maps in the same ear. This could lead to some confusion in the listening experience, especially for speech recognition where the lowest frequency (acoustic stimulation) portions of speech are presented to the normal tonotopic place, and the mid- and high frequency speech frequencies are presented to locations much more basal than normal. This discrepancy could lead to difficulty in understanding speech sounds and possibly require a long time to adapt to the device. In addition, the lower success rate of the Hybrid, short-electrode implant for very elderly patients may reflect their reduced ability to adapt to such distortions.

Another interesting issue that has been revealed by A+E stimulation is also related to distorted place-frequency mapping. The use of a shorter electrode in an attempt to preserve acoustic hearing necessitates mapping the speech processor to present a fairly wide range of speech frequencies assigned to electrodes that occupy only a narrow space in the base of the cochlea. In the Hybid (10-mm) electrode, the speech frequencies from 700 to 8000 Hz are presented to six electrodes that extend only over 4 mm of cochlear distance. Yet some of these patients can understand speech at very high levels of performance when listening via electric-only stimulation (80%-correct for difficult consonant tests); scores are equivalent to long-electrode patients where the same speech range is presented to 22 electrodes spaced across 17 mm of cochlear space. This is an interesting finding, in that it suggests that the physical interaction of closely spaced electrodes may not limit performance in the Hybrid patients as is suspected for long-electrode patients. It seems that that this

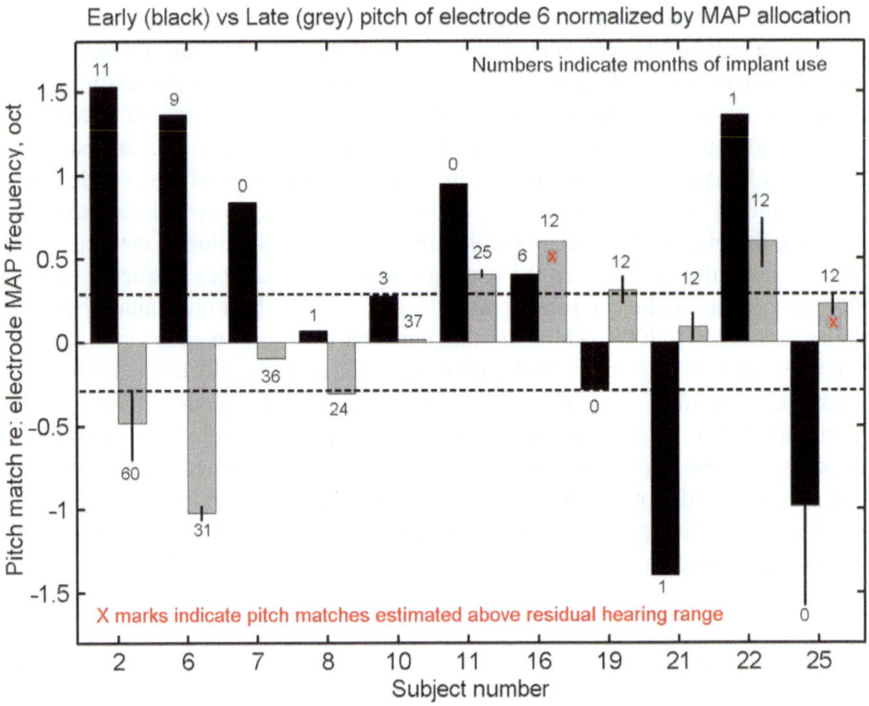

Fig. 3.12 A demonstration of pitch changes over time in short-electrode patients. Electric stimulation to the most apical electrode in the "short" Hybrid electrode is matched in pitch to an acoustically presented pure tone in the contralateral ear. The dark bars represent pitch matches made soon after the implant is initially hooked up; the gray bars represent pitch matches made after a long period of adaptation to the implant. The y-axis represents the pitch match frequency normalized to the center of the MAP frequency assigned to this apical electrode by the speech processor. Over time, most patients' pitch matches shifted to a frequency within the range of the speech processor MAP frequency. (Figure from Reiss et al. 2008)

new generation of A+E cochlear implant patients, who tend to have much better residual hearing than the previous generation of long-electrode patients, will be more successful users of electrical stimulation than the previous long-electrode implant candidates. This may reflect higher rates of nerve survival for these patients. Thus electrical stimulation in general may be capable of even higher levels of performance when it is provided to new categories of patients.

10 Summary

It has been shown clearly that the addition of residual acoustic hearing to electrical stimulation via a cochlear implant provides advantages to many patients as compared to electric stimulation alone. In addition, the technique of hearing preservation surgery has potentially widened candidacy to patients who previously were not considered appropriate for cochlear implantation. Future work will include research on the most appropriate methods for preserving residual hearing, as well as improved guidelines for which patients are the most appropriate candidates for which type of device. Optimal combinations of residual hearing in either the contralateral or ipsilateral ear with electric stimulation have not yet been fully explored. In addition, the provision of electrical stimulation to patients with levels of residual hearing much greater than in the past certainly will expand our knowledge of the capabilities of cochlear implants in general.

Acknowledgments This work was supported in part by NIDCD grants RO1DC000377 and 2P50 DC00242, and by GCRC/NCRR grant RR00059.

References

Adunka, O., Gstoettner, W., Hambek, M., Unkelbach, M. H., Radeloff, A., & Kiefer, J. (2004). Preservation of basal inner ear structures in cochlear implantation. *Otorhinolaryngology-Head and Neck Surgery*, 66(6), 306–312.

Balkany, T. J., Connell, S. S., Hodges, A. V., Payne, S. L., Telischi, F. F., Eshraghi, A. A., Adrien, A., Angeli, S. I., Germani, R., Messiah, S., & Arheart, K. L. (2006). Conservation of residual acoustic hearing after cochlear implantation. *Otology & Neurotology*, 27, 1083–1088.

Briggs, R. J., Tykocinski, M., Stidham, K., & Robinson J. B. (2005). Cochleostomy site: implications for electrode placement and hearing preservation. *Acta Oto-Laryngologica*, 125, 870–876.

Brown, C., & Bacon, S. (2009). Achieving electric-acoustic benefit with a modulated tone. *Ear and Hearing*, 30(5), 489–493.

Chang, J. E., Bai, J. Y., & Zeng F. G. (2006). Unintelligible low-frequency sound enhances simulated cochlear-implant speech recognition in noise. *IEEE Transactions on Biomedical-Engineering*, 53, 2598–2601.

Ching, T. Y. (2005). The evidence calls for making binaural-bimodal fitting routine. *Hearing Journal*, 58, 32–41.

Ching, T. Y. C., Incerti, P., & Hill, M. (2004). Binaural benefits for adults who use hearing aids and cochlear implants in opposite ears. *Ear and Hearing*, 25, 9–21.

Ciocca, V., Frabcusm, A.L., Aisha, R., & Wong, L. (2002). The perception of Cantonese lexical tones by early-deafened cochlear implantees. *Journal of the Acoustical Society of America*, 111, 2250–2256.

Coco, A., Epp, S. B., Fallon, J. B., Xu, J., Millard, R. E., & Shepherd, R. K. (2007). Does cochlear implantation and electrical stimulation affect residual hair cells and spiral ganglion neurons? *Hearing Research*, 225, 60–70.

Dorman, M. F., & Gifford, R. H. (2010). Combining acoustic and electric stimulation in the service of speech recognition. *International Journal of Audiology*, 49, 912–919.

Dorman, M. F., Smith, L. S., Smith, M., & Parkin, J. L. (1996). Frequency discrimination and speech recognition by patients who use the Ineraid and continuous interleaved sampling cochlear-implant processors. *Journal of the Acoustical Society of America*, 99, 1174–1184.

Dorman, M. F., Gifford, R. H., Spahr, A. J., McKarns, S. A. (2008). The benefits of combining acoustic and electric stimulation for the recognition of speech, voice and melodies. *Audiology & Neurotology*, 13(2), 105–112.

Dubno, J. R., & Dirks, D. D. (1989). Auditory filter characteristics and consonant recognition for hearing-impaired listeners. *Journal of the Acoustical Society of America*, 85, 1666–1675.

Duqeusnoy, A. J. (1983). Effect of a single interfering noise or speech source upon the binaural sentence intelligibility of aged persons. *Journal of the Acoustical Society of America*, 74, 739–743.

Fishman, K., Shannon, R. V., & Slattery, W. H. (1997). Speech recognition as a function of the number of electrodes used in the SPEAK cochlear implant speech processor. *Journal of Speech, Language, and Hearing Research*, 40, 1201–1215.

Fu, Q. J., & Shannon R. V. (1999). Effects of electrode configuration and frequency allocation on vowel recognition with the Nucleus-22 cochlear implant. *Ear and Hearing*, 20(4), 332–344.

Fu, Q. J., Shannon, R. V., & Wang, X. (1998). Effects of noise and spectral resolution on vowel and consonant recognition: acoustic and electric hearing. *Journal of the Acoustical Society of America*, 104, 3586–3596.

Fu, Q.-J., Zeng, F.-G., Shannon, R. V., & Soli, S. D. (1998). Importance of tonal envelope cues in Chinese speech recognition. *Journal of the Acoustical Society of America*, 104, 505–510.

Gantz, B. J., & Turner, C. W. (2003). Combining acoustic and electric hearing. *Laryngoscope*, 113, 1726–1730.

Gantz, B. J., Rubinstein J. T., Tyler, R. S., Teagle, H., Cohen, N. L., Waltzman, S. B., Miyamoto, R. T., & Kirk. I. (2000). Long-term results of cochlear implants in children with residual hearing. *Annals of Otology, Rhinology, and Laryngology*, 109(Suppl. 185, 12, Pt. 2), 33–36.

Gantz, B.J., Turner, C., & Gfeller, K.E. (2005). Preservation of hearing in cochlear implant surgery: advantages of combined electrical and acoustical speech processing. *Laryngoscope*, 115, 796–802.

Gantz, B. J., Turner, C. W., & Gfeller, K. (2006). Acoustic plus electric speech processing: results of a multicenter clinical trial of the Iowa/Nucleus Hybrid Implant. *Audiology & Neurotology*, 11(Suppl. 1), 63–68.

Gantz, B. J., Hansen, M. R, Turner, C. W., Oleson, J. J., Reiss, L. A., & Parkinson, A. J. (2009). Hybrid 10 clinical trial: preliminary results. *Audiology & Neurotology*, 14, 32–38.

Gfeller, K., Turner, C., Woodworth, G., Mehr, M., Fearn, R., Witt, S., & Stordahl, J. (2002). Recognition of familiar melodies by adult cochlear implant recipients and normal hearing adults. *Cochlear Implants International*, 3, 29–53.

Glasberg, B., & Moore, B. C. J. (1986). Auditory filter shapes in subjects with unilateral and bilateral cochlear impairments. *Journal of the Acoustical Society of America*, 79, 1020–1033.

Gstoettner, W., Kiefer, J., Baumgartner, W.-D., Pok, S., Peters, S., & Adunka, O. (2004). Hearing preservation in cochlear implantation for electric acoustic stimulation. *Acta Oto-Laryngologica*, 124, 348–352.

Henry, B. A., Turner, C. W., & Behrens, A. (2005). Spectral peak resolution and speech recognition in quiet: normal hearing, hearing impaired and cochlear implant listeners. *Journal of the Acoustical Society of America*, 118, 1111–1121.

Hodges, A. V., Schloffman, J., and Balkany, T. (1997). Conservation of residual hearing with cochlear implants. *American Journal of Otology*, 18, 179–183.

James, C., Albegger, K., Battmer, R., Burdo, S., Deggouj, N., Deguine, O., Dillier, N., Gersdorff, M., Laszig, R., Lenarz, T., Rodriguez, M. M., Mondain, M., Offeciers, E., Macías, Á. R., Ramsden, R., Sterkers, O., Von Wallenberg, E., Weber, B., Fraysse, B. (2005). Preservation of residual hearing with cochlear implantation: how and why. *Acta Oto-Laryngologica*, 125(5), 481–491(11).

Kiefer, J., Pok, M., Adunka, O., Sturzbecher, E., Baumgartner, W., Schmidt, M., Tillein, J., Yue, Q., & Gstoettner, W. (2005). Combined electric and acoustic stimulation of the auditory system: results of a clinical study. *Audiology & Neurootology*, 10, 134–144.

Kong, Y. Y., & Carlyon, R. P. (2007). Improved speech recognition in noise in simulated binaurally combined acoustic and electric stimulation. *Journal of the Acoustical Society of America*, 121, 3717–3727.

Kong, Y., Cruz, R., Jones, J., & Zeng, F. (2004). Music perception with temporal cues in acoustic and electric hearing. *Ear and Hearing*, 25, 173–185.

Kong, Y. Y., Stickney, G. S., & Zeng, F. G. (2005). Speech and melody recognition in binaurally combined acoustic and electric hearing. *Journal of the Acoustical Society of America*, 117, 1351–1361.

Kwon, B. J., & Turner, C. W. (2001). Consonant identification under maskers with sinusoidal modulation. *Journal of the Acoustical Society of America*, 110, 1130–1140.

Lehnhardt, E. (1993). Intracochlear placement of cochlear implant electrodes in soft surgery technique. *HNO*, 41(7), 356–359.

Lenarz, T., Stöver, T., Buechner, A., Lesinski-Schiedat, A., Patrick, J., Pesch, J. (2009). Hearing conservation surgery using the Hybrid-L electrode. Results from the first clinical trial at the Medical University of Hannover. *Audiology & Neurotology*, 14(Suppl. 1), 22–31.

Liberman, M. C., & Dodds, L. W. (1984). Single-neuron labeling and chronic cochlear pathology. III. Stereocilia damage and alterations of threshold tuning curves. *Hearing Research*, 16, 55–74.

Litovsky, R. Y., Parkinson, A., & Arcaroli, J. (2009). Spatial hearing and speech intelligibility in bilateral cochlear implant users. *Ear and Hearing*, 30, 419–431.

Long, C. J., Portnuff, C., Muralimanohar, R., & Litovsky, R. (2009). Binaural cues via acoustic and electric stimulation: simulations. Paper presented at the Convergence of Hearing Aid and Cochlear Implant Technology Workshop, Miami, FL.

Miller, G. A., & Licklider, J. C. R. (1950). The intelligibility of interrupted speech. *Journal of the Acoustical Society of America*, 22, 167–173.

Miller, C. A., Abbas, P. J., Robinson, B. K., Nourski, K. V., Zhang, F., & Jeng, F-C. (2006). Electrical excitation of the acoustically sensitive auditory nerve: single-fiber responses to electric pulse trains. *Journal of the Association of Research in Otolaryngology*, 7, 195–210.

Miller, C. A., Abbas, P. J., Robinson, B. K., Nourski, K. V., Zhjang, F., & Jeng, F.-C. (2009). Auditory nerve fiber responses to combined acoustic and electric stimulation. *Journal of the Association of Research in Otolaryngology*, 10, 425–445.

Moxon, E. C. (1971). *Neural and mechanical responses to electric stimulation of the cat's inner ear* (Unpublished doctoral dissertation). Massachusetts Institute of Technology.

Nelson, P., Jin, S.-H., Carney, A., & Nelson, D. A. (2003). Understanding speech in modulated interference: cochlear implant users and normal-hearing listeners. *Journal of the Acoustical Society of America*, 113, 961–968.

Ni, D., Shepard, R. K., Seldon, H. L., Xu, S., Clark, G. M., & Millard, R. E. (1992). Cochlear pathology following chronic electrical stimulation of the auditory nerve. I: normal hearing kittens. *Hearing Research*, 62, 63–81.

Nittrouer, S., & Chapman, C. (2009). The effects of bilateral electric and bimodal electric-acoustic stimulation on language development. *Trends in Amplification*, 13, 190–205.

Nourski, K. V., Abbas, P. J, Miller, C. A., Robinson, B. K, & Jeng, F.-C. (2007). Acoustic–electric interactions in the guinea pig auditory nerve: simultaneous and forward masking of the electrically evoked compoundaction potential. *Hearing Research*, 232, 87–103.

Peng, S. P., Tomblin, B., & Turner, C. W. (2008). Production and perception of speech intonation in pediatric cochlear implant recipients and individuals with normal hearing. *Ear and Hearing*, 29(3), 336–351.

Qin, M. K., & Oxenham, A. J. (2003). Effects of simulated cochlear implant processing on speech reception in fluctuating maskers. *Journal of the Acoustical Society of America*, 114, 446–454.

Reiss, L. R., Turner, C. W., Erenberg, S. R., & Gantz, B. (2007). Changes in pitch with a cochlear implant over time. *Journal of the Association of Research in Otolaryngology*, 8(2), 241–257.

Reiss, L. R., Gantz, B., & Turner, C. W. (2008). Cochlear implant speech processor frequency allocations may influence pitch perception. *Otology & Neurology*, 29, 160–167.

Rubinstein, J. T., Parkinson, W. S. Tyler, R. S., & Gantz, B. J. (1999a). Residual speech recognition and cochlear implant performance: effects of implantation criteria. *American Journal of Otolaryngology*, 20, 445–452.

Rubinstein, J. T., Wilson, B. S., Finley, C. C., & Abbas, P. J. (1999b). Pseudospontaneous activity: stochastic independence of auditory nerve fibers with electrical stimulation. *Hearing Research*, 127, 108–118.

Shallop, J., Arndt, P., & Turnacliff, K. (1992). Expanded indications for cochlear implantation: perceptual results in seven adults with residual hearing. *Journal of Speech-Language Pathology & Applied Behavior Analysis*, 16, 141–148.

Shannon, R.V., Zeng, F. G., Kamath, V., Wygonski, J., & Ekelid, M. (1995). Speech recognition with primarily temporal cues. *Science*, 270, 303–304.

Shepherd, R. K., Clark, G. M., & Black, R. C. (1983). Chronic electrical stimulation of the auditory nerve in cats. Physiological and histopathological results. *Acta Oto-Laryngologica*, Suppl. 399, 19–31.

Skarzynski, H., Lorens, A., & Piotrowska, A. (2003). A new method of partial deafness treatment. *Medical Science Monitor*, 9, 26–30.

Skarzynski, H., Lorens, A., & Piotrowska, A. (2007). Preservation of low frequency hearing in partial deafness cochlear implantation using the round window surgical approach. *Acta Oto-Laryngologica*, 127, 41–48.

Snyder, R. L, Bierer, J. A, & Middlebrooks, J. C. (2004). Topographic spread of inferior colliculus activation in response to acoustic and intracochlear electric stimulation. *Journal of the Association of Research in Otolaryngology*, 5, 305–322.

Stickney, G. S., Zeng, F. G., Litovsky, R. V., & Assmann, P. F. (2004). Cochlear implant speech recognition with speech maskers. *Journal of the Acoustic Society of America*, 116, 1081–1091.

Stypulkowski, P. H., & Van Den Honert, C. (1984). Physiological properties of the electrically stimulated auditory nerve. I. Compound action potential recordings. *Hearing Research*, 14, 205–223.

Trees, D. A., & Turner, C. W. (1986). Spread of masking in normal subjects and in subject with hearing loss. *Audiology*, 25, 70–83.

Turner, C. W. (2006). Hearing loss and the limits of amplification. *Audiology & Neurotology*, 11(Suppl. 1), 2–5.

Turner, C. W., Gantz, B. J., Vidal, C., Behrens, A., & Henry, B. A. (2004). Speech recognition in noise for cochlear implant listeners: benefits of residual acoustic hearing. *Journal of the Acoustical Society of America*, 115, 1729–1735.

Turner, C. W., Reiss, L. A., & Gantz, B. J. (2008). Combined acoustic and electric hearing: preserving residual acoustic hearing. *Hearing Research*, 242, 164–171.

Van Den Honert, C., & Stypulkowski, P. H. (1984). Physiological properties of the electrically stimulated auditory nerve. II. Single fiber recordings. *Hearing Research*, 14, 225–243.

Von Ilberg, C., Kiefer, J., Tillein, J., Pfenningdorff, T., Hartmann, R., Sturzebacher, E., & Klinke, R. (1999). Electro-acoustic stimulation of the auditory system. *Journal for Oto-Rhino-Laryngology*, 61, 334–340.

Wilson, B., Wolford, R., Lawson, D., & Schatzer, R. (2002). Speech processors for auditory prostheses. Third Quarterly Progress Report, Neural Prosthesis Program (contract N01-DC-2-1002 NIH). Washington, DC.

Xu, J., Shepard,, R. K., Milllard, R. E., & Clark, G. M. (1997). Chronic electrical stimulation of the auditory system at high stimulus rates: a physiological and histopathological study. *Hearing Research*, 105, 1–29.

Xu, L., Tsai, Y., & Pfingst, B. E. (2002). Features of stimulation affecting tonal-speech perception: implications for cochlear implants. *Journal of the Acoustic Society of America*, 112, 247–258.

Zhang, T., Spahr, A., & Dorman, M. (2010). Information from the voice fundamental frequency (F0) accounts for the majority of the benefit when acoustic stimulation is added to electric stimulation. *Ear and Hearing*, 31, 63–69.

Chapter 4
Implantable Hearing Devices for Conductive and Sensorineural Hearing Impairment

Ad Snik

1 General Introduction

Some degree of hearing loss affects 10 to 15% of the general population. The prevalence increases with age. More than one third of 70-year-old persons have hearing loss in excess of 30 dB HL (Hearing Loss), which makes them eligible for hearing aids. Although introduction of digital signal processing in acoustic hearing aids in the late 1990s has led to wider acceptance, owing to improved sound quality and comfort, only 20 to 25% of hearing impaired subjects are satisfied with their hearing aids. There are several medical and technological limitations that are responsible for this low acceptance rate of the acoustic hearing aids. The medical conditions limiting the use of acoustic hearing aids include chronic external otitis, chronic middle ear disease, or presence of profound hearing loss. The technological limitations are often related to the use of ear molds in these aids, including occlusion of the ear canal, leading to complaints about the user's own voice, annoying feedback as a result of high amplification and acoustic leakage, and undesirable blockage of residual hearing at low frequencies. A final reason for the low acceptance rate of acoustic hearing aids is social stigma associated with being perceived old or handicapped.

There are two different approaches to address the low acceptance rate issue. One approach is to continue to improve the design of acoustic hearing aids that, for example, adopt open ear mold fittings, feedback cancelation, and advanced miniaturization technology that can completely conceal the hearing aid in the ear canal (Dillon 2001). The other approach is the development of implantable auditory prostheses, including cochlear implants, direct bone-conduction devices, and middle ear implants. Cochlear

A. Snik (✉)
Department of Otorhinolaryngology, Radboud University Medical Centre,
PO Box 9101, 6500 HB, Nijmegen, the Netherlands
e-mail: a.snik@kno.umcn.nl

F.-G. Zeng et al. (eds.), *Auditory Prostheses: New Horizons*, Springer Handbook of Auditory Research 39, DOI 10.1007/978-1-4419-9434-9_4,

implants have been developed for patients with profound deafness and who cannot benefit from acoustic aids. A cochlear implant provides sound perception by means of electrical stimulation of the auditory nerve via an electrode array surgically placed in the cochlea. Cochlear implants are highly successful and are used by over 150,000 adults and children worldwide. Alternatively, direct bone-conduction devices have been developed for patients with aural atresia or chronic draining ears who cannot be fitted with acoustic aids. These patients typically have conductive or mixed hearing loss. Direct bone-conduction devices transform the incoming sound signals into vibrations that are transmitted to the cochlea by means of mechanical coupling to the skull (Stenfelt and Goode 2005). For example, the Baha (Bone anchored hearing aid) is a well established direct bone-conduction device that is being used by more than 65,000 people worldwide (Cochlear BAS Ltd, Göteborg, Sweden). The third and most recent type of implantable devices are middle ear implants that have been developed for those of moderately to severely hearing impaired patients with normal outer and middle ears, who need amplification but desire an invisible or well concealed solution. Different from conventional hearing aids, the acoustic output of the middle ear implants is replaced by a transducer that has to be surgically coupled to one of the middle ear ossicles, or directly to the cochlea (Magnan et al. 2005). In addition to cosmetics, the middle ear implants avoid most ear mold-related problems such as occlusion and feedback while also improving sound fidelity (Fredrickson et al. 1995; Ball et al. 1997). However, the middle ear implants have also generated a host of new technological and medical problems, from transducer placement and coupling efficiency to surgical risk and relatively high financial cost. The present chapter reviews audiological, surgical, and technological issues for implantable hearing devices other than cochlear implants. Subdivision is made into devices for patients with sensorineural hearing loss and devices for patients with conductive or mixed hearing loss.

2 Implantable Hearing Devices for Sensorineural Hearing Loss

2.1 Introduction

Several different types of implantable hearing aid have been developed for patients with sensorineural hearing loss (see Table 4.1). Initially, these concerned semi-implantable systems with the output transducer connected to the ossicular chain or tympanic membrane, while the electronics, battery, and microphone were worn externally.

Figure 4.1 presents a typical example. The external amplifier, often referred to as the audioprocessor (AP in the figure), transmits the amplified signal via a conduction link (SR) to the implanted transducer (T). Two types of transducer have been used, piezoelectric and electromagnetic. In a piezoelectric transducer, a voltage across piezoelectric materials causes deformation or bending. An alternating voltage will result in vibrations that can be used to drive the ossicular chain. Electromagnetic

Table 4.1 Middle ear implant systems that have been used in patients

Device name	Company	Principal investigator	Devices sold	Current status
TICA	Implex	Zenner, Leysieffer	>20	Off
Esteem	Envoy	Kraus	>70	FDA approval (3/17/2010)
Soundtec	Soundtec	Hough	>600	Off
Otologics MET	Otologics	Fredrikson	>800	Only in Europe
Carina	Otologics	Kasic, Jenkins	>500	Phase II
Vibrant Soundbridge	Med-El	Ball	>5000	

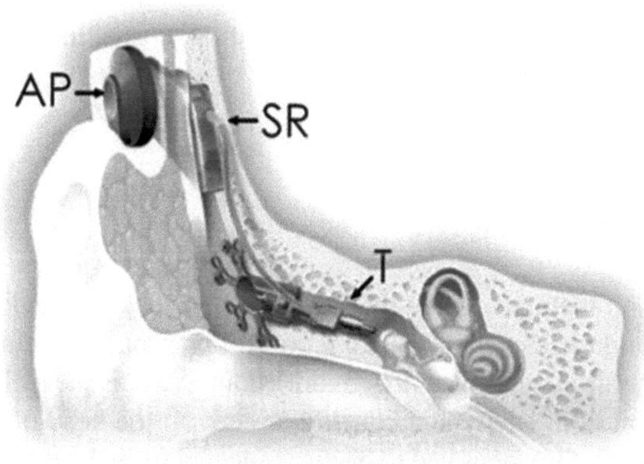

Fig. 4.1 Diagram of a semi-implantable device (Otologics MET). AP refers to the externally worn audio processor, SR to the subcutaneously placed receiving coil, and T to the transducer which is coupled to the incus. Courtesy of Otologics LLC, Boulder

transducers are available in two subtypes. The first subtype is a contactless set-up. A permanent magnet is connected to the tympanic membrane or the middle ear ossicles. The magnet is driven by a remote coil, mostly packed in an ear mold, placed in the ear canal. Passing an alternating current through the coil produces a fluctuating magnetic field that will cause the magnet to vibrate. The second subtype, called an electromechanical transducer, includes a magnet that moves within the coil. This transducer is encased in a special housing, mostly made of titanium and implanted. The movements of the magnet within the coil are transduced to the middle ear ossicles by means of direct coupling, for example, by a rod that can move through an opening in the transducer's housing.

2.2 *Outcomes of Implantable Hearing Aids*

This section examines the main audiological outcome measures including speech recognition at normal conversational level (65 dB SPL) and "functional gain." Speech recognition can be measured using different materials (and languages) and different scoring methods, such as at word or phoneme level. The "functional gain" is defined as unaided thresholds minus aided thresholds (Dillon 2001). However, when an air-bone gap occurs after surgery, this erroneously leads to extra "functional gain." Therefore, presurgery unaided thresholds have to be used. Another consideration is that several implantable devices make use of non-linear amplification. This means that the gain is input-dependent and largest with soft sounds near threshold level (Lenarz et al. 2001; Snik and Cremers 2001). "Functional gain" of these systems only expresses the gain for soft sounds and overestimates the gain at louder, yet normal conversational level. Snik and Cremers (2001) studied implant systems with non-linear amplification and reported that the gain at conversational levels was up to 10 dB less than the "functional gain" at the threshold level. As most of the older implant systems use linear amplification, corrections do not need to be made. However, the conditions used to measure the functional gain are specified in the more recent devices.

2.2.1 Piezoelectric Systems

The first system on the market was the Japanese piezoelectric device developed by Suzuki and co-workers (Suzuki 1988). Figure 4.2 presents a schematic diagram of this system. Its transducer (V in the figure) comprised a piezoelectric bimorph, anchored in the mastoid cavity and coupled directly to the head of the stapes. This coupling implied that the stapes had to be disconnected from the incus. Therefore, the system was only used in the case of disrupted ossicular chain, thus in patients with conductive or mixed hearing loss. Sect. 4.2.1 addresses this system in more detail.

Ohno and Kajiya (1988) showed that the output and gain of this piezoelectric device were limited. It is important to note that the power output is directly related to the size of the crystal. Owing to anatomical and technical limitations, it was fairly difficult to increase the size of the piezoelectric crystal. Leysieffer et al. (1997) introduced a modified piezoelectric system, in which the piezoelectric transducer was placed in a titanium housing, anchored in the mastoid cavity, and connected to the incus by means of a rod. The modified system achieved a maximum equivalent sound pressure level that was about 10 dB higher than that of the Japanese system (Leysieffer et al. 1997; Zenner and Leysieffer 2001). An additional appealing feature of the system, named the TICA, was that it was fully implantable. The microphone was placed subcutaneously in the ear canal, indicating that the ear canal and pinna were functionally involved in hearing with the TICA. However, there were problems with feedback because the tympanic membrane radiated the mechanical vibrations induced by the TICA transducer as an acoustic signal into the ear canal (Zenner and Leysieffer 2001). To minimize feedback without reducing the gain, the developers introduced the "reversible malleus neck dissection technique" to disrupt the feedback loop (Zenner and Leysieffer 2001). Although reversible, it is not generally accepted to disrupt the ossicular chain when applying a middle ear system (Magnan et al. 2005). Results from a clinical trial revealed limited gain. The manufacture of the TICA dissolved in 2001.

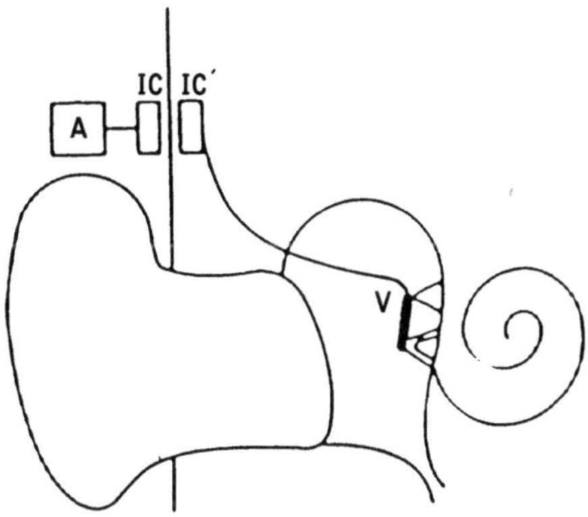

Fig. 4.2 Schematic drawing of the Japanese, piezoelectric middle ear device. A refers to the amplifier, IC the transcutaneous transmitter/receiver system, and V to the transducer. Courtesy of Prof. Dr. J. Suzuki

Another fully implantable piezoelectric system, formerly known as the St. Croix system, has recently been renamed as the Envoy Esteem (Chen et al. 2004). The piezoelectric transducer is connected to the head of the stapes, while a second sensor-transducer is coupled to the malleus and acts as the microphone of the system, in conjunction with the tympanic membrane. In principle, this seems to be the most natural solution, as the external ear canal and tympanic membrane are functionally involved. However, the ossicular chain is disrupted as a direct consequence of this set-up, because the incus has to be removed. It has been shown that the maximum output of this device is equivalent to 110 dB SPL. Furthermore, the surgery is complicated. Chen et al. reported several technical and surgical complications in their Phase I clinical study on 7 patients. Functional gain and speech perception in quiet were no better than the scores obtained with conventional devices. After modifications, a phase II study was started. In early 2010, the Envoy system received FDA approval. Nevertheless, clinical data regarding the Envoy system are limited. Barbara et al. (2009) showed that in 3 patients with an activated Envoy device, mean functional gain was only 19 dB while the mean hearing loss was 60 dB HL. Neither speech perception nor control data to the conventional devices were mentioned.

2.2.2 Contactless Electromagnetic Devices

As an alternative to the piezoelectric transducer, Heide et al. (1988) proposed an electromagnetic system that comprised a small magnet glued onto the eardrum at the level of the malleus. The microphone, battery, electronics, and the driving coil were placed in a special housing and inserted into the ear canal in the form of a deeply fitted in-the-ear hearing aid. Although the coupling between the transducer and middle ear was contactless, the ear canal was occluded by the driver. Heide et al. (1988) published audiometric data on 6 patients with mild to severe sensorineural hearing loss and found that the mean functional gain was 10 dB higher than the patients' own conventional hearing aid. Surprisingly, there was no improvement in speech recognition at 65 dB SPL, suggesting that the (linear) device amplified soft sound effectively but not sounds at normal conversation levels. A second experimental study with Heide's system was conducted in patients with conductive or mixed hearing loss (Kartush and Tos 1995). Because any slight shift would result in significant changes in power and the magnet had the tendency to dislocate, the device was taken off the market.

A major problem with the electromagnetic coupling between a permanent magnet and a remote driving coil is much lower efficiency than that of the piezoelectric coupling (Leysieffer et al. 1997). Efficiency decreases with the cube of the distance between the magnet and driving coil. Maniglia and co-workers proposed a solution to minimise the distance (Abbass et al. 1997; Maniglia et al. 1997). In their set-up, the magnet was glued to the body of the incus and the driving coil was implanted in the mastoid at only 0.5 to 1 mm away from the magnet. This distance was much smaller than that in the set-up of Heide et al. The developers found that the gain and output of this device were tens of decibels higher. However, the frequency response was poor below 2 kHz and the development effort has since stopped (Abbass et al. 1997).

Hough et al. (2001) also introduced an improved version of Heide's device called the Soundtec that used a permanent magnet and a driving coil. After separating the incudo-stapial joint, a small ring holding the magnet was placed around the head of the stapes, after which the ossicles were reconnected. The driver was placed deep into the ear, causing occlusion of the ear canal. The occlusion is considered to be a disadvantage (Magnan et al. 2005). Postoperatively, air-conduction thresholds were found to have deteriorated by about 5 dB. Several patients reported side effects from the magnet in the middle ear. Silverstein et al. (2005) introduced techniques to provide additional support for the magnet; they did not discuss whether or not this influenced the effectiveness of the whole system. Hough et al. (2002) reported results from 102 patients with moderate high frequency hearing loss who produced slightly better mean speech recognition score with the Soundtec (82% correct) than with conventional devices (77% correct). Using disability-specific questionnaires, Roland et al. (2001) did not find any difference in subjective benefit between the Soundtec device and acoustic hearing aids in a group of 23 patients. Silverstein et al. (2005) concluded that the Soundtec device might be an improvement over conventional devices in well selected patients, provided that they have favourable anatomy of the ear canal and middle ear, realistic expectations, and are fully aware of the side effects of the magnet in the middle ear. In 2004, the device was withdrawn from the market. It remains to be seen whether this marks the end of magnetic, contactless middle ear implants.

2.2.3 Devices with Electromechanical Transducers

Other electromagnetic contact systems have been put forward, for instance, the Otologics MET by Fredrickson et al. (1995). In their set-up, there is no air gap between the magnet and coil; the magnet moves in the coil and is attached to the incus by means of a connecting rod (see Fig. 4.1). Owing to the absence of an air gap between the magnet and coil, this device is much more powerful than the contactless types described above, with possible output up to an equivalent of 130 dB SPL. The transducer (T) is anchored in the surgically enlarged mastoid cavity, while the tip of its moving rod is placed in a small hole, made by laser, in the body of the incus. This transducer is connected electrically to a receiving coil (SR) placed subcutaneously (see Fig. 4.1). The rest of the device (microphone, amplifier, transmitter) is worn externally in a button-type of processor (AP) that is kept in place by means of magnets (one implanted as part of the receiver and the other in the button). This device has now been on the market in Europe for more than 10 years. No more than 3 dB changes were reported in hearing thresholds at most frequencies as a result of the surgery and/or the coupling between the transducer and incus (Jenkins et al. 2004). Jenkins et al. reported multicenter audiological results of 282 Otologics MET users. They studied two groups of patients: one group fitted in the US following a strict protocol that included comparisons with an acoustical behind-the-ear device (or BTE) with the same electronics as the Otologics audio processor (see Sect. 2.4 for details), and another group of patients from Europe with moderate to profound hearing loss from whom only "functional gain" data were available.

Because the audio processor used non-linear amplification, their gain data (up to 40 dB) should be considered as gain for soft sounds. Snik et al. (2004) measured the real functional gain as well as the maximum output in 5 experienced users of the Otologics MET with severe hearing loss. They put the sound processor in linear amplification mode and did not limit the maximum output. In situ maximum output was derived from aided input–output functions, measured objectively (with the microphone in the ear canal). The input level at which the output levelled off plus the patient's functional gain determined the maximum output. Highest mean functional gain (0.5–4 kHz) was 43 dB, while the individual maximum output values were between 103 and 120 dB SPL (mean 111 dB SPL). When compared to the popular NAL prescription rule, a rule that prescribes desired target gain and output (Dillon 2001), the measured gain and maximum output values were adequate for all the patients (Snik et al. 2004). In principle, the individual maximum output values are independent of the patient's hearing loss. This suggests that the observed range of 17 dB in maximum output was caused by variations in the effectiveness of the coupling between the transducer and incus.

Since the introduction of the semi-implantable Otologics MET in the late 1990s, development has been directed towards a totally implantable device. In 2005, such a device, called the Otologics Carina, was released for phase I testing. To deal with feedback, the microphone was placed subcutaneously behind the auricle, at a relatively long distance from the auditory meatus. Apart from acoustic feedback, sensitivity to chewing noises and vocalisation played important roles in determining the best microphone position (Jenkins et al. 2007b). Initial testing showed that the gain of the Carina was up to about 20 to 25 dB (Lefebvre et al. 2006), about half of the gain of the semi-implantable version. Feedback was the limiting factor. Jenkins et al. (2007a) presented the phase I trial data as well as the follow-up results at 12 months. Compared with the patients' acoustic device, Carina produced better aided thresholds and speech perception scores in quiet, but similar speech scores in noise. A phase II study was introduced after significant improvements in technical

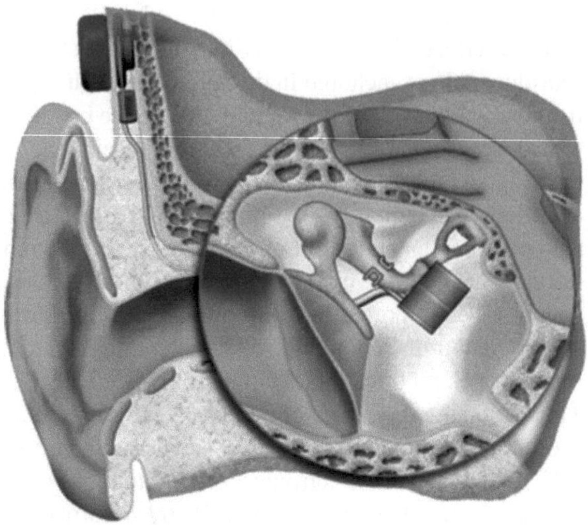

Fig. 4.3 Diagram of the Vibrant Soundbridge device, with enlarged projection of the middle ear to show the floating mass transducer, connected to the incus. Courtesy of Med-El Company, Innsbruck

and surgical aspects (Jenkins et al. 2007a). Bruschini et al. (2009) reported results in 5 Carina users with hearing loss between 60 and 70 dB HL. Mean gain was comparable to that reported by Lefebvre et al. (2006). Mean speech recognition improved from 16% in the unaided condition to 56% with the Carina device, suggesting non-optimal gain. In the case of conductive hearing loss, limited gain is less of a problem. Therefore, the Carina device has also been adapted for this group of patients (see Sect. 4.2.3).

A third type of electromagnetic middle ear implant was developed with the output transducer comprising a coil and magnet, mounted together in a 2 × 1.5 mm, hermetically sealed cylinder (Ball et al. 1997; Gan et al. 1997). When a current is applied, the magnet moves. As the magnet is relatively heavy, the cylinder vibrates in the opposite direction. This so-called floating mass transducer (FMT) is coupled directly to the ossicular chain to vibrate in parallel with the stapes. Figure 4.3 shows a diagram of this system, called the Vibrant Soundbridge.

The main advantage of the Vibrant Soundbridge over the transducer of the Otologics MET is that its housing is hermetically sealed and thus less vulnerable to body fluids. However, physiological dimensions determine the size of the transducer and thus the weight of the mass, which in turn determines the power output. Measurements on temporal bones have shown that the maximum output is equivalent to 110 dB SPL and that the frequency range is broad (Gan et al. 1997).

Luetje et al. (2002) showed safe results from a successful clinical trial involving 53 patients. Preoperative versus postoperative bone-conduction thresholds was within 3 dB at all frequencies. The (unspecified) functional gain was reported to be better with the Vibrant Soundbridge than with the patients' own acoustic devices. Fraysse et al. (2001; $n=25$) and Lenarz et al. (2001; $n=34$) reported similar improvements and generally overall preference for the Vibrant Soundbridge over the acoustic device. Snik et al. (2007) studied patient satisfaction in a unique group of 22 patients who could not wear an acoustic device because of therapy-resistant external otitis that worsened when an occluding ear mold was used. Given this patient population, the subjective satisfaction level was less than that for the Vibrant Soundbridge users reported in the other studies. It was suggested that selection bias might have played a role; the patients in the Snik et al. study could not wear an acoustic device, while those in the other studies could wear it but did not like their acoustic device.

Snik et al. (2004) measured real functional gain in 7 Vibrant Soundbridge users with severe sensorineural hearing loss between 65 and 75 dB HL. Sound processor was set in linear amplification mode and the output was unlimited. Measurements were the same as those reported above in the Otologics MET users. The highest mean functional gain was 40 dB, with the individual mean maximum output varying between 92 and 112 dB SPL (mean 102 dB SPL), about 10 dB below that with the Otologics MET device. According to the NAL prescription rule, the gain was adequate, but the maximum output was too low in several patients. This 10 dB difference in individual maximum output might be attributable to variance in the coupling efficiency of the FMT to the incus. The Vibrant Soundbridge and the Otologics MET have been on the market in Europe for longer than 10 years, with the former having more than 5,000 users. Extensive clinical data have been obtained with these two devices and will be discussed in Sect. 2.4 and 2.5.

2.3 Surgical Issues and Complications

This section deals with surgical issues and complications related to the devices available on the European market, such as the Envoy Esteem, Otologics devices, and the Vibrant Soundbridge. Only phase I data have been published on the two fully implantable devices, the Envoy Esteem and Otologics Carina. Serious technical and surgical problems have been described (Chen et al. 2004 and Jenkins et al. 2007a, respectively). After appropriate improvements, phase II studies have been started, but the results are not yet available.

With regard to the semi-implantable Otologics MET device, Jenkins et al. (2004) did not elaborate on surgical problems or device failures. They discussed how to achieve effective coupling to the incus and later introduced an intra-operative monitoring technique (Jenkins et al. 2007c).

Several studies have addressed surgical and technical problems with the Vibrant Soundbridge, including too narrow posterior tympanotomy and, in a number of cases, concomitant chorda problems. Sterkers et al. (2003; $n=125$) reported such problems in 5.6% of their patients, whereas Schmuziger et al. (2006; $n=22$) reported a higher percentage of 15%. Crimping of the FMT clip around the incus also led to problems when the diameter of the incus was small (Fraysse et al. 2001; Lenarz et al. 2001; Sterkers et al. 2003, Cremers et al. 2008). Sterkers et al. (2003) reported an incidence of 6.4%. Lenarz et al. (2001) proposed the use of bone cement to achieve better fixation and found that it was effective. Snik and Cremers (2004) compared the post-surgery results of five patients with the normal FMT coupling to six patients who received additional bone cement. No positive or negative effects could be detected at that time. However, 2 years later, one of their bone cement patients showed deterioration of the aided thresholds. Additional surgery revealed that the incus was necrotic (Cremers et al. 2008) and that the bone cement had disappeared (Serenocem, Corinthian Medical, Nottingham). Later on, a second patient from this group was found to have similar problems (J. Mulder, 2009, personal communication). This negative result suggests that bone cement should be used with caution. So far, no other groups have reported incus necrosis, but crimping of the FMT clip over the incus was found to cause erosion of incus, comparable to that seen in stapes revision surgery (Todt et al. 2004). Device failure rates were between 4 and 5% (Schmuziger et al. 2006; Mosnier et al. 2008). Mosnier et al. (2008) further suggested that device failures were only encountered with the first generation of Vibrant Soundbridge implants.

A disadvantage of electromagnetic devices over piezoelectric devices is their susceptibility to magnetic fields. MRI is not recommended in patients with electromagnetic devices. Nevertheless, Todt et al. (2004) reported on two Vibrant Soundbridge users who had undergone 1.5-Tesla MRI. No adverse effects were found. In contrast, Schmuziger et al. (2006) reported dislocation of the FMT after MRI was performed in one of their patients.

To place the FMT, a transcanal approach has been advocated. This approach is less invasive than the transmastoid approach. Truy et al. (2006) reported good results without any complications. However, Cremers and Snik (2005) reported serious long-term complications in 3 out of their 9 patients in whom the transcanal approach had been used (perforation of the tympanic membrane, wire loose in the auditory canal). Some of these patients suffered from severe external otitis, suggesting that the complications were the result of the patients' poor skin condition. More long-term evaluations are needed to determine whether this approach is as safe as the classic transmastoid approach in patients with normal skin condition in the ear canal.

2.4 Middle Ear Implant or Not?

To study whether middle ear implants are more beneficial than acoustic devices, several studies have compared the results obtained with these two types of devices in a within-subjects, repeated measurement design. Mostly, the patient's own acoustic device was used. Such studies have been reviewed in Sect. 2.2.3. In other studies, newly fitted acoustic devices with the same electronics as those of the implant's audioprocessors were used for comparison (Kasic and Fredrickson 2001; Uziel et al. 2003; Jenkins et al. 2004; Saliba et al. 2005; Truy et al. 2008). The latter comparison is preferred, because it minimizes sound processing as a variable. Thus, any difference can be ascribed primarily to the coupling of the amplifier to the ear. Such a comparison was made using the Otologics MET device in a large group of patients (Kasic and Fredrickson 2001). The results obtained with the two types of device were found to be "virtually identical," but no details were revealed (Jenkins et al. 2004).

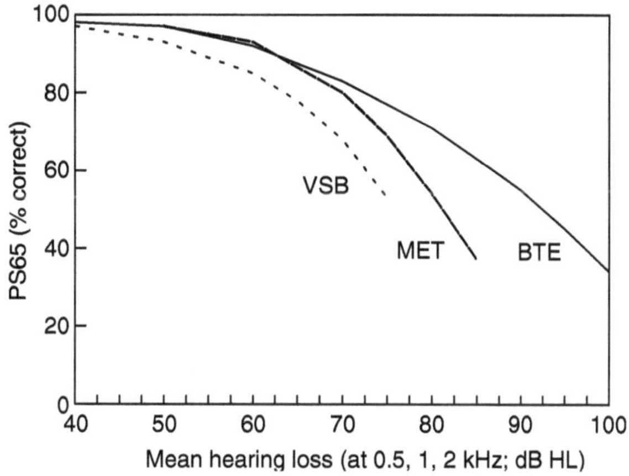

Fig. 4.4 Best-fit regression curves relating aided phoneme score obtained at 65 dB SPL presentation level (PS65) and the patient's mean hearing loss (PTA at 0.5, 1, and 2 kHz). BTE refers to acoustic device users, VSB refers to the Vibrant Soundbridge users, and MET to the Otologics MET users. Redrawn from Verhaegen et al. (2008)

For the Vibrant Soundbridge, Saliba et al. (2005) found that aided thresholds with the BTE were better than with the Vibrant Soundbridge, whereas speech recognition in noise was better with the Vibrant Soundbridge.

Uziel et al. (2003; $n=6$) and Truy et al. (2008; $n=6$) studied only patients who predominantly had high frequency hearing loss. These studies found that the Vibrant Soundbridge is the better option. Not only were its aided thresholds comparable or better than the control acoustic device, but its speech recognition in noise was also significantly better.

The Nijmegen group compared the results of two groups of patients with both middle ear implants (Vibrant Soundbridge, $n=22$, or Otologics MET, $n=10$) as well as a state-of-the art acoustic devices ($n=50$) (Verhaegen et al. 2008). Figure 4.4 shows their main outcome measure, the (unilateral) aided phoneme recognition score in quiet at a presentation level of 65 dB SPL, plotted against the individual mean hearing loss of the patients. To avoid crowdedness, only the best-fit second-order regression curves to the individual data points were shown. Verhaegen et al. (2008) concluded that, in general, the middle ear implants are not better than acoustic devices. For patients with severe hearing loss, acoustic devices are even the best option.

In summary, from an audiological point of view, today's semi-implantable middle ear devices can be considered as an effective amplification option when the hearing loss is not too severe. State-of-the-art acoustic devices are competitive, but they are not suitable for patients with chronic external otitis and might be less effective for patients with a predominantly high frequency hearing loss.

2.5 Cost-Effectiveness of Middle Ear Implantation

Many hearing impaired patients who need amplification ask for an invisible device. Visible hearing aids are often associated with handicap or old age. In such patients, the fully implantable devices might be an option.

In contrast with acoustic devices, middle ear devices involve surgery and much higher hardware cost. To justify the higher risk and cost, health authorities have been asking questions about treatment effectiveness of these middle ear devices. Snik and colleagues (2006, 2010) assessed changes in quality of life after middle ear implantation and concluded that semi-implantable devices are cost-effective for patients with sensorineural hearing loss and comorbid chronic external otitis, but probably not for patients who are just unwilling to accept acoustic devices.

3 Implantable Bone Conduction Devices for Conductive and Mixed Hearing Loss

3.1 Introduction

In patients with a significant air-bone gap, reconstructive surgery is the first option. However, surgery is not always possible as in the case of chronic middle ear disease or in patients with aural atresia (Declau et al. 1999). Bone-conduction devices become the next option. A bone-conduction device is composed of a powerful amplifier and a vibrating bone-conduction transducer. The vibrating transducer is pressed against the skin in the mastoid region and is mostly held in place by means of a steel band over the head. The amplified sounds are transmitted to the skull transcutaneously, stimulating the cochlea via bone conduction (Stenfelt and Goode 2005). Dillon (2001) showed that calculated sensation levels obtained with a bone-conduction transducer were 50 to 60 dB less effective than a powerful BTE coupled acoustically to the ear. Therefore, in general, bone-conduction devices must have powerful amplifiers.

A major drawback of bone-conduction devices is that pressure is needed to achieve effective coupling of the vibrating transducer. Even then, the skin and subcutaneous layers between the transducer and the skull attenuate the vibrations considerably (Håkansson et al. 1985).

Direct bone-conduction hearing aids are partially implantable. The main reason for implantation is to optimize energy transfer from the amplifier to the skull by avoiding the attenuation of the skin and subcutaneous layers. Two bone-conduction devices have been introduced, the Xomed Audiant device (Hough et al. 1986) and the bone-anchored hearing aid, the Baha device (Håkansson et al. 1985). In each case, surgery is required to achieve efficient coupling between the vibrating transducer and the skull.

3.2 Outcomes

Hough et al. (1986, 1994) developed the temporal bone stimulator (TBS) or the Xomed-Audiant device. A permanent magnet was implanted in the temporal bone and covered by a thin layer of skin. The magnet was driven by an external coil positioned

Fig. 4.5 Standard Baha device (with the latest digital soundprocessor BP 100) showing both the abutment and the implanted fixture. Courtesy of Cochlear Company

on the skin and kept in place by a coupling magnet. The external coil was powered by an amplifier in a BTE housing or in a larger, body-worn housing. One main disadvantage of the TBS was the relatively wide distance between the implanted magnet and the external driving coil, which resulted in low efficiency (Gatehouse and Browning 1990; Håkansson et al. 1990). Other problems included insufficient gain, medical issues, and high failure rate (Browning 1990; Wade et al. 1992; Snik et al. 1998). As a consequence, the TBS was taken off the market in mid 1990s.

Håkansson and co-workers (Håkansson et al. 1985; Tjellström and Håkansson 1995; Tjellström et al. 2001) developed the Baha device, which has been applied to more than 65,000 patients with conductive and mixed hearing loss (Cochlear BAS Ltd, Göteorg, Sweden). Figure 4.5 shows the Baha device, including its percutaneous coupling. The titanium fixture is implanted in the skull behind the auricle and will gradually undergo osseo-integration. A percutaneous titanium abutment is connected to the fixture, and an audio processor can be attached to the abutment. Håkansson et al. (1985) demonstrated that the percutaneous transmission is 10 to 15 dB more effective than the conventional transcutaneous transmission. Indeed, better sound field thresholds have been reported with Baha than with conventional bone conductors, making it possible to set a higher volume without saturating the amplifiers by loud sounds. Cremers et al. (1992) showed that in the sound field, harmonic distortion was significantly less with Baha than with conventional bone conductors. Because the improved thresholds were mainly in the high frequency range, significant improvement could be achieve in speech-in-noise tests. In the "Consensus statements on the Baha" (Snik et al. 2005), it was concluded that the

Baha can be applied in conductive hearing loss and mixed hearing loss up to a 65 dB HL sensorineural hearing loss component. If the sensorineural hearing loss component exceeds 65 dB HL, alternative treatment would be either cochlear implantation (Verhaegen et al. 2009), or middle ear implantation (see Sect. 4).

The Baha has been applied bilaterally in patients with bilateral conductive or mixed hearing loss with success, enabling sound localization and improving speech recognition in a noisy environment (Snik et al. 2005). However, a drawback of any bone-conduction device is that a single device will inevitably stimulate both cochleae, referred to as cross-stimulation, because of the limited transcranial attenuation of sound vibrations in the skull (Stenfelt 2005). The Baha has also been applied to patients with pure sensorineural hearing loss and comorbid chronic external otitis, as well as to patients with high frequency sensorineural hearing loss. In each of these cases, a major consideration to apply the Baha was the opportunity to leave the ear canal unoccluded. When a standard Baha is used at its maximum settings, the air-bone gap can be virtually closed in the mid-frequencies with an additional maximum "compensation" of 5 to 10 dB for the sensorineural hearing loss component (Carlsson and Håkansson 1997). This limited "compensation" is the reason that the patients with sensorineural hearing loss did not benefit from a standard Baha (Snik et al. 1995; Stenfelt et al. 2000). The Baha can also be used as a CROS (Contralateral Routing of Signal) device in patients with total unilateral deafness. The Baha device placed near the deaf ear picks up sounds from the deaf side to produce vibrations that stimulate the contralateral (normal functioning) cochlea via bone conduction. The pioneering work of Vaneecloo et al. (2001) has popularized the application of the Baha device to treat unilateral deafness, but the jury is still out regarding the methodology and benefits of the Baha CROS application over the conventional CROS device (Baguley and Plydoropulou 2009).

3.3 Surgical Issues and Complications

The Baha surgery comprises two stages, including first installation of the titanium fixture, followed by placement of the skin-penetrating abutment (Tjellström and Håkansson 1995). In adults, the two stages are performed in one and the same surgical procedure, while in children, a waiting period of 3 months in between is advocated. It is not advisable to attempt implantation before the age of 3 to 4 years because of the thin skull (Snik et al. 2005). Even then, device failure is higher among children than in adults owing to poorer osseo-integration and trauma. Therefore, in children, a second, sleeping fixture is often placed. Longitudinal studies showed that the percutaneous implant is safe to use and only a limited number of serious complications have been reported in adults. However, second surgery owing to implant loss or appositional growth of bone may occur in up to 25% of the children (Snik et al. 2005).

3.4 *Cost-Benefit Analysis and Future Developments*

Johnson et al. (2006) reviewed the cost-effectiveness of the Baha device and found minor improvement in quality of life measures. They advised professionals to proceed with caution when counselling candidates. On the other hand, Arunachalam et al. (2001) and Dutt et al. (2002) reported subjective benefit scores that were comparable to those reported in cochlear implant studies (e.g., Vermeire et al. 2005). As Baha implantation is much less costly than cochlear implantation, it seems that Baha can drive significant cost-benefit.

An important consideration in improving the effectiveness of the Baha is the position of the titanium fixture (Eeg-Olofsson et al. 2008): the closer it is to the cochlea, the better the result. This result suggests that a transducer implanted in the mastoid might be advantageous. Håkansson et al. (2008) developed a totally implantable bone-conduction transducer that not only places the transducer near the cochlea but also avoids percutaneous coupling. Initial evaluation of such a device showed effective stimulation of the ipsilateral cochlea, and, surprisingly, limited crossover stimulation (Reinfeldt 2009). Transcutaneous implantable bone-conduction devices have potentials to increase both efficiency and safety over the present percutaneous devices.

4 Active Middle Ear Implants for Conductive and Mixed Hearing Loss

4.1 *Introduction*

Active middle ear implantation with direct coupling between the transducer and cochlea has recently become an alternative treatment option for patients with conductive or mixed hearing loss (Colletti et al. 2006). Five different middle ear implant systems for conductive or mixed hearing loss are discussed, in their order of appearance on the market. To compare the outcomes of the different studies, the "effective functional gain" is used again and defined as the difference between bone-conduction and implant-aided thresholds.

4.1.1 The Piezoelectric Device for Conductive or Mixed Hearing Loss

The first system on the market comprised a semi-implantable device with a piezoelectric bimorph transducer, coupled directly to the stapes (Suzuki 1988; Ko and Zhu 1995; see Fig. 4.2). The piezoelectric transducer is implanted and connected directly to the (isolated) stapes, but the electronics, battery and microphone are worn externally in a behind-the-ear housing. Ohno and Kajiya (1988) showed that this device's maximum output was 75 dB SPL at 500 Hz, and 90 dB SPL at 4 kHz, with corresponding maximum gain being less than 5 dB at 500 Hz and 20 dB at

4 kHz. In the majority of patients, the air-bone gap could be closed with hardly any compensation for the sensorineural component (Suzuki et al. 1994). Nevertheless, these patients showed significantly better speech recognition in noise with this device than with an acoustic device using exactly the same electronics (Gyo et al. 1996). This improvement was ascribed to superior performance of the piezoelectric transducer compared to the acoustic coupling of the conventional hearing aid. Owing to anatomical and technical limitations, it was difficult to increase the gain and output of the device (Ko and Zhu 1995). Furthermore, aided thresholds deteriorated over time (Yanagihara et al. 2004). The device was first introduced in 1984 but was taken off the market in the late 1990s.

4.1.2 Contactless Electromagnetic Devices

As described in Sect. 2.2.2, Heide et al. (1988) were probably the first to use a contactless electromagnetic system in patients. Their device comprised a small magnet glued onto the eardrum. Kartush and Tos (1995) applied the same device in a modified form. In 10 patients with mixed hearing loss, the magnet was incorporated into the TORP (Total Ossicular Replacement Prosthesis) that had been used to reconstruct the middle ear. Results indicated that this system could "close" the air-bone gap and was at least as effective as acoustic hearing aids (Kartush and Tos 1995). Cayé-Thomasen et al. (2002) reported the long-term results of 9 out of the 10 patients implanted by Tos. All 9 patients had stopped using the device after a mean duration of 2 years because of problems with the driver and/or TORP dislocation. This contactless device is not available in the market.

4.1.3 Devices with Electromagnetic Transducers

The Vibrant Soundbridge was developed for patients with sensorineural hearing loss (see Sect. 2.2.3). Colletti and co-workers (2006) coupled the transducer of the Vibrant Soundbridge, the FMT, directly to the round window membrane of the cochlea, making it suitable to treat patients with conductive and mixed hearing loss. Huber et al. (2006) coined the term "vibroplasty" for this new treatment option. Colletti et al. (2006) placed the FMT in the enlarged round window niche in 7 patients and showed remarkable post-device fitting results that produced 0 to 28 dB "effective functional gain," averaged over frequencies of 0.5, 1, and 2 kHz (see Fig. 4.6). Several follow-up studies have been conducted since, with most of them placing the FMT in the round window niche (Linder et al. 2008; Beltrame et al. 2009; Cude et al. 2009; Frenzel et al. 2009). Patients with various types of hearing loss were also implanted, including congenital onset (aural atresia; Frenzel et al. 2009) and acquired conductive or mixed loss (Linder et al. 2008; Beltrame et al. 2009; Cude et al. 2009). Additionally, the FMT was coupled to the oval window with the aid of adapted TORPs (Huber et al. 2006; Hüttenbrink et al. 2008).

Figure 4.6a and b presents the effective gain data from these studies, namely of patients with (predominantly) conductive hearing loss (mean bone-conduction

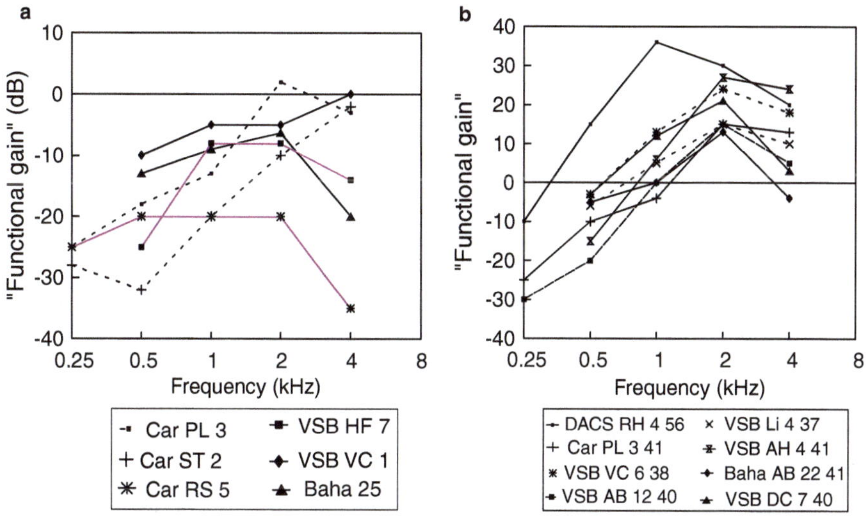

Fig. 4.6 The mean effective functional gain versus frequency in patients with predominantly conductive hearing loss (**a**) and mixed hearing loss (**b**). The labels per curve refer to the type of device used (VSB is Vibrant Soundbridge, Car is Otologics Carina) followed by the initials of the first author of the study, followed by the number of tested patients. In (**b**), the last number indicates the mean sensorineural hearing loss component of the study group (dB HL)

thresholds at 0.5, 1, and 2 kHz of <25 dB HL) and of those with mixed hearing loss, respectively. The labels to the curves indicate the device used, initials of the first author, number of patients in the study, and, in cases of mixed hearing loss, the mean sensorineural hearing loss component.

Siegert et al. (2007) used the Otologics Carina device in 5 patients with conductive hearing loss as a result of congenital aural atresia. After removal of the incus, the moving rod of the transducer was coupled directly to the head of the stapes with the aid of an adapted TORP. The air-bone gap was not entirely eliminated but reduced by 36 dB on average. Speech recognition scores at 65 dB SPL improved to approximately 70%. The mean effective functional gain as a function of frequency is presented in Fig. 4.6a (curve Car RS 5).

Tringali et al. (2008, 2009) also used the Carina in two patients with conductive hearing loss. In the first patient with congenital onset, the stapes proved to be fixed. After stapedotomy, a titanium stapes prosthesis was placed and connected to the moving rod of the transducer. In the second patient who had acquired hearing loss, a modified TORP was used, coupled to the transducer, which was attached to the round window membrane using a piece of facia. In each case, partial closure of the air-bone gap was reported. Speech recognition scores at 65 dB SPL were 80% and 95% in the two patients, respectively. Figure 4.6a (curve Car ST 2) presents the mean effective functional gain.

Using a similar approach to Tringali et al. (2009), Lefebvre et al. (2009) reported results from 6 patients with acquired conductive or mixed hearing loss. The effective functional gain remained stable over 12 months (Car PL 3 in Fig. 4.6a and Car PL 341 in 6b), but speech recognition scores were unstable. At 3 to 6 months after surgery, aided speech scores at 65 dB SPL were between 80% and 100%, whereas at 12 months follow-up, the average score was only 36%. The reason for this unstable speech performance was not clear.

Hausler et al. (2008) developed a semi-implantable DACS (Direct Acoustical Cochlear Stimulation) device especially for patients with advanced otosclerosis. A stapes prosthesis was placed surgically and connected to the electromechanical transducer, which was implanted in the mastoid. The DACS transducer was driven by a standard BTE device, coupled by means of a percutaneous plug. Four patients have been implanted, producing the highest effective functional gain (DACS RH 4 56 in Fig. 4.6b). However, the DACS patients also had the highest sensorineural hearing loss component. Evaluation at 2 years showed stable results.

4.2 Middle Ear implant or Baha?

To compare the effectiveness of middle ear implants versus Baha, the mean effective functional gain was measured in 25 subjects with conductive hearing loss from Nijmegen, who had been using the digital, standard Baha device (the Baha Divino) for at least 3 months. Figure 4.6a (curve Baha 25) shows clearly that, although the air-bone gap was not closed (negative gain values) in these Baha users, their mean effective gain was better than or comparable to that observed in various middle ear implant studies, except at 4000 Hz.

Except for the DACS users, all subjects with mixed hearing loss had a sensorineural component at about 40 dB HL (see labels to the curves in Fig. 4.6b). Still, significant variation in the functional gain existed between these studies, which might be the result of patient selection, surgical approach, effectiveness of the coupling, type of middle ear implant, and audio processor settings. For example, the recently introduced powerful Baha Intenso had a somewhat lower functional gain than the middle ear implants, but these two types of devices used different amplifications (Bosman et al. 2009; see curve Baha AB 22 41 in Fig. 4.6b). The Baha devices used linear amplification, while the middle ear implant devices used nonlinear compressive amplification. As a result, the effective gain of the middle ear implants at normal conversational levels would be 5 to 10 dB lower than that measured at threshold level (Snik and Cremers 2001). Taking the amplification difference into account, the functional gain was generally comparable between the middle ear implant and Baha devices.

4.3 Current Status

Middle ear implantation for the treatment of conductive and mixed hearing loss is relatively new and many issues need to be resolved. There is still debate about the best fixation of the FMT to the ossicular chain and its coupling to the cochlea (Gan et al. 1997; Hüttenbrink et al. 2008; Colletti et al. 2009). If the stapes is intact, it is not yet clear whether the FMT should be connected to the stapes or placed in the round or oval window. Another problem is the device noise that is audible to patients with pure conductive hearing loss but normal cochlear function. Third, long-term evaluations are needed to show whether the transducer couplings are stable over time. Finally, it is important to determine whether the intervention is cost-effective, as compared to the conventional solutions like Baha. At present, clinical utilization seems to precede technological development.

5 Summary

Compared with conventional acoustic devices, the middle ear implants have advantages and disadvantages in treating patients with sensorineural hearing loss. Because of the direct coupling of the hearing device to the middle ear ossicles, it was claimed that sound quality would be superior to acoustic devices. However, they produced generally similar speech recognition scores to acoustic devices, except for patients with high frequency hearing loss. The external part of semi-implantable middle ear devices, the audio processor, can be easily concealed by hair, making it cosmetically appealing. Several drawbacks are also apparent, including the need for surgery with general anesthesia, repeated surgery in case of updates, the non-compatibility with MRI, and the high financial cost. However, these middle ear implants are well suited for special patient populations such as those with comorbid external otitis.

Baha devices have been widely and successfully applied to patients with conductive or mixed hearing loss. The Bahas are especially suited for patients with atretic ears or chronic middle ear infection. Although the complication rate is low, the percutaneous coupling is still considered a serious drawback. This drawback can be addressed by new development in totally implantable bone-conduction transducer, which will make the percutaneous coupling obsolete. If the sensorineural hearing loss component exceeds 65 dB HL, the Baha is no longer effective; instead, a powerful middle ear implant coupled directly to one of the cochlear windows becomes a viable option. The use of middle ear implants to treat conductive or mixed hearing loss is promising but still experimental.

Acknowledgements I would like to express my sincere thanks to my colleagues Cor Cremers, Emmanuel Mylanus, Arjan Bosman, Jef Mulder, Myrthe Hol, and Carine Hendriks for their support through the years.

References

Abbass, H. A., Kane, M., Gaverick, S., Falk, T. J., Frenz, W., Ko, W. H., & Maniglia, A. J. (1997). Mechanical, acoustical and electromagnetic evaluation of the semi-implantable middle ear hearing device. *Ear Nose & Throat Journal,* 76, 321–327.

Arunachalam, P. S., Kilby, D., Meikle, D., Davison, T., & Johnson, I. J. (2001). Bone anchored hearing aid quality of life assessed by Glasgow Benefit. *Laryngoscope,* 111, 1260–1263.

Baguley, D. M., & Plydoropulou, V. (2009). Prevost AT Bone-anchored hearing aids for single-sided deafness. *Clinical Otolaryngology*, 34, 176–177.

Ball, G. R., Huber, A., & Goode, R. L. (1997). Scanning laser Doppler Vibrometry of the middle ear ossicles. *Ear Nose & Throat Journal*, 76, 213–218.

Barbara, M., Manni, V., & Monini, S. (2009). Totally implantable middle ear device for rehabilitation of sensorineural hearing loss: preliminary experience with the Esteem, Envoy. *Acta Oto-Laryngologica*, 129, 429–432.

Beltrame, A. M., Martini, A., Prosser, S., Giarbini, N., & Streitberger, C. (2009). Coupling the Vibrant Soundbridge to cochlea round window: auditory results in patients with mixed hearing loss. *Otology & Neurotology,* 30, 194–201.

Bosman, A. J., Snik, A. F. M., Mylanus, E. A. M., & Cremers, C. W. R. J. (2009). Fitting of the Baha Intenso. *International Journal of Audiology*, 48, 346–352.

Browning, G. G. (1990). The British experience of an implantable, subcutaneous bone conduction hearing aid. *Journal of Laryngology and Otology*, 104, 534–538.

Bruschini, L., Foru, F, Santoro, A., Bruschini, P., & Berrettini, S. (2009). Fully implantable Otologics MET Carina device for the treatment of sensorineural hearing loss. Preliminary results. *Acta Otorhinolaryngologica Italica*, 29, 79–85.

Carlsson, P. U., & Håkansson, B. E. (1997). The bone-anchored hearing aid: reference quantities and functional gain. *Ear and Hearing,* 18, 34–41.

Cayé-Thomasen, P., Jensen, J. H., Bonding, P., & Tos, M. (2002). Long-term results and experience with the first generation semi-implantable electromagnetic with ossicular replacement device for mixed hearing loss. *Otology & Neurotology*, 23, 904–911.

Chen, D. A., Backous, D. D., Arriaga, M. A., Garvin, R., Kobylek, D., Littman, T., Walgren, S., & Luca, D. (2004). Phase 1 clinical trial results of the Envoy System: a totally implantable middle ear device for sensorineural hearing loss. *Otolaryngology-Head and Neck Surgery*, 131, 904–916.

Colletti, V., Soli, S. D., Carner, M., & Colletti, L. (2006). Treatment of mixed hearing losses via implantation of a vibratory transducer on the round window. *International Journal of Audiology*, 45, 600–608.

Colletti, V., Carner, M., & Colletti, L. (2009). TORP vs round window implant for hearing restoration of patients with extensive ossicular chain defect. *Acta Oto-Laryngologica*, 129, 449–452.

Cremers, C. W. R. J., & Snik, A. F. M. (2005). The endaural approach for the Vibrant Soundbridge middle ear implant. *Journal of the Dutch ENT Society*, 11, 104.

Cremers, C. W. R. J., Snik, A. F. M., & Beynon, A. J. (1992). Hearing with the bone-anchored hearing aid (Baha HC 2000) compared to a conventional bone-conduction hearing aid. *Clinical Otolaryngology*, 17, 275–279.

Cremers, C. W. R. J., Verhaegen, V. J., & Snik, A. F. M. (2008). The floating mass transducer of the Vibrant Soundbridge interposed between the stapes and tympanic membrane after incus necrosis. *Otology & Neurotology*, 30, 76–78.

Cude, D., Murri, A., & Tinelli, N. (2009). Piezoelectric round window osteoplasty for Vibrant Soundbridge Implant. *Otology & Neurotology* 30, 782–786.

Declau, F., Cremers, C. W. R. J., & Van de Heyning, P. (1999). Diagnosis and management strategies in congenital atresia of the external auditory canal. Study Group on Otological Malformations and Hearing Impairment. *British Journal of Audiology*, 33, 313–327.

Dillon, H. (2001). *Hearing Aids*. New York: Thieme Verlag.

Dutt, S. N., McDermott, A. L., Jelbert, A., Reid, A. P., & Proops, D. W. (2002). The Glasgow Benefit Inventory in the evaluation of patient satisfaction with the bone anchored hearing aid: quality of life issues. *Journal of Laryngology and Otology Supplement*, 28, 7–14.

Eeg-Olofsson, M., Stenfelt, S., Tjellström, A., & Granström, G. (2008). Transmission of bone-conducted sound in the human skull measured by cochlear vibrations. *International Journal of Audiology*, 47, 761–769.

Fraysse, B., Lavieille, J. P., Schmerber, S., Enée, V., Truy, E., Vincent, C., Vaneecloo, F. M., & Sterkers, O. (2001). A multicenter study of the Vibrant Soundbridge middle ear implant: early clinical results and experience. *Otology & Neurotology*, 22, 952–961.

Fredrickson, J. M., Coticchia, J. M., & Khosla, S. (1995). Ongoing investigations into an implantable electromagnetic hearing aid for moderate to severe sensorineural hearing loss. *Otolaryngologic Clinics of North America*, 28, 107–120.

Frenzel, H., Hanke, F., Beltrame, M., Steffen, A., Schonweiler, R., & Wollenberg, B. (2009). Application of the Vibrant Soundbridge to unilateral osseus atresia cases. *Laryngoscope*, 119, 67–73.

Gan, R. Z., Wood, M. W., Ball, G. R., Dietz, T. G., & Dormer, K. J. (1997). Implantable hearing device performance measured by laser Doppler interferometry. *Ear Nose and Throat Journal*, 76(5), 297–309.

Gatehouse, S., & Browning, G. G. (1990). The output characteristics of an implanted bone-conduction prosthesis. *Clinical Otolaryngology*, 15, 503–513.

Gyo, K., Saiki, T. & Yanagihara, N. (1996). Implantable hearing aids using a piezoelectric ossicular vibrator: a speech audiometric study. *Audiology*, 35, 271–276.

Håkansson, B., Tjellström, A., Rosenhall, U., & Carlsson, P. (1985). The bone-anchored hearing aid: principles design and a psycho-acoustical evaluation. *Acta Oto-Laryngologica*, 100, 229–239.

Håkansson, B., Tjellström, A., & Carlsson, P. (1990). Percutaneous versus transcutaneous transducers for hearing by direct bone conduction. *Otolaryngology-Head and Neck Surgery*, 102, 339–344.

Håkansson, B., Eeg-Olofsson, M., Reinfeldt, S., Stenfelt, S., & Granström, G. (2008). Percutaneous versus transcutaneous bone conduction implant system: a feasibility study on a cadaver head. *Otology & Neurotology*, 29, 1132–1139.

Hausler, R., Stieger, C., Bernhard, H., & Kompis, M. (2008). A novel implantable hearing system with direct acoustic cochlear stimulation. *Audiology & Neurotology*, 13, 247–256.

Heide, J., Tatge, G., Sander, T., Gooch, T., & Prescott, T. (1988). Development of a semi-implantable hearing device. *Advances in Audiology*, 4, 32–43.

Hough, J. V., Vernon, J., Johson, B., Dormer, K., & Himelick, M. A. (1986). Experiences with implantable hearing devices and a presentation of a new device. *Annals of Otology, Rhinology, and Laryngology,* 95, 60–65.

Hough, J. V., Wilson, N., Dormer, K. J., & Rohrer, M. (1994). The Audiant bone conductor: update of patient results in North America. *American Journal of Otology*, 15, 189–197.

Hough, J. V., Dyer, R. K., Matthews, P., & Wood, M. W. (2001). Early clinical results: SOUNDTEC implantable hearing device phase II study. *Laryngoscope,* 111, 1–8.

Hough, J. V., Matthews, P., Wood, M. W., & Dyer, R. K. (2002). Middle ear electromagnetic semi-implantable hearing device: results of the phase II SOUNDTEC direct system clinical trial. *Otology & Neurotology*, 23, 895–903.

Huber, A. M., Ball, G. R., Veraguth, D., Dillier, N., Bodmer, D., & Sequeira, D. (2006). A new implantable middle ear hearing device for mixed hearing loss: A feasibility study in human temporal bones. *Otology & Neurotology*, 27, 1104–1109.

Hüttenbrink, K. B., Zahnert, T., Bornitz, M., & Beutner, D. (2008). TORP-vibroplasty: a new alternative for the chronically disabled middle ear. *Otology & Neurotology,* 29, 965–971.

Jenkins, H. A., Niparko, J. K., Slattery, W. H., & Neely, J. G., Fredrickson, J. M. (2004). Otologics Middle Ear Transducer Ossicular Stimulator: performance results with varying degrees of sensorineural hearing loss. *Acta Oto-Laryngologica*, 124, 391–394.

Jenkins, H. A., Atkins, J. S., Horlbeck, D., Hoffer, M. E., Balough, B., Arigo, J. V., Alxiades, G., & Garvis, W. (2007a). Phase I preliminary results of use of the Otologics MET Fully-Implantable Ossicular Stimulator. *Otolaryngology–Head and Neck Surgery*, 137, 206–212.

Jenkins, H. A., Pergola, N., & Kasic, J. (2007b). Anatomical vibrations that implantable microphones must overcome. *Otology & Neurotolology*, 28, 579–588.

Jenkins, H. A., Pergola, N., & Kasic, J. (2007c). Intraoperative ossicular loading with the Otologics fully implantable hearing device. *Acta Oto-Laryngologica*, 127, 360–364.

Johnson, C., Danhauer, J. L., Reith, A. C., & Latiolais, L. N. (2006). A systematic review of the nonacoustic benefits of bone-anchored hearing aids. *Ear and Hearing*, 27, 703–713.

Kartush, J. M., & Tos, M. (1995). Electromagnetic ossicular augmentation device. *Otolaryngologic Clinics of North America*, 28, 155–172.

Kasic, J. F., & Fredrickson, J. M. (2001). The Otologics MET Ossicular Stimulator. *Otolaryngologic Clinics of North America*, 34, 501–514.

Ko, W. H., & Zhu, W. L. (1995). Engineering principles of mechanical stimulation of the middle ear. *Otolaryngologic Clinics of North America,* 28, 29–42.

Lefebvre, P. P., DeCat, M., Cremers, C. W. R. J., Snik, A. F. M., & Maier, H. (2006). Otologics fully implantable hearing device results. *Wiener Medizinische Wochenschrift,* 156(Suppl. 119), 84.

Lefebvre, P. P., Martin, C., Dubreuil, C., Decat, M., Yazbeck, A., Kasic, J., & Tringali, S. (2009). A pilot study of the safety and performance of the Otologics fully implantable hearing device: transducing sounds via the round window membrane to the inner ear. *Audiology & Neurotology,* 14, 172–180.

Lenarz, T., Weber, B. P., Issing, P. R., Gnadeberg, D., Ambjørnsen, K., Mack, K. F., & Winter, M. (2001). Vibrant Sound Bridge System. A new kind hearing prosthesis for patients with sensorineural hearing loss. 2. Audiological results. *Laryngorhinootologie*, 80, 370–380.

Leysieffer, H., Baumann, J. W., Müller, G., & Zenner, H. P. (1997). Ein implantierbarer piezoelektrischer Hörgerätewandler für Innenohrschwerhörige. Teil II: Klinisches Implantat. *Zeitschrift für Hals Nasen und Ohrenheilkunde,* 45, 801–815.

Linder, T., Schlegel, C., Demin, N., & van der Westhuizen, S. (2008). Active middle ear implants in patients undergoing subtotal petrosectomy: new application for the Vibrant Soundbridge device and its implication for lateral cranium base surgery. *Otology & Neurotology*, 30, 41–47.

Luetje, C., Brackman, D., Balkany, T. J., Maw, J., Baker, R. S., Kelsall, D., Backous, D., Miyamoto, R., Parisier, S., & Arts, A. (2002). Phase III clinical trial results with the Vibrant Soundbridge implantable middle ear hearing device: a prospective controlled multicenter study. *Otolaryngology–Head and Neck Surgery*, 126, 97–107.

Magnan, J., Manrique, M., Dillier, N., Snik, A., & Häusler, R. (2005). International consensus on middle ear implants. *Acta Oto-Laryngologica*, 125, 920–921.

Maniglia, A. J., Ko, W. H., Gaverick, S. L., Abbass, H., Kane, M., Rosenbaum, M. L., & Murray, G. (1997). Semi-implantable middle ear electromagnetic hearing device for sensorineural hearing loss. *Ear Nose & Throat Journal*, 76, 333–341.

Mosnier, I., Sterkers, O., Bouccara, D., Labassi, S., Bebear, J. P., Bordure, P., Dubreuil, C., Dumon, T., Frachet, B., Fraysse, B., Lavielle, J.F., & Magnan, J. (2008). Benefit of the Vibrant Soundbridge device in patients implanted for 5 to 8 years. *Ear and Hearing,* 29, 281–284.

Ohno, T., & Kajiya, T. (1988). Performance of middle ear implants. *Advance in Audiology,* 4, 85–96.

Reinfeldt, S. (2009). *Bone-conduction hearing in human communication* (Unpublished thesis). Chalmers University of Technology, Göteborg, Sweden.

Roland, P. S., Shoup, A. G., Shea, M. C., Jones, D. B., & Richey, H. S. (2001). Verification of improved patient outcomes with a partially implantable hearing aid, the SOUNDTEC direct hearing system. *Laryngoscope,* 111, 1682–1686.

Saliba, I., Calmels, M. N., Wanna, G., Iversenc, G., James, C., Deguine, O., & Fraysse, B. (2005). Binaurality in middle ear implant recipients using contralateral digital hearing aids. *Otology & Neurotology*, 26, 680–685.

Silverstein, H., Atkins, J., Thompson, J. H., & Gilman, N. (2005). Experience with the SOUNDTEC implantable hearing aid. *Otology & Neurotology,* 26, 211–217.

Schmuziger, N., Schimmann, F., Wengen, D., Patscheke, J., & Probst, R. (2006). Long-term assessment after implantation of the Vibrant Soundbridge device. *Otology & Neurotology*, 27, 183–188.

Siegert, R., Mattheis, S., & Kasic, J. (2007). Fully implantable hearing aids in patients with congenital auricular atresia. *Laryngoscope,* 117, 336–340.

Snik, A. F. M., & Cremers, C. W. R. J. (2001). Vibrant semi-implantable hearing device with digital sound processing: effective gain and speech perception. *Archives of Otolaryngology–Head & Neck Surgery,* 127, 1433–1437.

Snik, A. F. M., & Cremers, C. W. R. J. (2004). Audiometric evaluation of an attempt to optimize the fixation of the transducer of a middle-ear implant to the ossicular chain with bone cement. *Clinical Otolaryngology*, 29, 5–9.

Snik, A. F. M., Mylanus, E. A. M., & Cremers, C. W. R. J. (1995). Bone-anchored hearing aids in patients with sensorineural hearing aids loss and persistent otitis media. *Clinical Otolaryngology*, 20, 31–35.
Snik, A. F. M., Dreschler, W. A., Tange, R. A., & Cremers, C. W. R. J. (1998). Short and longterm results with implantable transcutaneous and percutaneous bone-conduction devices. *Archives of Otolaryngology–Head & Neck Surgery*, 134, 265–268.
Snik, A. F. M., Noten, J., & Cremers, C. W. R. J. (2004). Gain and maximum output of two electromagnetic middle ear implants: are real ear measurements helpful? *Journal of American Academy of Audiology*, 15, 249–257.
Snik, A. F. M., Mylanus, E. A. M., Proops, D. W., Wolfaardt, J. F., Hodgetts, W. E., Somers, T., Niparko, J. K., Wazen, J. J., Sterkers, O., Cremers, C. W., & Tjellström, A. (2005). Consensus statements on the BAHA system: where do we stand at present? *Annals of Otology, Rhinology, and Laryngology*, 195, 2–12.
Snik, A. F. M., van Duijnhoven, N. T., Mylanus, E. A., & Cremers, C. W. R. J. (2006). Estimated cost-effectiveness of active middle-ear implantation in hearing-impaired patients with severe external otitis. *Archives of Otolaryngology–Head & Neck Surgery*, 132, 1210–1215.
Snik, A. F. M., van Duijnhoven, N. T., Mulder, J. J., & Cremers, C. W. (2007). Evaluation of the subjective effect of middle ear implantation in hearing-impaired patients with severe external otitis. *Journal of the American Academy of Audiology*, 18, 496–503.
Snik, A. F. M., Verhaegen, V., Mulder, J. J, & Cremers, C. W. R. J. (2010). Cost-effectiveness of implantable middle ear hearing devices. *Advances in Otorhinolaryngology*, 69, 14–19.
Stenfelt, S. (2005). Bilateral fitting of BAHAs and BAHA fitted in unilateral deaf persons: acoustical aspects. *International Journal of Audiology*, 44, 178–189.
Stenfelt, S., & Goode, R.L. (2005). Bone-conducted sound: physiological and clinical aspects. *Otology & Neurotology*, 26, 1245–1261.
Stenfelt, S., Håkansson, B., Jonsson, R., & Granström, G. (2000). A bone-anchored hearing aid for patients with pure sensorineural hearing impairment: a pilot study. *Scandinavian Audiology*, 29, 175–185.
Sterkers, O., Boucarra, D., Labassi, S., Bebear, J. P., Dubreuil, C., Frachet, B., Fraysse, B., Lavielle, J. P., Magnan, J., Martin, C., Truy, E., Uziel, A., & Vaneecloo, F. M. (2003). A middle ear implant, the Symphonix Vibrant Soundbridge: retrospective study of the first 125 patients implanted in France. *Otology & Neurotology*, 24, 427–436.
Suzuki, J. I. (1988). *Middle ear implants: implantable hearing aids. Advances in Audiology,* 4. Basel, Switzerland: Karger.
Suzuki, J. I., Kodera, K., Nagai, K., & Yabe, T. (1994). Long-term clinical results of the partially implantable piezoelectric middle ear implant. *Ear Nose & Throat Journal*, 73, 104–107.
Tjellström, A., & Håkansson, B. (1995). The bone-anchored hearing aids: design principles, indications and long-term clinical results. *Otolaryngologic Clinics of North America*, 28, 53–72.
Tjellström, A., Håkansson, B., & Granström, G. (2001). Bone-anchored hearing aids. Current status in adults and children. *Otolaryngologic Clinics of North America*, 34, 337–364.
Todt, I., Seidl, R. O., Mutze, S., & Ernst, A. (2004). MRI scanning and incus fixation in Vibrant Soundbridge implantation. *Otology & Neurotology*, 25, 969–972.
Tringali, S., Pergola, N., Ferber-Viart, C., Truy, E., Berger, P., & Dubreuil, C. (2008). Fully implantable hearing device as a new treatment of conductive hearing loss in Franceschetti syndrome. *International Journal of Pediatric Otorhinolaryngology*, 72, 513–517.
Tringali, S., Pergola, N., Berger, P., & Dubreuil, C. (2009). Fully implantable hearing device with transducer on the round window as a treatment of mixed hearing loss. *Auris Nasus Larynx,* 36, 353–358.
Truy, E., Eshraghi, A. A., Balkany, T. J., Telishi, F. F., Van de Water, T., & Lavielle, J. P. (2006). Vibrant Soundbridge surgery: evaluation of transcanal surgical approaches. *Otology & Neurotology*, 27, 887–895.

Truy, E., Philibert, B., Vesson, J. F., Labassi, S., & Collet, L. (2008). Vibrant Soundbridge versus conventional hearing aid in sensorineural high-frequency hearing loss: a prospective study. *Otology & Neurotology*, 29, 684–687.

Uziel, A., Mondain, M., Hagen, P., Dejean, F., & Doucet, G. (2003). Rehabilitation for high-frequency sensorineural hearing impairment in adults with the Symphonix Vibrant Soundbridge: a comparative study. *Otology & Neurotology*, 24, 775–783.

Vaneecloo, F. M., Ruzza, I., Hanson, J. N., Gerard, T., Dehaussy, J., Cory, M., Arrouet, C., & Vincent, C. (2001). Appareillage mono pseudo stereophonique par Baha dans les cophoses unilaterales: à propos de 29 patients. *Revue de Laryngologie Otologie Rhinologie*, 122, 343–350.

Verhaegen, V. J. O., Mylanus, E. A. M., Cremers, C. W. R. J., & Snik, A. F. M. (2008). Audiological application criteria for implantable hearing aid devices: a clinical experience at the Nijmegen ORL clinic. *Laryngoscope*, 118, 1645–1649.

Verhaegen, V. J. O., Mulder, J. J. S., Mylanus, E. A. M., Cremers, C. W. R. J., & Snik, A. F. M. (2009). Profound mixed hearing loss: bone-anchored hearing aid system or cochlear implant? *Annals of Otology, Rhinology, and Laryngology*, 118, 693–697.

Vermeire, K., Broks, J. P. L., Wuyts, F. L., Cochet, E., Hofkens, A., & Van de Heyning, P. H. (2005). Quality-of-life benefit from cochlear implantation in the elderly. *Otology & Neurotology*, 26, 188–195.

Wade, P. S., Halik, J. J., & Chasin, M. (1992). Bone-conduction implants: transcutaneous versus percutaneous. *Otolaryngology–Head and Neck Surgery*, 106, 68–74.

Yanagihara, N., Honda, N., Sato, H., & Hinohira, Y. (2004). Piezoelectric semi-implantable middle ear hearing device: Rion device E-type, long-term results. *Cochlear Implants International*, Suppl. 1, 186–188.

Zenner, H. P., & Leysieffer, H. (2001). Total implantation of the implex TICA hearing amplifier implant for high-frequency sensorineural hearing loss. *Otolaryngologic Clinics of North America*, 34, 417–446.

Chapter 5
Vestibular Implants

Justin S. Golub, James O. Phillips, and Jay T. Rubinstein

1 Introduction

1.1 Overview of the Vestibular System

The vestibular system is responsible for detecting and relaying three-dimensional spatial information to the central nervous system. It acts in an exquisitely sensitive manner to control several critical physiologic functions, including stabilizing visual images during head movement, maintenance of balance, and postural control. Unlike the five traditional senses, input from the vestibular system is not normally processed at a conscious level. In the pathologic state, however, the importance of the vestibular system becomes readily apparent.

1.1.1 Anatomy and Physiology

The vestibular system consists of five paired sensory organs located in the temporal bone adjacent to the cochlea. Rotational motion is detected by three semicircular canals: the superior (or anterior) canal, the posterior canal, and the lateral (or horizontal) canal. Each of these canals lies orthogonal to the other, such that all three spatial dimensions are anatomically represented. Linear acceleration is detected by

J.T. Rubinstein (✉)
Virginia Merrill Bloedel Hearing Research Center, University of Washington, Box 357923, Seattle, WA 98195-7923, USA

Department of Otolaryngology-Head and Neck Surgery, University of Washington, Seattle, WA 98195-6515, USA

Department of Bioengineering, University of Washington, Seattle, WA, USA
e-mail: rubinj@uw.edu

F.-G. Zeng et al. (eds.), *Auditory Prostheses: New Horizons*, Springer Handbook of Auditory Research 39, DOI 10.1007/978-1-4419-9434-9_5,

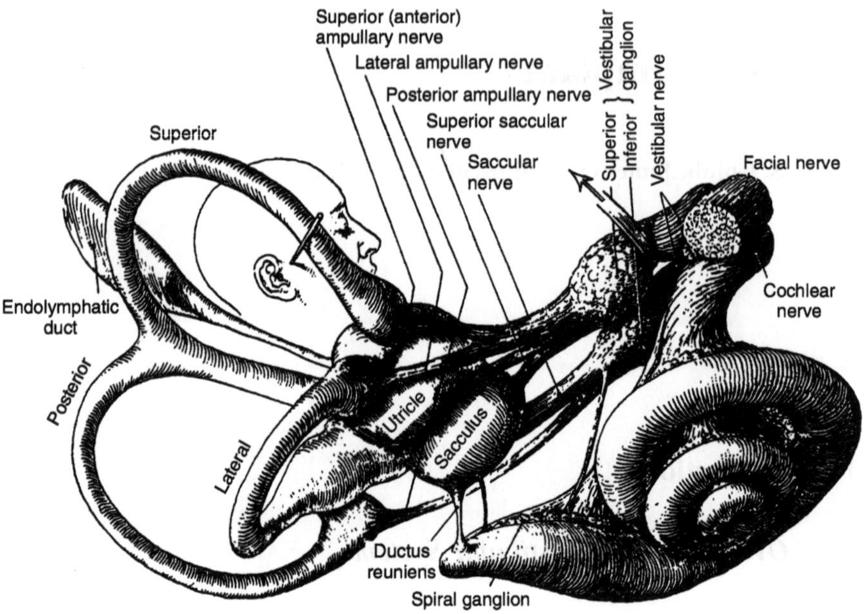

Fig. 5.1 Anatomy of the vestibular system. Three semicircular canals (superior [or anterior], posterior, and lateral [or horizontal]) detect changes in angular motion. Two otolith organs (the utricle and saccule [or sacculus]) detect changes in linear motion. Nerves from each organ converge to form the vestibular nerve. The close proximity of the cochlea, cochlear nerve, and facial nerve is illustrated. From Brodel (1946). *Three unpublished drawings of the anatomy of the human ear.* Philadelphia: WB Saunders (Lysakowski 2010)

the two otolith organs: the utricle and the saccule. The utricle lies in approximately the horizontal plane and the saccule lies in the vertical plane (Fig. 5.1).

The bony semicircular canals are filled with perilymphatic fluid. Within them are membranous semicircular canals filled with endolymphatic fluid. Angular acceleration induces movement of these fluids. This is detected at the *ampullae*, which contain the sensory elements of the semicircular canals, the *cristae*. Each *crista* contains mechanosensory hair cells lined with stereocilia and a kinocilium on their apical surface. Endolymphatic fluid motion displaces a gelatinous cupula overlying the hair cells of the ampulla. The stereocilia are then deflected. If the deflection occurs towards the kinocilium, the result is a depolarization of the hair cell (Fig. 5.2a, b). Conversely, if the deflection is away from the kinocilium, the result is hyperpolarization. Synaptic transmission to the primary afferents codes these polarizations as an increase or decrease, respectively, in afferent spike rate.

The *maculae* are the sensory elements of the utricle and saccule. Like the cristae of the semicircular canals, the maculae of the utricle and saccule contain mechanosensory hair cells. Instead of a cupula, however, the otolith organs contain a gelatinous layer embedded with dense otoliths (or otoconia) composed of calcium carbonate crystals. The hair cells of the otolith organs deflect with movement of the overlying otolith layer.

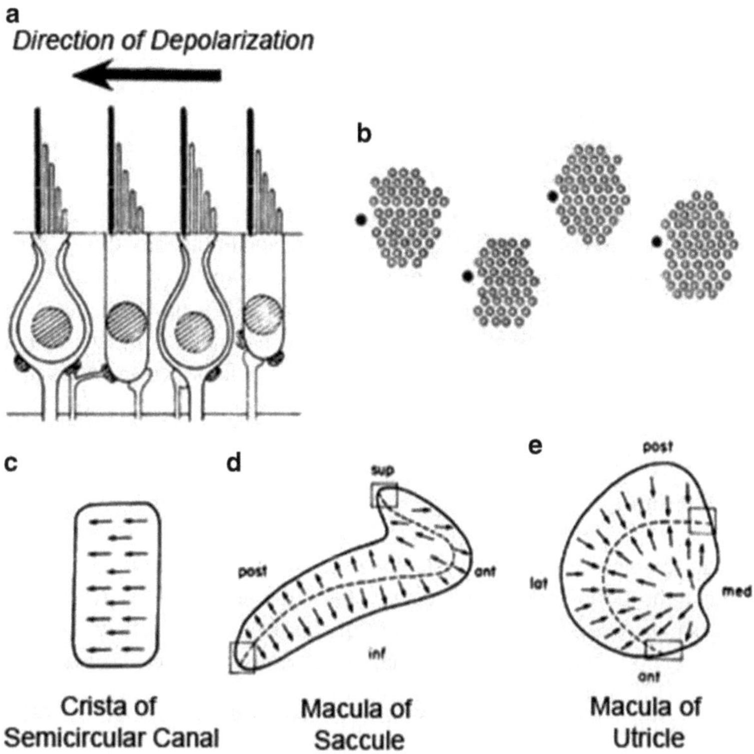

Fig. 5.2 Sensory elements of the vestibular organs. (**a**) Vestibular sensory hair cells with kinocilium (*black*) and adjacent "stair-step" arrangement of stereocilia (*white*). Depolarization occurs when deflection of hair cells is towards the kinocilium. The hair cells depicted in this image are thus said to be oriented towards the left. (**b**) Axial view of the four hair cells in (**a**). (**c**) Orientation of hair cells in a semicircular canal crista. Note that all hair cells are oriented in the same direction. (**d**) Orientation of hair cells in the macula of the saccule. The region in which the orientation of hair cells changes 180° is termed the striola (*dotted line*). (**e**) Orientation of hair cells in the macula of the utricle. Modified from Lindemann, H. H. (1969). *Ergeb Anat Entwicklungsgesch,* 42, 1–113 (Lysakowski 2010)

One key difference between the otolith organs and the semicircular canals lies in the hair cell orientation. In the cristae of the semicircular canals, all hair cells are oriented in approximately the same direction. In contrast, hair cells in the maculae of the otolith organs are oriented in many different directions (Fig. 5.2c–e). For this reason, each semicircular canal encodes motion in only one axis, whereas each otolith organ encodes motion from numerous directions. Because of this simpler signal output, the semicircular canals are typically chosen as the target for vestibular implants.

Hair cells of the vestibular organs synapse with branches of the vestibular nerve, whose cell body lies in the vestibular (Scarpa's) ganglion. The axons then travel proximally within the superior and inferior vestibular nerves before separating and terminating in the vestibular nuclei in the brainstem. From there, second order neurons project to several important tracts, including the vestibulospinal and the vestibulo-ocular pathways (Goebel and Sumer 2007; Goldberg 2000).

It is important to note that there is a *tonic neural firing rate* between a few to over 200 spikes/second in afferents originating from the vestibular organs, even in the absence of motion (Gong and Merfeld 2000). Acceleration will either raise or decrease the firing frequency from this baseline rate depending on its direction. The neural signal is therefore said to be *rate encoded* (Goldberg and Fernandez 1971). The paired contralateral vestibular organ outputs a similar signal in the opposite direction. Thus the vestibular system on each side of the body can independently encode motion information from all directions. The physiological redundancy of this "push-pull" system potentially provides the ability to restore vestibular function with a unilateral implant (Della Santina et al. 2007; Gong et al. 2008; Gong and Merfeld 2002).

1.1.2 Function

One critical role of the vestibular system is to maintain accurate visual image stabilization during head movement. This occurs through a neural reflex arc known as the vestibulo-ocular reflex (VOR). Movement of the head to the left will cause rapid (approximately 5 to 9 ms delay) and equal movement of the eyes to the right (Tabak et al. 1997). This enables objects of interest to remain stationary on the fovea of the retina for maximal visual resolution. From an evolutionary standpoint, one can appreciate many essential situations that would require maintaining objects in sharp focus (e.g., prey) while the head is constantly moving (e.g., running after prey). The smooth pursuit and optokinetic pathways can serve as two backup systems if the vestibulo-ocular system is nonfunctional; however, their capacity is limited by the relatively long time required for visual motion processing (Carey and Della Santina 2005).

The other essential role of the vestibular system is maintenance of balance and posture. This is largely mediated through the vestibulospinal system. Proprioception and visual information are two other means for achieving balance control. These three systems ordinarily work together. The more systems that are impaired, however, the more challenging balance becomes. For example, visual balance inputs require adequate light and relatively simple, stable images. Proprioceptive balance inputs require a reliable relationship between the support surface and limb position. Scenarios with multiple impaired systems (e.g., vestibular patients in the dark or on uneven/compliant surfaces where proprioception is challenging) are unfortunately common and result in falls and injury (Wall et al. 2002; White 2007).

1.2 Peripheral Vestibular Disorders

The vital role of the vestibular system becomes apparent when examining the effect of vestibular pathology. Symptoms range from incapacitating vertigo in the acute state to postural instability and reduced visual acuity in the chronic state.

Vestibular disorders can be categorized into groups depending on their time course (acute versus chronic versus recurrent or episodic) and their location (unilateral versus bilateral). This organizational rubric is helpful because symptoms depend more on timing and sidedness than the underlying etiology. Because current vestibular implants are targeted towards treating disorders peripheral to the vestibular nerve, the discussion is limited to peripheral vestibular disorders and does not include central disorders that cause vestibular symptoms.

1.2.1 Acute Unilateral Disorders

In acute, unilateral vestibular pathology, the predominant symptom is vertigo. This occurs because of a sudden imbalance in the neural output from the vestibular system. The signal from the diseased side is reduced, resulting in a mismatch between the two sides of the body. The brain interprets this as vertigo: a sensation of whirling or spinning. Nausea and vomiting often occur, resulting in difficulty performing basic activities until symptoms abate. Examples of acute, unilateral disease include vestibular neuronitis, labyrinthitis, or trauma to the vestibular organ (Lempert and Neuhauser 2009; Neuhauser et al. 2008).

1.2.2 Chronic Unilateral Disorders

Over time, the brain will adapt to asymmetric vestibular input. This process is referred to as central compensation and is discussed further in Sect. 3.3. Thus the symptom profile of chronic disorders does not typically include vertigo. In addition, a functional vestibular system on only one side of the body is required for maintenance of relatively good vestibular function. As a result, chronic, unilateral vestibular pathology is often asymptomatic. An example is a slowly growing vestibular schwannoma (acoustic neuroma). Acute, unilateral processes that progress to impair the vestibular system permanently, such as a severe bacterial labyrinthitis or a temporal bone fracture through the labyrinth, would also fall into this category.

On occasion, however, appropriate central compensation does not occur. This can happen as a rare result, for example, of viral neuronitis. Such patients do suffer from imbalance and brief vertigo during head turns.

1.2.3 Recurrent Acute (Episodic) Unilateral Disorders

Unfortunately, in diseases marked by recurrent episodes, the brain does not have time to adapt. Vertigo is then experienced during attacks. Benign paroxysmal positional vertigo (BPPV), the most common peripheral vestibular disorder, is one such example of an episodic disorder. Treatment consists of simple repositioning maneuvers (Bhattacharyya et al. 2008; Epley 1992).

Another well-known example is Ménière's disease. The classic symptomatic triad includes fluctuating sensorineural hearing loss, tinnitus, and paroxysmal disabling vertigo. Aural fullness is often an accompanying symptom. Vertiginous episodes typically last 20 min to 12 h. Patients are often incapacitated during attacks, unable to work, and are sometimes relegated to lying down on the floor at onset. The disorder is relatively common with a prevalence of 43 per 100,000 (Gates 2006; Thorp and James 2005; Van de Heyning et al. 2005).

1.2.4 Chronic Bilateral Disorders

In chronic bilateral vestibular pathology, symptoms are dominated by vestibular hypofunction. Impairment of the vestibulo-ocular reflex (VOR) results in oscillopsia, a tremendously incapacitating condition where head movement causes retinal image motion. This results in "visual bobbing" and reduced visual acuity. In addition, impairment of the vestibulospinal tract causes postural instability. Patients with impaired proprioception (such as diabetics, the elderly, or when walking on uneven or compliant surfaces) or impaired vision (such as in the dark) are especially symptomatic. Examples of such disorders include congenital or genetic anomalies, exposure to ototoxic medication, or age-related decline of vestibular function, i.e., presbystasis (Merchant et al. 2000; Rauch et al. 2001).

1.3 Clinical Need for Vestibular Neurostimulation

Vestibular disorders are widely prevalent and can cause significant morbidity and mortality. Because of the aging population, the prevalence is only anticipated to increase. The resulting vertigo and postural imbalance can lead to falling. Of those over 65 years old, fall-related complications are one of the leading causes of death (CDC 2007; Sattin 1992). In addition, acute vertiginous symptoms are often incapacitating, causing missed work, and even the inability to leave one's home (Lempert and Neuhauser 2009; Neuhauser et al. 2008).

Advancements in treatment of vestibular disorders are sorely needed. With the exceptions of vestibular rehabilitation, superior canal dehiscence syndrome, and the popularization of intratympanic gentamicin for Ménière's disease, the medical and surgical treatment for vestibular pathology has remained unchanged in the past two decades. Indeed, destructive procedures such as labyrinthectomy and intratympanic gentamicin are still commonly the best option for patients with severe Ménière's disease. For those with chronic, bilateral, and fixed or with fluctuating vestibular loss, there is no effective means to restore function.

At the same time, numerous types of neural prostheses have been developed over the past several decades. The most successful example is the cochlear implant. With continued refinement, hearing abilities have greatly improved and indications for implantation continue to expand (Gantz et al. 2006; Shkel and Zeng 2006; Won et al. 2008; Zeng 2004). Implantable retinal prostheses are under active investigation for those with sensory blindness (Weiland et al. 2005). Neural prostheses have even been developed that substitute one modality for another in order to restore a missing sense (Danilov and Tyler 2005). In the past decade, research in vestibular neurostimulation has finally placed us close to the clinical reality of treating a wide range of vestibular disorders with electric stimulation.

1.4 Disease Processes that May be Ameliorated by a Vestibular Implant

Vestibular prosthesis implantation could be useful in various pathologic states, particularly those characterized by recurrent acute attacks (in which central compensation does not occur) and for chronic bilateral hypofunction. Chronic, unilateral disorders are frequently asymptomatic because of central compensation and the contralateral redundancy of the vestibular system. Thus an implant would only be appropriate if a lack of compensation were to cause severe residual symptoms. Acute, unilateral disorders are typically transient and in most cases would not be appropriate for surgical intervention. Non-implanted vestibular prostheses that facilitate rehabilitation, however, may be ideal in these situations (Wall et al. 2002; Weinberg et al. 2006).

1.4.1 Recurrent Acute (Episodic) Unilateral Disorders

Ménière's disease is the second-most common cause of episodic vertigo after BPPV. Yet, the pathology is poorly understood. Hydrops or expansion of the endolymphatic space is often seen on autopsy; however, an animal model with surgically induced hydrops does not manifest signs of clinical Ménière's disease (Kimura 1982). The lack of a clinically pertinent animal model has significantly hampered research into diagnostic and therapeutic interventions for this poorly understood disease. One widely hypothesized etiologic mechanism is that hydrops leads to microperforations and admixture of endolymph and perilymph. This pathologic change neutralizes the endolymphatic potential and results in vestibular hypofunction. The abrupt unilateral loss of vestibular afferent input then causes symptomatic vertigo (Schuknecht 1982b, 1984; Schuknecht and Gulya 1983).

The majority of Ménière's patients improve their symptoms with medical and dietary management. Approximately 15%, however, will progress and require surgery. Creation of an endolymphatic shunt is the only non-destructive surgical procedure; however, its long-term efficacy may be poor (Gates 2006; Thorp and James 2005). Other surgical therapies aim at ablating or partially ablating the diseased vestibular organ, including intratympanic gentamicin, vestibular nerve section, and labyrinthectomy. These procedures all carry significant risks, including hearing loss and chronic unsteadiness because of poor central compensation (Gates 2006; Thorp and James 2005; Van de Heyning et al. 2005).

Because vertiginous attacks may be precipitated by an acute loss of vestibular tone, Ménière's disease could be an ideal candidate for a "pacemaker"-style vestibular implant that simply replaces the missing neural impulses during attacks. Patients would activate the device at the onset of vertigo. A human clinical trial involving a vestibular implant for Ménière's disease is discussed further in Sect. 4.1.

1.4.2 Chronic Bilateral Disorders

Patients with severe chronic bilateral disorders have no vestibular function and thus suffer from oscillopsia and chronic postural instability. Regardless of the etiology, the resulting deficit is the same. Afflicted patients would likely benefit from sensor-based vestibular implants to restore function. This topic is discussed further in Sect. 2.2.2.

1.4.3 Uncompensated Chronic Unilateral Disorders

Unilateral disorders that are not followed by appropriate central compensation of the imbalance in tonic vestibular output from the two sides of the body may result in chronic symptoms. This could occur, for example, as a sequela of vestibular neuronitis. While it is a relatively rare complication of vestibular neuronitis, the incidence of presumed viral vestibulopathy is so high that there is a high prevalence of such patients with poorly compensated unilateral lesions. Like in episodic disorders, patients might benefit from a "pacemaker"-based vestibular implant but the mechanism of efficacy would be different. The goal of electrical stimulation in these cases would be to raise the gain between angular eye movement and angular head movement of the affected semicircular canals through tonic electrical stimulation. Sensor-based implants might also be employed to increased gain. For such a treatment to be effective, the implant procedure must not further damage rotational sensitivity.

2 Vestibular Implant Design and Function

A vestibular implant bypasses the vestibular end organs to directly stimulate the vestibular nerve much as a cochlear implant bypasses the cochlea to stimulate the cochlear nerve directly. Prior to development of implant prototypes, it was known that electrical stimulation of the vestibular nerve could induce the vestibulo-ocular reflex (Cohen and Suzuki 1963; Suzuki et al. 1964) and modulate balance (Hlavacka and Njiokiktjien 1985; Nashner and Wolfson 1974; Pavlik et al. 1999). Human studies have even shown that balance abilities can be enhanced during platform perturbations using transcutaneous stimulation of vestibular afferents (galvanic vestibular stimulation) (Scinicariello et al. 2001). The goal of vestibular implantation is to provide vestibular functionality and/or reduce symptomatology by programmed or sensory-mediated stimulation of the vestibular nerve.

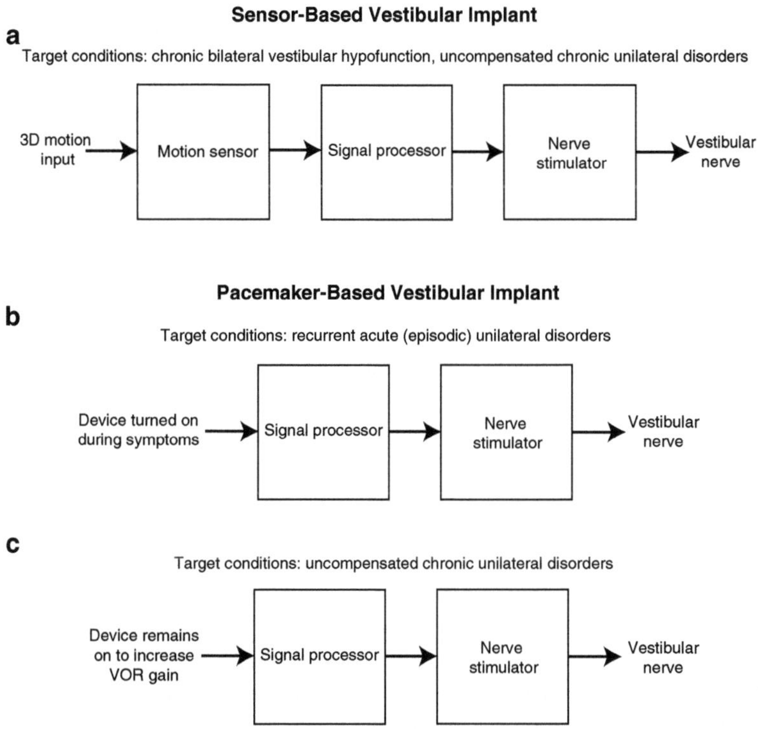

Fig. 5.3 Two designs for a vestibular implant. (**a**) A sensor-based vestibular implant would restore function in chronic bilateral vestibular hypofunction or increase VOR gain in uncompensated chronic unilateral disorders. (**b**) A pacemaker-based vestibular implant would aim to halt disabling symptoms caused by acute episodes of recurrent unilateral conditions or (**c**) increase VOR gain in uncompensated chronic unilateral disorders

2.1 Design of a Vestibular Implant

There are two basic designs of vestibular implants, each of which can be broken down into several components. The first design would attempt to optimally reproduce natural physiology using built-in motion sensors to replace missing positional information. Figure 5.3a shows such a device that consists of three elements: a motion sensing unit, a signal processor, and a nerve stimulator/electrode (Shkel and Zeng 2006). This design would be analogous to the external microphone, speech processor, and stimulator/electrode of the standard cochlear implant (Gates and Miyamoto 2003; Rubinstein 2004a; Zeng 2004). A simpler design could forgo the motion sensing unit and consist only of a signal processor and a nerve stimulator/electrode (Rubinstein et al. 2011). Figure 5.3b, c shows this latter design that would serve as a vestibular "pacemaker," replacing missing afferent information with a pre-programmed algorithm and restoring tonic activity to the afferents. The following section compares and contrasts these two designs.

In either design, the stimulator generates a biphasic charge-balanced pulsatile current that is delivered to branches of the vestibular nerve via individual channels and electrodes. Increasing either the amplitude of the current (amperage) or the frequency of the pulse (pulse rate, Hz) may result in greater stimulation of the nerve. The velocity of eye movements evoked by pulsatile stimulation appears to increase as a near linear function of pulse frequency (Gong and Merfeld 2000). If more afferents are recruited with increasing current, a similar relationship between stimulation current and eye velocity may be obtained.

As mentioned in Sect. 1.1.1, each semicircular canal encodes information regarding a specific axis of rotation. In contrast, the utricle and saccule each encode motion in multiple directions, and the resulting neural signal from an individual end organ is quite complex. Two recent reports demonstrated horizontal and vertical eye movements when the utricle and saccule, respectively, were stimulated with fine needles in cats (Goto et al. 2003, 2004). Earlier reports, however, have inconsistently shown expected eye movements by stimulating macular regions where hair cells are differentially oriented (Curthoys 1987; Curthoys and Oman 1986; Fluur and Mellstrom 1971). For this reason, efforts for developing a vestibular implant have focused on stimulating the individual ampullary nerves of the semicircular canals rather than the utricular or saccular nerve (Della Santina et al. 2007).

Because of the push-pull redundancy of the vestibular system, only one functional vestibular system is necessary for functioning of the VOR and for maintenance of posture. Thus bilateral vestibular implantation might confer little advantage over unilateral implantation (Della Santina et al. 2007; Gong and Merfeld 2002). One study showed a modest improvement in gain with bilateral implantation, but concluded that unilateral implantation was a more logical avenue for research and development (Gong et al. 2008).

2.2 *Pacing Versus Sensor-Based Modulation*

2.2.1 Pacemaker-Based Vestibular Implant

Vestibular disorders characterized by acute or recurrent unilateral hypofunction primarily cause vertiginous symptoms because of asymmetric vestibular input to the brain. The diseased ear outputs a reduced signal whereas the healthy, contralateral ear outputs a normal tonic signal. In this situation, simple replacement of the reduced signal could restore symmetric input to the brain and halt symptoms. A motion sensing component is thus not required for this indication. Rather, the implant need only "pace" the vestibular nerve during acute symptomatic episodes. The processor of such an implant would be programmed with an algorithm to output a baseline, constant-rate, pulsatile stimulation when activated. This could be customized and "mapped" through software modification just as is done after cochlear implantation.

Uncompensated chronic unilateral disorders may also benefit from a pacemaker-based implant. Tonic electrical stimulation could be used to raise the gain of the affected semicircular canals.

While true prosthetic replacement of vestibular function with a motion sensing component may remain as an ultimate goal, this pacemaker-based design could still be helpful in several diseases. Furthermore, the more basic design allows for earlier commencement of human clinical trials (Rubinstein et al. 2011).

2.2.2 Sensor-Based Vestibular Implant

For patients with chronic bilateral vestibular hypofunction, the predominant deficit comes from an inability to maintain postural stability and utilize the VOR to stabilize vision during head motion. Restoring this deficit requires a sensor-based system that will convey head velocity and acceleration information. Uncompensated chronic unilateral disorders may also benefit from a sensor-based design in order to raise the gain of the diseased side.

Such a system would consist of a baseline, constant-rate, pulsatile stimulation analogous to the natural output of the vestibular system. During head acceleration the signal would be modified to convey information about the change in motion. For example, the electrode at the ampullary nerve of the horizontal semicircular canal could be programmed to increase the stimulation frequency if the head were rotated in one direction in the yaw plane and to decrease the stimulation frequency if the head were rotated in the opposite direction. The rate of stimulation would largely reflect head velocity in each case.

To allow for greater dynamic range in the direction of decreased frequency, the baseline stimulation rate can be set to levels higher than physiologic (Della Santina et al. 2007; Merfeld et al. 2007). For example, when implanting guinea pigs, Merfeld and Gong set the baseline stimulation pulse frequency of their prosthesis to 150 Hz, which is more than twice as high as the normal guinea pig baseline of ~60 Hz. Central compensation allows for rapid (<1 day) adaptation to this supraphysiologic baseline.

Increasing the dynamic range by raising the baseline stimulation rate may be particularly important for unilateral implantation in patients with bilateral vestibular hypofunction. The normal baseline pulse rate is relatively low, leaving "unlimited" room to increase, but little room (only 60 Hz in the guinea pig) to decrease. With normal bilateral vestibular physiology, the side that has little dynamic range to decrease will always be complemented by the paired contralateral vestibular organ that has ample dynamic range to increase. In unilateral vestibular implantation for bilateral vestibular hypofunction, however, there is no contralateral side to provide an increased signal dynamic range (Gong and Merfeld 2002).

2.3 Preservation of Residual Vestibular Function

Not all vestibular disorders that would benefit from implantation are characterized by total loss of function of the candidate side. In Ménière's disease, for example, vestibular physiology is often normal or near normal between attacks (Gates 2006;

Thorp and James 2005). For this reason, it is advantageous to develop a vestibular implant that allows continued functioning of the implanted vestibular system in addition to superimposed function of the device. This would be analogous to so-called electro-acoustic stimulation or "hybrid" cochlear implants (see Turner and Gantz, Chap. 3).

A "hybrid" vestibular implant would be possible by implanting electrodes within the perilymphatic space adjacent to the ampullary nerves while maintaining a patent fluid space inside the membranous labyrinth. Unlike the cochlea, which contains nerve terminals along the entire two and a half turns of the organ, the nerve terminals in the semicircular canal are confined to the ampulla. Thus the electrode need only enter a small segment of the organ, lying adjacent to the ampulla. Local current spread may allow for robust activation of the ampullary nerve. Because of the limited trauma of this technique, native vestibular function could be preserved.

Implantation of such vestibular devices would employ minimally traumatic "soft-surgical" techniques learned from hybrid cochlear implantation (Gantz et al. 2005). Likewise, it would be important to follow hearing preserving techniques employed during vestibular surgery such as semicircular canal plugging (Agrawal and Parnes 2001, 2008; Limb et al. 2006). In a recent study, rotational sensitivity was preserved after semicircular canal implantation of 5 rhesus macaques using a device designed by Rubinstein, Phillips, and colleagues and fabricated under subcontract by Cochlear Ltd (Rubinstein et al. 2011). This device was recently approved by the FDA for use in a human feasibility study and is discussed further below and in Sect. 4.1.

2.4 *Examples of Devices*

2.4.1 Vestibular Implant Prototypes

Over the past decade, several groups have developed vestibular implant prototypes that have been implanted in animal models. The authors' group is currently investigating a system intended for human implantation developed at the University of Washington with funding from NIDCD and manufactured under subcontract by Cochlear Ltd. The device consists of a modified Nucleus Freedom system (Fig. 5.4). The receiver-stimulator is attached to a modified trifurcating array of 9 electrodes. Each electrode array is implanted 2.5 mm into the perilymphatic space adjacent to the ampullary nerve. This first generation device is intended to function as a vestibular pacemaker and so lacks a motion sensor. The external signal ("speech") processor contains modified software that can be programmed to provide the appropriate pacing algorithm for each patient. As mentioned in the preceding section, this device is intended to preserve the native physiology of the semicircular canals. A human clinical trial of this device for patients with Ménière's disease is currently underway (see Sect. 4.1).

Other groups have developed sensor-based vestibular implants for testing in animal models. Merfeld et al. developed the first prototype of a vestibular implant prosthesis (Gong and Merfeld 2000), which uses a similar design consisting of a single

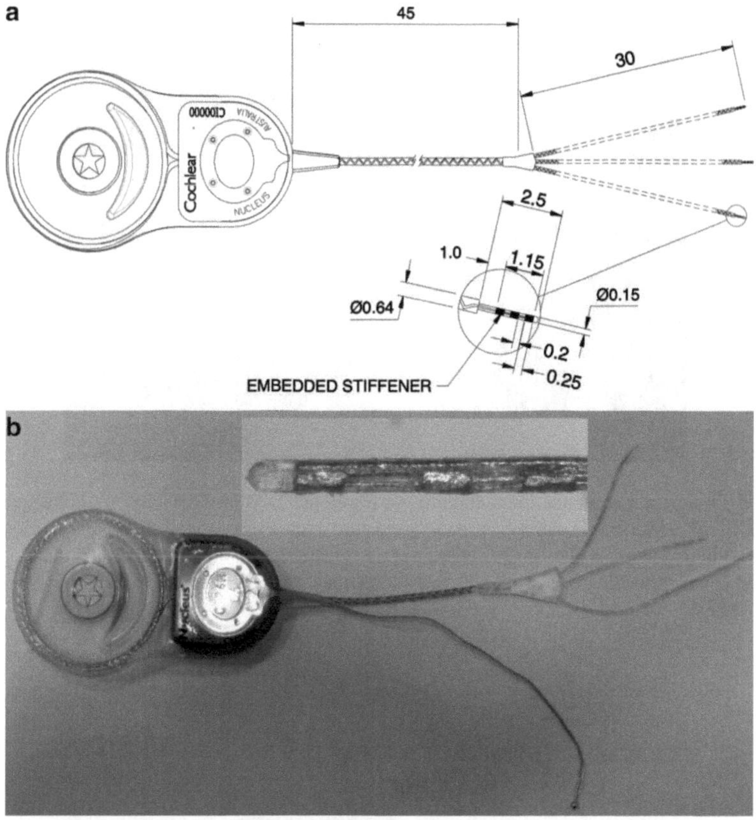

Fig. 5.4 Design of the pacemaker-based UW/Nucleus vestibular implant. The implant consists of three channels with three electrodes each (nine electrodes total). (**a**) Dimensional schematic of the receiver/stimulator and electrode array. (**b**) Photograph of the receiver/stimulator and electrode array. Inset is of magnified electrode array. Courtesy of Cochlear, Ltd

rotational sensor (Fig. 5.5). A piezoelectric vibrating gyroscope is integrated onto a circuit board with a microprocessor. The system contains 3 platinum electrodes on a single channel that are implanted into a single semicircular canal adjacent to the ampulla. The footprint of the device is slightly larger than two quarters. This device has been employed in numerous experiments involving both monkeys (Gong et al. 2008; Lewis et al. 2002; Merfeld et al. 2007) and guinea pigs (Gong and Merfeld 2000, 2002; Merfeld et al. 2006). For these preliminary animal studies, the prosthesis was surgically attached to the top of the animal's head rather than being internally implanted. The electrodes were connected to the prosthesis via a percutaneous wire.

Della Santina and colleagues have developed an integrated device consisting of three micromachined angular gyro-rate sensors built into a circuit board with a microprocessor (Fig. 5.6). An external lithium battery serves as the power source. The authors envision that refinement and size reduction could ultimately allow for human subcutaneous implantation. The first generation device is roughly the size of

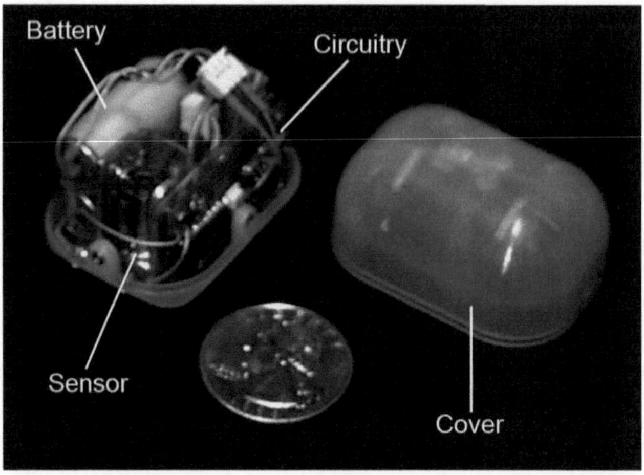

Fig. 5.5 Design of a vestibular implant prototype developed by Merfeld, Gong, and colleagues. This prosthesis includes a single piezoelectric vibrating gyroscope to transmit angular motion information to one semicircular canal. The impact contains three electrodes on the single channel (not depicted). A nylon cover measuring 4.3 × 3.1 × 2.5 cm encloses the circuitry (Gong and Merfeld, 2000)

Fig. 5.6 Design of a vestibular implant prototype circuitry developed by Della Santina and colleagues (Johns Hopkins Multichannel Vestibular Prosthesis, version MVP1). The prosthesis includes three micromachined angular gyro rate sensors (*arrows* indicate the plane of detection). Each gyro transmits angular motion information to one semicircular canal. All three canals are thus stimulated via a 3-channel design, where each channel contains a single electrode (not depicted) (Della Santina et al. 2005, 2007). Reprinted with permission from Charles Della Santina, PhD, MD

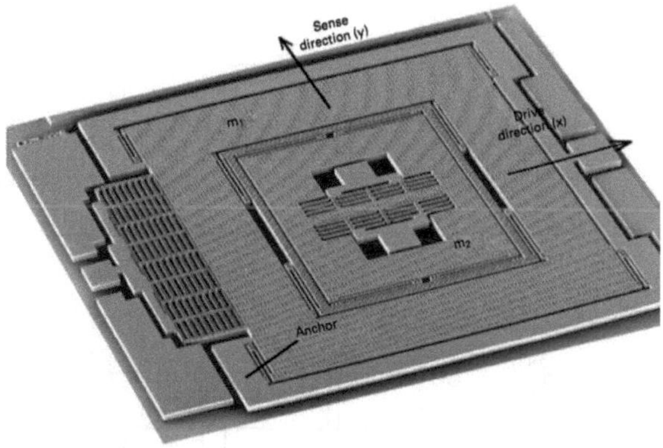

Fig. 5.7 Design of microelectromechanical (MEMS) sensor prototype developed by Shkel, Zeng, and colleagues. The depicted silicon chip is 5×5 mm and contains three gyroscopes and three linear and angular accelerometers. Reprinted with permission from S. Karger AG, Basel (Shkel and Zeng 2006)

a modern cochlear implant receiver-stimulator. The 3 electrode leads are then implanted into each semicircular canal adjacent to the ampullae. This device has been used in several experiments in a chinchilla animal model. Because of size constraints, the device was externally head mounted. A partly compensatory EVOR was observed in multiple planes. One of the key challenges in this small animal model was attaining specific canal stimulation because of current spread.

Shkel and Zeng have developed a microelectromechanical system (MEMS)-based prototype that contains 3 gyroscopes and 3 linear and angular accelerometers on a 5×5 mm silicon chip (Shkel and Zeng 2006) (Fig. 5.7). The same MEMS technology could ultimately be applied to the signal processor, allowing for significant miniaturization, reduction of fabrication costs, and creation of a "balance on-a-chip" system.

2.4.2 Nonimplanted Vestibular Prostheses

For acute vestibular pathology or non-surgical candidates, implantation of a permanent device would be unwarranted. Various non-implanted vestibular prostheses have been developed that may aid in vestibular rehabilitation to overcome acute symptoms or chronic deficits. Such devices typically relay balance information through different sensory modalities, including vibrotactile sensation of the trunk (Weinberg et al. 2006) and electrotactile stimulation of the tongue, e.g., the BrainPort device (Danilov and Tyler 2005; Danilov et al. 2006, 2007; Robinson et al. 2009). Further discussion is beyond the scope of this chapter.

3 Topics in Vestibular Implant Research

3.1 Evoked Vestibulo-ocular Reflex (EVOR)

Electronic stimulation of branches of the vestibular nerve has been show to produce nystagmus in a variety of animal models (Cohen and Suzuki 1963; Cohen et al. 1964; Suzuki et al. 1964, 1969). Furthermore, the eye movements

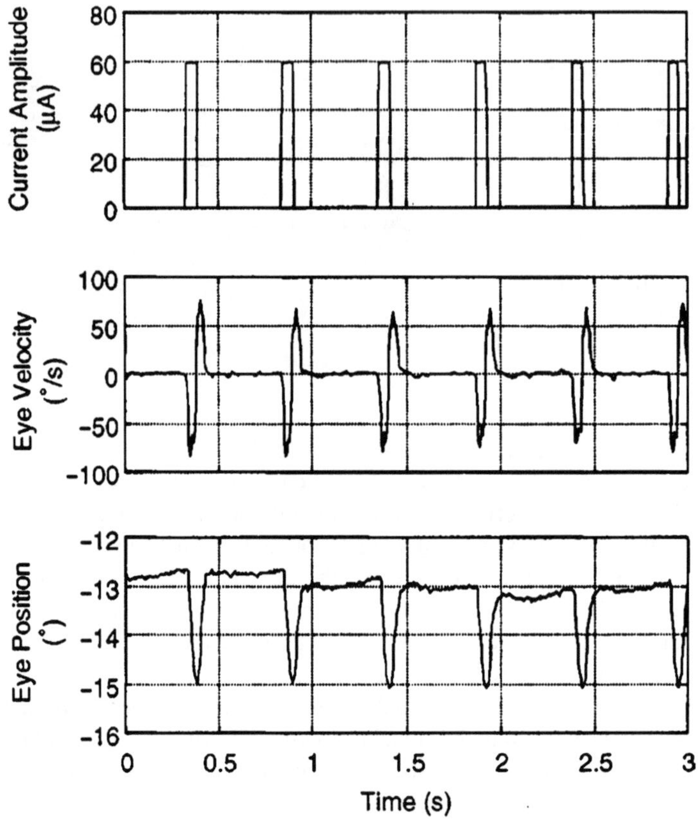

Fig. 5.8 Evoked vestibulo-ocular reflex (EVOR) from prosthetic stimulation of the superior semicircular canal in a guinea pig model. Six pulse trains from the implant (*top*) induced changes in eye velocity (*middle*) and position (*bottom*). Each pulse train consisted of 20 biphasic current pulses (Gong and Merfeld 2000)

correlated with the orientation of the plane of the stimulated semicircular canal. Figure 5.8 shows an example of this electrically induced VOR or evoked VOR (EVOR). An EVOR was recently reproduced experimentally in three human subjects undergoing cochlear implant surgery, in which the posterior ampullary nerve was stimulated and a primarily vertical nystagmus was observed (Wall et al. 2007).

Because VOR is considered to be one of the best objective measures of vestibular function (Gong and Merfeld 2000), EVOR serves a powerful analytic role in vestibular implant research and design. EVOR has several important applications, including assisting with optimal electrode placement intra-operatively, monitoring implant functionality postoperatively, and studying central nervous system adaptation and plasticity. Clinically, sensor-based vestibular implants would replace an absent physiological VOR during head movement with an EVOR.

3.1.1 Intra-operative EVOR to Facilitate Electrode Placement

Animal studies have employed EVOR to aid in placement of prototype vestibular implant electrodes intra-operatively. A brisk EVOR (i.e., nystagmus) after an implanted electrode is activated confirms appropriate placement (Merfeld et al. 2006, 2007). In addition, an electrically evoked compound action potential (ECAP) can be obtained from the vestibular nerve just as it is from the cochlear nerve with cochlear implantation (Rubinstein 2004b). Indeed, ECAP was found to correlate highly to EVOR (Nie et al. 2011). The ECAP was measured using standard clinical ECAP software to assist in correct electrode position when implanting rhesus macaques with a vestibular implant. Like EVOR, ECAP will likely play an important role in confirming correct electrode placement during clinical vestibular implantation (Nie et al. 2011; Rubinstein et al. 2011).

3.1.2 Postimplantation EVOR

EVOR has also been used in animal studies to evaluate device functionality post implantation (Della Santina et al. 2005, 2007; Gong and Merfeld 2000, 2002; Lewis et al. 2002; Merfeld et al. 2006, 2007). For example, sinusoidal modulation of the pulse train frequency has resulted in physiologically appropriate sinusoidal eye movements in both monkey and guinea pig models (Merfeld et al. 2006; Rubinstein et al. 2011). Evoked eye movements have also been shown to vary with the current amplitude (Merfeld et al. 2007). Recently, a deaf patient with bilateral vestibular loss received a modified cochlear implant with an additional electrode that was implanted near the posterior ampullary nerve. Postoperatively, frequency or current modulation of the electrical stimulation resulted in smooth conjugated eye movements (Guyot 2010).

For clinical purposes, EVOR is desired in sensor-based implants during head rotation in order to replace the physiologic VOR that is absent in the diseased state. Animal studies employing rotational chair testing with video and magnetic coil oculography have demonstrated that EVOR is able to approximate normal VOR with potentially near normal gain and phase (Merfeld et al. 2007).

In pacemaker-based implants, the goal is not to restore VOR functionality, but rather to abate symptoms, such as vertigo and nystagmus, caused by an imbalance in the vestibular output on the two sides of the body. Thus EVOR is primarily desired not for visual tracking purposes but rather to "neutralize" pathologic nystagmus occurring during vertigo episodes.

EVOR at undesired times would result in unwanted nystagmus and is a potential side effect of vestibular implantation. For example, in sensor-based implants, there is a tonic baseline level of stimulation that occurs when the head is not moving. This is analogous to the natural baseline output of the vestibular system. In these situations, the vestibular implant should not result in an EVOR. However, animal studies have demonstrated that when vestibular implants are turned on or off, EVOR (i.e., nystagmus) immediately occurs. Fortunately, the brain is able to adapt quickly to the new baseline signal and unwanted evoked nystagmus ceases within 1 day in both guinea pig (Merfeld et al. 2006) and monkey models (Lewis et al. 2002; Merfeld et al. 2007) and within 30 min in one human study (Guyot 2010).

An important study by Merfeld et al. recently demonstrated that despite rapid adaptation to a tonic baseline stimulation continued for over 90 days, monkeys were still able to have an EVOR in response to deviations from this baseline by modulating the pulse rate with head rotation (Merfeld et al. 2007). This study demonstrates that head-motion induced EVOR can still occur despite disappearance of unwanted EVOR caused by the chronic baseline (i.e., non-motion induced) signal.

3.2 Canal-Specificity and Current Spread

Another important goal of vestibular implantation research is the avoidance of current spread from the stimulating electrodes beyond their intended target. If this occurs within the vestibular system, "cross talk" could result in an unintended vestibular organ being stimulated (Della Santina et al. 2007; Weinberg et al. 2006). An example is inducing a vertical eye movement when the horizontal canal ampulla is stimulated because of current spread to the nearby superior canal ampulla (Rubinstein et al. 2011). This phenomenon would result in erroneous vestibular input, reducing the efficacy of the implant, and potentially inducing or exacerbating vestibular symptoms.

Spread beyond the vestibular system carries the danger of auditory side effects. In fact, much of the concern about current spread comes from the converse situation; a well established side effect of cochlear implantation is vestibular symptoms (Brey et al. 1995; Enticott et al. 2006).

Adjusting the properties of the electrical stimulus can reduce current spread. For example, high frequency pulsatile electrical stimulation yields less current spread than low frequency stimulation (Merfeld et al. 2006; Rubinstein 1991; Rubinstein and Spelman 1988) and bipolar stimulation produces less spread than monopolar stimulation.

Several studies have demonstrated the ability to stimulate the semicircular canal ampullae selectively without measurable hearing effects. In a proof-of-concept experiment, Tang and colleagues implanted all 3 semicircular canals of 6 chinchillas. Hearing was preserved in 2 animals. The authors acknowledged the challenge of implantation in a small animal model and predicted better results with more refined surgical technique and device design (Tang et al. 2009). Rubinstein and colleagues conducted auditory brainstem responses (ABRs) in 4 rhesus macaques implanted with electrodes in 2 semicircular canals. Hearing was preserved in 3 of 4 animals (Rubinstein et al. 2011).

A third and less likely target of current spread is the facial nerve. Avoidance of this consequence has been demonstrated in animal models (Merfeld et al. 2006). In one study, the implant current amplitude was decreased until facial nerve stimulation was not observed. Subsequent modification of the signal was performed only through frequency adjustment (Merfeld et al. 2007). In cochlear implant surgery this complication is seen frequently in congenital anomalies of the ear as well as in cases of otosclerosis and is usually easily addressed by manipulating stimulation parameters in the speech processor (Bigelow et al. 1998; Graham et al. 2000; Papsin 2005).

3.3 Adaptation

Adaptation occurs when the central nervous system appropriately adjusts to a change in neural input. *Plasticity* is a common term used to describe the ability to retain this adaptive state (Jones 2000). In contrast, *habituation* is a process whereby repeated exposure to a sensory stimulus causes a response decline. Adaptation is a complex and dynamic process, whereas habituation is a simple tendency towards zero. The remarkable capacity for adaptation was demonstrated in a landmark study where the direction of VOR was reversed in human subjects whose vision was optically flipped (Gonshor and Jones 1973).

As discussed above, stimulation of the vestibular nerve when turning on a vestibular implant for the first time results in evoked nystagmus (EVOR). This side effect is not desirable because the implant should simply mimic the missing normal tonic output from the vestibular system. Within several hours, however, the central nervous system will adapt to this acute stimulus, and the nystagmus will cease (Lewis et al. 2002; Merfeld et al. 2007). Rapid adaptation will also occur to prosthesis pulse rates more than double that of an animal's normal physiologic baseline (Gong and Merfeld 2002).

A concern is that in chronic stimulation of the vestibular nerve, the brain will broadly habituate, rather than adapt, to the signal. If this were the case, any stimulation over baseline would also decay, and the EVOR desired during head movement will eventually not occur. Evidence from several animal studies suggests that there is central adaptation, but not habituation, to vestibular implants. Monkeys that were exposed to continual tonic vestibular implant stimulation for over 90 days were still able to have robust EVORs in response to modulation of this baseline signal during head rotation. This occurred despite complete adaptation (lack of EVOR) from the tonic baseline stimulation (Merfeld et al. 2007).

Another important observation is that after adaptation to the baseline stimulation, turning the device off for the first time also leads to nystagmus in the opposite direction. These so-called after-effects are characteristic of adaptation, but not habituation (Gong and Merfeld 2002; Merfeld et al. 2007). In addition, the evoked nystagmus upon turning the device on or off will also fatigue to zero. In a guinea pig study, the EVOR decayed to 20% of the original value after only three on-off cycles. Acclimation to either device state (on or off) may represent a form of so-called dual-state adaptation (Merfeld et al. 2006). This finding has important consequences because vestibular implant users may need to abruptly deactivate their device in several situations, e.g., showering or changing batteries. Evoked nystagmus would clearly be undesirable, particularly when it is accompanied by other substantial effects such as spatial disorientation. It remains to be seen whether humans implanted with a vestibular prosthesis may benefit from periodically turning on and off the device to maintain dual-state adaptation.

The central vestibular system has the ability to adapt to the EVOR gain. Merfeld and colleagues showed that, in a monkey model, when the device was intentionally set to understimulate and thus result in a low gain, the gain increased over time. However, the increase was modest and the gain did not approach the normal value (normal gain = 1) (Merfeld et al. 2007).

Lastly, adaptation has been shown to minimize effects of current spread beyond the desired canal ampulla. In the same study, undesired vertical EVOR that resulted during horizontal canal stimulation eventually disappeared (Merfeld et al. 2007). Even more dramatically, intentionally stimulating the posterior canal during yaw rotation has been shown to result in adjustment of the EVOR to the correct horizontal direction (Lewis et al. 2002).

4 Clinical Trials

4.1 UW/Nucleus Vestibular Implant Trial for Ménière's Disease

As of this writing, there is one current human clinical trial of a vestibular implant, which has been developed by the authors' group at the University of Washington and built under subcontract by Cochlear Ltd. The trial is a feasibility study and allows implantation of up to 10 human subjects. The device is described in Sect. 2.4.1 and consists of a modified Nucleus Freedom cochlear implant system designed to act as a sensorless vestibular pacemaker. The standard FDA-approved receiver/stimulator is attached to customized trifurcating electrode arrays, with each being implanted into one of the three semicircular canals. The external processor contains programmable algorithms to control the electrical stimulation. Preliminary studies demonstrated preservation of rotational sensitivity, thus allowing native rotational and device-induced stimulation of the vestibular nerve (Rubinstein et al. 2011).

Ménière's disease will serve as an ideal candidate disease for this initial feasibility trial of a pacemaker-style implant. When subjects experience a vestibular attack, they will simply activate the device, which will replace the lost afferent signal by directly stimulating the vestibular nerve. The encoded signal can be adjusted just as cochlear implant patients can adjust the speech processing strategy of their sound processor.

The current study is for Ménière's patients who meet "definitive" disease criteria under the AAO-HNS classification system (Committee 1995). Patients must have failed medical management and be considering surgical options. While preliminary animal studies have documented the ability of this implant system to preserve hearing (Rubinstein et al. 2011), this initial feasibility study will only involve patients with speech discrimination <80% and pure tone averages >50 dB above 2 kHz in the involved ear. Patients must also have a functioning contralateral vestibular system. This trial has recently been granted FDA approval.

4.2 *Future Human Trials*

Initial clinical trials, such as the trial of the UW/Nucleus device for Ménière's disease, will provide information regarding the safety and functionality of vestibular implantation. Such experience should allow further device refinement and potential expansion of indications to include other vestibulopathies such as chronic unilateral lesions with poor compensation.

4.3 *Animals Studies Still Needed*

While a pacemaker-based vestibular implant has already been fabricated for intended human use, all sensor-based implants that have been studied thus far were designed for animal experimentation. Preclinical animal studies are needed on future sensor-based implants designed in conjunction with device manufacturers intended for human implantation.

Animal studies should also continue to investigate adaptation mechanisms and the neurophysiologic changes that result after implantation. This information could result in refinements in vestibular implant design and functionality.

Lastly, more basic research is needed on the neural encoding from the utricle and saccule. If the goal is to mimic normal vestibular physiology as closely as possible, this may be an unrealistic quest with prosthetic implantation of only three of five vestibular organs.

5 The Future

The clinical field of vestibular implantation is just now emerging. Numerous opportunities exist for future avenues of human and preclinical research.

While EVOR has been repeatedly demonstrated in animal models, the gain in sensor-based implants needs to be optimized. Even if a subnormal gain were tolerated in early generation implants, efforts in signal optimization should attempt to approach more normal levels.

Research should also be conducted on creating a totally implantable vestibular implant. Like cochlear implants, all the components of a vestibular implant could be completely internalized. Advances in miniaturization, microelectromechanical systems (MEMS), nanotechnology, battery technology, and circuit energy efficiency could help realize this goal. Ongoing research in totally implantable cochlear implants makes such advances likely.

Vestibular implants in the form described in this chapter would only be helpful for disorders peripheral to the vestibular nerve; however, for some central disorders, vestibular implantation might still prove beneficial. Patients with neurofibromatosis 2, for example, develop bilateral vestibular schwannomas (acoustic neuromas) and may ultimately require resection of both vestibulocochlear nerves (see McCreery and Otto, Chap. 8). A vestibular brainstem implant may prove helpful for bilateral vestibular hypofunction due to this condition. Such a device could potentially be combined with an auditory brainstem device (Schwartz 2008). Obviously such a consideration would require detailed understanding of the surgical anatomy of the vestibular nuclei.

Ultimately, it may prove impossible to achieve completely normal vestibular physiology with prosthetic electrical stimulation. Efforts in tissue engineering and regenerative medicine strive to replace lost function with biologic, rather than electronic, products. While total regeneration of the vestibular apparatus may be decades away, a more realistic near-term goal would be to combine the two technologies. For example, it is known that vestibular neurons will slowly degenerate following hair cell death (Schuknecht 1982a). This could reduce implant efficacy in some patients, although electrical stimulation could have neurotrophic effects. Degenerative effects may be even more pronounced than with cochlear implantation since Scarpa's ganglion lies further away from the vestibular end organs than the spiral ganglion lies from the cochlea (Wall et al. 2002). An electrode array impregnated with neurotrophic growth factors may help solve this problem by augmenting the ability to neurally transmit the implant-delivered signal.

References

Agrawal, S. K., & Parnes, L. S. (2001). Human experience with canal plugging. *Annals of the New York Academy of Sciences,* 942, 300–305.

Agrawal, S. K., & Parnes, L. S. (2008). Transmastoid superior semicircular canal occlusion. *Otology & Neurotology,* 29(3), 363–367.

Bhattacharyya, N., Baugh, R. F., Orvidas, L., Barrs, D., Bronston, L. J., Cass, S., Chalian, A. A., Desmond, A. L., Earll, J. M., Fife, T. D., Fuller, D. C., Judge, J. O., Mann, N. R., Rosenfeld, R. M., Schuring, L. T., Steiner, R. W., Whitney, S. L., Haidari, J. (2008). Clinical practice guideline: benign paroxysmal positional vertigo. *Otolaryngology-Head and Neck Surgery,* 139(Suppl. 4), 47–81.

Bigelow, D. C., Kay, D. J., Rafter, K. O., Montes, M., Knox, G. W., & Yousem, D. M. (1998). Facial nerve stimulation from cochlear implants. *American Journal of Otology,* 19(2), 163–169.

Brey, R. H., Facer, G. W., Trine, M. B., Lynn, S. G., Peterson, A. M., & Suman, V. J. (1995). Vestibular effects associated with implantation of a multiple channel cochlear prosthesis. *American Journal of Otology,* 16(4), 424–430.

Carey, J. P., & Della Santina, C. P. (2005). Principles of applied vestibular physiology. In C. W. Cummings, P. W. Flint, B. H. Haughey, K. T. Robbins, J. R. Thomas, L. A. Harker, M. A. Richardson, & D. E. Schuller (Eds.), *Cummings otolaryngology-head & neck Surgery* (4th ed., pp. 3115–3159). Philadelphia: Elsevier.

Centers for Disease Control and Prevention. (2007). WISQARS Leading Causes of Death Reports. http://webappa.cdc.gov/sasweb/ncipc/leadcaus10.html. Accessed December 20, 2009.

Cohen, B., & Suzuki, J. I. (1963). Eye movements induced by ampullary nerve stimulation. *American Journal of Physiology,* 204, 347–351.

Cohen, B., Suzuki, J. I., & Bender, M. B. (1964). Eye movements from semicircular canal nerve stimulation in the cat. *Annals of Otology, Rhinology, and Laryngology,* 73, 153–169.

Committee on Hearing and Equilibrium (1995). Guidelines for the diagnosis and evaluation of therapy in Meniere's disease. *Otolaryngology-Head and Neck Surgery*, 113(3), 181–185.

Curthoys, I. S. (1987). Eye movements produced by utricular and saccular stimulation. *Aviation Space and Environmental Medicine,* 58(9, pt. 2), A192–197.

Curthoys, I. S., & Oman, C. M. (1986). Dimensions of the horizontal semicircular duct, ampulla and utricle in rat and guinea pig. *Acta Oto-Laryngologica,* 101(1–2), 1–10.

Danilov, Y., & Tyler, M. (2005). Brainport: an alternative input to the brain. *Journal of Integrative Neuroscience,* 4(4), 537–550.

Danilov, Y. P., Tyler, M. E., Skinner, K. L., & Bach-y-Rita, P. (2006). Efficacy of electrotactile vestibular substitution in patients with bilateral vestibular and central balance loss. *Conference Proceedings–IEEE Engineering in Medicine and Biology Society,* Suppl., 6605–6609.

Danilov, Y. P., Tyler, M. E., Skinner, K. L., Hogle, R. A., & Bach-y-Rita, P. (2007). Efficacy of electrotactile vestibular substitution in patients with peripheral and central vestibular loss. *Journal of Vestibular Research,* 17(2–3), 119–130.

Della Santina, C., Migliaccio, A., Patel, A. (2005). Electrical stimulation to restore vestibular function development of a 3-d vestibular prosthesis. *Conference Proceedings–IEEE Engineering in Medicine and Biology Society,* 7, 7380–7385.

Della Santina, C. C., Migliaccio, A. A., & Patel, A. H. (2007). A multichannel semicircular canal neural prosthesis using electrical stimulation to restore 3-d vestibular sensation. *IEEE Transactions on Biomedical Engineering,* 54(6, pt. 1), 1016–1030.

Enticott, J. C., Tari, S., Koh, S. M., Dowell, R. C., & O'Leary, S. J. (2006). Cochlear implant and vestibular function. *Otology & Neurotology,* 27(6), 824–830.

Epley, J. M. (1992). The canalith repositioning procedure: for treatment of benign paroxysmal positional vertigo. *Otolaryngology-Head and Neck Surgery,* 107(3), 399–404.

Fluur, E., & Mellstrom, A. (1971). The otolith organs and their influence on oculomotor movements. *Experimental Neurology,* 30(1), 139–147.

Gantz, B. J., Turner, C., Gfeller, K. E., & Lowder, M. W. (2005). Preservation of hearing in cochlear implant surgery: advantages of combined electrical and acoustical speech processing. *Laryngoscope,* 115(5), 796–802.

Gantz, B. J., Turner, C., & Gfeller, K. E. (2006). Acoustic plus electric speech processing: preliminary results of a multicenter clinical trial of the Iowa/Nucleus Hybrid implant. *Audiology & Neurotology,* 11 (Suppl. 1), 63–68.

Gates, G. A. (2006). Meniere's disease review 2005. *Journal of the American Academy of Audiology,* 17(1), 16–26.

Gates, G. A., & Miyamoto, R. T. (2003). Cochlear implants. *New England Journal of Medicine,* 349(5), 421–423.

Goebel, J. A., & Sumer, B. (2007). Vestibular physiology. In G. B. Hughes & M. L. Pensak (Eds.), *Clinical otology* (3rd ed.). New York: Thieme.

Goldberg, J. M., & Fernandez, C. (1971). Physiology of peripheral neurons innervating semicircular canals of the squirrel monkey. I. Resting discharge and response to constant angular accelerations. *Journal of Neurophysiology,* 34(4), 635–660.

Goldberg, M. E. (2000). The vestibular system. In E. R. Kandel, J. H. Schwartz, & T. M. Jessell (Eds.), *Principles of neural science*. New York: McGraw-Hill.

Gong, W., Haburcakova, C., & Merfeld, D. M. (2008). Vestibulo-ocular responses evoked via bilateral electrical stimulation of the lateral semicircular canals. *IEEE Transactions on Biomedical Engineering,* 55(11), 2608–2619.

Gong, W., & Merfeld, D. M. (2000). Prototype neural semicircular canal prosthesis using patterned electrical stimulation. *Annals of Biomedical Engineering,* 28(5), 572–581.

Gong, W., & Merfeld, D. M. (2002). System design and performance of a unilateral horizontal semicircular canal prosthesis. *IEEE Transactions on Biomedical Engineering,* 49(2), 175–181.

Gonshor, A., & Jones, G. M. (1973). Proceedings: Changes of human vestibulo-ocular response induced by vision-reversal during head rotation. *The Journal of Physiology,* 234(2), 102P–103P.

Goto, F., Meng, H., Bai, R., Sato, H., Imagawa, M., Sasaki, M., et al. (2003). Eye movements evoked by the selective stimulation of the utricular nerve in cats. *Auris Nasus Larynx,* 30(4), 341–348.

Goto, F., Meng, H., Bai, R., Sato, H., Imagawa, M., Sasaki, M., et al. (2004). Eye movements evoked by selective saccular nerve stimulation in cats. *Auris Nasus Larynx,* 31(3), 220–225.

Graham, J. M., Phelps, P. D., & Michaels, L. (2000). Congenital malformations of the ear and cochlear implantation in children: review and temporal bone report of common cavity. *Journal of Laryngology and Otology. Supplement,* 25, 1–14.

Guyot, J. P. (2010). Vestibular implantation: from the concept to the first studies in the human. Presented at the 11th International Conference on Cochlear Implants and Other Implantable Auditory Technologies, Stockholm, Sweden.

Hlavacka, F., & Njiokiktjien, C. (1985). Postural responses evoked by sinusoidal galvanic stimulation of the labyrinth. Influence of head position. *Acta Oto-Laryngologica,* 99(1–2), 107–112.

Jones, G. M. (2000). Posture. In E. R. Kandel, J. H. Schwartz, & T. M. Jessell (Eds.), *Principles of neural science*. New York: McGraw-Hill.

Kimura, R. S. (1982). Animal models of endolymphatic hydrops. *American Journal of Otolaryngology,* 3(6), 447–451.

Lempert, T., & Neuhauser, H. (2009). Epidemiology of vertigo, migraine and vestibular migraine. *Journal of Neurology,* 256(3), 333–338.

Lewis, R. F., Gong, W., Ramsey, M., Minor, L., Boyle, R., & Merfeld, D. M. (2002). Vestibular adaptation studied with a prosthetic semicircular canal. *Journal of Vestibular Research,* 12(2–3), 87–94.

Limb, C. J., Carey, J. P., Srireddy, S., & Minor, L. B. (2006). Auditory function in patients with surgically treated superior semicircular canal dehiscence. *Otology & Neurotology,* 27(7), 969–980.

Lysakowski, A. (2010). Anatomy of the vestibular system. *Cummings otolaryngology-head & neck surgery* (5th ed., pp. 1850–1865). Philadelphia: Elsevier.

Merchant, S. N., Velazquez-Villasenor, L., Tsuji, K., Glynn, R. J., Wall, C., 3rd, & Rauch, S. D. (2000). Temporal bone studies of the human peripheral vestibular system. Normative vestibular hair cell data. *Annals of Otology, Rhinology, and Laryngology. Supplement,* 181, 3–13.

Merfeld, D. M., Gong, W., Morrissey, J., Saginaw, M., Haburcakova, C., & Lewis, R. F. (2006). Acclimation to chronic constant-rate peripheral stimulation provided by a vestibular prosthesis. *IEEE Transactions on Biomedical Engineering,* 53(11), 2362–2372.

Merfeld, D. M., Haburcakova, C., Gong, W., & Lewis, R. F. (2007). Chronic vestibulo-ocular reflexes evoked by a vestibular prosthesis. *IEEE Transactions on Biomedical Engineering,* 54(6, Pt. 1), 1005–1015.

Nashner, L. M., & Wolfson, P. (1974). Influence of head position and proprioceptive cues on short latency postural reflexes evoked by galvanic stimulation of the human labyrinth. *Brain Research,* 67(2), 255–268.

Neuhauser, H. K., Radtke, A., von Brevern, M., Lezius, F., Feldmann, M., & Lempert, T. (2008). Burden of dizziness and vertigo in the community. *Archives of Internal Medicine,* 168(19), 2118–2124.

Nie, K., Bierer, S. M., Ling, L., Oxford, T., Rubinstein, J. T., & Phillips, J. O. (2011). Characterization of the electrically-evoked compound action potential of the vestibular nerve. [in press].

Papsin, B. C. (2005). Cochlear implantation in children with anomalous cochleovestibular anatomy. *Laryngoscope,* 115(1, Pt. 2, Suppl. 106), 1–26.

Pavlik, A. E., Inglis, J. T., Lauk, M., Oddsson, L., & Collins, J. J. (1999). The effects of stochastic galvanic vestibular stimulation on human postural sway. *Experimental Brain Research,* 124(3), 273–280.

Rauch, S. D., Velazquez-Villasenor, L., Dimitri, P. S., & Merchant, S. N. (2001). Decreasing hair cell counts in aging humans. *Annals of the New York Academy of Sciences,* 942, 220–227.

Robinson, B. S., Cook, J. L., Richburg, C. M., & Price, S. E. (2009). Use of an electrotactile vestibular substitution system to facilitate balance and gait of an individual with gentamicin-induced bilateral vestibular hypofunction and bilateral transtibial amputation. *Journal of Neurologic Physical Therapy,* 33(3), 150–159.

Rubinstein, J. T. (1991). Analytical theory for extracellular electrical stimulation of nerve with focal electrodes. II. Passive myelinated axon. *Biophysical Journal,* 60(3), 538–555.

Rubinstein, J. T. (2004a). How cochlear implants encode speech. *Current Opinion in Otolaryngology & Head and Neck Surgery,* 12(5), 444–448.

Rubinstein, J. T. (2004b). An introduction to the biophysics of the electrically evoked compound action potential. *International Journal of Audiology,* 43 (Suppl. 1), S3–9.

Rubinstein, J. T., Bierer, S., Fuchs, A. F., Kaneko, C., Ling, L., Nie, K., et al. (2011). Prosthetic implantation of the semicircular canals with preservation of rotational sensitivity: a "hybrid" vestibular implant (in preparation).

Rubinstein, J. T., & Spelman, F. A. (1988). Analytical theory for extracellular electrical stimulation of nerve with focal electrodes. I. Passive unmyelinated axon. *Biophysical Journal,* 54(6), 975–981.

Sattin, R. W. (1992). Falls among older persons: a public health perspective. *Annual Review of Public Health,* 13, 489–508.

Schuknecht, H. F. (1982a). Behavior of the vestibular nerve following labyrinthectomy. *Annals of Otology, Rhinology, and Laryngology. Supplement,* 97, 16–32.

Schuknecht, H. F. (1982b). Meniere's disease, pathogenesis and pathology. *American Journal of Otolaryngology,* 3(5), 349–352.

Schuknecht, H. F. (1984). The pathophysiology of Meniere's disease. *American Journal of Otology,* 5(6), 526–527.

Schuknecht, H. F., Gulya, A. J. (1983). Endolymphatic hydrops. An overview and classification. *Annals of Otology, Rhinology, and Laryngology. Supplement,* 106, 1–20.

Schwartz, M. S., Otto, S. R., Shannon, R. V., Hitselberger, W. E., & Brackmann, D. E. (2008). Auditory brainstem implants. *Neurotherapeutics,* 5(1), 128–136.

Scinicariello, A. P., Eaton, K., Inglis, J. T., & Collins, J. J. (2001). Enhancing human balance control with galvanic vestibular stimulation. *Biological Cybernetics,* 84(6), 475–480.

Shkel, A. M., & Zeng, F. G. (2006). An electronic prosthesis mimicking the dynamic vestibular function. *Audiology & Neurotology,* 11(2), 113–122.

Suzuki, J. I., Cohen, B., & Bender, M. B. (1964). Compensatory eye movements induced by vertical semicircular canal stimulation. *Experimental Neurology,* 9, 137–160.

Suzuki, J. I., Goto, K., Tokumasu, K., & Cohen, B. (1969). Implantation of electrodes near individual vestibular nerve branches in mammals. *Annals of Otology, Rhinology, and Laryngology,* 78(4), 815–826.

Tabak, S., Collewijn, H., Boumans, L. J., & van der Steen, J. (1997). Gain and delay of human vestibulo-ocular reflexes to oscillation and steps of the head by a reactive torque helmet. I. Normal subjects. *Acta Oto-Laryngologica,* 117(6), 785–795.

Tang, S., Melvin, T. A., & Della Santina, C. C. (2009). Effects of semicircular canal electrode implantation on hearing in chinchillas. *Acta Oto-Laryngologica,* 129(5), 481–486.

Thorp, M. A., & James, A. L. (2005). Prosper Meniere. *Lancet,* 366(9503), 2137–2139.

Van de Heyning, P. H., Wuyts, F., & Boudewyns, A. (2005). Surgical treatment of Meniere's disease. *Current Opinion in Neurology,* 18(1), 23–28.

Wall, C., 3rd, Merfeld, D. M., Rauch, S. D., & Black, F. O. (2002). Vestibular prostheses: the engineering and biomedical issues. *Journal of Vestibular Research,* 12(2–3), 95–113.

Wall, C., 3rd, Kos, M. I., & Guyot, J. P. (2007). Eye movements in response to electric stimulation of the human posterior ampullary nerve. *Annals of Otology, Rhinology, and Laryngology,* 116(5), 369–374.

Weiland, J. D., Liu, W., & Humayun, M. S. (2005). Retinal prosthesis. *Annual Review of Biomedical Engineering,* 7, 361–401.

Weinberg, M. S., Wall, C., Robertsson, J., O'Neil, E., Sienko, K., & Fields, R. (2006). Tilt determination in MEMS inertial vestibular prosthesis. *Journal of Biomechanical Engineering,* 128(6), 943–956.

White, J. A. (2007). Laboratory tests of vestibular and balance functioning. In G. B. Hughes & M. L. Pensak (Eds.), *Clinical otology* (3rd ed.). New York: Thieme.

Won, J. H., Schimmel, S. M., Drennan, W. R., Souza, P. E., Atlas, L., & Rubinstein, J. T. (2008). Improving performance in noise for hearing aids and cochlear implants using coherent modulation filtering. *Hearing Research,* 239(1–2), 1–11.

Zeng, F. G. (2004). Trends in cochlear implants. *Trends in Amplification,* 8(1), 1–34.

Chapter 6
Optical Stimulation of the Auditory Nerve

Claus-Peter Richter and Agnella Izzo Matic

1 Introduction

Improvements in cochlear implant devices during the last decade have mainly been achieved through novel coding strategies rather than through improvement of the neural interface. The neural interface, however, is a bottleneck for transferring information from the cochlear implant to the auditory nerve. Electric current spreads in the tissue and neighboring electrode contacts cannot be considered independent stimulation sources. Simultaneous transfer of information at adjacent electrodes may lead to deleterious interactions. Therefore, contemporary coding strategies use sequential stimulation paradigms that avoid simultaneous stimulation at neighboring electrode contacts. These coding strategies provide good speech recognition in quiet listening environments but fail in noisy backgrounds. It has been argued that an increase in the number of independent channels that transfer information to the auditory nerve could improve patient performance in noisy listening environments. Therefore, an important objective in implant electrode design is to maximize the spatial selectivity of stimulation.

1.1 Number of Frequency Bands and Speech Recognition

The number of frequency bands required to transmit speech information accurately is an important measurement used in optimization of multi-electrode stimulation of the cochlea. Shannon et al. (2004) and Turner et al. (1995) used acoustic models to

C.-P. Richter (✉)
Department of Otolaryngology, Feinberg School of Medicine,
Northwestern University, 303 E. Chicago Avenue,
Searle 12-561, Chicago, IL 60611, USA
e-mail: cri529@northwestern.edu

F.-G. Zeng et al. (eds.), *Auditory Prostheses: New Horizons*, Springer Handbook of Auditory Research 39, DOI 10.1007/978-1-4419-9434-9_6,

study the speech information transmitted by fixed-filter speech processing schemes. They assessed the optimal number of filter bands to be used as well as the number of cochlear implant electrodes. For quiet listening conditions, a normal-hearing listener could obtain near normal speech recognition with a 4-channel processor. These results were confirmed by other groups (Hochmair and Hochmair-Desoyer 1983; Hochmair-Desoyer et al. 1983); however, for noisy listening conditions 4 channels are not sufficient. Previous work with cochlear implants has demonstrated that speech recognition scores increase with an increasing number of electrodes (Holmes et al. 1987; Dorman et al. 1989, 1997, 1998; Geier and Norton 1992; Kilney et al. 1992; Lawson 1993, 1996; McKay et al. 1994; Collins et al. 1997; Eddington et al. 1997; Fishman et al. 1997). The puzzling aspect of the aforementioned data is that even the best performing cochlear implant users appear to be limited to the equivalent of 7 to 10 "spectral channels." Although there are factors such as coding strategy, the brand of the cochlear implant device, and warping in the spectral-tonotopic mapping, the primary factor limiting speech recognition scores seems to be electrode interaction (Friesen et al. 2001; O'Leary et al. 2009). More selective stimulation of spiral ganglion cells could allow for more discrete stimulation of the auditory system, thus providing an increased number of independent sub-populations of spiral ganglion cells for speech processing.

1.2 Strategies to Increase Selectivity of Stimulation with Electrical Current

Measurements of potential distributions in the cochlea have been conducted (Black et al. 1981; O'Leary et al. 1985; Kral et al. 1998; Vanpoucke et al. 2004). The measurements demonstrated that with monopolar stimulation, current spreads widely in the cochlea. More selective stimulation was possible if the current was steered using multipolar stimulation paradigms (van den Honert and Stypulkowski 1987; Kral et al. 1998). However, multipolar stimulation was less power efficient. More recently, it has been suggested to use thin-film penetrating electrodes and place them directly into the auditory nerve to increase the selectivity of stimulation (see Chap. 7) (Middlebrooks and Snyder 2007).

1.3 Near-Infrared Radiation, an Alternative for Neural Interfaces

A recently developed method to stimulate neurons uses optical radiation. Wells et al. (2005a, b) have used a pulsed infrared laser to stimulate the rat sciatic nerve and have pioneered a new field of neural stimulation. The transient deposition of optical radiation into the neural tissue directly evokes an action potential from the rat sciatic nerve in response to each laser pulse. To stimulate neural activity artificially, the foremost advantage of using optical radiation over electrical current is the spatial resolution of stimulation. Only neural tissue that is directly in the optical path will be stimulated when the radiation is absorbed.

This technique should not be confused with optogenetics, which is a technique for photochemical neural activation in mammalian neurons. Channel rhodopsins have to be expressed in the cell membrane of the neuron, after which the cells can be stimulated upon irradiation with light pulses (e.g., Boyden et al. 2005; Zhang et al. 2006). Readers are referred to a recent review by Kramer, Fortin, and Trauner (Kramer et al. 2009).

2 Stimulation of Neural Tissue with an Infrared Pulsed Laser

Inspired by the experiments by Wells and co-workers, experiments have been conducted to stimulate the auditory nerve with near-infrared radiation. Optical radiation may afford the possibility to independently stimulate small populations of neurons, increasing the number of perceptual channels for a cochlear implant user, which are required for music perception and speech recognition in noise.

2.1 Proof of Principle

In several animal models, it has been shown that the auditory nerve can be stimulated with optical radiation. With the optical fiber placed in front of the round window or through a cochleostomy into the scala tympani, compound action potentials (CAPs) could be evoked with optical radiation (Fig. 6.1). In gerbils (*Meriones unguiculatus*), average stimulation thresholds were 18 mJ/cm^2. Radiation energy could be increased by a factor of 30 to 100 until drastic changes were seen in cochlear function (Izzo et al. 2006). Optical radiation evoked CAPs in normal hearing and in deaf gerbils, in which the cochleae lacked outer and inner hair cells. Acoustically evoked compound action potential thresholds were elevated by more than 40 dB after neomycin application in acutely deaf and more than 60 dB in chronically deaf animals. Optically evoked compound action potential thresholds were not significantly elevated in acutely deaf animals (Fig. 6.2). However, in chronically deaf animals optically evoked CAP thresholds were elevated. The change in CAP amplitude was correlated with the number of surviving spiral ganglion cells and the optical parameters that were used for stimulation (Richter et al. 2008).

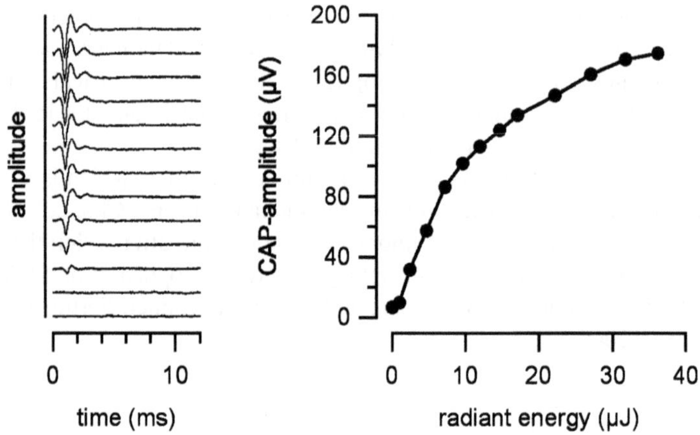

Fig. 6.1 Compound action potentials recorded with a large electrode from the gerbil round window, while stimulating with optical radiation (λ = 1860 nm, pulse duration = 100 µs, pulse repetition rate = 10 Hz). *Left panel* shows individual recorded traces for increasing radiant energy from *bottom* to *top*. *Right panel* shows the corresponding compound action potential peak-to-peak amplitudes

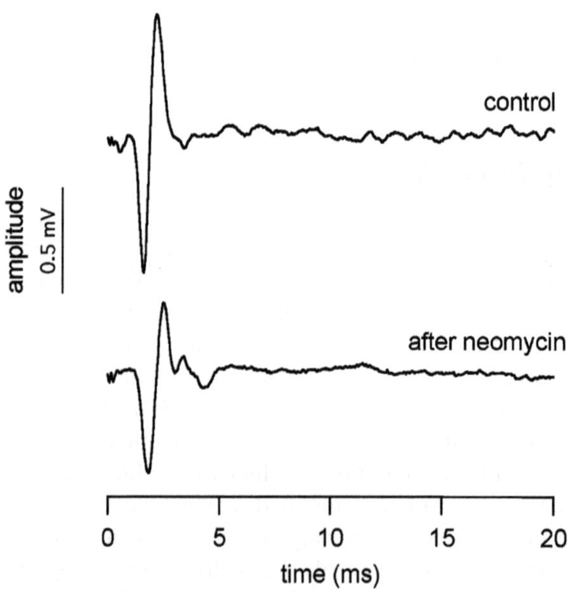

Fig. 6.2 Traces recorded from the guinea pig round window during stimulation with optical radiation (λ = 1860 nm, pulse duration = 100 µs, pulse repetition rate = 10 Hz, radiation energy is fixed). Shown is an example of a compound action potential (CAP) before and after deafening with neomycin (20 mM in buffered lactated Ringer's solution) injected into scala tympani. Few changes of the CAP were observed

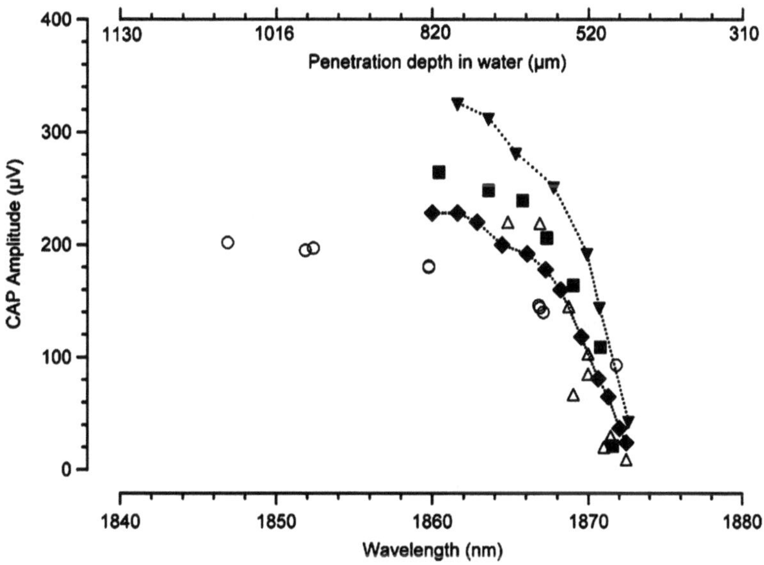

Fig. 6.3 Wavelength variability of laser evoked neural response. At longer wavelengths (shorter penetration depths), the CAP amplitude is at a minimum. When the wavelength is decreased (penetration depth increases), the CAP amplitude grows, until it reaches a plateau. Wavelengths between 1844 and 1873 nm were tested. Each type of data marker represents a single data point measured on a different animal. The axis representing penetration depth serves as a guide and is not meant to indicate a linear change

2.2 *Radiation Wavelength and Pulse Duration*

Various optical parameters have been studied for stimulation with optical radiation. The radiation sources were different lasers, including the Ho:YAG and different Aculight diode lasers. Wavelengths were between 1840 and 1888 nm, and discrete values of 1950 and 2120 nm. Energy per pulse was varied between 4 and 150 μJ; pulse durations were between 5 and 2000 μs. The lasers were coupled to an optical fiber, 50 to 600 μm in core diameter. Optical stimulation of the auditory nerve was possible at all wavelengths. At 1950 nm, the penetration depth of the radiation in water is approximately 100 μm. When irradiated with 1950-nm radiation, the tip of the optical fiber had to be placed closer to the target structure to achieve stimulation. The effect of the radiation wavelength on the compound action potential amplitude was examined systematically between 1840 and 1888 nm (Izzo et al. 2007). Figure 6.3 shows that the CAP amplitude increases with increasing penetration depth of the radiation until a plateau is reached. The results suggest that more neurons were recruited with increasing penetration depth of the radiation. Saturation of the CAP amplitude indicated that all the neurons in the beam path were activated and no further neurons could be recruited if the penetration depth was increased (Izzo et al. 2007).

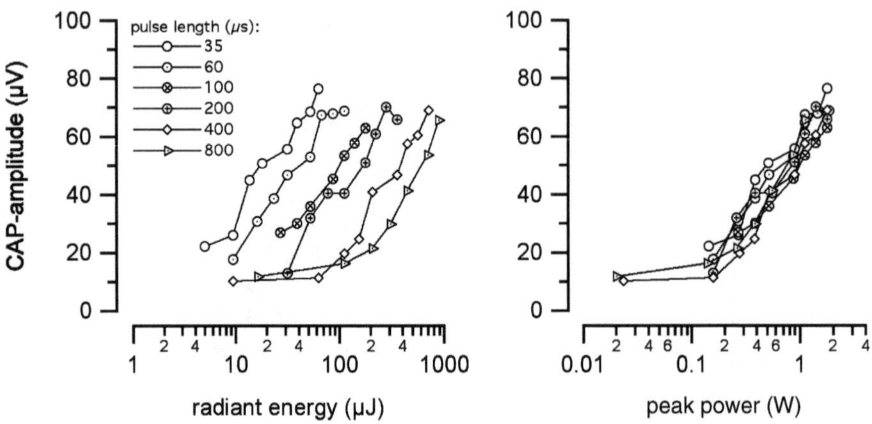

Fig. 6.4 The traces show the compound action potential (CAP) amplitudes recorded with a large electrode from the guinea pig round window during stimulation with optical radiation ($\lambda = 1860$ nm, pulse repetition rate = 10 Hz). The different traces reflect different pulse durations between 35 and 800 µs. Plotting the CAP amplitudes versus radiant energies suggest that shorter pulses are more efficient (*left panel*). Converting the radiant energies into peak power and replotting the data (*right panel*) shows that there are no differences among the different pulse durations

Threshold radiant exposures varied with pulse duration. The values were smaller for shorter pulse durations. Initial experiments were conducted in gerbils and the pulse duration was varied between 5 and 300 µs. Average radiant exposures were 1.6 ± 0.2 mJ/cm^2 for 5 µs and were 15.1 ± 2.2 mJ/cm^2 for 300 µs pulse duration (Izzo et al. 2007). Similar experiments were repeated in mice (C57BL/6J strain) and pigmented guinea pigs (*Cavia porcellus*). Threshold radiant exposures were similar among the different animal species. Furthermore, for all experiments, shorter pulses appeared more efficient than longer pulses. Threshold radiant exposures required to evoke an equal-amplitude CAP were smaller for shorter pulses. However, if the CAP amplitudes were plotted versus the peak power, the threshold differences between the pulse durations disappear (Fig. 6.4).

2.3 Stimulation Thresholds

Thresholds varied for neural stimulation with optical radiation in different neural systems. Radiant exposure thresholds were about 2 orders of magnitude smaller for stimulating the auditory neurons (~15 mJ/cm^2) than for stimulating other peripheral nerves, e.g., ~0.7 J/cm^2 for the gerbil facial nerve (Teudt et al. 2007), ~0.3 J/cm^2 (Wells et al. 2005b) and ~1.7 J/cm^2 (Duke et al. 2009) for the rat sciatic nerve, and 1 J/cm^2 for rat cavernous nerves (Fried et al. 2008). There are, however, differences in the outcomes measured in each of these experiments. In the cochlear optical stimulation experiments, a compound nerve action potential was used as the measurement (CAP).

However, for stimulation of the facial nerve or the sciatic nerve a functional muscle contraction was measured. Furthermore, differences could be attributed to the determination of the spot size, which is included in the calculation of the radiant exposure. While for the auditory nerve experiments the spot size was generally assumed to be the area of the optical fiber, Wells used the distance and the numerical aperture of the optical fiber to calculate the spot size. More recent experiments by Duke et al. (2009) reported radiant exposure values that included the experimentally determined spot size.

2.4 The Beam Path

Optical tissue properties, which are wavelength dependent, determine the interaction between the radiation and the tissue of interest. For the present experiments, the radiation wavelength was in the near infrared, 1840 to 1880 nm. The relevant tissues were the cochlear fluids, lipids, blood, and bone. For the given wavelengths, it has been shown that absorption governs the optical extinction while scatter plays a lesser role. At 1840 to 1880 nm, the corresponding penetration depth of the radiation in water is 977 to 382 μm (Hale and Querry 1973). In other words, the radiation energy decreases for each penetration depth to 1/e (1/Euler number) or to 37% of its value at the tip of the optical fiber. Because there is little scatter of these wavelengths in water, the beam diverges very little over 1 to 2 penetration depths through the fluids. The divergence angle is about 1°. In contrast to the fluids, bone scatters the radiation. For modiolar bone, the divergence angle is approximately 26° (measurements were made in thick tissue sections using the knife-edge technique).

Experiments conducted in guinea pigs have demonstrated that the optical path in the cochlea varies according to the orientation of the optical fiber (Moreno et al. in press). The results also demonstrated that best frequencies of stimulation depend on the orientation of the optical fiber. The access selected (basal turn cochleostomy) allowed stimulation of the spiral ganglion cells that encode best frequencies between 1 and 15 kHz in the guinea pig cochlea. However, the experiments also demonstrated that stimulation likely occurs at sites that are further away from the tip of the optical fiber and that it is difficult to irradiate the section of the spiral ganglion directly opposite to the optical fiber (Fig. 6.5). The latter corresponds to frequencies between ~10 and 20 kHz. The current surgical access does not allow stimulation of the very base of the cochlea, corresponding to best frequencies above 20 kHz.

With a multi-channel electrode array inserted in the inferior colliculus, a frequency range between ~1 and 25 kHz could be monitored in the aforementioned experiments. The frequency range for the current experiments was similar to that published previously for the guinea pig (Snyder et al. 2004, 2008). With the results from auditory single nerve fiber recordings (Robertson 1984; Tsuji and Liberman 1997), a frequency place map of the guinea pig cochlea can be constructed using the function suggested by Greenwood (1990). From this, the lowest frequency for the guinea pig is calculated to be 0.052 kHz and the highest frequency is 43.76 kHz (Müller 1996).

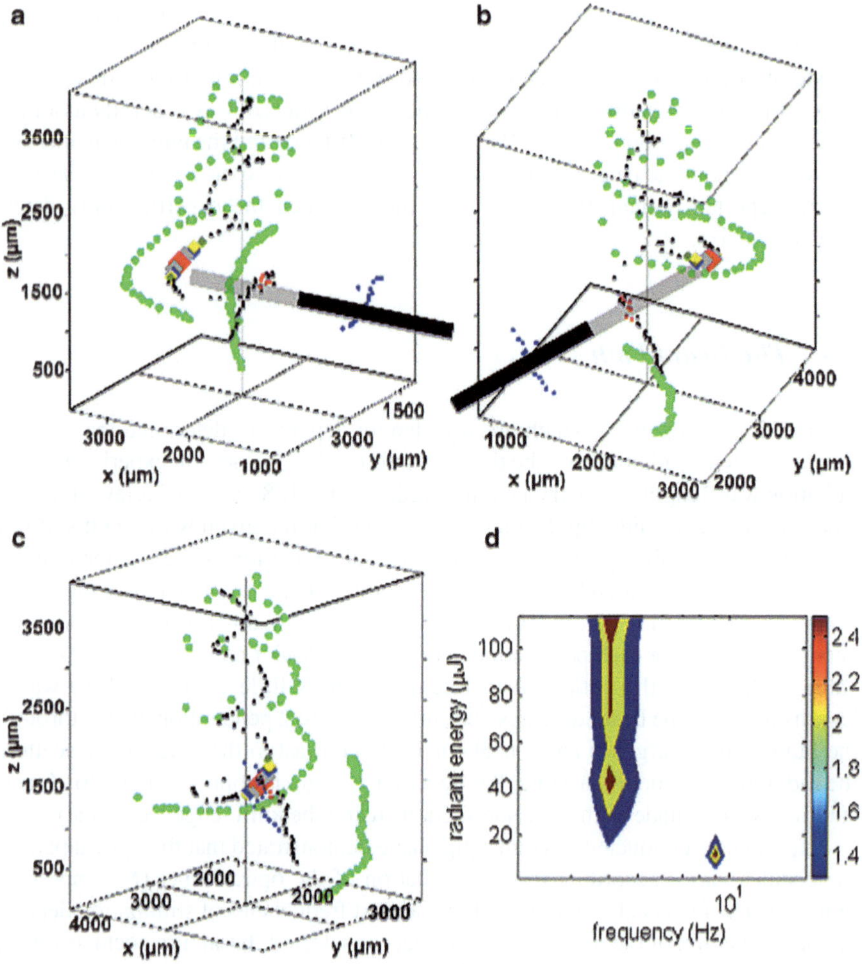

Fig. 6.5 Reconstruction of the basal section of a guinea pig cochlea. *Green circles* show the location of the outer pillar feet, *black circles* show the center location of the spiral ganglion, *yellow circles* indicate the sites of stimulation as determined from the inferior colliculus recordings, *red circles* indicate the site at the modiolus marked by the CO_2 laser as the center of the beam path, and *blue circles* show the location of the cochleostomy

2.5 The Target Structure

Reconstructions of the optical path in the cochlea demonstrated that spiral ganglion cells and the auditory nerve fibers in the center of the modiolus could be irradiated, depending on the orientation of the optical fiber. However, responses to optical stimulation could only be recorded when spiral ganglion cells were in the beam path (Moreno et al. in press). The latter finding suggested that stimulation occurs at the

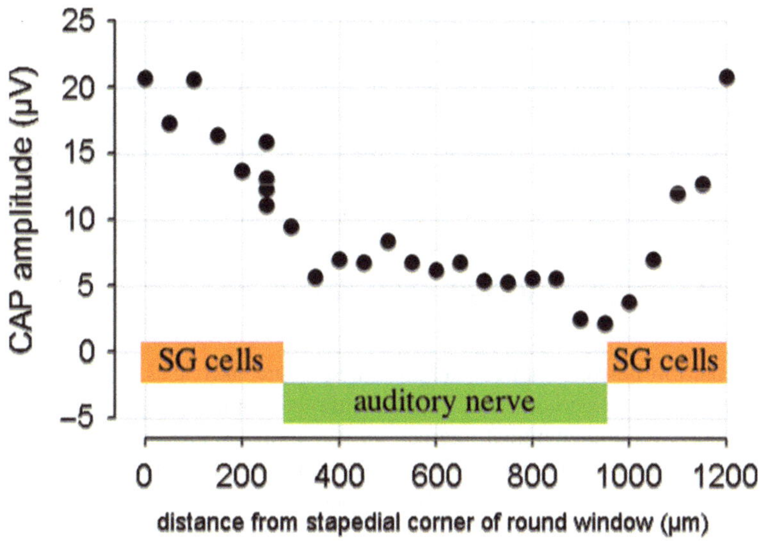

Fig. 6.6 Compound action potential amplitudes recorded from the gerbil round window. The optical fiber was moved for each measurement across the opening of the round window. The starting location was over the spiral ganglion as well as was the end location. The track of the optical fiber was across the modiolus. Largest CAP amplitudes were recorded while the optical fiber was oriented towards the spiral ganglion cells

spiral ganglion cell bodies. This view is supported by results obtained from recording compound action potentials; the responses were largest when the optical fiber was oriented towards the spiral ganglion cells but decreased when the optical fiber was oriented towards the nerve fibers in the center of the modiolus (Fig. 6.6).

2.6 Stimulation of the Central Auditory System

Lee and coworkers (2009) have demonstrated that neurons of the central auditory system can be stimulated using similar techniques. In acute animal experiments using rats, a 400-µm diameter optical fiber was placed on the surface of the cochlear nucleus and was used to irradiate the tissue with an Aculight diode laser ($\lambda = 1849$–1865 nm; $\tau_p = 5$ µs–10 ms; f = 2–1000 Hz). An evoked response was recorded from vertex to ear-level electrodes. This optically evoked auditory brainstem response (oABR) was a multi-peaked waveform reminiscent of an ABR evoked by acoustic stimulation. The oABR waveform peaks had latencies between 3 and 8 ms, longer than ABRs evoked by direct electrical stimulation of the same region. The discrepancy between oABR and eABR latencies may be explained by different mechanisms of neural depolarization by optical and electrical stimulation. Reproducible oABRs were detected at radiant exposure thresholds as low as 169 mJ/cm^2, with 50 μs pulse duration, and 5 Hz repetition rate. The oABR was stable during continuous optical stimulation for 30 min. In control experiments, after the optical path was blocked and after the rat was euthanized, the oABR disappeared. No thermal tissue damage was found on histological examination when pulse durations were less than 1 ms and radiant exposure levels were less than 2.05 J/cm^2.

3 The Mechanism of Optical Stimulation

The interaction between optical radiation and tissue may lead to photochemical, photomechanical, or photothermal effects. Although some regimes of optical stimulation occur via photochemical interaction, it is unlikely that pulsed, mid-infrared lasers, such as the FEL, the Ho:YAG, or the Aculight diode lasers, could be involved in a photochemical reaction. The individual photon energies emitted by these lasers are significantly lower than the energies required to move an electron to an excited state, as is needed for a photochemical reaction. The energy in individual photons from the Ho:YAG is ~0.58 eV. The value corresponds to a ~52 kJ/mol ion bond energy. Typical ion bond energies are in the range of 100 to 1000 kJ/mol. Furthermore, data from Wells et al. (2007b) did not identify a particular wavelength between 2000 and 10,000 nm at which the laser stimulation was significantly enhanced, which would have suggested a photochemical reaction (Wells et al. 2005a, b).

The energy required for a primary photomechanical mechanism is also much higher than that used by Wells et al. and Izzo et al., so it is unlikely that this would be the primary mechanism of pulsed, mid-infrared laser stimulation of nerves (Izzo et al. 2006, 2007; Wells et al. 2007b, c). Nevertheless, the possibility that a photomechanical mechanism is responsible for optical neural stimulation has been investigated. For photomechanical effects, both laser-induced stress waves and volumetric thermal expansion were considered as a mechanism for the stimulation. Wells et al. excluded a photomechanical effect from stress wave generation because the stimulation threshold for an action potential was independent of the laser pulse duration (Wells et al. 2007a). Moreover, the parameter space of optical stimulation is outside the region of stress confinement. Stress confinement occurs when the optical energy accumulates in the tissue before a laser-induced stress wave can propagate out of the irradiated area, leading to large stress waves (Jacques 1992). Theoretically, for stress confinement to occur when irradiating with these wavelengths, the pulse lengths should be shorter than 500 ns, which is orders of magnitude shorter than the pulses used for optical stimulation. Wenzel et al. used a Q-switched, frequency doubled Nd:YAG laser (λ = 532 nm; τ_p = 10 ns; f = 2–10 Hz) with laser parameters for stress confinement to generate pressure waves and directly stimulate hair cells in the cochlea (Wenzel et al. 2009). In contrast to generating a pressure wave, infrared neural stimulation is used to generate neural responses through direct interaction between the radiation and the nerve.

Wells et al. (2007a) concluded that transient optical neural stimulation occurred via a thermal mechanism and measured the temperature rise associated with stimulation of the sciatic nerve. The radiation is likely absorbed by water in the tissue and, subsequently, most of the radiation energy is transferred to heat. Wells et al. also concluded that thermal confinement is necessary to achieve optical stimulation. Thermal confinement exists when the thermalized optical energy delivered by a single pulse accumulates in the irradiated tissue before any of the heat can dissipate through conduction or convection (Jacques 1992). For thermal confinement to be achieved at these wavelengths, the pulse duration of the stimuli are typically longer than 500 ns and shorter than 200 ms.

It is currently unknown how the temperature rise induces neural depolarization. Potential mechanistic candidates include thermal activation of a particular ion channel, thermally induced biophysical changes at the membrane (increase ion channel conductance, decrease in Nernst equilibrium potential), or through general expansion of the neural membrane that facilitates the flux of ions. Wells et al. (2007b) measured an expansion of the sciatic nerve membrane of approximately 300 nm during laser stimulation. Although generation of a pressure wave through volumetric thermal expansion is possible in thermal confinement, there is no evidence yet that volumetric thermal expansion is the mechanism by which optical stimulation occurs in peripheral or sensory nerves (Wells et al. 2007b; Richter et al. 2008).

An alternative mechanism for INS could be the direct activation of heat-sensitive ion channels, which are present in nerves. The channels are termed the transient receptor potential (vanilloid) or TRPV channels (Caterina et al. 1997; Harteneck et al. 2000; Güler et al. 2002; Montell 2005). TRPV1 is most well-known for being activated by the chemical capsaicin, the main ingredient in hot chili peppers that produces a burning sensation. TRPV1 channels are also opened in the presence of other vanilloid compounds, acid ($pH \leq 5.9$) and heat ($\geq 43°C$), making it a key channel in peripheral nociception. TRPV channels are found in small neurons, including the sciatic nerve, the dorsal root and trigeminal ganglia of rats (Caterina et al. 1997), as well as cochlear structures of both the rat and guinea pig (Balaban et al. 2003; Zheng et al. 2003; Takumida et al. 2005). TRPV ion channels are expressed centrally at the cell body and in the neuron's periphery.

Immunostaining data show that the TRPV1 channel is present in gerbil and mouse spiral ganglion cells (Richter and Matic, in preparation). Moreover, preliminary *in vivo* experiments in gerbils showed that the infusion of the TRPV1 antagonist capsazepine in the cochlea reversibly reduces the amplitude of optically evoked compound action potentials. Contributions of the TRPV1 channel to the generation of action potentials were examined in mice lacking the gene for the TRPV1 channel. While most of the TRPV1-knockout mice showed no ABR response to optical stimulation, all control animals did (Suh et al. 2007, 2009). In addition to the elevation of the ABR thresholds for optical stimulation, the thresholds for acoustical stimulation were elevated as well. TRPV channels apparently are important for normal auditory function. Although there is evidence that TRPV channels are involved in optical stimulation, no definitive mechanism has been identified at present.

4 Selectivity of Neural Stimulation

Optical stimulation of peripheral nerves was selective. A pulsed, near-infrared laser could be used to map the sciatic or the facial nerve (Teudt et al. 2007; Wells et al. 2007a). Experiments in pigmented guinea pigs were used to determine whether optical stimulation of the cochlea was as selective as stimulation with acoustic tone pips. The optical radiation was delivered via a 200-μm diameter optical fiber positioned in front of the spiral ganglion neurons of the basal cochlear turn. The spread of activation of cochlear optical stimulation was estimated from neural responses in the central nucleus of the inferior colliculus (ICC). ICC responses were recorded using a multi-channel penetrating electrode array in acutely deafened guinea pigs. For the measurement, either the frequency of the tone pip or the location of the optical fiber was kept constant while the intensity of the stimulus was varied. For acoustical stimulation, the sound level was varied and for optical stimulation the radiant energy. The neural responses in the ICC were compared at the different contacts along the array. Activity profiles were constructed. Most commonly, tuned activation profiles, or spatial tuning curves (STCs), occurred with a single minimum at one of the array's contacts (Fig. 6.7). However, in a limited number of cases two or more separate peaks were observed. Closer inspection of the response properties on electrodes between the peaks suggests that these STCs should be treated as one broad STC. The differences between activation profiles likely depend upon the orientation of the optical fiber in the cochlea and the neural activities at the electrode contacts (Moreno et al. in press; Richter et al. in review).

To quantify and compare the response areas obtained with optical and acoustical stimulation, the single-peaked response profiles were selected. The sharpness of the response was measured using a two-step procedure: determine the radiation energy for $d'=2$ at the most sensitive electrode contact and then measure the width of the $d'=1$ iso-contour for this energy level (Fig. 6.7). The widths were normalized to the length of the multi-channel electrode array. Optically evoked STCs were on average 357 ± 206 μm ($d'=2$, N = 28) wide and acoustically evoked STCs were 383 μm ± 131 μm (N = 61) wide. The results indicated that the spread of activation as a result of optical stimuli is comparable to that produced with acoustic pips.

STC width measurements for electrical stimulation have been published for ball and for banded electrodes. Ball electrodes were placed visually, and the resulting STCs widths for bipolar stimulation were 429 μm when placed radially and were about 700 μm when placed longitudinally. STCs widths of 548 and 684 μm were reported for banded tripolar and banded bipolar electrodes, respectively (Snyder et al. 2004). For monopolar stimulation the widths of the STCs were larger than 1500 μm for both the banded and the ball electrode. The corresponding pure tone STCs were 382 μm wide (Snyder et al. 2004).

For optical stimulation, the energy range over which neural activity increased above spontaneous rates was up to factor of 10. However, the dynamic range in this study is biased by the limitations of the stimulation source, which had a maximum output of 130 μJ. In addition to the total range over which a neuron increases its rate of action potentials, it would be important to know whether optical stimulation

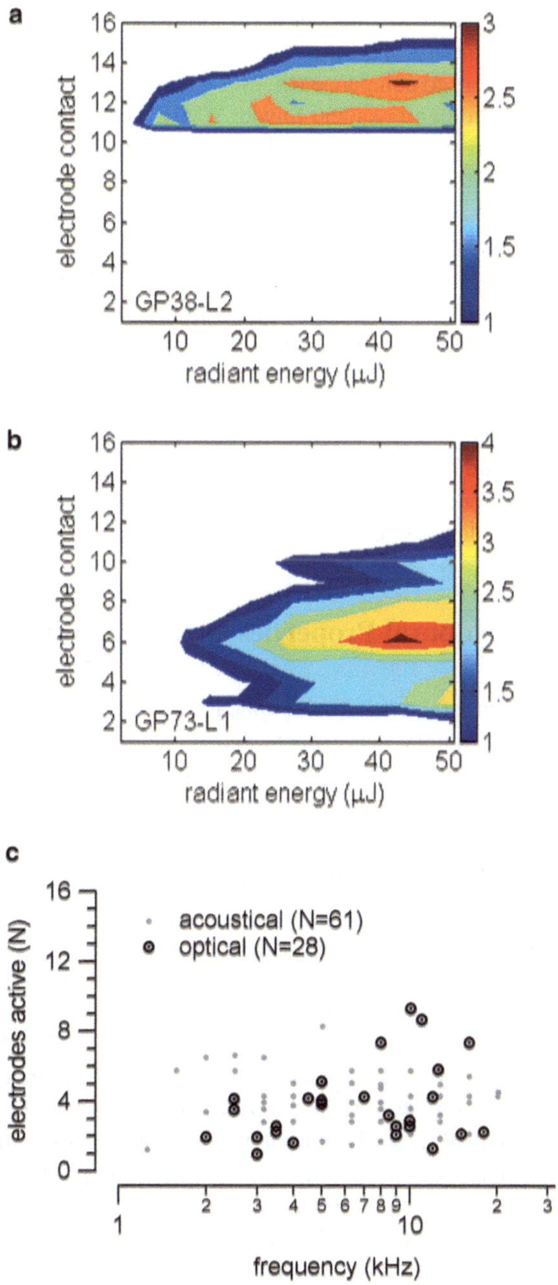

Fig. 6.7 Shown are two examples for spatial tuning curves (STCs) obtained from neural activity recorded from the inferior colliculus (**a, b**). For each figure the traces are recorded for different radiant energies with increasing energy from left to right. The width of the STCs were measured at an energy level which results in a d'=2 at the most sensitive electrode contact, (**c**) shows the widths of the STCs obtained with optical and with acoustical stimulation expressed in number of active electrodes

provides a greater number of discriminable level steps compared to cochlear electrical stimulation. This may be assessed from the spike rate, by applying signal detection theory to compare each level with the next discriminable level step inducing a d' of 1 (Middlebrooks and Snyder 2007). The number of stimulus levels between the cumulative d' = 1 and d' = 3 contours were measured at the electrode contact in the ICC that showed the lowest threshold. The two d' units were than divided by the measured difference in stimulus levels and the "discrimination slope" resulted. For electrical stimulation this slope was expressed in d' units per decibel. For electrical stimulation with an intraneural electrode the slope was 0.73d'; for a monopolar and bipolar electrode configuration the values were 1.98 and 1.94d', respectively. A similar calculation can be done for optical stimulation. When the slope is expressed in units of d' per radiant energy (μJ), it is on average 0.08d' (the average energy difference is 25.3 ± 21 μJ, N = 10). Expressed in dB ($dB = (10*\log(energy_{d^1=3}/energy_{d^1=1}))$) the slope is 0.42d' (the average dB value is 4.8 ± 2.3, N = 10). It is not clear whether electrical and optical values can be compared directly because for optical stimulation the radiant energy is reported as a measure, while for electrical stimulation the current amplitude is used.

5 Temporal Properties of Optical Stimulation

The temporal resolution of optical stimulation was determined with recordings from single auditory nerve fibers and recordings from the ICC using a multi-channel electrode array, both while stimulating the basal turn of the cochlea.

5.1 Low Spontaneously Active Single Nerve Fiber Recordings

With increasing radiant energy, the rate of action potentials increased monotonically to a certain value. However, the rate of evoked action potentials was always less than the stimulation rate. The ratio of the stimulation rate and the spike rate increased if the stimulation rate was above 100 Hz. For sustained stimulation, the optically evoked response rates were slower than with acoustic stimulation and were close to 400 APs/s for only two neurons, while the average maximally driven rate with optical stimulation was 97 ± 52.5 APs/s (Littlefield et al. 2010).

Interspike time histograms (INTHs) were plotted and were used to calculate the entrainment, which describes the ability of a stimulus to evoke a neural response, i.e., an action potential. The entrainment index is calculated by counting all action potentials that occur in an interval equal to the time between successive stimuli (1/stimulation rate), with the interval centered around the time equivalent to the stimulus period, and by dividing this number by the number of spikes counted in the entire interval histogram. Inspection of the histograms shows that most of the action potentials occur directly after the laser pulse if the stimulus rate is below the maximum possible optically evoked response rate of the neuron. However, at higher stimulation rates, not every laser pulse evoked an action potential. For the selected neurons, the entrainment index was about 0.5 for stimulus repetition rates up to 50 Hz. The index decreased with increasing stimulus repetition rates and was below 0.2 for stimulus rates >200 Hz.

Firing efficiency (FE) is the probability of a neuron generating an action potential after a stimulus and was as high as 100% for some neurons at stimulus rates less than 100 Hz, but it decreased drastically at higher stimulus rates. When plotted versus the power, there was a trend for FE to increase with optical power. Neurons that were stimulated at different pulse rates were combined. It is apparent that the pulse rate affects the absolute value of the firing efficiency, and the data variation may be from these pulse rate differences.

Poststimulus time histograms (PSTHs) were constructed for different stimulation rates. At a 200-Hz optical stimulation rate, the majority of the action potentials occurred in the first time interval between pulses, but this became more random at higher stimulation rates. Neurons were unable to exactly follow the stimulus. To quantify the responses further, the PSTHs were used to calculate the delay time and delay jitter. If the pulse repetition rate was below 50 Hz, the average time between the stimulus and the first action potential (latency) was about 4 ms. The latency was influenced by the pulse rate. At higher rates, the stimulation time interval was shorter than the recovery time of the neuron.

If the time between two subsequent optical pulses is considered the time for one stimulus cycle, the time of occurrence for *n* action potentials can be quantified by the vector strength (VS). The VS is about 1 for stimulus frequencies <50 Hz, but decreases drastically for stimulus frequencies >100 Hz. Again, neurons stimulated with different pulse rates were combined when VS was plotted versus power. It is apparent that the pulse rate affects the absolute value of the vector strength. For electrical stimulation, it has been shown in a neuron model that high-rate electrical stimulation rates (about 5000 pps) produce stochastic responses that are similar to the activity patterns obtained from spontaneously active neurons (Rubinstein et al. 1999). It would be important to know whether a similar behavior occurs for optical stimulation at pulse repetition rates above 100 pps. Ongoing experiments are designed to determine whether the stochastic properties at higher rates of optical stimulation are similar to those obtained from spontaneously active neurons. For the single fiber study, one has to keep in mind that the neurons are all low spontaneously active neurons.

5.2 Inferior Colliculus Responses

Neural responses from the ICC were analyzed for number of maxima in the PSTH, the delay time and instantaneous response rate. Post stimulus time histograms show several maxima (Fig. 6.8). One to four maxima following the stimulus could be identified. The time between the stimulus and the response became shorter with increasing optical energy. At stimulation threshold, the first maximum occurred at 5.5 ms, the second maximum at 7.6 ms, the third maximum at 9.3 ms, and the

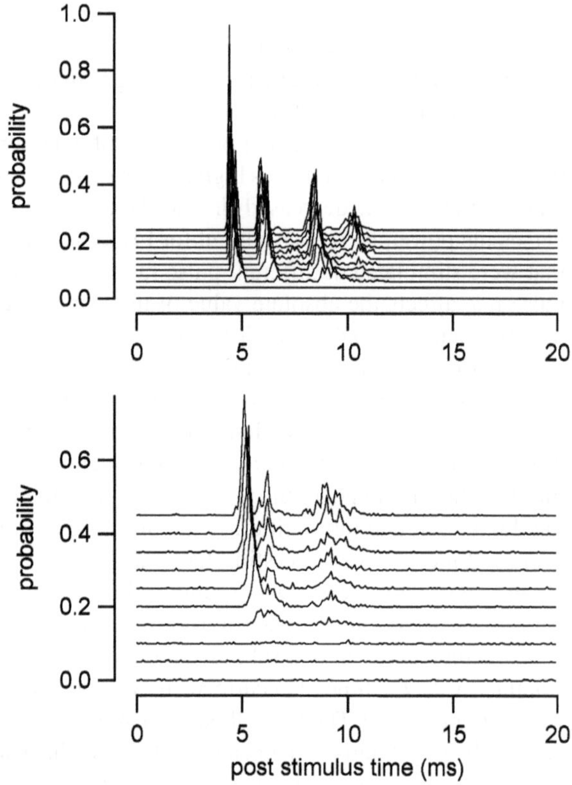

Fig. 6.8 Two examples for post stimulus histograms obtained from a guinea pig during stimulation with optical radiation ($\lambda = 1860$ nm, pulse duration = 100 μs, pulse repetition rate = 10 Hz). A typical response pattern can be seen with up to four maxima

fourth maximum at 11.7 ms. The delay time for the first maximum is similar at the different contacts of the recording electrode (different depths into the ICC). At low stimulus levels, only a single maximum can be detected. With increasing stimulus level, multiple maxima appear and the delay time for the first maximum decreases. Note that not all the recordings revealed a maximum at about 5 ms and fewer recordings had maxima at about 7, 9, and 11 ms. It is not clear which parameters determine the number of maxima in the recordings. In comparison, ICC recordings during pure tone stimulation before deafening the animal only showed a single peak in the PSTH.

All recordings from the ICC with clearly identifiable single neurons revealed spontaneous neural activities. With increasing energy, evoked neural responses could be recorded. The rate did not increase for all clusters monotonically with increasing energy. After the evoked activity reached a maximum, most often the activity decreased with further increases in energy. A similar finding of the increase in neural activity of the inferior colliculus has been reported for tonal stimulation (e.g., Ehret and Merzenich 1988).

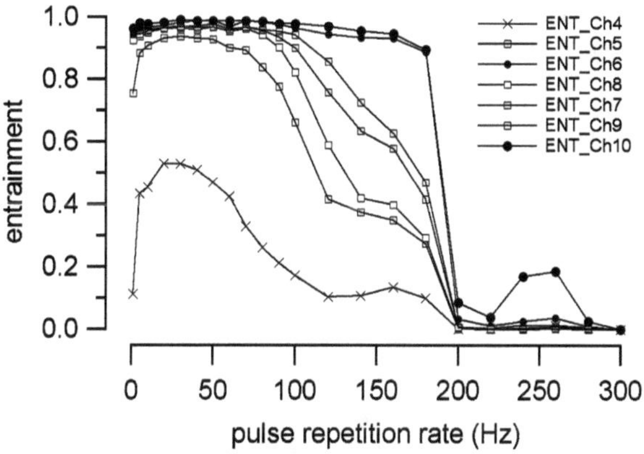

Fig. 6.9 Entrainment was calculated from the post stimulus histogram. The number of events in the first maximum of the histogram was divided by the number of stimulus presentations. A neural response at the time of the first maximum following each stimulus would correspond to and entrainment of 1. The different traces reflect the responses obtained at different contacts along the multi-channel array inserted into the inferior colliculus. The stimulation energy was minimum to evoke a response at array contact 10. At this contact the entrainment was best. If the location was moved to other contacts the entrainment decreased drastically. Note that for repetition rates at 1 and 5 Hz the entrainment was less than for faster repetition rates

Pulse durations were varied while the pulse rate (10 Hz) was kept constant. The responses for a selected electrode were examined. The response pattern was similar to the response pattern observed previously: a single maximum occurred about 5.5 ms after the optical pulse, a second maximum was at about 6.7 ms, and the third maximum at about 9 ms after the optical pulse. Changing the pulse duration did not change the response pattern at the ICC electrode.

Post stimulus histograms were obtained from 5 animals while irradiating at levels well above the stimulation threshold (at pulse duration of 100 μs). The pulse rate was changed from 1 to 300 Hz. The delay time for the first PSTH maximum at about 5 ms increases slightly with increasing pulse rate. Moreover, the width of the first maximum remains the same for pulse rates below 300 Hz. However, when increasing the optical stimulation rate, the second maximum disappears and the delay time for the third maximum increases. At 300-Hz pulse rate, the first maximum at 5 ms is broad and subsequent maxima are missing.

Figure 6.9 shows the entrainment for different contacts along an electrode array in the ICC. The entrainment index decreases with increasing pulse rate. The entrainment differs at the various contacts. While for some traces the entrainment index is close to 1 up to 200 Hz pulse rate, for other traces the entrainment index is barely 0.5, even at low stimulation rates. Interestingly, the trace with the entrainment index close to 1 for pulse rates up to 200 Hz corresponds to the contact at which the ICC STC shows a minimum. It indicates that the most robust responses are obtained for neurons in the center of the optical path.

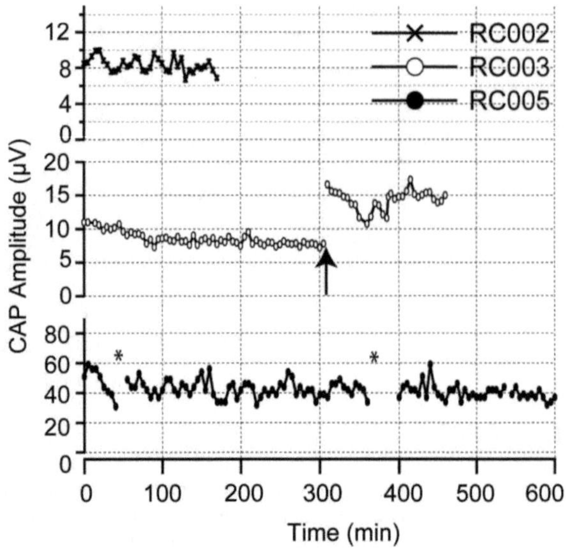

Fig. 6.10 Short-term safety was evaluated by continuous stimulation ($\lambda = 1860$ nm, pulse duration = 100 μs, pulse repetition rate = 200 Hz, radiation power = 250 mW/pulse). The compound action potential (CAP) amplitudes were recorded every 5 min. The CAP amplitudes remained stable over a 10-h duration

6 Chronic Experiments

At present, chronic experiments are underway but have not been completed. During the last 5 years, the size of the stimulation source has been reduced from the size of an entire building, as with the Free Electron Laser, to the experimental Ho:YAG laser, the table top Aculight diode laser, and the cell-phone-sized stimulator currently in use for chronic implantation in cats. Acute experiments have been conducted during which the cats have been continually stimulated for up to 10 h. As shown in Fig. 6.10, where CAP amplitude remained stable over 10 h of continual optical stimulation, the stimulation did not compromise cochlear function (Rajguru et al. 2010). Ongoing experiments will be conducted to verify that stimulation over extended periods of time is safe and effective.

7 Summary

In 2002, the Free Electron Laser was used to conduct initial experiments and to explore whether optical stimulation of a peripheral nerve, the sciatic nerve, is feasible. Experiments have shown that stimulation of the nerve is possible without acutely damaging the nerve. Subsequently, detailed studies were conducted with the Ho:YAG laser. It has been shown that optical stimulation is extremely selective. The concept was transferred to the cochlea in 2004, when a Ho:YAG laser was used to conduct the first experiments. With the development of a tabletop unit by Aculight, the optical parameters for stimulation have been optimized. With the optimized parameter set, the stimulation source has been further miniaturized and an implantable unit for chronic stimulation in the cat is available. Necessary safety studies are underway that will lead to the development of a unit that can be used in clinical trials.

However, there are some limitations of optical stimulation and some challenges that have yet to be addressed. While the radiation energy can be delivered through air in a non-contact mode, water and tissue along the optical path will reduce the incident energy. For optical stimulation, typical wavelengths were between 1840 and 2100 nm. Assuming that water is the dominant absorber, the radiation wavelengths correspond to penetration depths of 100 to 1500 μm. In other words, the radiation energy at the tip of the optical fiber or at the radiation source decreases with each penetration length by about 10 dB. Therefore, usable paradigms for optical stimulation have to consider the distance and the materials of the absorbers between the radiation source and the target neuron. The presence of highly scattering tissue, such as bone, in the optical path will increase spot size and decrease selectivity. In addition, because optical stimulation targets a restricted volume of tissue and the optical radiation spreads very little, the placement of the stimulating source needs to be accurate relative to the neurons to evoke a response. While this feature is desirable for some applications, it will be a detriment to others. In addition, the extremely selective stimulation can make it difficult to detect the neural responses.

Acknowledgements This project has been funded with federal funds from the National Institute on Deafness and Other Communication Disorders, National Institutes of Health, Department of Health and Human Services, under Contract No. HHSN260-2006-00006-C/NIH No. N01-DC-6-0006, NIH grant 1R41DC008515-01, NIH grant 1R41DC008515-02, NIH grant F31 DC008246-01, E.R. Capita Foundation.

References

Balaban, C. D., Zhou, J., & Li, H. (2003). Type 1 vanilloid receptor expression by mammalian inner ear ganglion cells. *Hearing Research*, 175, 165–170.

Black, R. C., Clark, G. M., & Patrick, J. F. (1981). Current distribution measurements within the human cochlea. *IEEE Transactions on Biomedical Engineering*, 28(10), 721–725.

Boyden, E. S., Zhang, F., Bamberg, E., Nagel, G., & Deisseroth, K. (2005). Millisecond-timescale, genetically targeted optical control of neural activity. *Nature Neuroscience*, 8(9), 1263–1268.

Caterina, M. J., Schumacher, M. A., Tominaga, M., Rosen, T. A., Levine, J. D., & Julius, D. (1997). The capsaicin receptor: a heat-activated ion channel in the pain pathway. *Nature*, 389, 816–824.

Collins, L. M., Zwolan, T. A., & Wakefield, G. H. (1997). Comparison of electrode discrimination, pitch ranking, and pitch scaling data in postlingually deafened adult cochlear implant subjects. *Journal of the Acoustical Society of America*, 101(1), 440–455.

Dorman, M. F., Dankowski, K., McCandless, G., & Smith, L. M. (1989). Consonant recognition as a function of the number of channels of stimulation by patients who use the Symbion cochlear implant. *Ear and Hearing*, 10, 288–291.

Dorman, M. F., Loizou, P. C., & Rainey, D. (1997). Speech intelligibility as a function of the number of channels of stimulation for signal processors using sine-wave and noise-band outputs. *Journal of the Acoustical Society of America*, 102, 2403–2411.

Dorman, M. F., Loizou, P. C., Fitzke, J., & Tu, Z. (1998). The recognition of sentences in noise by normal-hearing listeners using stimulations of cochlear-implant signal processors with 6–20 channels. *Journal of the Acoustical Society of America*, 104, 3583–3585.

Duke, A. R., Cayce, J. M., Malphrus, J. D., Konrad, P., Mahadevan-Jansen, A., & Jansen, E. D. (2009). Combined optical and electrical stimulation of neural tissue in vivo. *Journal of Biomedical Optics Letters*, 14(6), 060501–060501 - 060501–060503.

Eddington, D. K., Rabinowitz, W. R., Tierney, J., Noel, V., & Whearty, M. (1997). Speech processors for auditory prostheses. *8th Quaterly progress report, NIH Contract N01-DC-6-2100*.

Ehret, G., & Merzenich, M. M. (1988). Neuronal discharge rate is unsuitable for encoding sound intensity at the inferior-colliculus level. *Hearing Research*, 35(1), 1–7.

Fishman, K. E., Shannon, R. V., & Slattery, W. H. (1997). Speech recognition as a function of the number of electrodes used in the SPEAK cochlear implant speech processor. *Journal of Speech, Language, and Hearing Research*, 40(5), 1201–1215.

Fried, N. M., Lagoda, G. A., Scott, N. J., Su, L. M., & Burnett, A. L. (2008). Noncontact stimulation of the cavernous nerves in the rat prostate using a tunable-wavelength thulium fiber laser. *Journal of Endourology*, 22(3), 409–413.

Friesen, L. M., Shannon, R. V., Baskent, D., & Wang, X. (2001). Speech recognition in noise as a function of the number of spectral channels: comparison of acoustic hearing and cochlear implants. *Journal of the Acoustical Society of America*, 110(2), 1150–1163.

Geier, L. V., & Norton, S. (1992). The effect of limiting the number of Nucleus 22 cochlear implant electrodes programmed on speech perception. *Ear and Hearing*, 13, 340–348.

Greenwood, D. D. (1990). A cochlear frequency-position function for several species–29 years later. *Journal of the Acoustical Society of America*, 87(6), 2592–2605.

Güler, A. D., Lee, H., Iida, T., Shimizu, I., Tominaga, M., & Caterina, M. (2002). Heat-evoked activation of the ion channel, TRPV4. *Journal of Neuroscience*, 22(15), 6408–6414.

Hale, G. M., & Querry, M. R. (1973). Optical constants of water in the 200 nm to 200 μm region. *Applied Optics*, 12, 555–563.

Harteneck, C., Plant, T. D., & Schulz, G. u. n. (2000). From worm to man: three subfamilies of TRP channels. *Trends in Neuroscience*, 23, 159–166.

Hochmair, E. S., & Hochmair-Desoyer, I. J. (1983). Percepts elicited by different speech-coding strategies. *Annals of the New York Academy of Sciences*, 405, 268–279.

Hochmair-Desoyer, I. J., Hochmair, E. S., Burian, K., & Stiglbrunner, H. K. (1983). Percepts from the Vienna cochlear prosthesis. *Annals of the New York Academy of Sciences*, 405, 292–306.

Holmes, A., Kemker, F. J., & Merwin, G. (1987). The effects of varying the number of cohlear implant electrodes on speech perception. *American Journal of Otology*, 8, 240–246.

Izzo, A. D., Richter, C. P., Jansen, E. D., & Walsh Jr., J. T. (2006). Laser stimulation of the auditory nerve. *Lasers in Surgery and Medicine*, 38(8), 745–753.

Izzo, A. D., Walsh, J. T., Jr., Jansen, E. D., Bendett, M., Webb, J., Ralph, H., & Richter, C. P. (2007). Optical parameter variability in laser nerve stimulation: a study of pulse duration, repetition rate, and wavelength. *IEEE Transactions on Biomedical Engineering*, 54(6, Pt. 1), 1108–1114.

Jacques, S. L. (1992). Laser-tissue interactions. Photochemical, photothermal, and photomechanical. *Surgical Clinics of North America*, 72(3), 531–558.

Kilney, P., Zimmerman-Phillips, S., Zwolan, T., & Kemink, J. (1992). Effects of channel number and place of stimulation on performance with the cochlear corporation multichannel implant. *American Journal of Otology*, 13, 117–123.

Kral, A., Hartmann, R., Mortazavi, D., & Klinke, R. (1998). Spatial resolution of cochlear implants: the electrical field and excitation of auditory afferents. *Hearing Research*, 121(1–2), 11–28.

Kramer, R. H., Fortin, D. L., & Trauner, D. (2009). New photochemical tools for controlling neuronal activity. *Current Opinion in Neurobiology*, 19(5), 544–552.

Lawson, D. (1993). New processing strategies for multichannel cochlear prosthesis. *Progress in Brain Research*, 97, 331–321.

Lawson, D. (1996). Speech processors of auditory prostheses. *Third quarterly progress report, NIH Contract No1-DC-5-2103*.

Lee, D. J., Hancock, K. E., Mukerji, S., & Brown, M. C. (2009). Optical stimulation of the central auditory system. *Abstracts of the Association for Research in Otolaryngology*, 32, 314.

Littlefield, P. D., Vujanovic, I., Mundi, J., Matic, A. I., & Richter, C.-P. (2010). Laser stimulation of single auditory nerve fibers. *The Laryngoscope* (in press).

McKay, C. M., McDermott, H. J., & Clark, G. M. (1994). The beneficial use of channel interactions for improvement of speech perception for multichannel cochlear implants. *Australian Journal of Audiology*, 15(Suppl. 2), 20–21.

Middlebrooks, J. C., & Snyder, R. L. (2007). Auditory prosthesis with a penetrating nerve array. *Journal of the Association for Research in Otolaryngology*, 8(2), 258–279.

Montell, C. (2005). The TRP superfamily of cation channels. *Science Signaling: The Signal Transduction Knowledge Environment*, 272, re3.

Moreno, L. E., Rajguru S. R., Matic A. I., Yerram N., Robinson A., Hwang M., Stock SR., & Richter C. P. Infrared neural stimulation: beam path in the guinea pig cochlea, Hear Res (in press).

Müller, M. (1996). Frequenz- und Intensitätsanalyse im Innenohr der Säuger (Unpublished *Habilitationsschrift*). Johann Wolfgang Goethe-Universität, Frankfurt/Main, Germany.

O'Leary, S. J., Black, R. C., & Clark, G. M. (1985). Current distributions in the cat cochlea: a modelling and electrophysiological study. *Hearing Research*, 18(3), 273–281.

O'Leary, S. J., Richardson, R. R., & McDermott, H. J. (2009). Principles of design and biological approaches for improving the selectivity of cochlear implant electrodes. *Journal of Neural Engineering*, 6(5), 055002.

Rajguru S. M., Matic A. I., Robinson A. M., Fishman A. J., Moreno L. E., Bradley A., Vujanovic I., Breen J., Wells J. D., Bendett M., Richter C. P. (2010). Optical cochlear implants: Evaluation of surgical approach and laser parameters in cats, Hear Res, 269 (1-2): 102–111.

Richter, C. P., Bayon, R., Izzo, A. D., Otting, M., Suh, E., Goyal, S., Hotaling, J., & Walsh, J. T., Jr. (2008). Optical stimulation of auditory neurons: effects of acute and chronic deafening. *Hearing Research*, 242(1–2), 42–51.

Richter, C-P., Rajguru, S. M., Matic, A. I., Moreno E. L., Fishman, A. J., Robinson, A. M., Suh, E., Walsh, Jr. JT. Spread of cochlear excitation during stimulation with pulsed infrared radiation: inferior colliculus measurements, (in review).

Robertson, D. (1984). Horseradish peroxidase injection of physiologically characterized afferent and efferent neurones in the guinea pig spiral ganglion. *Hearing Research*, 15(2), 113–121.

Rubinstein, J. T., Wilson, B. S., Finley, C. C., & Abbas, P. J. (1999). Pseudospontaneous activity: stochastic independence of auditory nerve fibers with electrical stimulation. *Hearing Research*, 127(1–2), 108–118.

Shannon, R. V., Fu, Q. J., & Galvin, J., 3 rd. (2004). The number of spectral channels required for speech recognition depends on the difficulty of the listening situation. *Acta Oto-Laryngologica Supplement*, 552, 50–54.

Snyder, R. L., Bierer, J. A., & Middlebrooks, J. C. (2004). Topographic spread of inferior colliculus activation in response to acoustic and intracochlear electric stimulation. *Journal of the Association for Research in Otolaryngology*, 5(3), 305–322.

Snyder, R. L., Middlebrooks, J. C., & Bonham, B. H. (2008). Cochlear implant electrode configuration effects on activation threshold and tonotopic selectivity. *Hearing Research*, 235(1–2), 23–38.

Suh, E., Izzo, A. D., Walsh Jr., J. T., & Richter, C.-P. (2007). The role of Transient Receptor Potential channels in neural activation. *Abstracts of the Association for Research in Otolaryngology*, 30, 109.

Suh, E., Matic, A. I., Otting, M., Walsh Jr., J. T., & Richter, C.-P. (2009). Optical stimulation in mice which lack the TRPV1 channel. *Proceedings of SPIE, Volume 7180*, 71801–71805.

Takumida, M., Kubo, N., Ohtani, M., Suzuka, Y., & Anniko, M. (2005). Transient receptor potential channels in the inner ear: Presence of transient receptor potential channel subfamily 1 and 4 in the guinea pig inner ear. *Acta Oto-Laryngologica*, 125, 929–934.

Teudt, I. U., Nevel, A., Izzo, A. D., Walsh, J. J. T., & Richter, C.-P. (2007). Optical stimulation of the facial nerve–a new monitoring technique? *The Laryngoscope*, 117, 1641–1647.

Tsuji, J., & Liberman, M. C. (1997). Intracellular labeling of auditory nerve fibers in guinea pig: central and peripheral projections. *Journal of Comparative Neurology*, 381(2), 188–202.

Turner, C. W., Souza, P. E., & Forget, L. N. (1995). Use of temporal envelop cues in speech recognition by normal and hearing-impaired listeners. *Journal of the Acoustical Society of America*, 97, 2568–2576.

van den Honert, C., & Stypulkowski, P. H. (1987). Single fiber mapping of spatial excitation patterns in the electrically stimulated auditory nerve. *Hearing Research*, 29(2–3), 195–206.

Vanpoucke, F., Zarowski, A., Casselman, J., Frijns, J., & Peeters, S. (2004). The facial nerve canal: an important cochlear conduction path revealed by Clarion electrical field imaging. *Otology & Neurotology*, 25(3), 282–289.

Wells, J., Kao, C., Jansen, E. D., Konrad, P., & Mahadevan-Jansen, A. (2005a). Application of infrared light for in vivo neural stimulation. *Journal of Biomedical Optics*, 10, 064003.

Wells, J. D., Kao, C., Mariappan, K., Albea, J., Jansen, E. D., Konrad, P., & Mahadevan-Jansen, A. (2005b). Optical stimulation of neural tissue in vivo. *Optics Letters*, 30(5), 504–506.

Wells, J., Konrad, P., Kao, C., Jansen, E. D., & Mahadevan-Jansen, A. (2007a). Pulsed laser versus electrical energy for peripheral nerve stimulation. *Journal of Neuroscience Methods*, 163(2), 326–337.

Wells, J., Kao, C., Konrad, P., Milner, T., Kim, J., Mahadevan-Jansen, A., & Jansen, E. D. (2007b). Biophysical mechanisms of transient optical stimulation of peripheral nerve. *Biophysical Journal*, 93(7), 2567–2580.

Wells, J. D., Thomsen, S., Whitaker, P., Jansen, E. D., Kao, C. C., Konrad, P. E., & Mahadevan-Jansen, A. (2007c). Optically mediated nerve stimulation: identification of injury thresholds. *Lasers in Surgery and Medicine*, 39(6), 513–526.

Wenzel, G. I., Balster, S., Zhang, K., Lim, H. H., Reich, U., Massow, O., Lubatschowski, H., Ertmer, W., Lenarz, T., & Reuter, G. (2009). Green laser light activates the inner ear. *Journal of Biomedical Optics*, 14(4), 044007.

Zhang, F., Wang, L. P., Boyden, E. S., & Deisseroth, K. (2006). Channelrhodopsin-2 and optical control of excitable cells. *Nature Methods*, 3(10), 785–792.

Zheng, J., Dai, C., Steyger, P. S., Kim, Y., Vass, Z., Ren, T., & Nuttall, A. L. (2003). Vanilloid receptors in hearing: altered cochlear sensitivity by vanilloids and expression of TRPV1 in the organ of corti. *Journal of Neurophysiology*, 90(1), 444–455.

Chapter 7
A Penetrating Auditory Nerve Array for Auditory Prosthesis

John C. Middlebrooks and Russell L. Snyder

1 Introduction

Many chapters of this volume attest to the success of contemporary cochlear implants in bringing hearing to the severely or profoundly deaf. These devices consist of arrays of electrodes inserted into the scala tympani of the cochlea. In many cases, such an implant can deliver impressive levels of speech recognition. Nevertheless, there are too many cases in which performance with a cochlear implant is disappointing, either as a result of known pathology of the cochlea or auditory nerve or for other, unknown reasons. Moreover, even in the best cases, cochlear implant users show poor speech recognition in complex auditory environments, poor pitch recognition, and poor spatial hearing, even in cases of bilateral implantation. Some of these limitations are inherent in the basic biophysics of intrascalar stimulation. The scala-tympani stimulating electrodes lie within a volume of electrically conductive fluid (the perilymph) and are separated from auditory nerve fibers and cell bodies by a bony wall (the osseous spiral lamina). Those aspects of electrode placement result in a complex and attenuated current path from the electrodes to excitable neural elements, thereby elevating excitation thresholds and frustrating efforts to stimulate restricted neural populations. Frequency-specific activation of low frequency fibers originating in the cochlear apex is especially challenging, because most scala-tympani electrodes are located within the basal turn of the cochlea. Not even the most apical scala-tympani electrodes are likely to achieve selective stimulation of low frequency fibers because of the compact geometry of the apical portion of the spiral ganglion (Stakhovskaya et al. 2007); also, there is in

J.C. Middlebrooks (✉)
Departments of Otolaryngology, Neurobiology & Behavior, and Cognitive Science,
404D Medical Sciences D, University of California at Irvine,
Irvine, CA 92697-5310, USA
e-mail: j.midd@uci.edu

F.-G. Zeng et al. (eds.), *Auditory Prostheses: New Horizons*, Springer Handbook of Auditory Research 39, DOI 10.1007/978-1-4419-9434-9_7,

contemporary devices a high incidence of unintended deviation of apical electrodes into the scala vestibuli (Skinner et al. 2007; Finley et al. 2008).

A seemingly obvious alternative to cochlear implants based on scala-tympani electrodes would be to place stimulating electrodes directly into the auditory nerve, i.e., in the same anatomical compartment as the neural elements that are to be stimulated. Indeed, intraneural electrodes were used in some of the first demonstrations of auditory prostheses for the profoundly deaf. In pioneering work by Simmons and colleagues (Simmons et al. 1965; Simmons 1966, 1979), 4 or 6 75-μm diameter wires were inserted into the auditory nerves of deaf human volunteers. Subjects could distinguish among stimulated electrodes and, with varying degrees of resolution, could discriminate among current levels and among rates of pulsatile stimulation. Speech recognition by those subjects was negligible, however, for reasons that possibly included non-optimal positioning of the wires, traumatic response of the nerve to insertion of the wires, and lack of the high-speed digital sound-processing technology that is available today. In contrast, contemporaneous research with improved scala-tympani stimulation was showing promising results. For that reason, research on intraneural stimulation for auditory prosthesis was set aside, and it was the scala-tympani electrode array that was favored for the development that has led to the present-day cochlear implant.

With advances in technology over the past decade, there has been a revival of interest in intraneural stimulation for auditory prosthesis. A group at the University of Utah has developed a stimulating array consisting of multiple 1- or 1.5-mm-long silicon spikes, each ending in a single platinum-plated electrode. That group demonstrated that the array could be implanted into a cat auditory nerve with preservation of the physiological viability of the nerve (Badi et al. 2002); that auditory brainstem responses (ABRs) could be elicited by auditory nerve stimulation through the array (Badi et al. 2003; Hillman et al. 2003); and, based on ABR measurements, that various electrodes could activate somewhat non-overlapping auditory nerve fiber populations (Badi et al. 2006). Recordings from the cat auditory cortex demonstrated activation of the ascending auditory pathway by auditory nerve stimulation with the Utah electrode array (Kim et al. 2007). The range of the cortical frequency representation that was activated in the latter study was surprisingly narrow, however, possibly because the spikes of the electrode array were too short to reach the fibers from the middle and apical cochlear turns, which lie deep in the auditory nerve. Another group, at the University of Michigan, tested a stimulating array having 5 electrodes distributed along a single silicon-substrate shank (Niparko et al. 1989a, b; Zappia et al. 1990). Implantation of that device in guinea pig auditory nerves showed excellent histocompatibility and functional viability for periods of 5 weeks, as monitored with middle latency responses.

A recent series of experiments has undertaken study of the feasibility of intraneural stimulation as a mode of auditory prosthesis, using anesthetized cats as an animal model (Middlebrooks and Snyder 2007, 2008, 2010). The stimulating-electrode array was similar to that used by the earlier Michigan group – in this case, a single-shank 16-electrode device. Frequency-specific activation of the ascending auditory pathway was monitored by recording with a similar single-shank device with electrodes at 32 depths along the tonotopic axis of the central nucleus of the inferior colliculus (ICC). In each animal, responses to sounds first were measured under normal-hearing conditions. Then, the animal was deafened and responses were recorded to electrical stimulation of the auditory nerve with a penetrating intraneural array and, in some animals, with a conventional scala-tympani cochlear implant. The results from those studies demonstrate that activation of the auditory pathway with intraneural electrodes is indeed feasible and that such electrodes offer several advantages compared to conventional scala-tympani electrodes. Those advantages include: frequency-specific stimulation of a broader range of the cochlear frequency representation; narrower tonotopic spread of excitation; lower excitation thresholds; reduced interference between simultaneously stimulated channels; and enhanced transmission of temporal fine structure. This chapter begins by summarizing the basic stimulating and recording approach that was used in the Middlebrooks and Snyder experiments and by describing the responses of ICC neurons to acoustical and electrical stimulation. Then, specific characteristics of intraneural and scala-tympani stimulation are compared. Finally, the animal results are interpreted with regard to potential clinical application in humans.

2 Tonotopic Activation of the ICC

The ICC contains a well characterized representation of frequency (a "tonotopic" map), in which the characteristic frequencies (CFs) of neurons increase with increasing depth in the ICC along a dorsolateral to ventromedial axis (e.g., Rose et al. 1963). In the recent studies of intraneural stimulation (Middlebrooks and Snyder 2007, 2008, 2010), a single-shank silicon-substrate recording array was positioned such that extracellular spike activity could be recorded simultaneously at 32 sites in 100-μm steps along the ICC tonotopic axis. The responses to pure tones were used to adjust the depth of the recording array to sample ICC neurons with CFs from ≤1 kHz to ≥32 kHz. Then, the recording array was fixed in place, responses to sounds at each of the 32 recordings sites were recorded, and the CF at each recording site was estimated. The map of CF versus ICC depth for one animal is shown in Fig. 7.1a. The tonotopic distributions of activity in that animal elicited by 7 representative pure tones are represented in Fig. 7.1b–h in the form of spatial tuning curves (STCs), which plot a measure of ICC activity (colors) as a function of stimulus level (abscissa) and ICC depth (ordinate). For any particular pure tone, activity was fairly restricted in depth (ranges of ~300 to 500 μm in depth at 10 dB above threshold in this example). Across all the tested frequencies, there was a steady progression of superficial-to-deep locus of activity associated with low-to-high tone frequency.

After recording of CFs in each animal, the animal was deafened by infusion of neomycin sulfate into the scala tympani, and an intraneural electrode array was implanted. In most cases, a scala-tympani electrode array was implanted, tested, and then explanted prior to implantation of the intraneural array. The scala-tympani array was an animal version of the Nucleus 22 banded electrode array from Cochlear

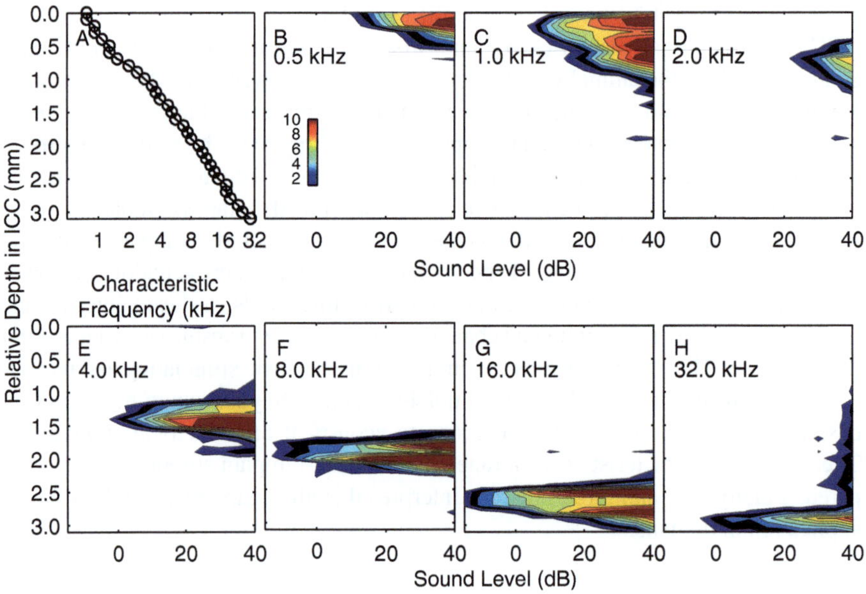

Fig. 7.1 Spread of activation along the tonotopic axis of the central nucleus of the inferior colliculus (ICC) in response to pure-tone stimulation in normal-hearing conditions. In each panel, the "Relative Depth" plotted on the ordinate is the depth along the 32-site recording array inserted along the tonotopic axis of the ICC, with depths plotted relative to the most superficial recording site. (**a**) Characteristic frequencies (CFs) at each of the 32 recording sites. (**b–h**) Spatial tuning curves (STCs) representing responses to pure tones at 7 frequencies. In each STC, the contours represent cumulative discrimination index (d') based on a receiver-operating-characteristic analysis of spike rates as a function of increasing stimulus level (Middlebrooks and Snyder 2007). The entire colored area in each panel represents the range of sound levels (abscissa) and relative depths (ordinate) at which tones at the specified frequency activated ICC neurons at or above a criterion of d' = 1. Cat 0509 (From Middlebrooks and Snyder 2007)

Ltd., differing only in that the animal version had only 8 electrodes, compared to the 22 electrodes of the human version. Scala-tympani electrodes were stimulated in two configurations: monopolar (one active scala-tympani electrode and a distant wire in a muscle as the return) and bipolar (one active scala-tympani electrode and an adjacent scala-tympani electrode as the return). The intraneural electrode array consisted of a single silicon-substrate shank, 15-μm thick, and a maximum of 240-μm wide, tapering toward the tip (NeuroNexus Technologies, Ann Arbor, MI). There were 16 iridium-plated stimulating sites spaced at 100-μm intervals along the shank. In most cases, and in all the examples illustrated in this chapter, a "modiolar trunk" placement of the intraneural electrode array was used. In those cases, the array was inserted through a small hole placed in the osseous spiral lamina of the cochlea, which was visualized through an enlarged round window. The electrode array penetrated the trunk of the auditory nerve basal to the basal turn, approximately perpendicular to the long axis of nerve fibers. In other cases (presented in Middlebrooks and Snyder 2008), the intraneural array was placed intracranially,

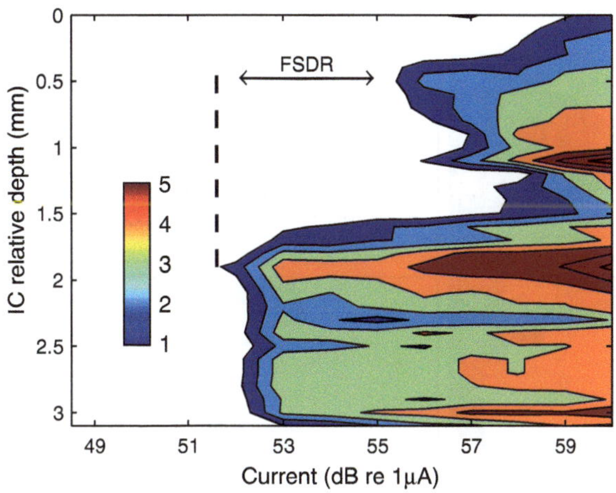

Fig. 7.2 Tonotopic spread of activation in the ICC in response to single electrical pulses presented through a scala-tympani electrode. Spread of activation is represented by a STC, as defined in Fig. 7.1. "FSDR" is the frequency-specific dynamic range, which is the range of current levels between the threshold at the most sensitivity ICC depth and the level at which activation spread to a tonotopically distant ICC depth. Cat 0509 (From Middlebrooks and Snyder 2007)

penetrating the auditory nerve as it exited the internal acoustic meatus. Electrical stimuli presented through both scala-tympani and intraneural electrodes were single biphasic pulses, 41 μs per phase with no interphase gap, or trains of such pulses.

An example of the spread of activation elicited by single pulses presented through one electrode of a scala-tympani array is presented by the STC in Fig. 7.2. The tonotopic spread of activation was broad. Stimuli within 1 dB of the minimum threshold activated neurons at nearly half of the recording sites in the ICC, corresponding to CFs from ~5 to 32 kHz. Current levels more than ~4 dB above the minimum threshold activated neurons along the entire recording array. The difference in currents between the minimum threshold and the level at which tonotopically distant neurons were activated was defined as the frequency-specific dynamic range (FSDR), which was 3.7 dB in this example. Lowest threshold activity restricted to the high frequency representation in the ICC was typical of many cases of scala-tympani stimulation. In other cases, activation spread from high- to low-CF regions with increasing stimulus level, but activity was never restricted to superficial low-CF regions in the ICC without concomitant lower threshold activity in higher-CF regions.

Figure 7.3 shows results of intraneural stimulation in the same animal as that represented in Fig. 7.2. Compared to stimulation with scala-tympani electrodes, intraneural stimulation produced more restricted spread of activation, lower thresholds, and wider FSDRs. The figure shows STCs from 6 of the 16 intraneural electrode sites, selected to show the range of tonotopic loci of activation. The STC elicited by single-pulse stimulation of the deepest electrode (Fig. 7.3a) lay in the

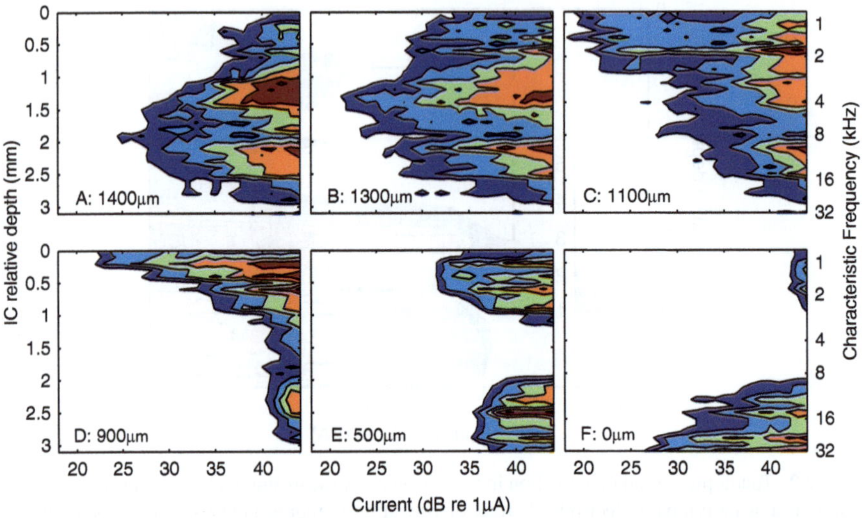

Fig. 7.3 Tonotopic spread of activation in the ICC in response to single electrical pulses presented through individual intraneural electrodes. Spread of activation is represented by a STC, as defined in Fig. 7.1. In each panel, the depth in the auditory nerve of the stimulating electrode is stated relative to the depth of the most superficial electrode. Cat 0509 (From Middlebrooks and Snyder 2007)

middle of the ICC tonotopic representation, indicating excitation of auditory nerve fibers from the middle cochlear turn. Successively more superficial electrodes showed a progression of STCs toward lower frequencies, indicating excitation of fibers from the apical turn (Fig. 7.3b–d). Electrodes located even more superficial produced bi-lobed STCs, as in Fig. 7.3e. Finally, the most superficial electrode elicited an STC that was restricted to the high-CF region of the ICC (Fig. 7.3f). This tonotopic progression of activation, from middle, to low, to bi-lobed, to high frequencies, was a consistent finding of intraneural array placements using the modiolar trunk approach. That progression accords with the geometry of the auditory nerve in the cat, as demonstrated anatomically by Arnesen and Osen (1978). That study showed a spiral organization of the nerve, with fibers from the cochlear apex lying innermost, middle-turn fibers lying more superficial along the medial surface of the nerve, and basal-turn fibers lying along its lateral surface. Bi-lobed STCs were observed consistently in the Middlebrooks and Snyder (2007) study and could be attributed to excitation spreading to both the deep fascicles of apical fibers and to the adjacent, superficial fascicles of basal fibers. In tests of intracranial placements of electrode arrays, some cases showed the "spiral" tonotopic pattern that was observed with the modiolar trunk placement (as in Fig. 7.3) and other cases showed a more monotonic progression of frequencies from middle frequencies (deep sites) to low frequencies (superficial) consistent with the nerve having rotated so that the trajectory of the stimulating array missed the basal-most fibers (Middlebrooks and Snyder 2008).

3 Characteristics of Intraneural Compared to Scala-Tympani Cochlear-Implant Stimulation

Consistent differences were observed between responses to intraneural and scala-tympani stimulation that appear to bode well for the application of intraneural stimulation to auditory prostheses for clinical application. Specific characteristics are described here.

3.1 Selective Access to a Wider Range of the Frequency Representation

Intraneural stimulation produced frequency-specific activation of restricted neural populations distributed throughout the entire spectrum of frequencies represented in the ICC. The responses to stimulation of single intraneural electrodes could produce STCs centered at ICC locations with CFs anywhere from 0.6 to 38 kHz. In some cases, STCs appeared to be centered above or below the span of the ICC recording array, suggesting that the range of activated CFs was underestimated because of limitations in the length of the recording array and in the range of calibrated frequencies in the audio system. Auditory nerve fibers originating throughout the entire cochlea exit through the base, so in principle the range of frequencies that could be stimulated with intraneural electrodes is limited only by knowledge of the locations of particular fiber groups.

In the case of scala-tympani electrodes, in contrast, frequency-specific excitation was limited to middle to high frequencies represented in the basal half of the cochlea. In the animal model, any activation of ICC regions with CFs below ~4 kHz was accompanied by lower threshold activation of higher-CF regions. This suggests that the low-CF activation by scala-tympani electrodes represented non-specific spread of excitation from basal electrodes to apical fibers as they passed by the basal turn of the cochlea. In the larger cochlea of humans, it might be possible to achieve greater specificity of activation because of specific access to spiral ganglion cells that are spread more widely around the basal turn. The apical portion of the human spiral ganglion is more compact than the basal turn, however (Stakhovskaya et al. 2007), so it is difficult to see how restricted low frequency regions could be stimulated selectively with scala-tympani electrodes, even if it were possible to insert an scala-tympani array well into the second turn of the cochlea. Indeed, psychophysical studies in cases of extraordinarily deep cochlear-implant placements in humans show frequency judgments by subjects that conflict with predictions based on the cochlear frequency map and on the longitudinal position of electrodes along the scala-tympani array (Baumann and Nobbe 2006); also, such deeply inserted arrays show only limited utility in speech recognition (e.g., Arnoldner et al. 2007).

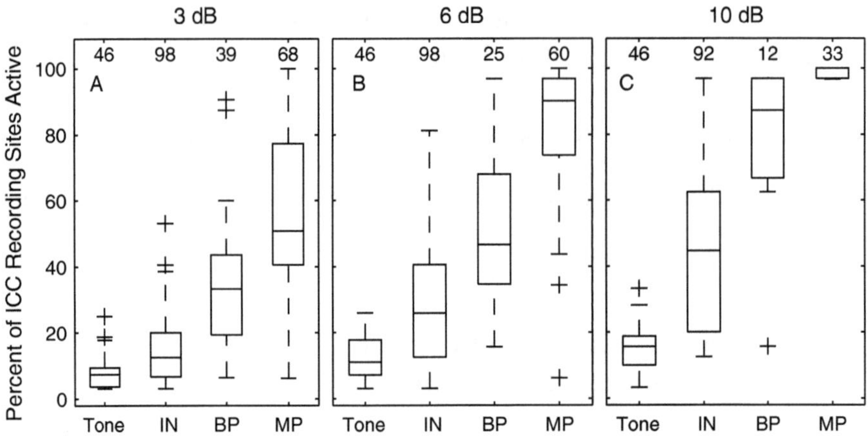

Fig. 7.4 Distributions of spread of ICC activation. These box-and-whisker plots represent the spread of ICC activation in response to various stimulus configurations and levels. In each case, the datum that is represented is the percentage of ICC recording sites that were active at a criterion of d' ≥ 1. Each panel represents distributions at 3, 6, or 10 dB above the threshold at the most sensitive ICC depth. Within each panel, each box has horizontal lines at the median and lower and upper quartile values. The whiskers show the extent of data lying within 1.5 times the interquartile distance from the upper and lower quartiles, and the plus signs represent outliers. The number above each box indicates the number of electrode sites (compiled across multiple cats) or the number of tone frequencies represented in the distribution. Stimulus configurations are indicated as Tone (pure tones in normal-hearing conditions), IN (single pulses through intraneural electrodes in deaf conditions), MP (scala-tympani electrodes in a monopolar configuration), and BP (scala-tympani electrodes in a bipolar configuration) (From Middlebrooks and Snyder 2007)

3.2 *Restricted Spread of Excitation*

A striking characteristic of responses to stimulation of intraneural electrodes compared to those obtained with scala-tympani electrodes was the restricted spread of activation along the ICC tonotopic axis, which can be assumed to reflect restricted spread of excitation within the auditory nerve. Spread of activation was quantified by the percentage of ICC recording sites that were activated by a particular electrode at a level 3, 6, or 10 dB above the minimum threshold for that level. The distributions of such measures across stimulation electrodes and across cats are represented by the box plots in Fig. 7.4 for conditions of pure-tone stimulation in normal-hearing conditions (Tone), intraneural stimulation (IN), and scala-tympani stimulation using bipolar (BP) or monopolar (MP) electrode configurations. Spread of activation in the Tone condition always was the most restricted, but spread in the IN condition was not much broader. Spread in the IN condition consistently was more restricted than that in conditions of MP or BP scala-tympani stimulation. It is somewhat misleading to compare spread of activation between acoustic and electrical stimulation at constant decibel levels above threshold because the dynamic range of electrical hearing is so much smaller than for acoustical hearing. A more reasonable comparison would be the spread of activation by sounds at 10 dB above threshold compared with the spread of activation by electrical pulses at 3 dB above threshold. In that comparison, the activation by intraneural stimulation actually is somewhat more restricted than that obtained with pure-tone stimulation.

3.3 Increased Frequency-Specific Dynamic Range

In addition to showing more restricted spread of activation at near-threshold current levels, intraneural stimulation typically exhibited broader dynamic ranges of current levels over which activation was restricted to contiguous frequency regions, i.e., larger FSDRs, as defined in Sect. 2. In each of 10 animals, examples of intraneural electrodes were selected that gave ICC activation indicative of restricted apical-, middle-, and basal-turn auditory nerve stimulation. In many cases, only the lower limits of FSDRs could be estimated for intraneural stimulation because there was no spread of activation to non-contiguous frequencies regions within the range of currents available on the stimulator. In the 10 animals, median FSDRs were 16.5, >15.5, and >12.5 dB for apical-, middle-, and basal-turn electrodes, respectively. In the same animals, the scala-tympani stimulating electrode or electrode pair was selected that gave the broadest FSDRs. Median FSDRs were considerably narrower for scala-tympani stimulation: 3.6 dB for monopolar and 6.2 for bipolar configurations. We assume that the breadth of FSDRs observed with intraneural stimulation reflects some electrotonic compartmentalization because of the fascicular organization of the auditory nerve. In contrast, the narrow FSDRs observed with scala-tympani stimuli likely reflect the low resistance milieu of the perilymph-filled scala tympani.

3.4 Lower Thresholds for Excitation

The positions of intraneural electrodes adjacent to auditory nerve fibers resulted in substantially lower threshold current levels than the thresholds that were obtained with scala-tympani stimulation, in which the electrodes lay in a separate anatomical compartment from the nerve. Again, in each animal, intraneural electrodes were selected for analysis that gave selective activation of apical-, middle-, and basal-turn fibers, and scala-tympani electrodes were selected that exhibited the lowest thresholds. Median thresholds were 26, 29, and 31 dB re 1 μA for apical-, middle-, and basal-turn intraneural electrodes, respectively. A current threshold of 26 dB, given a phase duration of 41 μs, corresponds to a charge threshold of <1 nanoCoulomb. Current thresholds were more than an order of magnitude higher for scala-tympani stimulation, with medians of 50.0 and 59.4 dB re 1 μA for monopolar and bipolar scala-tympani configurations, respectively. These values indicate that most or all of the currents needed to cover the dynamic range of a typical intraneural electrode, calculated as the threshold plus the FSDR, would be less than the current at threshold for typical intrascalar electrodes.

3.5 Reduced Interference Between Simultaneously Stimulated Channels

A major limitation of conventional multi-channel cochlear implants is the possibility of between-channel interference resulting from summation of electrical current among simultaneously stimulated electrodes. Such electrical summation would lead to unintended shifts in operating current ranges and summation of loudness, both of which would lead to a loss of functional independence among channels of information. The problem of current summation is largely circumvented in most contemporary cochlear implant systems by the use of pulsatile stimulation strategies in which the pulses delivered to nearby electrodes are interleaved in time, never arriving simultaneously. A disadvantage of this strategy, however, is that the temporal fine structure in the audio input signal is lost, replaced by the fine structure of the fixed-rate pulse trains.

Experiments in the animal model confirmed that simultaneous stimulation of pairs of scala-tympani electrodes resulted in substantial between-channel interference (Middlebrooks and Snyder 2007). A measure of such interference was the amount by which the threshold for activation by one electrode was reduced by simultaneous stimulation of a second electrode. When two nearby scala-tympani electrodes were stimulated with simultaneous pulses at equal currents, the threshold for the more sensitive of the two electrodes was reduced by nearly 6 dB, which is close to the reduction that would be expected if the currents were to be summed and presented through the same electrode. Smaller threshold interactions were observed for larger scala-tympani electrode separations, but even so, the thresholds reductions were consistent with a model based on simple summation of electrical currents

Between-channel interference was substantially lower in conditions of intraneural stimulation. In the example shown in Fig. 7.5, three intraneural electrodes stimulated individually showed restricted, well separated STCs (Fig. 7.5a, c, e; the electrodes were located 0, 1400, and 400 μm deep to the most superficial electrode). When those electrodes were stimulated simultaneously in pairs, the resulting STCs showed only superposition (Fig. 7.5b, d, f). That is, the individual contribution of each of the constituent single-channel STCs was evident and, in each pair, the current on one electrode had little or no effect on the threshold or spread of activation of the second electrode. A large part of this apparent independence of simultaneously stimulated electrodes is the result of the absence of overlap of STCs, implying lack of overlap of electrical fields in the auditory nerve. For 23 pairs of electrodes in which the individual STCs showed little or overlap, threshold reductions during simultaneous stimulation averaged only 0.3 dB. In contrast, 15 intraneural electrode pairs that showed substantial STC overlap exhibited threshold reductions averaging 3.6 dB. A quantitative estimate of STC overlap and threshold reduction confirmed the contribution of STC overlap to between-channel interference for both scala-tympani and intraneural stimulation (Middlebrooks and Snyder 2007). Nevertheless, for any given amount of STC overlap, intraneural stimulation showed significantly less threshold reduction, and presumably less inter-channel interference, than did

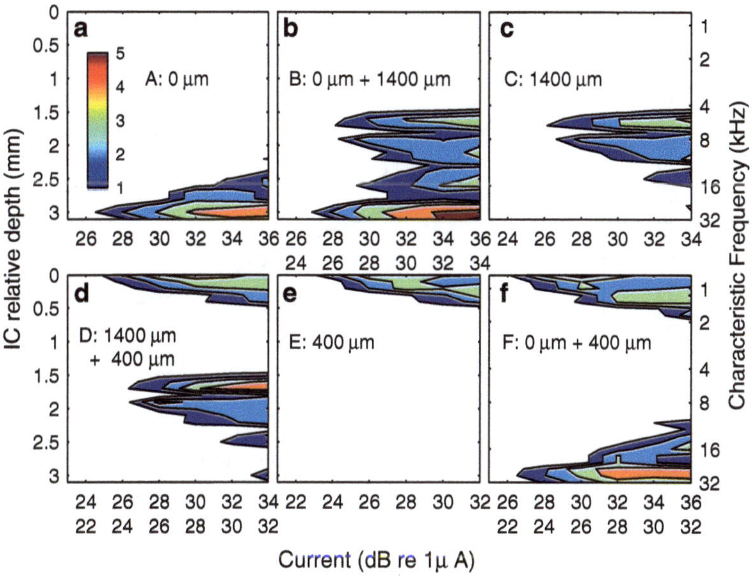

Fig. 7.5 Superposition of ICC activation under conditions of single-pulse stimuli presented simultaneously to pairs of intraneural electrodes. Spread of activation is represented by STCs as defined in Fig. 7.1. Panels (**a**), (**c**), and (**e**) represent activation by single electrodes and panels (**b**), (**d**), and (**f**) represent activation by pairs of electrodes. (From Middlebrooks and Snyder 2007)

stimulation with scala-tympani electrodes. A likely explanation for the reduced interference among intraneural electrodes is similar to that advanced for the broader FSDRs seen with intraneural stimulation, namely that compartmentalization within the fascicles of the auditory nerve restricts the tonotopic spread of electrical current more so than does the perilymph that fills the scala tympani.

3.6 Enhanced Transmission of Temporal Fine Structure

The temporal fine structure of a sound consists of the detailed timing of the waveform, in contrast to the envelope, which is the slower waxing and waning of waveform amplitude. Users of conventional cochlear implants show surprisingly poor sensitivity to temporal fine structure. For instance, implant users require changes of ~10% or more in the rates of pulse trains in order to perceive a change in rate pitch when base rates are below~200 to 300 pps, and rate sensitivity is lost entirely at higher pulse rates (Shannon 1983; Tong and Clark 1985; Townshend et al. 1987; McKay et al. 2000; Zeng 2002; Landsberger and McKay 2005; van Hoesel 2007; Carlyon et al. 2008). In contrast, normal-hearing listeners can discriminate pure tone frequencies with difference limens of ~1% up to frequencies of ~2000 Hz. One might speculate that the intimate contact of intraneural electrodes with auditory

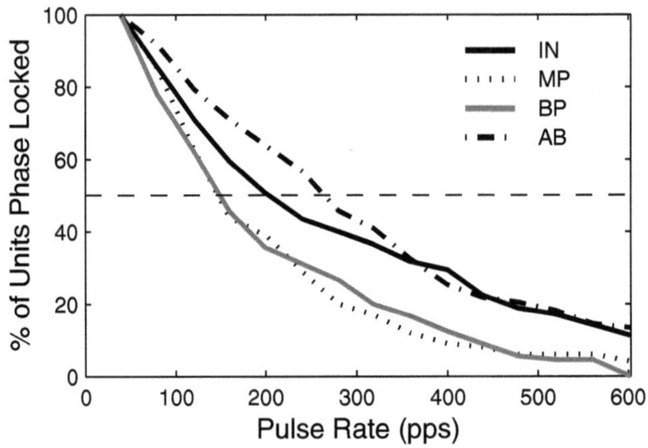

Fig. 7.6 Cumulative distribution of limiting rates of ICC units. The limiting rate of an ICC neuron is the highest pulse rate to which the neuron showed statistically significant phase locking to the stimulus. *Lines* represent the percentage of ICC neurons showing limiting rates at or above each pulse rate. *Line types* represent various electrode types: *IN* intraneural, *MP* monopolar intrascalar, *BP* bipolar intrascalar, *AB* apical ball (From Middlebrooks and Snyder 2010)

nerve fibers might permit transmission of temporal fine structure that is superior to that obtained with conventional scala-tympani electrodes. A recent study tested that hypothesis in an animal model by presenting pulse-train stimuli at various rates using both intraneural and scala-tympani electrodes and monitoring the precision of phase locking by ICC neurons (Middlebrooks and Snyder 2010). A statistical test was used to test for significant phase locking to various pulse rates, and the *limiting rate* was defined as the maximum pulse rate to which ICC neurons showed significant phase locking.

A substantially larger proportion of ICC neurons showed phase locking to high-rate pulse trains presented through intraneural electrodes than to such pulse trains presented through scala-tympani electrodes. The cumulative distributions of limiting rates for various stimulation conditions are shown in Fig. 7.6. The median values of limiting rate for scala-tympani electrodes stimulated in MP and BP configurations were 120 pulses per second (pps), whereas the median limiting rate for intraneural stimulation was 200 pps. At the highest pulse rate tested, 600 pps, phase locking to scala-tympani pulse trains was nearly absent, whereas 13% of ICC neurons showed significant phase locking to pulse trains presented through intraneural electrodes. The "AB" condition refers to an experimental stimulation condition described below.

Although the results of Middlebrooks and Snyder (2010) support the hypothesis that transmission of temporal fine structure is enhanced under conditions of intraneural stimulation, the conclusion of that study was that there is no specific evidence for a causal relationship between close proximity of electrodes to auditory nerve fibers and enhanced temporal transmission. Examination of the limiting rates of ICC neurons as a function of their CFs (as determined with tonal stimulation prior to deafening) revealed that most of the neurons having high limiting rates tended to

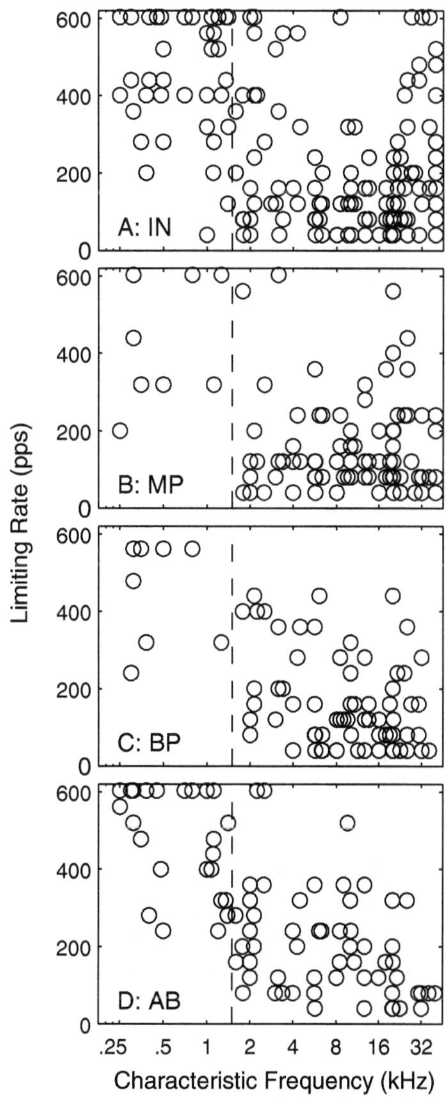

Fig. 7.7 Limiting rates as a function of characteristic frequency. *Each panel* represents a particular electrode type or configuration, as indicated (From Middlebrooks and Snyder 2010)

have low CFs, generally <1.5 kHz. That relationship is shown in Fig. 7.7 for four stimulation conditions. The sparse sampling of low-CF neurons with scala-tympani electrodes did not permit a determination of whether poor transmission of temporal fine structure through scala-tympani electrodes was the result of a property of the electrode-neural interface or simply an inability to excite low frequency auditory pathways selectively, which might have some as yet unknown intrinsic adaptation for high temporal resolution. That logical confound was resolved by tests of ICC phase locking to pulse trains presented through a ball electrode placed directly on the osseous spiral lamina of the apical turn of the cat's cochlea. This "apical ball" (AB) electrode was intended to simulate a scala-tympani electrode positioned much further apically than can be accomplished with the usual basal-turn insertion of a scala-tympani array. As expected, stimulation through the apical ball electrode activated a larger percentage of low-CF ICC neurons than was possible with conventional scala-tympani electrodes (Fig. 7.7d). Moreover, many of those neurons with CF's <1.5 kHz exhibited phase locking even to the highest pulse rates that were tested. A 2-way analysis of variance, with CF and electrode configuration (IN, MP, BP, and AB) as factors, indicated a strong dependence of limiting rate on CF ($p<.0001$) but no significant dependence of limiting rate on electrode configuration after accounting for CF ($p=0.25$).

The results of Middlebrooks and Snyder (2010) suggest that the larger percentage of ICC neurons showing high limiting rates under conditions of intraneural stimulation is the result primarily of the enhanced ability of intraneural electrodes to activate low frequency pathways rather than any special property of the electrode-neural interface. That study also demonstrated that ICC neurons with low CFs tended to have shorter latencies from electrical pulses to ICC-neuronal spikes than do neurons with higher CFs. The results indicate that the fibers originating in the cochlear apex that can be activated selectively by intraneural electrodes feed into brainstem pathways that are specialized for high temporal acuity, which is evident as high limiting rates and short latencies. It seems likely that the anatomical and/or physiological specializations in this pathway have evolved to preserve the temporal fine structure that, in normal hearing, is carried by low- but not by high-CF auditory nerve fibers.

4 Implications for Clinical Application

In the recent studies by Middlebrooks and Snyder (2007, 2008, 2010), intraneural stimulation exhibited improvements over scala-tympani stimulation in essentially every parameter that could be tested in that acute animal model, suggesting that intraneural stimulation could form the basis of an improved auditory prosthesis. There are many hurdles to cross, however, before intraneural stimulation could be implemented in a clinical auditory prosthesis, most notably the demonstration of safety of long-term implantation of a penetrating electrode array into the auditory nerve. Nevertheless, it is natural to speculate on benefits that might be realized by clinical application of intraneural stimulation. One can imagine improvements in transmission of both spectral and temporal aspects of sound stimuli as well as several other possible benefits that are of a more practical nature. Those issues will be considered separately here.

4.1 Representation of Sound Spectra

In normal hearing, the spectral content of sound is decomposed by cochlear mechanics, resulting in a tonotopic representation of frequency components at specific locations along the cochlear spiral. In sound processors for auditory prostheses, the frequency decomposition normally performed by the cochlea is replaced by digitally implemented banks of bandpass filters, and the output from each filter is directed to a single scala-tympani electrode at a tonotopically appropriate location. Present-day cochlear implants have as many as 22 scala-tympani electrodes, but the number of functionally independent channels of information delivered to the brain is considerably lower than the number of physical electrodes. In one study, for instance, speech perception in noise was measured as varying numbers of scala-tympani electrodes were disabled (Friesen et al. 2001). Performance was unaffected by decreasing the number of active electrodes down to 4 (for implant users with poor speech recognition) or 7 to 10 (for the best users). In contrast, normal-hearing listeners in the same study showed decrements in recognition of vocoder-processed speech (Shannon et al. 1995) when channel numbers were decreased below 20. Those results suggest that cochlear-implant users are limited in the spectral resolution that they can achieve through a multi-channel scala-tympani array, presumably because of interference among channels and/or the spread of excitation from individual channels.

Several characteristics of responses to intraneural stimulation demonstrated in the cat model (Middlebrooks and Snyder 2007) suggest that intraneural stimulation might offer, in general, a higher-resolution representation of sound spectrum as cochlear place and, more specifically, an increase in the number of functional independent channels of information from the electrode array to the brain compared to that offered by scala-tympani stimulation. *First*, intraneural electrodes can access restricted populations of low-CF fibers from the cochlear apex, which broadens the spectrum of frequencies that can be stimulated selectively, compared to the mid- to high frequency spectrum that is available to scala-tympani electrodes. In the animal model, intraneural stimulation provided frequency-specific stimulation over roughly double the length of the ICC tonotopic axis compared to that provided by scala-tympani stimulation. *Second*, the spread of excitation by individual intraneural electrodes was substantially more restricted than that obtained with scala-tympani electrodes, resulting in greater frequency specificity by single electrodes and less overlap of neural populations activated by simultaneously stimulated pairs of electrodes. *Third*, tests of simultaneous stimulation of pairs of intraneural or of pairs of scala-tympani electrodes offer a direct demonstration of reduced between-channel interference in conditions of intraneural stimulation.

Finally, the greater FSDR obtained with intraneural electrodes would increase the dynamic range of current levels that could be applied to each electrode without ectopic activation of distant frequency regions. That increased dynamic range would permit a decrease in the amount of amplitude compression needed to map the broad dynamic range of acoustic hearing into the more limited dynamic range of electric hearing, thereby decreasing the amount of distortion of speech signals. Also, if one assumes that perceptual resolution of electrical current in decibels is at least as good with intraneural stimulation as it is with scala-tympani stimulation, an increased dynamic range might produce an increase in the number of perceptually discriminable amplitudes on each channel, thereby enhancing the representation of the amplitude dimension of frequency-amplitude spectra.

Enhanced spectral resolution that might be obtained using an auditory prosthesis based on intraneural stimulation presumably would aid in all aspects of sound recognition, particularly in the presence of background noise. Improved low frequency spectral resolution would enhance sensitivity to low frequency sounds and would enhance place-based pitch perception. Improved high frequency spectral resolution might provide sufficient resolution for users to exploit the spatial cues to sound-source elevation and front-back location that are provided by the direction-specific filtering by the pinna (Middlebrooks and Green 1991); such cues would be available, however, only if the sound-processor microphone were to be positioned within the pinna.

4.2 Representation of the Temporal Structure of Sounds

The temporal envelope of speech sounds carries important information about manner of articulation, voicing, vowel identity, and prosody (reviewed by Rosen 1992). The most important frequencies for speech envelope information range from ~2 to 16 Hz or higher (Van Tasell et al. 1987; Rosen 1992; Drullman et al. 1994; Shannon et al. 1995; Fu and Shannon 2000; Xu et al. 2005; Xu and Zheng 2007). At higher frequencies, temporal fine structure carries information needed for pitch recognition, up to frequencies around 1000 Hz. In the minority of present-day cochlear-implant processors that use continuous-waveform stimulation (e.g., Simultaneous Analog Signal; Battmer et al. 1999; Zimmerman-Phillips and Murad 1999), the temporal waveform at the output of each bandpass filter is delivered more or less directly to each scala-tympani electrode. In the majority of processors, however, a low-pass-filtered envelope is extracted from the output of each bandpass filter, and that envelope is used to modulate a constant-rate electrical pulse train (e.g., Wilson et al. 1991; Skinner et al. 1994). In either case, the implant user's ability to utilize the temporal information is limited by the ability of the auditory brainstem to transmit the timing of auditory nerve spikes to central auditory structures. The recent work by Middlebrooks and Snyder (2010) demonstrates that the greatest temporal fidelity is found in auditory pathways with low CFs originating in the cochlear apex. At least in the available animal studies, those pathways can be activated selectively by intraneural electrodes but not by scala-tympani electrodes.

The ability to stimulate high temporal fidelity, low frequency, auditory pathways with intraneural electrodes would be expected to enhance temporal aspects of hearing in a future auditory prosthesis that employed such electrodes. In turn, improved temporal hearing presumably would improve rate-based pitch perception, which is relatively poor in present-day implant users. Improved pitch perception would enhance many auditory tasks, including discrimination of lexical tones in tonal languages (Xu and Zhou, Chap. 14), recognition of prosody in non-tonal languages, music appreciation (McDermott, Chap. 13), and segregation of voices in complex auditory scenes on the basis of vocal pitch contours. Also, improved transmission of temporal fine structure in cases of bilateral implantation might help in discrimination of interaural time differences, which in normal-hearing listeners are the dominant cues for the location of sounds in the horizontal dimension (Wightman and Kistler 1992; Macpherson and Middlebrooks 2002). Users of bilateral conventional cochlear implants show moderate sensitivity to interaural time differences in low pulse rate stimuli, but that sensitivity is lost at pulse rates >100 pps (van Hoesel and Tyler 2003; van Hoesel 2007; van Hoesel, Chap. 2).

In the majority of present-day cochlear-implant systems, the problem of interference among channels is mitigated somewhat by use of pulsatile stimulation strategies in which pulses on nearby channels are interleaved in time, which eliminates direct summation of stimulus currents. The disadvantage of such strategies, however, is that the temporal fine structure in the native audio input is lost and replaced by the fine structure of the fixed-rate pulse trains. Also, the limited duty cycle of the brief electrical pulses, necessary to avoid temporal overlap on adjacent channels, requires high peak currents in order to achieve sufficient charge transfer per pulse. The greatly reduced between-channel interference observed with intraneural stimulation in animal studies raises the possibility of dispensing with pulsatile stimulation altogether in favor of continuous-waveform "analog" stimulation. Analog stimulus would preserve the temporal fine structure of the audio signal and would offer an increased duty cycle, thereby reducing peak currents. Tests of continuous-waveform stimulation are in progress.

4.3 Additional Implications for Clinical Application

In addition to possible improvements in transmission of spectral and temporal information, there would be several additional benefits that intraneural stimulation might offer for a next-generation auditory prosthesis. One such benefit would be reduced power consumption. The placement of intraneural electrodes immediately adjacent to auditory nerve fibers results in a greater efficiency of stimulation, with reductions in the current levels needed to cover the dynamic range from threshold to saturating levels at which activation spreads to non-contiguous frequency regions. Reduced power consumption would offer longer battery life, which might eliminate the bulky behind-the-ear battery pack that is needed for contemporary cochlear-implant systems, in turn introducing the potential of a totally implantable prosthetic.

An auditory prosthesis based on an intraneural electrode array might offer options in certain situations in which a scala-tympani array would not be appropriate. One such situation is the case in which the scala tympani is occluded by bone, precluding the insertion of a scala-tympani electrode array. Such bone growth might occur as a sequelum of meningitis or severe otosclerosis. Another such situation is the case in which there is residual hearing to be preserved, typically at low frequencies. A growing number of such patients is being implanted with "short" scala tympani arrays intended to preserve native auditory function in the cochlear apex. There is a concern that such implantation might result in some loss of the low frequency hearing immediately after implantation (e.g., Gifford et al. 2008) or might hasten such loss in the long term, leaving the patient with no low frequency acoustic hearing and a scala-tympani array with even less than the usual limited coverage of apical sites. In principle, a full frequency intraneural electrode array could be inserted without disrupting middle ear sound conduction or cochlear mechanics. Indeed, Middlebrooks and Snyder have implanted intraneural arrays in normal-hearing cats and have preserved normal auditory thresholds at all but the highest frequencies (unpublished observations).

5 Summary and Conclusions

Auditory nerve stimulation with penetrating intraneural electrodes were employed in some of the first efforts at prosthetic treatment of sensorineural deafness, but that approach was abandoned in favor of stimulation using scala-tympani electrodes. The precision of stimulation of restricted auditory nerve fiber populations by scala-tympani electrodes is ultimately limited, however, by the location of the electrodes in an anatomical compartment separate from the compartment containing the intended targets of stimulation. Advances in the micromachining of implantable devices have revived interest in intraneural stimulation and have enabled new experiments in animal models. Experiments have been conducted in anesthetized cats, with the precision of activation of the ascending auditory pathway monitored by recording at multiple sites along the frequency map in the ICC. Those experiments have demonstrated the following benefits of auditory nerve stimulation with intraneural electrodes compared to conventional scala-tympani electrodes: frequency-selective excitation across the entire audible spectrum; spread of excitation comparable to that obtained with tones in normal-hearing conditions; stimulus current requirements reduced by over an order of magnitude; greatly reduced interference among simultaneously stimulated channels; and enhanced transmission of temporal fine structure.

The enhancement in transmission of sound-related information that could be demonstrated in an animal model leads to predictions of benefits that might be obtained if intraneural stimulation were to be implemented for a clinical auditory prosthesis. Of course, many technical challenges lie between the present animal experiments and clinical trials, including development of a chronically implantable intraneural electrode array and development of a speech processor that exploits all the advantages of intraneural stimulation. Questions of safety that must be addressed include the possibility of damage to the auditory nerve, either from physical trauma from the implanted electrode array or a foreign-body reaction, the possibility of cerebrospinal fluid leak, and concerns about the long-term stability of the intraneural electrode array relative to the tonotopic organization of the auditory nerve. If such challenges are overcome, however, it is reasonable to consider a new generation of auditory prostheses offering improved speech recognition in complex auditory scenes, improved music perception, improved recognition of vocal prosody and lexical tones, and improved spatial hearing.

Acknowledgements The authors thank James Wiler for expert technical assistance, Chris Ellinger and Dwayne Vailliencourt for engineering and machining support, and Zekiye Onsan and Russell Adrian for help with illustrations and preparation of the text. The authors' work was supported by NIH contracts NO1-DC-5-0005 and N263-2009-00024C.

References

Arnesen, A. R., & Osen, K. K. (1978). The cochlear nerve in the cat: topography, cochleotopy, and fiber spectrum. *Journal of Comparative Neurology*, 178, 661–678.

Arnoldner, C., Riss, D., Baumgartner, W. D., Kaider, A., & Hamzavi, J. S. (2007). Cochlear implant channel separation and its influence on speech perception–implications for a new electrode design. *Audiology & Neurotology*, 12, 313–324.

Badi, A. N., Hillman, T., Shelton, C., & Normann, R. A. (2002). A technique for implantation of a 3-dimensional penetrating electrode array in the modiolar nerve of cats and humans. *Archives of Otolaryngology–Head & Neck Surgery*, 128, 1019–1025.

Badi, A. N., Kertesz, M. D., Gurgel, R. K., Shelton, C., & Normann, R. A. (2003). Development of a novel eighth-nerve intraneural auditory neuroprosthesis. *Laryngoscope*, 113, 833–842.

Badi, A. N., Owa, A. O., Shelton, C., & Normann, R. A. (2006). Electrode independence in intraneural cochlear nerve stimulation. *Otology & Neurotology*, 28, 16–24.

Battmer, R. D., Zilberman, Y., Haake, P., & Lenarz, T. (1999). Simultaneous analog stimulation (SAS)–Continuous Interleaved Sampler (CIS) pilot comparison study in Europe. *Annals of Otology, Rhinology, and Laryngology Supplement*, 177, 69–73.

Baumann, U., & Nobbe, A. (2006). The cochlear implant electrode-pitch function. *Hearing Research*, 213, 34–42.

Carlyon, R. P., Long, C. J., & Deeks, J.M. (2008). Pulse-rate discrimination by cochlear-implant and normal-hearing listeners with and without binaural cues. *Journal of the Acoustical Society of America*, 123, 2276–2286.

Drullman, R., Festen, J.M., & Plomp, R. (1994). Effect of temporal smearing on speech reception. *Journal of the Acoustical Society of America*, 95, 1053–1064.

Finley, C. C., Holden, T. A., Holden, L. K., Whiting, B. R., Chole, R. A., Neely, G. J., Hullar, T. E., & Skinner, M. W. (2008). Role of electrode placement as a contributor to variability in cochlear implant outcomes. *Otology & Neurotology*, 29, 920–928.

Friesen, L. M., Shannon, R. V., Baskent, D., & Wang, X. (2001). Speech recognition in noise as a function of the number of spectral channels. Comparison of acoustic hearing and cochlear implants. *Journal of the Acoustical Society of America*, 110, 1150–1163.

Fu, Q. J., & Shannon, R. V. (2000). Effect of stimulation rate on phoneme recognition by Nucleus-22 cochlear implant listeners. *Journal of the Acoustical Society of America*, 107, 589–597.

Gifford, R. H., Dorman, M. F., Spahr, A. J., Bacon, S.P., Skarzynski, H., & Lorens, A.(2008). Hearing preservation surgery: psychophysical estimates of cochlear damage in recipients of a short electrode array. *Journal of the Acoustical Society of America*, 124, 2164–2173.

Hillman, T., Badi, A. N., Normann, R. A., Kertesz, T., & Shelton, C. (2003). Cochlear nerve stimulation with a 3-dimensional penetrating electrode array. *Otology & Neurotology*, 24, 764–768.

Kim, S. J., Badi, A. N., & Normann, R. A. (2007). Selective activation of cat primary auditory cortex by way of direct intraneural auditory nerve stimulation. *Laryngoscope*, 117, 1053–1062.

Landsberger, D. M., & McKay, C. M. (2005). Perceptual differences between low and high rates of stimulation on single electrodes for cochlear implantees. *Journal of the Acoustical Society of America*, 117, 319–327.

Macpherson, E. A., & Middlebrooks, J. C. (2002). Listener weighting of cues for lateral angle: the duplex theory of sound localization revisited. *Journal of the Acoustical Society of America*, 111, 2219–2236.

McKay, C. M., McDermott, H. J., & Carlyon, R. P. (2000). Place and temporal cues in pitch perception: are they truly independent? *Acoustical Research Letters Online*, 1, 25–30.

Middlebrooks, J. C., & Green, D. M. (1991). Sound localization by human listeners. *Annual Review of Psychology*, 42, 135–59.

Middlebrooks, J. C., & Snyder, R. L. (2007). Auditory prosthesis with a penetrating nerve array. *Journal of the Association for Research in Otolaryngology*, 8, 258–279.

Middlebrooks, J. C., & Snyder, R. L. (2008). Intraneural stimulation for auditory prosthesis: modiolar trunk and intracranial stimulation sites. *Hearing Research*, 242, 52–63.

Middlebrooks, J. C., & Snyder R. L. (2010). Selective electrical stimulation of the auditory nerve activates a pathway specialized for high temporal acuity. *Journal of Neuroscience*, 30, 1937–1946.

Niparko, J. K., Altschuler, R. A., Xue, X. L., Wiler, J. A., & Anderson, D. J. (1989a). Surgical implantation and biocompatibility of central nervous system auditory prostheses. *Annals of Otology, Rhinology, and Laryngology*, 98, 965–70.

Niparko, J. K., Altschuler, R. A., Evans, D. A., Xue, X. L., Farraye, J., & Anderson, D. J. (1989b). Auditory brainstem prosthesis: biocompatibility of stimulation. *Archives of Otolaryngology–Head & Neck Surgery*, 101, 344–52.

RoseJ, Greenwood DD, Goldberg JM, Hind JE (1963). Some discharge characteristics of single neurons in the inferior colliculus of the cat. I. Tonotopical organization, relation of spike-counts to tone intensity, and firing patterns of single elements. *Journal of Neurophysiology*, 26, 294–320.

Rosen, S. (1992). Temporal information in speech: acoustic, auditory and linguistic aspects. *Philosophical Transactions of the Royal Society London. B Biological Sciences*, 336, 367–73.

Shannon, R. V. (1983). Multichannel electrical stimulation of the auditory nerve in man. II. Channel interaction. *Hearing Research*, 12, 1–16.

Shannon, R. V., Zeng, F.-G., Kamath, V., Wygonski, J., & Ekelid, M. (1995). Speech recognition with primarily temporal cues. *Science* 270, 303–304.

Simmons, F. B. (1966). Electrical stimulation of the auditory nerve in man. *Annals of Otology, Rhinology, and Laryngology*, 84, 24–76.

Simmons, F. B. (1979). Electrical stimulation of the auditory nerve in cats: long term electrophysiological and histological results. *Annals of Otology, Rhinology, and Laryngology*, 88, 533–539.

Simmons, F. B., Epley, J. M., Lummis, R. C., Guttman, N., Frishkopf, L. S., Harmon, L. D., & Zwicker E. (1965). Auditory nerve: electrical stimulation in man. *Science*, 148, 104–106.

Skinner, M. W., Clark, G. M., Whitford, L. A., Seligman, P. M., Staller, S. J., Shipp, D. B., Shallop, J. K., Everingham, C., Menapace, C. M., Arndt, P. M., Antogenelli, T., Brimacombe, J., Pijl, S., Daniels, P., George, C., McDermott, H., & Beiter, A. L. (1994). Evaluation of a new spectral peak coding strategy for the Nucleus 22 Channel Cochlear Implant System. *American Journal of Otology*, 15 (Suppl. 2), 15–27.

Skinner, M. W., Holden, T. A., Whiting, B. R., Voie, A. H., Brunsden, B., Neely, J. G., Saxon, E. A., Hullar, T. E., & Finley, C. C. (2007). In vivo estimates of the position of advanced bionics electrode arrays in the human cochlea. *Annals of Otology, Rhinology, and Laryngology Supplement*, 197, 2–24.

Stakhovskaya, O., Sridhar, D., Bonham, B. H., & Leake, P. A. (2007). Frequency map for the human cochlear spiral ganglion: implications for cochlear implants. *Journal of the Association for Research in Otolaryngology*, 8, 220–233.

Tong, Y. C., & Clark, G. M. (1985). Absolute identification of electric pulse rates and electrode positions by cochlear implant patients. *Journal of the Acoustical Society of America*, 77, 1881–1888.

Townshend, B., Cotter, N., van Compernolle, D., & White, R. L. (1987). Pitch perception by cochlear implant subjects. *Journal of the Acoustical Society of America*, 82, 106–115.

van Hoesel, R. J. (2007). Sensitivity to binaural timing in bilateral cochlear implant users. *Journal of the Acoustical Society of America*, 121, 2192–2206.

van Hoesel, R. J., & Tyler, R. S. (2003). Speech perception, localization, and lateralization with bilateral cochlear implants. *Journal of the Acoustical Society of America*, 113, 1617–1630.

van Tasell, D. J., Soli, S. D., Kirby, V. M., & Widin, G. P. (1987). Speech waveform envelope cues for consonant recognition. *Journal of the Acoustical Society of America*, 82, 1152–61.

Wightman, F. L., & Kistler, D. J. (1992). The dominant role of low-frequency interaural time differences in sound localization. *Journal of the Acoustical Society of America*, 91, 1648–1661.

Wilson, B. S., Finley, C. C., Lawson, D. T., Wolford, R. D., Eddington, D. K., & Rabinowitz, W. M. (1991). Better speech recognition with cochlear implants. *Nature*, 352, 236–238.

Xu, L., Thompson, C. S., & Pfingst, B. E. (2005). Relative contributions of spectral and temporal cues for phoneme recognition. *Journal of the Acoustical Society of America*, 117, 3255–67.

Xu, L., & Zheng, Y. (2007). Spectral and temporal cues for phoneme recognition in noise. *Journal of the Acoustical Society of America*, 122, 1758.

Zappia JJ, Hetke JF, Altschuler RA, Niparko JK (1990). Evaluation of a silicon-substrate modiolar eighth nerve implant in a guinea pig. *Archives of Otolaryngology–Head & Neck Surgery*, 103, 575–582.

Zeng, F.-G. (2002). Temporal pitch in electric hearing. *Hearing Research*, 174, 101–106.

Zimmerman-Phillips, S., & Murad, C. (1999). Programming features of the CLARION Multi-Strategy Cochlear Implant. *Annals of Otology, Rhinology, and Laryngology Supplement*, 177, 17–21.

Chapter 8
Cochlear Nucleus Auditory Prostheses

Douglas B. McCreery and Steven R. Otto

1 The Cochlear Nucleus ABI as a Treatment for Profound Hearing Loss

1.1 Clinical Indications for the Cochlear Nucleus Auditory Brainstem Implant

Persons lacking a functional auditory nerve cannot benefit from a cochlear implant but may benefit from a prosthesis that employs an array of stimulating electrodes adjacent to, or within, the cochlear nucleus. Worldwide, more than 700 patients have received such cochlear nucleus auditory brainstem implants, usually described as "auditory brainstem implants" or ABIs (Otto et al. 2002; Schwartz et al. 2008; Colletti et al. 2009b). The most frequent indication for an ABI is Type 2 Neurofibromatosis (NF2), a genetic disorder with an autosomal dominant inheritance pattern in which the patient often develops life-threatening tumors (benign vestibular schwannomas) on both eighth nerves. Typically, the vestibular and auditory components of the nerve are destroyed by the invasive tumor or during its surgical resection. The prevalence of NF2 is approximately 1 in 40,000 live births with a high probability of bilateral acoustic tumors (Evans et al. 1992, 1999). There is a 50% chance of offspring inheriting the disorder, and many individuals with NF2 continue to have children. Thus, while NF2 is quite rare, worldwide there are many thousands of persons afflicted with this disorder.

D.B. McCreery (✉)
Huntington Medical Research Institutes, Neural Engineering Program,
734 Fairmount Avenue, Pasadena, CA 91105, USA
e-mail: dougmc@hmri.org

F.-G. Zeng et al. (eds.), *Auditory Prostheses: New Horizons*, Springer Handbook of Auditory Research 39, DOI 10.1007/978-1-4419-9434-9_8,

In recent years, the indications for the cochlear nucleus ABI have expanded beyond NF2, to include persons with bilateral deafness of other etiologies that preclude a cochlear implant. These etiologies include bilateral cochlear ossification (usually as a complication of meningitis), unilateral vestibular schwannomas in combination with deafness in the contralateral ear that is not amenable to a cochlear implant, congenital cochlear nerve aplasia or hypoplasia, malformations of the inner ear, and bilateral traumatic avulsions of the cochlear nerve, which often is accompanied by damage to the bony and membranous labyrinth (Kuchta 2007; Seki et al. 2007; Colletti et al. 2009a). In persons with ossified cochlea, a cochlear implant may be attempted after partial removal of the obstructing ossification, but often these people do not derive acceptable benefit from their cochlear implants, and an ABI may then provide them with improved hearing (Colletti et al. 2004). It is noteworthy that many of these "non-tumor" ABI users have been reported to perform significantly better with their ABIs than the "tumor" patients. Some of the best results have been achieved by users with traumatic avulsion of the cochlear nerves and short duration of hearing loss prior to implantation of the ABI (Colletti et al. 2009a). Possible explanations for the marked differences in performance between the "non-tumor" and "tumor" ABI users are discussed in Sect. 1.3.

In children with severe or profound sensorineural hearing loss, a cochlear implant has become a standard of care, but a subgroup of these children achieves little or no benefit from their cochlear implants. Most of these children have congenital bilateral cochlear nerve aplasia and/or agenesis, often with associated cochlear malformations. At the University of Verona as of the end of 2006, 24 such children, aged 14 months to 16 years, have received auditory brainstem implants (Colletti 2007). All have used their devices for more than 75% of waking hours, and all have achieved an awareness of environmental sounds and have developed the ability to utter words and simple sentences. They also exhibited significant development of non-auditory cognitive skills (Colletti and Zoccante 2008). It has been recommended that all children who receive cochlear implants but do not subsequently develop hearing-related skills should undergo CT or MRI imaging studies of the cerebellopontine angle and the cochlea. If the imaging studies reveal cochlear nerve degeneration or malformations of inner ear structures, an auditory brainstem implant is recommended (Carner et al. 2009). At the House Ear institute and Clinic in Los Angeles, CA (HEI), 21 pre-teen children and teenagers (age 12 to 18 years) who were deafened by NF2 have undergone implantation with multichannel ABIs. The experiences with these young people has served to emphasize the importance of strong motivation and strong family support in achieving a successful outcome and in maximizing the benefits that the patients receive from their implants (Otto et al. 2004).

Colletti and his associates (Colletti et al. 2002) have discussed the risks and benefits, and some of the ethical considerations attendant to implanting cochlear nucleus auditory prostheses. They noted that in patients with NF2, for whom removal of the tumors is a life-saving procedure, the prevalence and severity of the complications after tumor removal has not increased in those who have received an ABI. However, for "non-tumor" patients, for whom surgical access to the brainstem is not necessary in order to save their lives, a different equation may apply. The authors noted

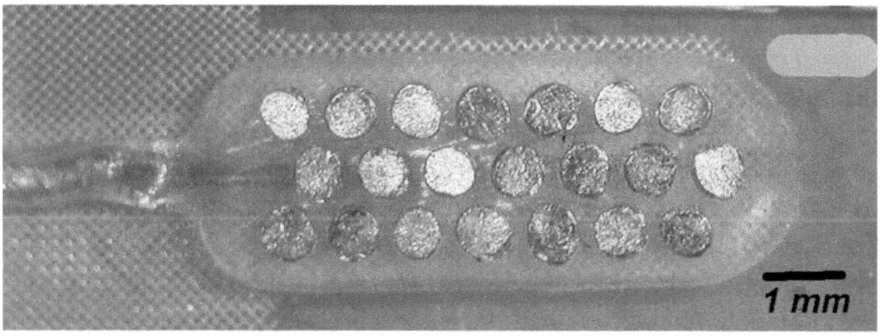

Fig. 8.1 The 21-electrode array of the Nucleus ABI24 Auditory brainstem implant system

that cerebellopontine angle surgery is currently used worldwide in neuro-otology as an elective procedure for several disabling but not life-threatening disorders, including vestibular neurectomy in Meniere's disease, neurovascular decompression of the cranial nerves for relief of trigeminal neuralgia, hemifacial spasm, disabling positional vertigo, tinnitus, and spasmodic torticolis; total deafness is no less incapacitating than some of the above mentioned disorders, particularly when it occurs in childhood.

1.2 The Evolution of the Cochlear Nucleus ABI

The first implantation of an ABI was performed by Hitselberger and House in 1979 (Edgerton et al. 1982). It was a single spherical electrode implanted adjacent to the cochlear nucleus of a patient with NF2, following removal of bilateral vestibular schwannomas. Subsequently, an array of 2, and later 3, electrodes affixed to a polyester mesh backing was developed by the Huntington Medical Research Institute of Pasadena, CA. These were used in the HEI program in combination with modified 3M–House sound processors originally designed for cochlear implants, and the first patient used the device with benefit for over 20 years (House and Hitselberger 2001). In 1992, in conjunction with the Huntington Medical Research Institute and Cochlear Ltd. (Lane Cove, Australia), 8-electrode and 21-electrode arrays were developed, and these multichannel devices have in most cases provided the patients with greater benefits than the early devices (Edgerton et al. 1982; Schwartz et al. 2008). The Nucleus 24 ABI (ABI24) differs from the earlier Nucleus 22 ABI in that several sound processing strategies are supported, and the ABI24 also offers the opportunity for intra-operative neural response telemetry. The ABI24 supports the Spectral Peak (SPEAK), the Continuous Interleaved Sampling (CIS), and the Advanced Combination Encoder (ACE) processing strategies. However, in the USA, only the SPEAK strategy has been approved by the Food and Drug Administration. The ABI24 and ABI22 systems utilize 21 0.7-mm platinum disk electrodes aligned on a flexible silicone and polyester mesh backing (Fig. 8.1) for implantation into the lateral recess of the fourth ventricle.

At present, more than 260 persons with NF2 have received the ABI at HEI, making it the largest program in the world. A recent review (Schwartz et al. 2008) summarizes the HEI program, including the surgical procedures, the postsurgical management of the patients, and the benefits that the patients derive from their implants. The surgical resection of the vestibular schwannomas and subsequent implantation of the ABI electrode array is always via the translabyrinthine approach, which provides excellent visualization of the relevant anatomy (Brackmann et al. 1993; Fayad et al. 2006). However, the retrosigmoid approach has been used successfully in other programs (Colletti et al. 2000; Lenarz et al. 2001; Kuchta 2007). In the event that the tumor on the eighth nerve can be resected or reduced without complete destruction of the eighth nerve, the retrosigmoid approach preserves the structures of the inner ear, allowing for a cochlear implant that may provide the patient with greater benefits than an ABI (Huy et al. 2008).

1.3 Device Fitting and the Management of ABI Users

In the HEI ABI program, a great deal of attention is paid to assessing and accommodating individual differences in the patients' responses to their ABI. Several factors make the dynamics of ABI fitting and outcomes more variable and less predictable than that of cochlear implants and undoubtedly contribute to the variability in the benefits that different patients receive from their ABI. Variations in the anatomy of the cerebellopontine angle after surgical removal of the schwannomas (Brackmann et al. 1993) and differences in placement of the electrode array affect the patients' percepts from the electrical stimulation. These percepts often change over time and must be reassessed with appropriate adjustments to the sound processor in order to optimize performance (Otto and Staller 1995). The impact that NF2 typically has on the patients' general health and ability to function must always be considered. The hallmark of the disease is the development of bilateral vestibular schwannomas, and the most common presenting symptom in adults is progressive hearing loss. However, persons with NF2 also develop tumors on other cranial and spinal nerves as well as other nervous system tumors, primarily meningiomas. Presenile cataracts, ocular motility disorders, peripheral neuropathies, and skin tumors are other common findings. Many patients become severely disabled and life expectancy is reduced (Elvsashagen et al. 2009).

Previously, when electrical access to the electrode array was via a percutaneous connector, it had been possible to test patients very soon after the surgery, in order to obtain an initial assessment of their responses to the ABI (Portillo et al. 1993; Zeng and Shannon 1994). The completely implanted receiver/stimulator system used in current ABIs has led to delaying this testing for 4 to 6 weeks after the tumor removal surgery, until swelling around the surgical site has subsided and the patients have recovered sufficiently from the surgery. During this interval, patients are confronted with the psychological trauma of losing any remaining hearing and possibly a decline in other physical abilities (e.g., facial nerve function). Because of the chronic and progressive nature of NF2, this can be an ongoing process that influences all facets of patient management, including how they use and benefit from their ABI (Otto et al. 2002). An experienced multidisciplinary team is necessary to deal effectively with these issues. An essential part of this process is the preoperative counseling, during which the potential benefits and limitations of the ABI are thoroughly discussed with the patient, including the possibility that the implant might provide no hearing at all. A patient's unrealistic expectations can have pervasive negative effects on successful long-term use and benefit from their ABI. Despite the use of intraoperative monitoring to help to position the ABI array, the outcomes are not known until the device is actually activated several weeks later. Even neural response telemetry (NRT), used extensively in programming cochlear implants in infants (Hughes et al. 2000) and some ABIs (Colletti 2007), has not proven accurate in predicting how well the device will function. For example, at the HEI there was little if any correlation between the presence (or absence) of compound action potentials and auditory or non-auditory percepts in 15 awake ABI recipients (Otto et al. 2005). It is for this reason that a cautiously optimistic outlook is best maintained with patients prior to initial device activation.

One major change in the implantation protocol occurred fairly early in the ABI program. While bilateral vestibular schwannomas are a hallmark of NF2, the tumors on the left and right nerves tend to originate at different times and to develop at different rates, and so they are removed in separate surgeries that may be months or years apart. Originally, ABI devices were implanted only at the time of removal of the second tumor, and the patients not only had to adjust to sudden complete deafness, but simultaneously had also to adjust to the new sounds from their ABI. This seemed to complicate the adaptation process unnecessarily, so clearance was obtained from the Food and Drug Administration to implant an ABI at the time of first-side acoustic tumor removal, even if the patients retained useful hearing in the contralateral ear (Otto et al. 1997). At the HEI, about a third of the NF2 patients have received an ABI at the time of their first-side tumor removal. This protocol has allowed patients to gain experience with their ABIs prior to losing all acoustic hearing at the time of removal of the second tumor and appears to ease the psychological burden of transitioning to complete deafness. It also provides a second opportunity to achieve a functional and beneficial ABI system if the first device does not provide useful hearing, which occurs in about 9% of the patients at the HEI. Overall, the ABI has achieved a very good record of safety and efficacy that resulted in FDA approval of the Nucleus ABI24 system in 2000. However, during initial activation, the patient's vital signs are monitored because of the proximity of the electrode array to neural tracts and brain stem nuclei controlling respiratory and cardiovascular function. Emergency resuscitation equipment and experienced medical personnel are available during this process.

Fitting an ABI is more complicated and time-consuming than fitting a cochlear implant because of more frequent non-auditory percepts as well as the marked individual differences in electrode-specific pitch percepts (Otto and Staller 1995). The process can be facilitated by periodic breaks for patients whose ability to tolerate lengthy testing may be affected adversely by the NF2. In the HEI program, the initial activation and fitting of the ABI usually is spread over a two-day period. This allows new ABI recipients to use the device after the first session before returning home, and thus any difficulties can be addressed during the second day's testing. Subsequently, ABI recipients are re-evaluated every three months for the first year and then annually. This follow-up schedule has been shown to accommodate the changes in auditory and non-auditory percepts over time and to track and monitor performance (Otto et al. 1998).

The software used to program the Nucleus ABI24 allows many parameters to be set, including, but not limited to, stimulus pulse duration, stimulus amplitude, inter-pulse interval, sensitivity of the microphone pre-amplifier, the number of spectral maxima transmitted, the bandwidth of each spectral channel, level of power output from the transcutaneous transmitter coil, and overall frequency range of the encoded sound. It is by no means clear what the optimum *a priori* combination of settings will be for individual patients. It is clear, however, that the parameters that provide the greatest benefit to a particular patient can change significantly over time and that interactions occur between these parameters. For example, increasing the stimulus pulse duration may minimize non-auditory sensations but can result in a decrease in battery life, stimulation rate, and the number of spectral channels. By necessity, fitting ABIs involves subjective and objective processes to maximize performance within the perceptual limitations of the ABI recipients and the capabilities of the programming system.

Prior to the first activation of their ABI, the patients are again counseled about the potential outcomes and are provided with written instructions regarding the fitting process. They are informed that they may hear sound, or experience non-auditory percepts, including a tingling sensation, mild dizziness, or visual "jittering" (Otto et al. 2002). Patients who are both blind and deaf have been fitted with ABIs at the HEI, and these individuals require substantial modification to the fitting process, including establishing an unambiguous signal to the patient to cue when an electrode is being activated. The ABI usually provides the greatest benefit when used in combination with lip-reading, and patients with little or no vision tend to derive less benefit from their device. If possible, it is extremely helpful to discuss the ABI in detail, prior to the patients becoming completely deaf or blind.

The determination of the thresholds for auditory percepts for each electrode starts with a very low stimulus current that increases until the patient indicates detection of some percept. If it is an auditory percept, the stimulus amplitude is increased until the patient indicates that the percept is "comfortably loud." It is not uncommon for individual electrodes of an ABI to produce non-auditory percepts, and in some cases the patient finds these to be sufficiently uncomfortable such that the particular electrode cannot be used (Lenarz et al. 2001; Colletti et al. 2002; Otto et al. 2002). The locations of the non-auditory sensations tend to be equally distributed between the head and body. While they most often are described as tingling, mild jittering of vision often is reported, and occasionally a feeling of dizziness requires the deactivation of the offending electrodes. In one group of 61 patients of the HEI series, 24% of 404 electrodes could not be used because of non-auditory sensations. The incidence of non-auditory sensations was greater for the most medial and lateral electrodes, with the most medially located electrodes having the highest incidence. Non-auditory percepts were almost always ipsilateral to the side of the implant (Otto et al. 2002). These sensations often can be reduced by increasing the stimulus pulse duration and concurrently decreasing the stimulus current. However, this must be balanced against the finding that shorter pulses are most efficient in ABI stimulation and therefore extend battery life in the sound processor (Shannon and Otto 1990). Patients are queried regarding the occurrence of any non-auditory sensations, and they are asked to rate the intensity of these on a 4-point scale from "barely noticeable" to "intolerable," and electrodes can often be retained for use even if they generate mild non-auditory percepts (Otto and Staller 1995). During actual use of the ABI, stimulation with these electrodes typically occurs over such a brief interval that non-auditory sensations may not be detected. If the patient has only a few electrodes that are completely free of non-auditory sensations, the availability of these additional electrodes can be important. After the auditory threshold, comfort level, and presence and magnitude of any non-auditory sensations have been determined for all of the electrodes, an effort is made to balance the comfortable loudness levels across all useable electrodes, prior to the assessment of the pitch percepts from each electrode.

Many methods have been used for evaluating and ranking the pitch percepts produced by the ABI electrodes, and one approach uses custom computer software (Long et al. 2005). At the HEI, the following approach is used so that progressively more precise pitch rankings may be made as the patient gains experience with their implant. First, all the useable electrodes are activated in sequence and the patient is queried regarding whether they elicit the same or different pitch percepts. About 60% of patients demonstrate a somewhat ordered range of pitch sensation across the ensemble of electrodes, usually increasing in pitch in a lateral-to-medial direction across the array (Otto et al. 2002). The other patients have reported either a falling pitch pattern in the lateral-to-medial direction or a random pattern of pitch across the electrode array. Some patients report essentially no pitch differences for different electrodes, in which case the remainder of the pitch assessment and ranking procedure is deferred until a later session. Sometimes patients will begin to perceive subtle pitch differences after they obtain more experience with their implant. If the patient describes different pitch differences across the array, they are asked to scale the pitch on a scale of 1 to 100 (low to high pitch). This provides information about the distribution and overall range of their pitch percepts, which can differ greatly for different patients (Otto et al. 1998). If necessary, the pitch ranking across electrodes is refined by conducting paired comparisons between the adjacent electrodes from the initial ranking. Unfortunately, there is no way of altering the range and distribution of the pitch percepts, which may be at least partially related to the proximity of some of the electrodes to surviving neurons in the cochlear nucleus (Fayad et al. 2006). It would be desirable to be able to change the pitch percepts by steering the stimulation to discrete populations of surviving neurons, but this is not possible with the current hardware and software. For example, if a patient has two narrow groups of pitch percepts, each concentrated at extreme ends of the pitch range, this unusual pitch pattern would have a different effect on speech recognition than an evenly distributed pitch pattern. Combining bipolar pairs on the electrode array has been used in a few patients to achieve intermediate pitches (Otto and Staller 1995) but has yielded mixed results.

After a reliable relative pitch ranking of electrodes is achieved, the sound processor is programmed with the patient's threshold levels, comfortable loudness levels, and electrode pitch-rank ordering. As has been done since the early days of multi-channel cochlear implants, each electrode is assigned to be activated over a limited range of acoustic frequencies. With the ABI, it is adequate to cover an overall acoustic frequency range of approximately 160 to 6188 Hz, which encompasses the spectral range of speech and of the most important environmental sounds.

The FDA-approved Spectral Maxima (SPEAK) speech processing strategy searches for peaks in the sound spectrum and uses one or more electrodes to convey this acoustic activity to the user. In practice, the number of spectral maxima (up to 9) usually is 1 or 2 less than the number of usable electrodes. ABI recipients with NF2 can derive substantial benefits from a sound processor program which utilizes a least four electrodes that are pitch-rank ordered and convey at least 3 spectral maxima. More electrodes (up to about 8) have resulted in improved speech recognition (Otto et al. 2002). As few as 2 electrodes that are properly pitch-rank ordered often provided above chance levels of speech perception and in fact has provided better performance than 8 or more electrodes that are not pitch-ranked. This practice has provided reasonably good performance, as verified by hundreds of assessments of speech perception in ABI recipients with different numbers of useable electrodes (Kuchta et al. 2004). If the numbers of usable electrodes and spectral peaks are identical, the SPEAK processing strategy is essentially identical to the "continuous interleaved sampling" strategy (Wilson et al. 1991) that has been widely used by cochlear implant users, and in some ABI programs outside of the US (Colletti et al. 2005, 2009b; Behr et al. 2007). However, a study comparing speech perception with the SPEAK and ACE strategies in the best ABI performers found no significant difference in performance between the two strategies (Otto, personal communication).

For the patients with little or no ability to pitch-rank their electrodes, the policy at the HEI has been to program their processors according to the pitch-ranking pattern seen in about 60% of patients. Increasingly higher acoustic frequency bands are assigned to electrodes 2 through 22 on the electrode array (Otto and Staller 1995; Otto et al. 2002). The patent's pitch-ranking percepts are evaluated periodically. The default map is retained if the patient performs well and indicates a strong preference for the accustomed pitch-ranking, but they are encouraged to experiment with revised electrode maps. They may even experience a temporary decline (approximately 10%) in speech perception, but if the programming changes are appropriate, this decline is temporary, and higher levels of performance may be achieved.

1.4 Benefits that Users Derive from Cochlear Nucleus ABIs

With few exceptions, most ABI users do not perform as well as cochlear implant users on speech recognition and related tasks. However, most users do derive real benefits from their implants, most notably when the device is used in combination with lip-reading in a relatively quiet environment (Lenarz et al. 2002; Otto et al. 2002; Schwartz et al. 2008; Colletti et al. 2009b). Figure 8.2a, b summarizes speech

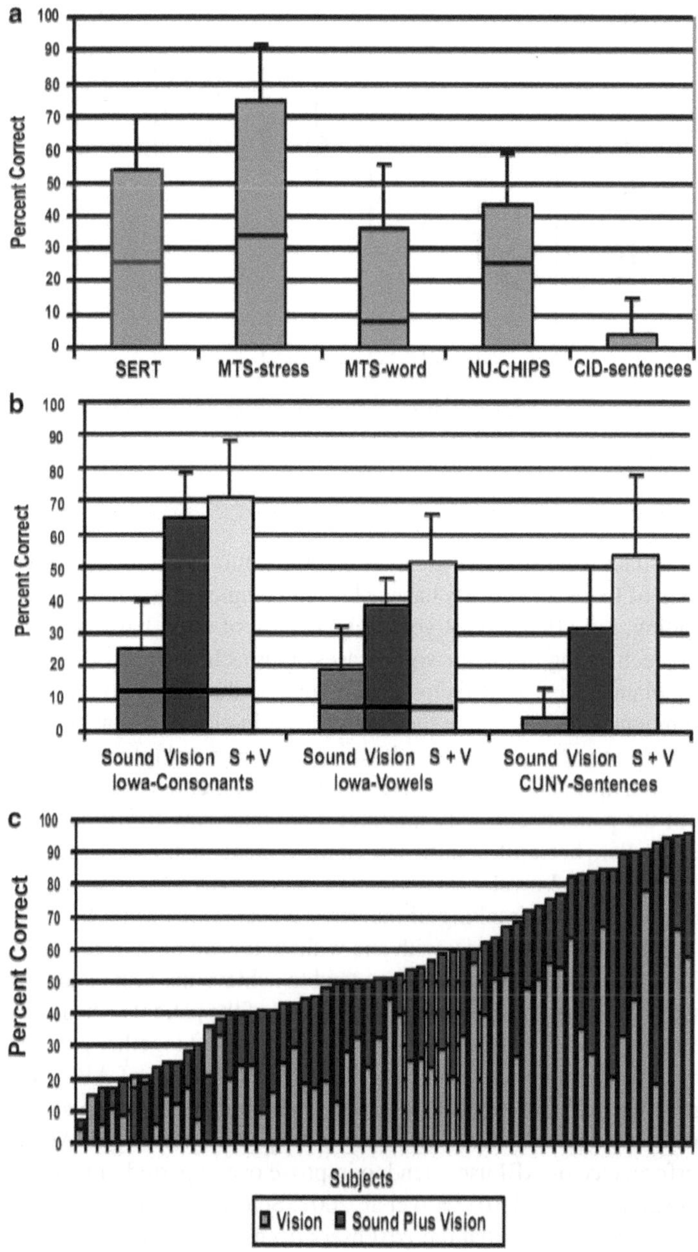

Fig. 8.2 Performance of 61 users of ABIs from the House Ear Institute series. (**a**) Performance on standard audiology tests without the aid of lip reading. (**b**) Their performance on three word recognition tests using the ABI alone ("sound"), using lip-reading alone ("vision") and when their ABI is used in conjunction with lip-reading ("S+V") (from Schwartz et al. 2008). (**c**) The performance of the 61 users ranked according to their ability to interpret CUNY sentences correctly when their ABI is used alone or in combination with lip-reading ("sound+vision") (from Schwartz et al. 2008, adapted from Otto et al. 2002). (**d**) Improvements over time in the recognition of MTS (monosyllable-trochee-spondee) words by ABI users with NF2. (Reproduced with permission from Schwartz et al. 2008)

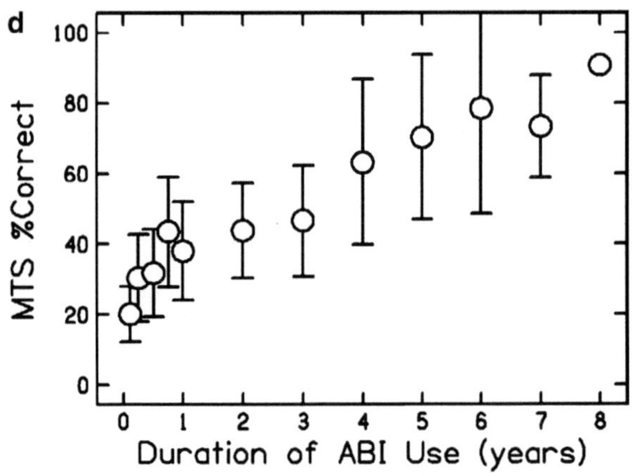

Fig. 8.2 (continued)

performance data from 61 ABI users at the HEI (Otto et al. 2002; Schwartz et al. 2008). None of these individuals have achieved complete speech recognition without lip-reading, but after several years, about 20% of them have developed some open-set speech recognition, in some cases approaching that of some users of cochlear implants. Most ABI recipients do develop the ability to recognize words and environmental sounds above the chance level in closed-set tasks. For recognition of CUNY sentences (sentences with minimal contextual information), the mean improvement in lip-reading when also using their ABI is about 30% (range: 0–66%), and 31% of the patients correctly interpret more than 70% of the sentences. They achieve a smaller, but still significant, improvement in the recognition of Iowa Consonants and Iowa Vowels.

Figure 8.2c shows the ranking of the performance of the same 61 ABI users on recognition of CUNY sentences, with and without the aid of lip-reading. ABI users who perform well with lip-reading alone tend to achieve the best results when their ABI is used to augment the lip-reading, but several of the patients who have achieved the highest scores when the ABI and lip-reading are used together performed quite poorly with lip-reading alone. It is clear that the performance of ABI users spans a very large range, and this variability across users must be taken in account in the development of future ABI that will improve the performance of all ABI users.

The performance of ABI users tend to improve over a period of many months or even years (Lenarz et al. 2001; Otto et al. 2002; Schwartz et al. 2008; Colletti et al. 2009b). Figure 8.2d shows that the ABI users from the HEI series with the longest experience achieve higher scores on the MTS (monosyllable-trochee-spondee) test. Speech recognition may continue to improve for as long as 10 or 15 years after implantation. However, the first few months after initial activation of the implant seem to be the most critical, and patients who do not persevere may deny themselves the eventual substantial long-term benefits. In the Department of Otolaryngology at the Medical University of Hannover (Lenarz et al. 2001) more

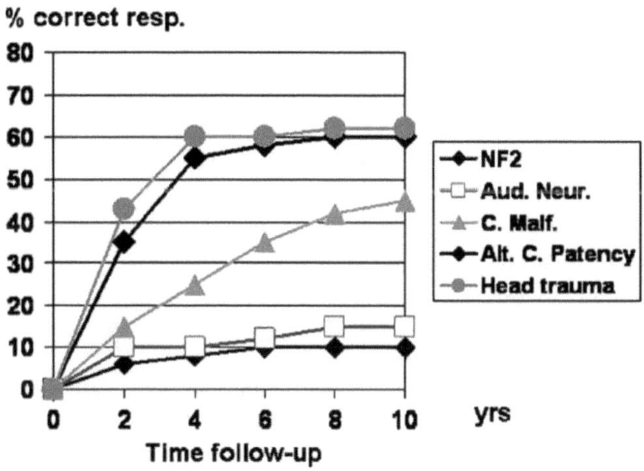

Fig. 8.3 Average performance (% correct on open-set speech) over time for patients at the University of Verona who were deafened by various pathologies (*NF2* Type 2 Neurofibromatosis, *Aud Neur* auditory neuropathy, *Alt, C. patency* altered cochlear patency and cochlear malformation, and *head trauma*). (Reproduced with permission of Wolters Kluwer, from Colletti et al. 2009a)

than half of the patients demonstrated improved performance during the first 3 months, and for some users, performance continued to improve during the first 6 months and then mostly reached a plateau, particularly in the closed-set vowel and consonant confusion tests. The European ABI multicenter clinical trial also showed that in a more realistic listening environment, the performance of ABI users improves over time, especially during the first 6 to 9 months after device activation (Sollmann et al. 2000). The findings are generally similar for patients whose deafness is of etiologies other than NF2 (Fig. 8.3).

In summary, the ABI provides persons with NF2 with sound awareness, discrimination of many environmental sounds, and an average 30% improvement in speech understanding when combined with lip-reading. Most of the NF2 patients find their ABI to be helpful, and they use the device during most waking hours. As with cochlear implants, performance is negatively impacted in a noisy environment. An additional benefit to many users is suppression of the tinnitus which often occurs after removal of vestibular schwannomas (Soussi and Otto 1994; Behr et al. 2007; Schwartz et al. 2008).

Etiology of deafness significantly affects the performance of ABI users. Colletti et al. (Colletti et al. 2009a) have documented the performance of 82 patients from the University of Verona ABI program, including 48 whose deafness is of etiologies other than NF2 ("non-tumor" patients). The non-tumor adult patients scored from 10% to 100% in open-set speech perception tests (average, 59%) while the patients with NF2 scored from 5% to 31% (Fig. 8.3). The non-tumor patients deafened by traumatic bilateral damage to the auditory nerves or cochlea achieved the highest scores, followed by those with post-meningitis cochlear ossifications, followed by

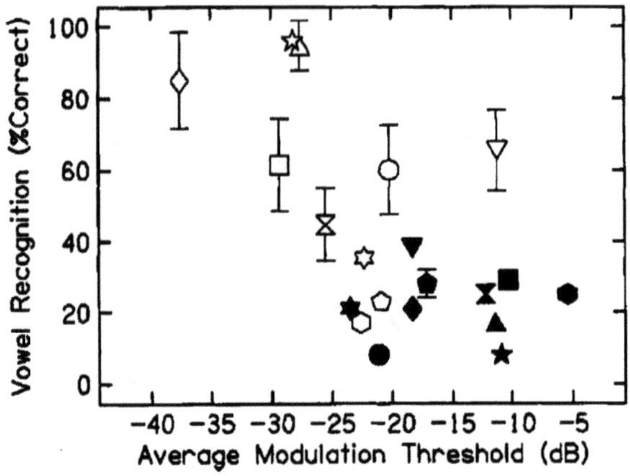

Fig. 8.4 Vowel recognition as a function of the average modulation threshold, expressed in dB re 100% modulation. *Open symbols* represent non-tumor patients; *filled symbols* represent patients with Type 2 Neurofibromatosis. (Reproduced with permission of John Wiley and Sons, Inc., from Colletti and Shannon 2005)

those with cochlear malformations. The average performance of patients with severe congenital cochlear malformations is no better than that of the patients with NF2.

It is not clear why many of the non-tumor patients achieve better performance than those with NF2. The duration of deafness prior to the implant may explain some the difference across the categories of non-tumor patients. Also, in the tumor patients, the cochlear nucleus may have been damaged either by the space-occupying tumors or during their surgical removal. As a group, the modulation detection thresholds (MDT) of the NF2 subjects were significantly higher than those of the non-tumor subjects (Colletti and Shannon 2005) and the MDTs were significantly correlated with vowel recognition scores (Fig. 8.4). The relatively poor performance of the NF2 subjects cannot be attributed entirely to their high MDT, because a few of the non-tumor subjects exhibited high MDT but attained vowel recognition scores that were higher than those of any of the NF2 subjects. However, the high MDT is the only psychoacoustic variable that has been found to distinguish the NF2 and non-tumor ABI users, and it is possible that all users of ABIs, including those with NF2, would benefit from a prosthesis that allows improved modulation detection. In cochlear-implant users, there is a strong correlation between phoneme recognition scores and subjects' mean modulation detection thresholds (Fu 2002).

2 Electrical Stimulation in the Cochlear Nucleus and Related Safety Issues

A neural prosthesis must activate the requisite neurons without causing injury to the neurons or to other tissue elements. Electrical stimulation is an effective, albeit unnatural, means of exciting neurons and axons, so certain precautions are necessary if tissue injury is to be avoided. There are several mechanisms by which electrical stimulation might inflict tissue injury (McCreery 2004). The review by Shepherd and McCreery (2006) addresses this topic with special reference to cochlear implants and cochlear nucleus auditory prostheses. The propensity for stimulation-induced neural injury is influenced by all of the stimulus parameters, including charge density and charge per phase. One of the cardinal principles of functional electrical stimulation is that the charge injection into the tissue should be by electrochemical processes that are completely reversible (Cogan 2008). It is generally accepted that the stimulus charge density should not exceed that for which the processes that mediate charge injection into the tissue are completely reversible (Cogan 2008). Adherence to this principle does not guarantee that stimulation-induced injury will not occur (McCreery et al. 1990), but it does minimize the risk of generating toxic products at the electrode-tissue interface, including the large pH changes and the accompanying generation of gas bubbles that are potentially injurious (Huang et al. 2001). Polished platinum disk electrodes used in the array shown in Fig. 8.1 can support a charge density of ~100 $\mu C/cm^2$ (Rose and Robblee 1990).

For the NF2 patients who have received auditory brainstem implants that employ an array of macroelectrodes (Fig. 8.1), the threshold for auditory percepts ranges from less than 0.01 μC/ph to more than 0.2 μC/ph (Shannon and Otto 1990; Colletti et al. 2005). Some patients require as much as 0.3 μC/phase in order for the auditory percepts to reach maximum comfortable loudness (Otto et al. 2002), which corresponds to a geometric charge density of 75 $\mu C/cm^2$. This is near the maximum reversible charge density recommended for smooth platinum. At least in the cat's cerebral cortex, charge per phase and charge density interacts to determine the propensity for tissue damage (McCreery et al. 1990). The analogous study has not been conducted for macroelectrodes implanted over the cochlear nucleus, but in the cerebral cortex, the combination of 75 $\mu C/cm^2$ and 0.3 μC/ph would not be expected to be injurious. This is consistent with the observation that the stimulus amplitude for threshold and maximum comfortable loudness levels for ABI users tend to be stable over many years (Otto et al. 2002; Schwartz et al. 2008).

The inclusion of penetrating microelectrodes in a cochlear nucleus auditory prosthesis (Sect. 3.3) will require some special precautions. There is a risk of tissue damage during insertion of the electrodes into the vascular brain tissue and then during their long-term residence in the brain. There also is the opportunity for injury from the electrical stimulation *per se*. Because of their small geometric surface areas (typically 1000 to 5000 μm^2), the stimulus charge density at the microelectrode-tissue interface typically exceeds that recommended for chronic stimulation with platinum or platinum alloys. However, other electrode materials, including several forms of activated (oxidized) iridium do support electrochemical processes not shared by platinum or its alloys, which increases their safe charge injection capacity (Robblee et al. 1983; Cogan 2008). The safety issues attendant to prolonged microstimulation in the cochlear nucleus have been investigated in a series of studies in a cat model, and protocols for safely exciting cochlear nucleus neurons have been developed (McCreery et al. 1994, 1997, 2000; McCreery 2004). Although stimulus parameters that do not induce histologically detectable neuronal damage may still produce prolonged depression of neuronal excitability, stimulus parameters have been identified which do not inflict tissue injury and which also preserve most of the dynamic range of the neuronal response to the electrical stimulation.

3 The Challenge of Improving the Performance of Cochlear Nucleus Auditory Prostheses

While nearly all persons with auditory brainstem implants do derive real benefits from their device, there clearly is need for much improvement, particularly for the patients afflicted with NF2. Also, while many patients whose deafness is of etiologies other than NF2 often do acquire better speech perception than the NF2 patients, their performance spans a wide range (Colletti et al. 2009a), and typically they require many months or even years for their performance to approach that of most users of multi-channel cochlear implants (Fig. 8.3). The challenges facing the development of improved ABIs include effective extraction and conveyance of auditory information to the appropriate neuronal populations within the cochlear nucleus; how best to encode sound information for a prosthesis that by its very nature will bypass some of the neural circuitry that normally encodes the temporal and spectral features of sound; and the development of novel electrode arrays. To date, ABIs have employed and adapted sound encoding strategies designed for multi-channel cochlear implants, in which the emphasis has focused on utilizing the surviving anatomical elements and the tonotopic organization of the cochlea in order to convey information about the spectral and the temporal features of sound. Future cochlear nucleus prostheses could be improved by processing strategies that better convey sound spectral information and expand the concept of information "channels" to include dimensions other than acoustic frequency.

3.1 The ABI as a Conveyer of Sound Spectral Information

As in a cochlear implant, the sound processor of an ABI partitions the sound spectrum into contiguous channels that are encoded into the electrical stimuli that are directed to individual electrodes. An array of macrostimulating electrodes implanted on the surface of the brainstem and separated from the cochlear nucleus by the pontobulbar body and the glia-pia membrane (Moore and Osen 1979) can activate neurons throughout much of the cochlear nucleus (Shannon et al. 1997). The relatively broad spread of the stimulus current within the nucleus certainly will limit the ability of different electrodes to access the tonotopic organization of the cochlear nucleus selectively and thus also their ability to convey unique sound spectral information. Clinical data has shown that in most cases, ABI users with more functional electrodes tend to perform better (Otto et al. 2002). However, while cochlear-implant users can achieve very high levels of speech recognition in quiet with as few as 4 usable electrodes (Shannon et al. 2004), ABI users rarely achieve the same level of performance. It is not clear to what extent the limited ABI performance derives from the diffuse stimulation provided by the surface electrodes or is a result of a loss of the capacity for the neuronal machinery of the cochlear nucleus to make full use of the spectral information that the implants can provide. However, it is clear that future ABIs will need to convey spectral information to the appropriate locations within the cochlear nucleus.

3.2 The ABI and the Functional Neuroanatomy of the Cochlear Nucleus

The opportunities and challenges for improved cochlear nucleus auditory prostheses can best be appreciated in the context of the anatomy and physiology of the cochlear nucleus, where many of the basic features of sound are extracted and encoded by intermixed, but partially segregated, populations of neurons.

In the cochlear nucleus, many of the basic temporal and spectral features of sound are extracted and processed into parallel channels of auditory information by morphologically and physiologically distinct neuronal populations that project from the cochlear nucleus to other structures in the auditory brainstem. In all mammals, there is an ordered spatial representation of acoustic frequencies within each subdivision of the cochlear nucleus (Powell and Cowen 1962; Leake and Snyder 1989; Snyder et al. 1997), whereby auditory nerve axons that originate from a small segment of the tonotopically ordered cochlea terminate in a thin laminar plexus abutting those originating from contiguous small regions of the cochlea. Of course, this tonotopically ordered innervation would be absent in persons without a functional auditory nerve, so for a cochlear nucleus auditory prosthesis that utilizes stimulating electrodes implanted on the surface or within the nucleus, the ability to convey the appropriate information into a particular part of the cochlear nucleus and to a particular population of neurons will be determined by the spatial distribution of the stimulating electrode sites, the spatial localization of the electrical stimulus from each electrode site, and the extent to which the neuronal populations are spatially segregated. It also will be essential that the manner in which sound is encoded into the electric stimulus is compatible with the "neural code" that the neurons employ to convey the temporal and spectral features of the sound.

The human cochlear nucleus spans approximately 3 mm along its dorsolateral and ventrolateral dimensions but much of the rostral portion of the ventral nucleus lies beneath the middle cerebellar peduncle (Moore and Osen 1979). In the nomenclature of Osen and Brawer (Osen 1969; Brawer et al. 1974), the ventral cochlear nucleus (Fig. 8.5) includes a rostral region of spherical cells; a central region where globular bushy cells, spherical cells, and a heterogeneous population of multipolar cells are intermingled; and a caudal region containing the morphologically unique octopus cells (Osen 1969; Moore 1987; Cant and Benson 2003).

The partial spatial segregation and unique physiologic properties of the multipolar and octopus cells illustrates how these features might be exploited in an improved cochlear nucleus auditory prosthesis. The multipolar cells of the ventral cochlear nucleus are especially numerous in humans and in other primates (Moore 1987). The "Type I multipolar cells" (Smith and Rhode 1989; Cant and Benson 2003) respond best when the acoustic signal is amplitude modulated, suggesting that they are specialized to amplify and extract envelope information from the sound (Moller 1974, 2006; Frisina 2001). The axons of Type 1 multipolar cells project to contralateral auditory nuclei via the ventral acoustic stria, the trapezoid body (Adams 1979; Smith and Rhode 1989; Cant and Benson 2003). The importance to speech perception

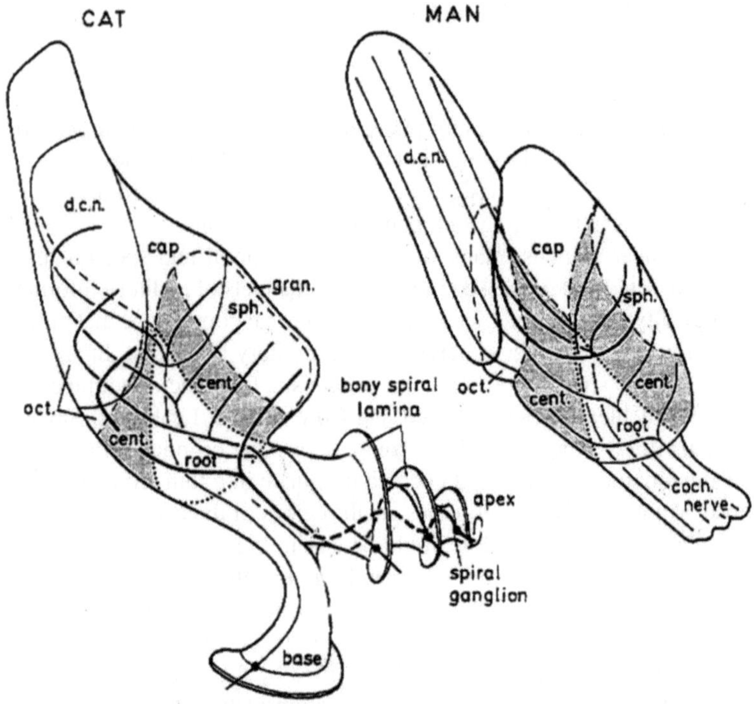

Fig. 8.5 A lateral view the human and feline cochlear nucleus complex. The regions of the central nucleus containing the multipolar cells is stippled. (Reproduced with permission of John Wiley and Sons, Inc., from Moore and Osen 1979)

of this contralateral projection is illustrated by a patient who lost all speech perception after a midline pontine tegmental hemorrhage in the region of the trapezoid body (Egan et al. 1996). The projection of the Type I multipolar cells is primarily to contralateral brainstem nuclei via the trapezoid body while the ipsilateral and contralateral projections of the bushy cells are more balanced (Warr 1972). Thus, the survival of the patient's ipsilateral pathways is consistent with her ability to perceive pure tones but her total loss of speech perception demonstrates the importance to speech perception of the information carried in the trapezoid body and by inference, the vital role of the Type I multipolar cells. When considering how cochlear nucleus auditory prostheses might be improved, especially for NF2 patients whose modulation detection tends to be poor, it is instructive to consider how the Type 1 multipolar cells and similar neurons in the cochlear nucleus extract and amplify sound envelope information, and how these mechanisms might be affected by loss of the auditory nerve and/or by damage to the cochlear nucleus by the disease or during surgical removal of the eighth nerve tumors. Each Type 1 multipolar neuron receives inputs from auditory nerve fibers that span a range of spontaneous firing rates, which imparts them with a wide dynamic range to acoustic stimuli and enhances their sensitivity to amplitude modulation (Smith and Rhode 1989; Frisina 2001), but this functionality would be lost in persons without a functional auditory nerve. Physiologic and modeling studies indicate that inhibitory inputs to the multipolar neurons can further enhance the multipolar neurons' sensitivity to amplitude modulation (Frisina et al. 1990; Dugue et al. 2007). Most of the GABA-ergic and glycinergic (inhibitory) neurons in the ventral cochlear nucleus are located in the small cell cap (Kolston et al. 1992; Moore et al. 1996), a cytoarchitecturally distinct region of the ventral cochlear nucleus that is especially prominent in humans (Fig. 8.5). It has been suggested that damage to this portion of the cochlea nucleus because of compression by the acoustic tumors of NF2 or during tumor removal could account for the poor modulation detection by ABI users with NF2 (Colletti and Shannon 2005).

The octopus cells are a homogeneous and morphologically distinct population of neurons located near the caudal pole of the posteroventral cochlear nucleus (Brawer et al. 1974; Moore 1987). They discharge action potentials with extremely precise timing and mainly at the onset of an acoustic stimulus, to which they quickly adapt (Godfrey et al. 1975; Rouiller and Ryugo 1984). Their response pattern suggests that they precisely encode broadband temporal timing information, which could be important for speech perception.

3.3 *The Role of Penetrating Microelectrodes in Future ABIs*

The fact that the octopus cells and to a lesser degree, the multipolar cells, are spatially segregated in the ventral cochlea nucleus suggests that they might be selectively accessible by an appropriately designed auditory prosthesis. Thus, stimuli conveying and accentuating sound amplitude modulation and spectral information could be directed to the central nucleus containing the multipolar cells, while a more broadly tuned representation of the precise timing of the onset of temporal features could be directed to the caudal part of the ventral nucleus containing the octopus cells.

Several groups have investigated microstimulation within the cochlear nucleus in animal models (McCreery et al. 2010). Ten patients of the HEI ABI series have received penetrating ABIs (PABIs), which includes a surface array similar to that shown in Fig. 8.1 and also an array of 8 or 10 activated iridium penetrating microelectrodes, each with an active surface area of 2000 or 5000 μm^2 (Fig. 8.6a). The penetrating array is implanted into the ventral cochlear nucleus with the hand-held inserter tool shown in Fig. 8.6c (McCreery 2008; Otto et al. 2008). Figure 8.6d depicts the locations of the surface and penetrating electrode arrays, adjacent to and within the cochlear nucleus.

Figure 8.7 shows the threshold and maximum comfort levels for the auditory percepts from PABI patient #2's penetrating and surface microelectrodes. The thresholds for auditory percepts for the penetrating electrodes were 1 or 2 orders in magnitude lower than for the surface array, reflecting the smaller distance between the penetrating electrode and the excitable neural elements. The thresholds from the

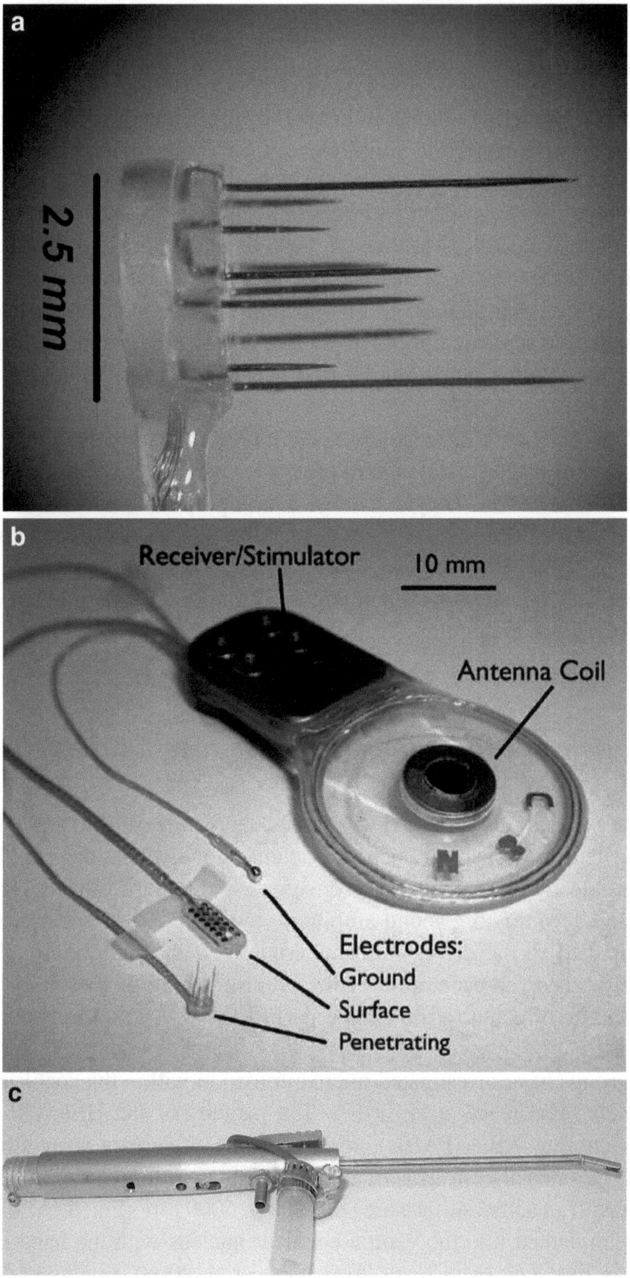

Fig. 8.6 (**a**) The array of penetrating microstimulating electrodes used in the clinical PABI system. (**b**) The implantable portion of the PABI system, showing the Nucleus 24 implantable receiver/transmitter and the surface and penetrating electrode arrays. (**c**) The hand-held inserter tool used to inject the penetrating array into the cochlea nucleus. (**d**) Schematic representation of the locations of the penetrating and surface electrode arrays, adjacent to and within the cochlear nucleus. (**a, c**) Reproduced with permission of Elsevier, from McCreery (2008). (**b, d**) Reprinted with permission of Wolters Kluwer, from Otto et al. 2008

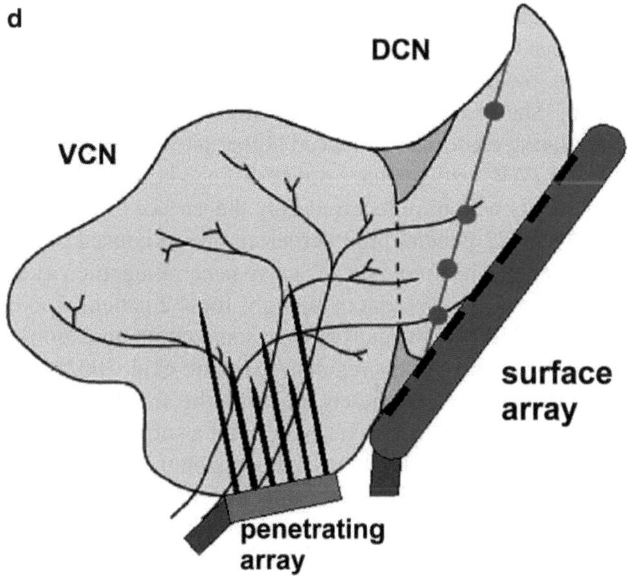

Fig. 8.6 (continued)

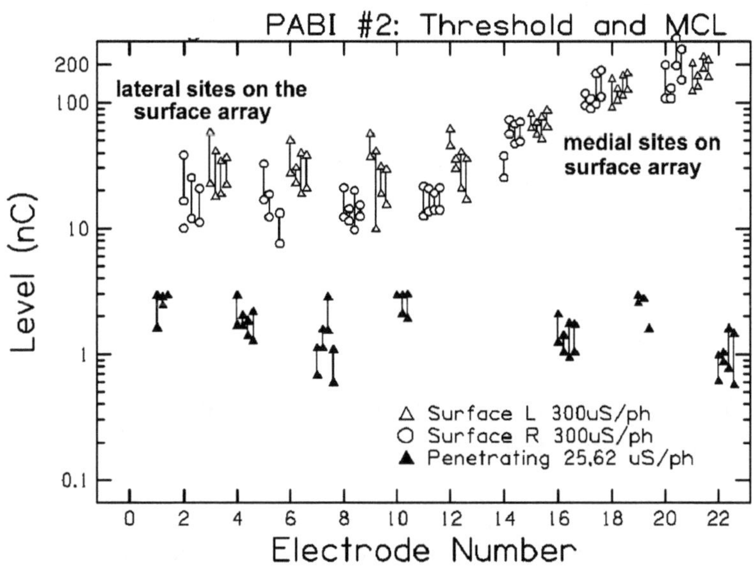

Fig. 8.7 Thresholds for auditory percepts (*lower symbol of each pair*) and maximum comfort level (*upper symbol*) for the surface and penetrating electrodes from PABI patient #2 measured over a span of 40 months after the implant surgery. Successive repeated measures for each electrode are displaced slightly to the right. L and R designate surface electrodes on the left (caudal) and right (rostral) side of the surface array shown in Fig. 8.1. (Reproduced with permission of Elsevier, from McCreery 2008)

surface and the penetrating electrodes have been quite stable for more than 3 years. This patient perceived different pitches from each of the penetrating microelectrodes, ranging from very low to very high, with the latter sounding similar to "the highest note on the piano." She describes the sound as similar to that of a calliope. In general, the shorter penetrating electrodes produced higher pitch percepts.

As a group, the PABI patients have not performed better on speech recognition tasks than the patients who have received only the surface array (Otto et al. 2008). However, only 14 of 72 penetrating microelectrodes produced auditory percepts, and only 2 PABI patients had more than 2 active penetrating microelectrodes (4 and 5 microelectrodes, respectively). Encouragingly, these 2 patients scored an average of 68% correct on NU-Chip words (Otto, personal communication) *vs.* a median score of about 38% for users of the regular ABI (Otto et al. 2002). This experience has revealed the difficulty of accurately targeting the small penetrating array into the small cochlear nucleus, which in humans is not a surface structure and must be located on the basis of often ambiguous landmarks on the surface of the brainstem, whose topology often is markedly distorted by the eighth nerve tumor (Brackmann et al. 1993). The difficulty of placing a sufficient number of electrode sites within the central nucleus of the VCN could be mitigated by using an array of many penetrating microelectrodes distributed over a much larger footprint. This strategy could best be implemented by an array of multisite microelectrodes, which offer the advantage of placing the maximum number of independent stimulating sites into the target nucleus with the minimum number of electrode shanks and thus the minimum risk of tissue injury. Arrays of such multisite silicon probes have been implanted into the ventral cochlear nucleus of cats for more than 300 days (McCreery et al. 2007; McCreery 2008). Compound neuronal responses evoked from the sites in the VCN were recorded in the contralateral inferior colliculus, and the threshold and growth of these responses were stable for at least 250 days after implantation.

Another finding was that PABI recipients with functioning penetrating electrodes preferred sound processor configurations using a combination of surface and penetrating electrodes, suggesting the penetrating and surface electrodes provide functionally independent auditory percepts.

3.4 The Opportunities for Novel Sound Processing Strategies for ABIs

Currently, the only sound processing strategy approved by the FDA for use in the USA by ABI recipients is the SPEAK (spectral peak) sound processing strategy, which also is used in other programs around the world that use the Cochlear Corporation ABIs. The ABI24 also supports the SPEAK, ACE, and CIS strategies. In the University of Verona program, patients may be transitioned to the ACE processing strategy after 6 months if their performance on speech-related tasks does not improve (Colletti et al. 2002), and the high-rate CIS processing strategy has been used routinely in at least one program (Behr et al. 2007). However, the SPEAK,

CIS, and ACE processing strategies were developed for cochlear implants, and the temporal features of sound that are encoded into the pattern of stimulus pulses may not be interpreted correctly by the deafferentated cochlear nucleus neurons. Efforts to develop improved ABIs should encompass how best to convey sound spectral information and also how to convey separate "channels" of information, including temporal information that is not directly related to the sound spectrum (Shannon and Otto 1990).

One facet of sound processing deserving of special attention, especially in light of the apparent deficit in modulation detection by ABI users with NF2 is how to convey information about sound amplitude modulation better. In the cochlear nucleus, the "neural code" for sound amplitude modulation appears to be the synchrony of the action potentials generated by an ensemble of neurons (Rhode and Greenberg 1994). The amplitude-modulated train of charge-balanced pulses used in all of the ABI sound processing strategies may not be optimal for the mechanism by which the neurons of the cochlear nucleus amplify and encode amplitude modulation, because their action potentials will tend to be synchronized with the stimulus pulses rather than with the modulation envelope. Finally, the strategies currently used to encode sound cannot convey the temporal fine structure of speech, a deficiency which appears to affect the performance of cochlear-implant users in noisy environments adversely (Zeng et al. 2005).

A compressed analog stimulus might circumvent these limitations of pulsatile stimulation, but an amplitude-modulated train of charge-balanced pulses interleaved across channels does reduce channel interaction (Wilson et al. 1991). Penetrating microelectrodes are better able to access the tonotopic organization of the ventral cochlea nucleus selectively than are macroelectrodes implanted on the surface of the nucleus (McCreery et al. 2010), and it may be possible to exploit this improved tonotopic specificity in order to use compressed analog stimuli without introducing excessive channel interaction. A compressed analog stimulus will require microelectrodes with greater charge injection capacity than what is required for pulsatile stimulation, but recently sputtered iridium oxide films with a charge injection capacity of at least 2000 $\mu C/cm^2$ have been demonstrated (Cogan et al. 2004).

ABI performance also might be improved by sound processing strategies that can better utilize the direct projection of the fusiform cells of the dorsal cochlear nucleus to the contralateral inferior colliculus, which overlaps with the projections from the ventral cochlear nucleus (Malmierca et al. 2005). The direct projections from the DCN to the inferior colliculus apparently cannot by itself support speech perception, because in the patient with the ventral pontine tegmental lesion described previously, the dorsal acoustic stria appears to have been spared, and yet all speech perception was lost. However, the dorsal cochlear nucleus (DCN), with its well ordered tonotopic organization, lies close to the medial part of an ABI's surface array, and it may be possible to alter and restrict the populations of DCN neurons that are activated by using combinations of electrodes to focus and "steer" the stimulus current (Firszt et al. 2007; Berenstein et al. 2008) and to create "virtual channels" as is presently being evaluated in cochlear implants.

3.5 Can Changes in the Clinical Management of Persons with NF2 Improve Their Performance with an ABI?

As noted previously, ABI users deafened by NF2 usually do not demonstrate the highest levels of sound-only speech recognition when compared to ABI users whose deafness is of other etiologies. This suggests that factors unique to NF2, including the removal of the acoustic tumor, may adversely affect the ability of the cochlear nucleus to process information provided by implants. In this respect, it is notable that some NF2 patients do retain quite good acoustic hearing immediately prior to the tumor removal surgery. However, the dorsal cochlear nucleus could be damaged by a tumor within the lateral recess or by surgical approaches to the floor of the lateral recess, and the ventral cochlear nucleus is vulnerable to damage by tumors within, or by surgical approaches to, the cerebellopontine angle (Abe and Rhoton 2006). Colletti and Shannon (2005) have speculated that transient vasospasm during tumor removal and/or the use of electric cautery to obtain hemostasis may damage neural elements in the CN that are vital for speech recognition. They also determined that persons with NF2 exhibit higher modulation detection thresholds than the non-tumor patients. The fact that the sound envelope information appears to be encoded strongly by multipolar cells supports the premise that during tumor removal, the multipolar cells either are lost or damaged, or their function is otherwise adversely affected. The latter might occur if there is damage to the inhibitory neurons of the small cell cap, whose inputs to the multipolar cells appears to enhance their sensitivity to amplitude modulation. It has been demonstrated in an animal model that several weeks of compression of the auditory nerve induces the loss of 50% of the neurons in the dorsal and posteroventral cochlear nuclei, and the latter contains a high density of multipolar cells (Sekiya et al. 2009). Most persons with NF2 do not undergo tumor removal surgery until the lesions have enlarged to the point of becoming life-threatening or until all useful hearing has been lost. If compression of the auditory nerve or the cochlear nucleus by the vestibular schwannomas is contributing to neuronal loss or to loss of neuronal function in the cochlear nucleus, it may be advantageous to remove the tumors surgically at an earlier stage, and the benefits that would accrue from earlier surgical removal would become greater as new technologies are developed that can better utilize the better preserved neuronal substrate. However, caution must be exercised when attributing damage in the cochlear nucleus to the tumor removal surgery or to compression of the cochlear nucleus by the enlarging tumor. In one study of 17 NF2 patients who received ABIs after removal of acoustic schwannomas, there was no significant correlation between tumor size and the patients' subsequent performance with their ABIs on device function (Otto et al. 1990). Also, some NF2 patients have received cochlear implants prior to removal of the vestibular schwannomas, at which time they received an ABI. Some of these people demonstrated similar or better speech recognition with their ABIs than with their cochlear implants (Otto, unpublished observations). While this could be the result of the patients' auditory nerve and/or the cochlea nucleus having sustained damage from the tumor prior to their receiving the cochlear implant, it does somewhat cloud the issue of the contribution of the tumor removal surgery to the relatively poor speech recognition by the NF2 patients.

4 Summary

Persons who lack functional auditory nerve cannot benefit from cochlear implants, but prostheses utilizing an electrode array implanted on the surface of the cochlear nucleus can restore some hearing. Worldwide, more than 700 persons have received these auditory brainstem implants or "ABIs," most frequently after surgical removal of the tumors that occur with Type 2 Neurofibromatosis (NF2). In recent years, ABIs have been used to treat hearing loss due to bilateral cochlear ossification, unilateral vestibular schwannomas in combination with deafness in the contralateral ear that is not amenable to a cochlear implant, congenital cochlear nerve aplasia or hypoplasia, malformations of the inner ear, and bilateral traumatic avulsions of the cochlear nerve. Typically, the ABI provides individuals with NF2 with improved speech understanding when combined with lip-reading and allows perception and discrimination of environmental sounds, but relatively few users have achieved significant open-set speech recognition. However, recent studies (Colletti and Shannon 2005) have shown that some ABI users with deafness of etiologies other than NF2 have achieved open-set speech recognition and in some cases, a level of performance approaching that of users of multi-channel cochlear implants.

The feasibility of supplementing an ABI array of surface electrodes with penetrating microstimulating electrodes has been demonstrated in animal studies. At present, 10 persons with NF2 have received a new type of cochlear nucleus implant that includes a surface array and an array of penetrating microelectrodes. Their speech perception is not significantly better than that of the NF2 patients who have only the surface arrays, but the technical advances have paved the way for incorporating penetrating microelectrodes into central auditory prostheses. In addition, improved understanding of the anatomy and physiology of the auditory system suggests opportunities for developing the next generation of cochlear nucleus auditory prostheses. These future implants might combine surface electrodes implanted over the cochlear nucleus and penetrating microstimulating electrodes within the nucleus, and employ sound processing strategies that are optimized for cochlear nucleus prostheses. There also is a need for animal models that can better elucidate how a space-occupying lesion in the cerebellopontine angle, and the surgical procedures that are employed to remove such a lesion, can affect the functionality of the cochlear nucleus.

References

Abe, H., & Rhoton, A. L., Jr. (2006). Microsurgical anatomy of the cochlear nuclei. *Neurosurgery*, 58(4), 728–739.

Adams, J. C. (1979). Identification of cochlear nucleus projections by removal of HRP reaction product. *Brain Research*, 177(1), 165–169.

Behr, R., Muller, J., Shehata-Dieler, W., Schlake, H. P., Helms, J., Roosen, K., Klug, N., Holper, B., & Lorens, A. (2007). The high rate CIS auditory brainstem implant for restoration of hearing in NF-2 patients. *Skull Base*, 17(2), 91–107.

Berenstein, C. K., Mens, L. H., Mulder, J. J., & Vanpoucke, F. J. (2008). Current steering and current focusing in cochlear implants: comparison of monopolar, tripolar, and virtual channel electrode configurations. *Ear and Hearing*, 29(2), 250–260.

Brackmann, D. E., Hitselberger, W. E., Nelson, R. A., Moore, J., Waring, M. D., Portillo, F., Shannon, R. V., & Telischi, F. F. (1993). Auditory brainstem implant: I. Issues in surgical implantation. *Archives of Otolaryngology–Head & Neck Surgery*, 108(6), 624–633.

Brawer, J. R., Morest, D. K., & Kane, E. C. (1974). The neuronal architecture of the cochlear nucleus of the cat. *Journal of Comparative Neurology*, 155, 251–300.

Cant, N. B., & Benson, C. G. (2003). Parallel auditory pathways: projection patterns of the different neuronal populations in the dorsal and ventral cochlear nuclei. *Brain Research Bulletin*, 60(5–6), 457–474.

Carner, M., Colletti, L., Shannon, R. V., Cerini, R., Barillari, M., Mucelli, R. P., & Colletti, V. (2009). Imaging in 28 children with cochlear nerve aplasia. *Acta Oto-Laryngologica*, 129(4), 458–461.

Cogan, S. F. (2008). Neural stimulation and recording electrodes. *Annual Review of Biomedical Engineering*, 10, 275–309.

Cogan, S. F., Plante, T. D., & Ehrlich, J. (2004). Sputtered iridium oxide films (SIROFs) for low-impedance neural stimulation and recording electrodes. *Proceedings of the 26th Annual International Conference of the IEEE EMBS* (pp. 4153–4156).

Colletti, L. (2007). Beneficial auditory and cognitive effects of auditory brainstem implantation in children. *Acta Oto-Laryngologica*, 127(9), 943–946.

Colletti, L., & Zoccante, L. (2008). Nonverbal cognitive abilities and auditory performance in children fitted with auditory brainstem implants: preliminary report. *Laryngoscope*, 118(8), 1443–1448.

Colletti, V., & Shannon, R. V. (2005). Open set speech perception with auditory brainstem implant? *Laryngoscope*, 115(11), 1974–1978.

Colletti, V., Sacchetto, L., Giarbini, N., Fiorino, F., & Carner, M. (2000). Retrosigmoid approach for auditory brainstem implant. *The Journal of Laryngology & Otology. Supplement*, (27), 37–40.

Colletti, V., Fiorino, F., Carner, M., Sacchetto, L., Miorelli, V., & Orsi, A. (2002). Auditory brainstem implantation: the University of Verona experience. *Archives of Otolaryngology–Head & Neck Surgery*, 127(1), 84–96.

Colletti, V., Fiorino, F. G., Carner, M., Miorelli, V., Guida, M., & Colletti, L. (2004). Auditory brainstem implant as a salvage treatment after unsuccessful cochlear implantation. *Otology & Neurotology*, 25(4), 485–496.

Colletti, V., Carner, M., Miorelli, V., Guida, M., Colletti, L., & Fiorino, F. (2005). Auditory brainstem implant (ABI): new frontiers in adults and children. *Archives of Otolaryngology–Head & Neck Surgery*, 133(1), 126–138.

Colletti, V., Shannon, R., Carner, M., Veronese, S., & Colletti, L. (2009a). Outcomes in nontumor adults fitted with the auditory brainstem implant: 10 years' experience. *Otology & Neurotology*, 30(5), 614–618.

Colletti, V., Shannon, R. V., Carner, M., Veronese, S., & Colletti, L. (2009b). Progress in restoration of hearing with the auditory brainstem implant. *Progress in Brain Research*, 175, 333–345.

Dugue, P., Le Bouquin Jeannes, R., & Faucon, G. (2007). Improving the dynamics of responses to amplitude modulated stimuli by modeling inhibitory interneurons in cochlear nucleus. *Conference Proceedings of the IEEE Engineering in Medicine and Biology Society, 2007* (pp. 1286–1289).

Edgerton, B. J., House, W. F., & Hitselberger, W. (1982). Hearing by cochlear nucleus stimulation in humans. *Annals of Otology, Rhinology, and Laryngology. Supplement*, 91(2, Pt. 3), 117–124.

Egan, C. A., Davies, L., & Halmagyi, G. M. (1996). Bilateral total deafness due to pontine haematoma. *Journal of Neurology, Neurosurgery & Psychiatry*, 61(6), 628–631.

Elvsashagen, T., Solyga, V., Bakke, S. J., Heiberg, A., & Kerty, E. (2009). [Neurofibromatosis type 2 and auditory brainstem implantation]. *Tidsskrift for den Nordke Laegeforening*, 129(15), 1469–1473.

Evans, D. G., Lye, R., Neary, W., Black, G., Strachan, T.,Wallace, A., & Ramsden, R. T. (1999). Probability of bilateral disease in people presenting with a unilateral vestibular schwannoma. *Journal of Neurology, Neurosurgery & Psychiatry*, 66(6), 764–767.

Evans, D. G., Huson, S. M., Donnai, D., Neary, W., Blair, V., Teare, D., Newton, V., Strachan, T., Ramsden, R., & Harris, R. (1992). A genetic study of type 2 neurofibromatosis in the United Kingdom. I. Prevalence, mutation rate, fitness, and confirmation of maternal transmission effect on severity. *Journal of Medical Genetics*, 29(12), 841–846.

Fayad, J. N., Otto, S. R., & Brackmann, D. E. (2006). Auditory brainstem implants: surgical aspects. *Advances in Oto-Rhino-Laryngology*, 64, 144–153.

Firszt, J. B., Koch, D. B., Downing, M., & Litvak, L. (2007). Current steering creates additional pitch percepts in adult cochlear implant recipients. *Otology & Neurotology*, 28(5), 629–636.

Frisina, R. D. (2001). Subcortical neural coding mechanisms for auditory temporal processing. *Hearing Research*, 158(1–2), 1–27.

Frisina, R. D., Smith, R. L., & Chamberlain, S. C. (1990). Encoding of amplitude modulation in the gerbil cochlear nucleus: II. Possible neural mechanisms. *Hearing Research*, 44(2–3), 123–141.

Fu, Q. J. (2002). Temporal processing and speech recognition in cochlear implant users. *Neuroreport*, 13(13), 1635–1639.

Godfrey, D. A., Kiang, N. Y., & Norris, B. E. (1975). Single unit activity in the posteroventral cochlear nucleus of the cat. *Journal of Comparative Neurology*, 162(2), 247–268.

House, W. F., & Hitselberger, W. E. (2001). Twenty-year report of the first auditory brain stem nucleus implant. *Annals of Otology, Rhinology, and Laryngology*, 110(2), 103–104.

Huang, C. Q., Carter, P. M., & Shepherd, R. K. (2001). Stimulus induced pH changes in cochlear implants: an in vitro and in vivo study. *Annals of Biomedical Engineering*, 29(9), 791–802.

Hughes, M. L., Abbas, P. J., Brown, C. J., & Gantz, B. J. (2000). Using electrically evoked compound action potential thresholds to facilitate creating MAPs for children with the Nucleus CI24M. *Advances in Oto-Rhino-Laryngology*, 57, 260–265.

Huy, P. T., Kania, R., Frachet, B., Poncet, C., & Legac, M. S. (2008). Auditory rehabilitation with cochlear implantation in patients with neurofibromatosis type 2. *Acta Oto-Laryngologica*, 129(9): 971–975.

Kolston, J., Osen, K. K., Hackney, C. M., Ottersen, O. P., & Storm-Mathisen, J. (1992). An atlas of glycine- and GABA-like immunoreactivity and colocalization in the cochlear nuclear complex of the guinea pig. *Anatomy and Embryology*, 186(5), 443–465

Kuchta, J. (2007). Twenty-five years of auditory brainstem implants: perspectives. *Acta Neurochirurgica. Supplement*, 97(Pt. 2), 443–449.

Kuchta, J., Otto, S. R., Shannon, R. V., Hitselberger, W. E., Brackmann, D. E. (2004). The multichannel auditory brainstem implant: how many electrodes make sense? *Journal of Neurosurgery*, 100(1), 16–23.

Leake, P. A., & Snyder, R. L. (1989). Topographic organization of the central projections of the spiral ganglion in cats. *Journal of Comparative Neurology*, 281(4), 612–629.

Lenarz, T., Moshrefi, M., Matthies, C., Frohne, C., Lesinski-Schiedat, A., Illg, A., Rost, U., Battmer, R. D., & Samii, M. (2001). Auditory brainstem implant: part I. Auditory performance and its evolution over time. *Otology & Neurotology*, 22(6), 823–833.

Lenarz, T., Matthies, C., Lesinski-Schiedat, A., Frohne, C., Rost, U., Illg, A., Battmer, R. D., & Samii, M. (2002). Auditory brainstem implant part II: subjective assessment of functional outcome. *Otology & Neurotology*, 23(5), 694–697.

Long, C. J., Nimmo-Smith, I., Baguley, D. M., O'Driscoll, M., Ramsden, R., Otto, S. R., Axon, P. R., & Carlyon, R. P. (2005). Optimizing the clinical fit of auditory brain stem implants. *Ear and Hearing*, 26(3), 251–262.

Malmierca, M. S., Saint Marie, R. L., Merchan, M. A., & Oliver, D. L. (2005). Laminar inputs from dorsal cochlear nucleus and ventral cochlear nucleus to the central nucleus of the inferior colliculus: two patterns of convergence. *Neuroscience*, 136(3), 883–894.

McCreery, D. B. (2004). Tissue reaction to electrodes: the problem of safe and effective stimulation of neural tissue. In K.W.. Horch & G.S. Dhillon (Eds.), *Neural prostheses: theory and practice* (pp. 592–607). River Edge, NJ; World Scientific Publishing.

McCreery, D. B. (2008). Cochlear nucleus auditory prostheses. *Hearing Research*, 242(1–2), 64–73.

McCreery, D. B., Agnew, W. F., Yuen, T. G., & Bullara, L. (1990). Charge density and charge per phase as cofactors in neural injury induced by electrical stimulation. *IEEE Transactions on Biomedical Engineering*, 37(10), 996–1001.

McCreery, D. B.,Yuen, T. G., Agnew, W. F., & Bullara, L. A. (1994). Stimulus parameters affecting tissue injury during microstimulation in the cochlear nucleus of the cat. *Hearing Research*, 77(1–2), 105–115.

McCreery, D. B., Yuen, T. G., Agnew, W. F., & Bullara, L. A. (1997). A characterization of the effects on neuronal excitability due to prolonged microstimulation with chronically implanted microelectrodes. *IEEE Transactions on Biomedical Engineering*, 44(10), 931–939.

McCreery, D. B.,Yuen, T. G., & Bullara, L. A. (2000). Chronic microstimulation in the feline ventral cochlear nucleus: physiologic and histologic effects. *Hearing Research*, 149(1–2), 223–238.

McCreery, D. B., Lossinsky, A., & Pikov, V. (2007). Performance of multisite silicon microprobes implanted chronically in the ventral cochlear nucleus of the cat. *IEEE Transactions on Biomedical Engineering*, 54(6), 1042–1052.

McCreery, D. B., Han, M., & Pikov, V. (2010). Neuronal activity evoked in the inferior colliculus of the cat by surface macroelectrodes and penetrating microelectrodes implanted in the cochlear nucleus. *IEEE Transactions on Biomedical Engineering*, 57(7), 1765–1773.

Moller, A. R. (1974). Responses of units in the cochlear nucleus to sinusoidally amplitude-modulated tones. *Experimental Neurology*, 45(1), 105–117.

Moller, A. R. (2006). History of cochlear implants and auditory brainstem implants. *Advances in Oto-Rhino-Laryngology*, 64, 1–10.

Moore, J. K. (1987). The human auditory brain stem: a comparative view. *Hearing Research*, 29(1), 1–32.

Moore, J. K., & Osen, K. K. (1979). The cochlear nuclei in man. *American Journal of Anatomy*, 154(3), 393–418.

Moore, J. K., Osen, K. K., Storm-Mathisen, J., & Ottersen, O. P. (1996). Gamma-Aminobutyric acid and glycine in the baboon cochlear nuclei: an immunocytochemical colocalization study with reference to interspecies differences in inhibitory systems. *Journal of Comparative Neurology*, 369(4), 497–519.

Osen, K. K. (1969). Cytoarchitecture of the cochlear nucleus in the cat. *Journal of Comparative Neurology*, 136, 453–483.

Otto, S. R., & Staller, S. (1995). Multichannel auditory brain stem implant: case studies comparing fitting strategies and results. *Annals of Otology, Rhinology, and Laryngology. Supplement, 166*, 36–39.

Otto, S. R., House, W. F., Brackmann, D. E., Hitselberger, W. E., & Nelson, R. A. (1990). Auditory brain stem implant: effect of tumor size and preoperative hearing level on function. *Annals of Otology, Rhinology, and Laryngology*, 99(10, Pt. 1), 789–790.

Otto, S. R., Brackmann, D. E., Staller, S., & Menapace, C. M. (1997). The multichannel auditory brainstem implant: 6-month coinvestigator results. *Advances in Oto-Rhino-Laryngology*, 52, 1–7.

Otto, S. R., Shannon, R. V., Brackmann, D. E., Hitselberger, W. E., Staller, S., & Menapace, C. (1998). The multichannel auditory brain stem implant: performance in twenty patients. *Archives of Otolaryngology–Head & Neck Surgery*, 118(3, Pt. 1), 291–303.

Otto, S. R., Brackmann, D. E., Hitselberger, W. E., Shannon, R. V., & Kuchta, J. (2002). Multichannel auditory brainstem implant: update on performance in 61 patients. *Journal of Neurosurgery*, 96(6), 1063–1071.

Otto, S. R., Brackmann, D. E., & Hitselberger, W. (2004). Auditory brainstem implantation in 12- to 18-year-olds. *Archives of Otolaryngology–Head & Neck Surgery*, 130(5), 656–659.

Otto, S. R., Waring, M. D., & Kuchta, J. (2005). Neural response telemetry and auditory/nonauditory sensations in 15 recipients of auditory brainstem implants. *Journal of the American Academy of Audiology*, 16(4), 219–227.

Otto, S. R., Shannon, R. V., Wilkinson, E. P., Hitselberger, W. E., McCreery, D. B., Moore, J. K., & Brackmann, D. E. (2008). Audiologic outcomes with the penetrating electrode auditory brainstem implant. *Otology & Neurotology*, 29(8), 1147–1154.

Portillo, F., Nelson, R. A., Brackmann, D. E., Hitselberger, W. E., Shannon, R. V., Waring, M. D., & Moore, J. K. (1993). Auditory brain stem implant: electrical stimulation of the human cochlear nucleus. *Advances in Oto-Rhino-Laryngology*, 48, 248–252.

Powell, T. B. S., & Cowen, W. M. (1962). An experimental study of the projection of the cochlear nucleus. *Journal of Anatomy*, 96, 269–284.

Rhode, W. S., & Greenberg, S. (1994). Encoding of amplitude modulation in the cochlear nucleus of the cat. *Journal of Neurophysiology*, 71(5), 1797–1825.

Robblee, L. S., Lefko, J., & Brummer, S. B. (1983). Activated Ir: an electrode suitable for reversible charge injection in saline solution. *Journal of the Electrochemical Society*, 130, 731–733.

Rose, T. L., & Robblee, L. S. (1990). Electrical stimulation with Pt electrodes. VIII. Electrochemically safe charge injection limits with 0.2 ms pulses. *IEEE Transactions on Biomedical Engineering*, 37(11), 1118–1120.

Rouiller, E. M., & Ryugo, D. K. (1984). Intracellular marking of physiologically characterized cells in the ventral cochlear nucleus of the cat. *Journal of Comparative Neurology*, 225(2), 167–186.

Schwartz, M. S., Otto, S. R., Shannon, R. V., Hitselberger, W. E., & Brackmann, D. E. (2008). Auditory brainstem implants. *Neurotherapeutics*, 5(1), 128–136.

Seki, Y., Samejima, N., & Komatsuzaki, A. (2007). Auditory brainstem implants: current state and future directions with special reference to the subtonsillar approach for implantation. *Acta Neurochirurgica. Supplement*, 97(Pt. 2), 431–435.

Sekiya, T., Canlon, B., Viberg, A., Matsumoto, M., Kojima, K., Ono, K., Yoshida, A., Kikkawa, Y. S., Nakagawa, T., & Ito, J. (2009). Selective vulnerability of adult cochlear nucleus neurons to de-afferentation by mechanical compression. *Experimental Neurology*, 218(1), 117–123.

Shannon, R. V., & Otto, S. R. (1990). Psychophysical measures from electrical stimulation of the human cochlear nucleus. *Hearing Research*, 47(1–2), 159–168.

Shannon, R. V., Fu, Q. J., & Galvin, J. (2004). The number of spectral channels required for speech recognition depends on the difficulty of the listening situation. *Acta Oto-Laryngolologica. Supplement* (552), 50–54.

Shannon, R. V., Moore, J. K., McCreery, D. B., & Portillo, F. (1997). Threshold–distance measures from electrical stimulation of human brainstem. *IEEE Transactions on Rehabilitation Engineering*, 5(1), 70–74.

Shepherd, R. K., & McCreery, D. B. (2006). Basis of electrical stimulation of the cochlea and the cochlear nucleus. *Advances in Oto-Rhino-Laryngology*, 64, 186–205.

Smith, P. H., & Rhode, W. S. (1989). Structural and functional properties distinguish two types of multipolar cells in the ventral cochlear nucleus. *Journal of Comparative Neurology*, 282(4), 595–616.

Snyder, R. L., Leake, P. A., & Hradek, G. T. (1997). Quantitative analysis of spiral ganglion projections to the cat cochlear nucleus. *Journal of Comparative Neurology*, 379(1), 133–149.

Sollmann, W. P., Laszig, R., & Marangos, N. (2000). Surgical experiences in 58 cases using the Nucleus 22 multichannel auditory brainstem implant. *Journal of Laryngology & Otology. Supplement* (27), 23–26.

Soussi, T., & Otto, S. R. (1994). Effects of electrical brainstem stimulation on tinnitus. *Acta Oto-Laryngologica*, 114(2), 135–140.

Warr, W. B. (1972). Fiber degeneration following lesions in the multipolar and globular cell areas in the ventral cochlear nucleus of the cat. *Brain Research*, 40(2), 247–270.

Wilson, B. S., Finley, C. C., Lawson, D. T., Wolford, R. D., Eddington, D. K., & Rabinowitz, W. M. (1991). Better speech recognition with cochlear implants. *Nature*, 352(6332), 236–238.

Zeng, F. G., & Shannon, R. V. (1994). Loudness-coding mechanisms inferred from electric stimulation of the human auditory system. *Science*, 264(5158), 564–566.

Zeng, F. G., Nie, K., Stickney, G. S., Kong, Y. Y., Vongphoe, M., Bhargave, A., Wei, C., & Cao, K. (2005). Speech recognition with amplitude and frequency modulations. *Proceedings of the National Academy of Sciences of the United States of America*, 102(7), 2293–2298.

Chapter 9
Midbrain Auditory Prostheses

Hubert H. Lim, Minoo Lenarz, and Thomas Lenarz

1 Introduction

In sorrowful reference to his deafness, Ludwig van Beethoven wrote in October of 1802 in his famous Heiligenstadt Testament (translated from the original German), "I was soon compelled to withdraw, to live life alone…[W]hat a humiliation for me when someone standing next to me heard a flute in the distance and I heard nothing or someone heard a shepherd singing and again I heard nothing…Such experiences brought me close to despair; a little more of that and I would have been at the point of ending my life. The only thing that held me back was my art. [so I] endure this wretched existence" (Lockwood 2003). If Beethoven were alive today, would he take the risk of being implanted with some computer chip into the brain to restore his hearing or even a crude form of hearing?

The field of auditory prostheses has made tremendous progress in the past 30 years, such that implanting devices into the head to restore hearing is no longer thought of as some radical or science-fiction phenomenon, but rather has become a standard treatment for many individuals suffering from hearing loss. There are many different types of implantable hearing devices, including middle ear implants (Snik, Chap. 4), cochlear implants (CIs) (Zeng, Chap. 1), and cochlear nucleus implants (McCreery, Chap. 8). There are also several alternative devices still under investigation that are designed to activate the auditory nerve using either traditional electric stimulation (Middlebrooks, Chap. 7) or novel optical stimulation (Richter, Chap. 6). Which device is provided to the deaf patient depends on the type of hearing loss as well as the expected success of hearing with that treatment. If the surgical risks are low and the benefits are high, then the decision to implant can be justified.

H.H. Lim (✉)
Department of Biomedical Engineering, University of Minnesota,
312 Church Street S.E., NHH 7-105, Minneapolis, MN 55455, USA
e-mail: hlim@umn.edu

F.-G. Zeng et al. (eds.), *Auditory Prostheses: New Horizons*, Springer Handbook of Auditory Research 39, DOI 10.1007/978-1-4419-9434-9_9,

For example, the cochlear implant has proven to be safe and cost-effective in that many CI patients can converse over the telephone after being completely deaf, and implanted deaf children can integrate into mainstream schools (Zeng, Chap. 1).

In contrast to peripheral implant devices, central auditory implants have experienced slower progress. As presented in Chap. 8 (McCreery), the first auditory brainstem implant (ABI) was implanted as early as 1979, which coincided with some of the first CIs, yet achieved performance levels dramatically lower than CIs. The rationale for the ABI and for deep brain surgery for array implantation stemmed from experiences surrounding a subpopulation of deaf patients, those with Neurofibromatosis Type 2 (NF2), who had to undergo tumor removal surgery. NF2 is associated with bilateral acoustic neuromas that develop along the eighth cranial nerve (consisting of the auditory and vestibular nerves). In most cases of tumor removal, both the auditory and vestibular nerves are compromised. Considering that open head surgery and access to the brainstem are required for tumor removal and the patient usually becomes completely deaf after the operation (assuming deafness in the contralateral ear because of prior tumor removal), it is possible to implant a flat electrode array on the surface of the cochlear nucleus (the auditory portion of the brainstem) with minimal added surgical risks to restore some auditory sensations. From 1979 to 1992 there were a total of 25 ABI patients (Schwartz et al. 2008) that increased to roughly 500 by 2005 (Colletti and Shannon 2005) and now exceeds 1000. NF2 ABI patients rarely achieve open-set speech perception and performance is usually limited to lip-reading enhancement and environmental awareness. However, they receive an emotional and psychological benefit from the device in feeling connected to the outside world that significantly improves their quality of life.

In order to improve hearing performance with central auditory implants, there has been a renewed interest in stimulating other central auditory regions that may provide higher levels of speech perception. The present chapter specifically describes the development of one type of central auditory prosthesis that targets midbrain regions beyond the cochlear nucleus, particularly the inferior colliculus (IC). Currently there is an ongoing clinical trial for penetrating stimulation within the central nucleus of the IC (ICC) with a device known as the auditory midbrain implant (AMI). Section 2 will provide the rationale for selecting the ICC as an alternative auditory prosthesis site. Section 3 presents two interesting cases of surface stimulation of the IC that date back to as early as some of the first attempts at cochlear and nerve stimulation for hearing (i.e., during the early 1960s). Section 4 will then cover the animal and cadaver studies that enabled translation of the AMI into clinical trials as well as the first human psychophysical and speech results. Finally, Sect. 5 will discuss new directions for improving auditory midbrain prostheses.

2 Rationale for the AMI

The ABI is the only clinically approved hearing device for those who cannot benefit from CIs (Fig. 9.1). In the U.S., the ABI is only available to NF2 patients. Because of the growing experience with and acceptance of the ABI surgery and implantation

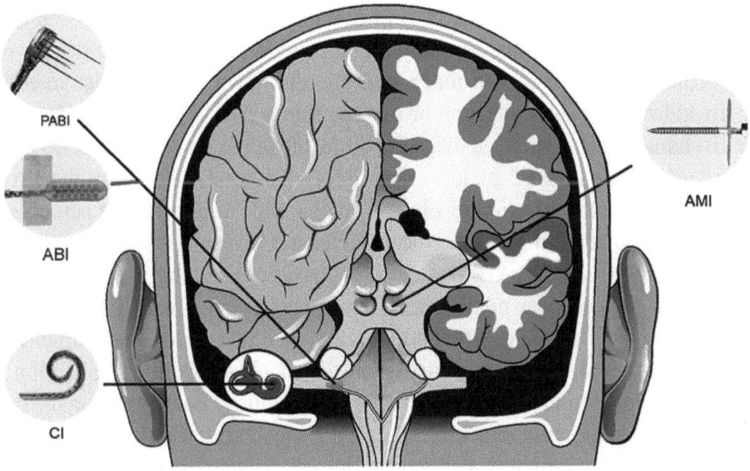

Fig. 9.1 Simplified brain schematic showing locations of different auditory implants. Both the penetrating auditory brainstem implant (PABI) and auditory midbrain implant (AMI) are in clinical trials. All the shown devices have been developed by Cochlear Ltd. though other auditory brainstem implants (ABIs) and cochlear implants (CIs) have been developed by various companies (Taken from Lenarz et al. (2006b) and reprinted with permission from Lippincott Williams and Wilkins)

worldwide, several countries have begun to allow ABI implantation into non-tumor patients who do not have functional auditory nerves (e.g., because of nerve avulsion or aplasia) or an implantable cochlea (e.g., because of ossification or malformations that prevent insertion of a cochlear electrode array). There has been a surprising difference in performance between the tumor and the non-tumor groups. In over 600 reported ABI NF2 patients, there have only been a few rare cases in which patients have achieved sufficient open-set speech perception (Lim et al. 2009a). However, a recent study (Colletti et al. 2009) showed that out of 48 non-tumor patients implanted with the ABI, more than half achieved sufficient open-set speech perception with a few reaching levels comparable to the top CI patients. These 48 non-tumor ABI patients obtained an average score of 59% on an open-set speech test compared to an average score of 10% across 32 NF2 ABI patients. Considering that similar implant technologies, stimulation strategies, and surgical approaches are used for both patient groups, these findings suggest that the limited performance observed in NF2 patients may be related to some form of damage induced at the level of the cochlear nucleus as a result of the tumor and/or tumor removal process (Colletti and Shannon 2005). This damage may result from tumor compression of the cochlear nucleus that can induce coding deficits in the central auditory system (Matthies et al. 2000; Crea et al. 2009) or even a compromised vasculature to the cochlear nucleus because of the tumor or its removal (Colletti and Shannon 2005).

Another proposed factor limiting speech perception with the ABI in NF2 patients and leading to the high performance variability across non-tumor patients is the diffuse nature of activation with surface stimulation (McCreery 2008). The cochlear nucleus is a complex neural structure consisting of distinct types of neurons specifically and tonotopically organized throughout its central region (Moore and Osen 1979; Cant and Benson 2003). Surface stimulation may not appropriately activate the deeper regions of the cochlear nucleus or even reach neurons that have not been damaged by the tumor. To address this issue, a new type of penetrating ABI (PABI; Fig. 9.1) consisting of 8 or 10 shanks, each with an activated iridium site at the tip (2000 or 5000 μm^2), has been implanted into 10 NF2 patients (McCreery, Chap. 8). Although the PABI can achieve low and stable activation levels as well as a wide range of pitch percepts across sites that were implanted into the cochlear nucleus, its overall performance has not yet exceeded that of the surface ABI device. It is possible that there is also tumor-related damage within the cochlear nucleus that limits performance with PABI stimulation. Considering the fact that many NF2 patients can still understand speech up until tumor removal (Slattery et al. 1998; Bance and Ramsden 1999; Colletti and Shannon 2005), even those with large tumors, suggests that at least auditory nuclei beyond the cochlear nucleus are still functionally intact for processing speech information. Therefore, stimulation of an auditory structure beyond the hypothesized damaged cochlear nucleus may provide improvements over ABI stimulation in NF2 patients.

2.1 ICC as a Potential Target for an Auditory Prosthesis

The IC is a three-dimensional structure that is somewhat spherical and consists of different regions associated with varying coding features (Ehret 1997; Oliver 2005). Figure 9.2 presents both an axial and parasagittal section of the human IC with the different labeled regions. The ICC (labeled as CN in Fig. 9.2) is the main ascending auditory region of the midbrain, which receives inputs originating from or passing through the lateral lemniscus. The dorsal cortex receives substantial descending projections from higher auditory and nonauditory centers and is designed for modulating information transmitted along the ascending pathway. The external nucleus of the IC, or lateral zone, has been associated with multi-modal information processing (i.e., auditory and non-auditory information integration across sensory and motor systems) as well as modulation of ascending and descending projections to and from lower auditory centers. Other regions within the IC have also been identified across species.

There are several properties of the ICC that make it a logical choice for an auditory prosthesis target. It is a converging center for almost all ascending auditory brainstem projections to higher perceptual centers (Casseday et al. 2002), which should provide access to pathways necessary for speech understanding. Anatomically, the ICC contains well defined laminated layers aligned roughly 45° from the parasagittal plane (Fig. 9.2) that correspond to a systematic frequency organization (Geniec and Morest 1971; Oliver 2005). Considering that speech perception performance for CI and normal hearing subjects has been correlated with the ability to provide sufficient frequency-specific information (Friesen et al. 2001; Shannon et al. 2004), the well defined tonotopic organization of the ICC compared to the other central auditory regions makes it a favorable target for an auditory prosthesis.

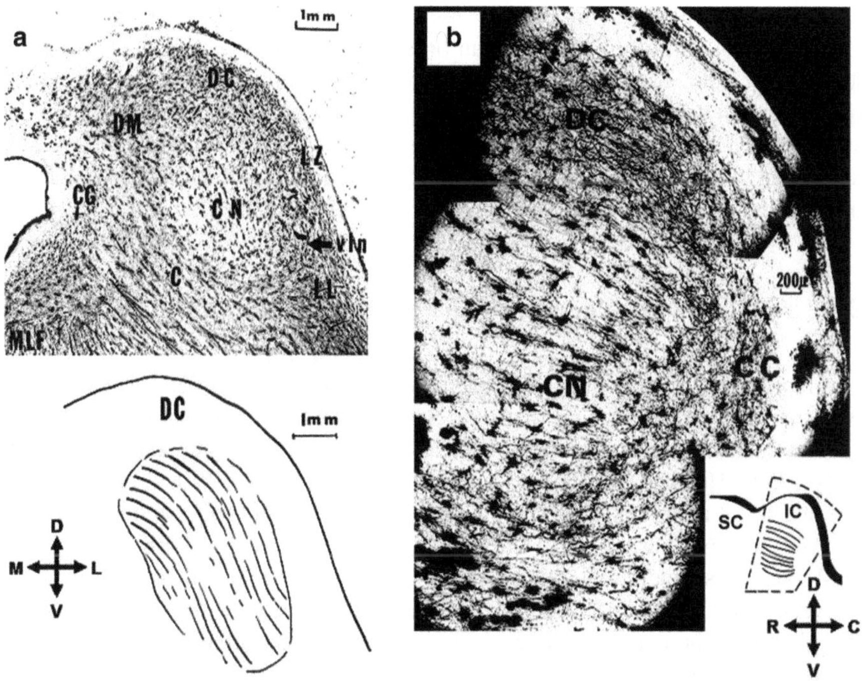

Fig. 9.2 Anatomy of inferior colliculus (IC).Histological sections of the human IC depicting its different subdivisions and layered structure using the Golgi-Cox method. (**a**) Axial section (*top*) at the junction of the caudal and middle thirds of the IC of a 55-year-old man, and its simplified schematic (*bottom*) showing the orientation of the dendritic laminae within the central nucleus. (**b**) Parasagittal section at the junction of the medial and middle thirds of the IC of a 53-year-old man; inset provides orientation of the dendritic laminae within the central nucleus and indicates the location of the section (*dashed lines*). *C* cuneiform area, *CC* caudal cortex, *CG* central gray, *CN* central nucleus, *DC* dorsal cortex, *DM* dorsomedial nucleus, *LL* lateral nucleus and dorsal nucleus of lateral lemniscus, *LZ* lateral zone, *MLF* medial longitudinal fasciculus, *SC* superior colliculus, vln ventrolateral nucleus. Anatomical directions: *C* caudal, *D* dorsal, *L* lateral, *M* medial, *R* rostral, *V* ventral (Taken from Geniec and Morest (1971) and reprinted with permission from Taylor and Francis Group)

The ICC also appears to exhibit some spatial organization along the isofrequency layers for other sound features important for speech perception. For example, in mice it has been shown that ICC neurons with lower pure-tone thresholds, sharper frequency tuning, and greater sensitivity to slower frequency sweep speeds are located more centrally within a lamina, and these properties systematically change in more concentrically outward regions (Stiebler 1986; Hage and Ehret 2003). In cats, it has been shown that a periodotopic (best modulation frequency) map exists along the dorsomedial-to-ventrolateral dimension of the ICC laminae with higher best modulation frequencies and shorter pure-tone latencies represented in more ventrolateral regions (Schreiner and Langner 1988; Langner et al. 2002). These findings suggest that frequency may be coded in one dimension while temporal, level, and even frequency interactions are coded along the other dimensions. From an engineering point of view, such an organization would be advantageous for a three-dimensional array in which appropriate spatial stimulation of the ICC could elicit different spectral, temporal, and level percepts, all features that make up the structure of a sound signal. From a surgical point of view, the IC is relatively easily accessible in humans and an array can be implanted with minimal added risk combined with NF2 tumor removal as achieved with ABI implantation (see Sect. 4.3).

2.2 The Growing Success of Deep Brain Stimulation (DBS)

The rapidly growing field of DBS has opened up the possibilities for penetrating ICC stimulation for hearing restoration. Some of the first reports of using DBS for alleviating neurological conditions, such as schizophrenia and pain, date back to the early 1950s (Heath 1954). There were also reports in the 1960s demonstrating the use of DBS of midbrain regions, including the IC, for severe pain patients (Nashold and Wilson 1966; Nashold et al. 1969). However, the rapid rise in DBS did not occur until after the serendipitous findings by Alim-Louis Benabid and colleagues at the University of Grenoble (France) in 1987 when they observed that high frequency stimulation of the thalamic nucleus ventralis intermedius could provide long-term suppression of tremor in Parkinson's disease patients (Benabid et al. 1987, 2009). Since then over 75,000 patients have been implanted with a DBS system; implanting penetrating arrays into neural tissue is now considered routine treatment for several neurological disorders, including Parkinson's disease and tremor (Breit et al. 2004; Wichmann and Delong 2006). Combined with the remarkable success of DBS for movement disorders, the continued improvements in safety of the implant is increasing the acceptance and potential of DBS for other neurological disorders, such as depression, obsessive compulsive disorder, and even hearing (e.g., tinnitus suppression). More importantly for the AMI, it should be possible to modify commonly used DBS stereotactic approaches to the midbrain for safe implantation of an electrode array into the ICC (Green et al. 2006; Wichmann and Delong 2006).

2.3 Consideration of Alternative Target Sites

There are other auditory regions that may be stimulated with an auditory prosthesis, though they appear less favorable than the IC. The auditory cortex is more superficially located and surgically accessible than the IC. However, it consists of a less defined functional organization (e.g., the tonotopic map is less consistent across animal subjects), in part because of its more plastic nature (Dahmen and King 2007; Keuroghlian and Knudsen 2007), and exhibits more complex coding of perceptual sound features. Some of the earliest attempts at brain stimulation for hearing applied electrical current to the auditory cortex (Penfield and Rasmussen 1950; Penfield 1958; Dobelle et al. 1973). However, cortical stimulation of different regions could elicit complex and abstract auditory percepts and generally required very high current levels (several milliamperes), which could be potentially dangerous for daily stimulation. Lower auditory nuclei, such as the superior olivary nuclei and the lateral lemnisci, may exhibit less complex processing compared to the IC because they are lower along the auditory pathway. However, these nuclei code sound in a more diffuse manner (i.e., no one nucleus serves as a converging center of information) and with a less defined and/or skewed tonotopic organization compared to the ICC (Ehret and Romand 1997; Nayagam et al. 2006). Although the medial geniculate body of the thalamus can be approached using stereotactic methods (Wichmann and Delong 2006; Owen et al. 2007) and provides access to most auditory projections ascending from lower centers to the auditory cortex, it may exhibit more complex processing compared to the IC since it is higher along the auditory pathway (Ehret and Romand 1997; Wang et al. 2008).

3 Surface Stimulation

The first reported attempt at IC stimulation for hearing restoration was in 1962 by Simmons and colleagues at Stanford University during a tumor removal operation (Simmons et al. 1964). Removal of a large tumor directly exposed both the right auditory nerve and surface of the IC. The day before the surgery, the patient was trained to various acoustic stimuli (e.g., sine and noise waveforms with varying rates and amplitudes) to enable the patient to better assess and compare the artificial percepts that would be induced by electrical activation of the auditory nerve and IC. Simmons' team was successful in eliciting auditory sensations to nerve stimulation for pulse widths between 0.1 and 1.0 ms and rates between 20 and 5000 pps for intensities much lower than 2 V. However, they were unable to elicit any auditory sensations when stimulating the surface of the IC up to intensities of 2 V. They repeated their stimulation paradigm in different locations across the IC surface but still were unable to elicit any auditory percepts. It is possible that surface IC stimulation may have activated predominantly modulatory and inhibitory neurons within the dorsal cortex and lateral zone (Jen et al. 2001; Kelly and Caspary 2005) rather than ascending auditory pathways within the ICC that are required for auditory perception.

Not until 2005 was there a second attempt at stimulating the surface of the IC to elicit auditory sensations (Colletti et al. 2007). Vittorio Colletti and colleagues at the University of Verona (Italy) implanted an NF2 patient with an ABI array onto the dorsal surface of the IC to assess if the limited performance with cochlear nucleus stimulation associated with tumor-related damage could be overcome by stimulating in a higher auditory center. Colletti's team implanted a Med-El Pulsar ci 100 ABI array (Innsbruck, Austria) that consists of 12 platinum disk electrodes (0.7 to 1.3 mm in diameter) organized on a 3-by-10 mm silicone carrier similar in concept to the ABI array shown in Fig. 9.1. Further details on the device and surgical approach are presented in Colletti et al. (2007) and Vince et al. (2010).

In contrast to the first attempt by Simmons' team, Colletti's group was able to elicit auditory sensations through surface stimulation of the IC across all 12 electrodes of the ABI array. However, thresholds were generally high for IC stimulation

(~10–50 nC corresponding to ~100–800 μA depending on the pulse width) in comparison to cochlear stimulation (~5–20 nC [Shannon 1985; Pfingst et al. 1997]). It is surprising that CI stimulation can achieve lower threshold ranges even though current must pass through a modiolar wall to activate distant neurons while the IC sites are in direct contact with the neurons. It is possible that higher current levels were necessary to activate excitatory and ascending neurons deeper within the IC to elicit auditory sensations. Nevertheless, these findings were encouraging for the use of IC stimulation to restore hearing. Surface IC stimulation could achieve systematic changes in loudness with current level and different pitch percepts across sites, both of which are important for speech understanding and stimulation strategies. Colletti's team also reported that the patient could obtain some speech understanding with significant improvements in lip-reading capabilities through continued use of the implant. Overall performance level for this patient is still significantly lower than what is achieved by CI patients and is within the general range achieved by a typical NF2 ABI patient. It is likely that surface IC stimulation does not effectively activate ascending auditory pathways necessary for speech understanding considering the extremely high current levels. The question remains whether penetrating stimulation along the tonotopic gradient of the ICC will provide better open-set speech perception compared to surface IC stimulation.

4 Penetrating Stimulation

During the early 2000s, Thomas Lenarz and Minoo Lenarz, along with several other colleagues at Hannover Medical University, collaborated with Cochlear Ltd. (Australia; led by Chief Scientist James Patrick) to develop a human prototype AMI array for penetrating stimulation across the tonotopic gradient of the ICC (Lenarz et al. 2006b). The AMI consisted of similar components as a typical CI except the electrode array was designed for the ICC. By 2003, this group had developed the first human prototype AMI and began testing its electrical activation and safety properties in a cat model (Lenarz et al. 2003; Reuter et al. 2004, 2005). During the early 2000s, Hubert Lim and David Anderson at the University of Michigan also began investigating the potential of ICC stimulation for hearing restoration (Lim and Anderson 2003; Lim and Anderson 2006). At that time, they used "Michigan" silicon substrate multi-site arrays for electrically stimulating multiple locations across the ICC and recording neural activity within the primary auditory cortex of guinea pigs. They were able to demonstrate that ICC stimulation could achieve lower thresholds, greater dynamic ranges, and more localized, frequency-specific activation compared to cochlear stimulation (Lim and Anderson 2006). They also identified distinct subregions along the isofrequency domain of the ICC that exhibited different electrical activation properties shown to be important for speech perception (Lim and Anderson 2007). In 2004, the two groups began a collaboration to translate the AMI into clinical application. The following sections present the AMI system as well as a summary of the collaborative animal and human surgical studies that led to the translation of the AMI into the first patients. The initial human AMI results are also summarized.

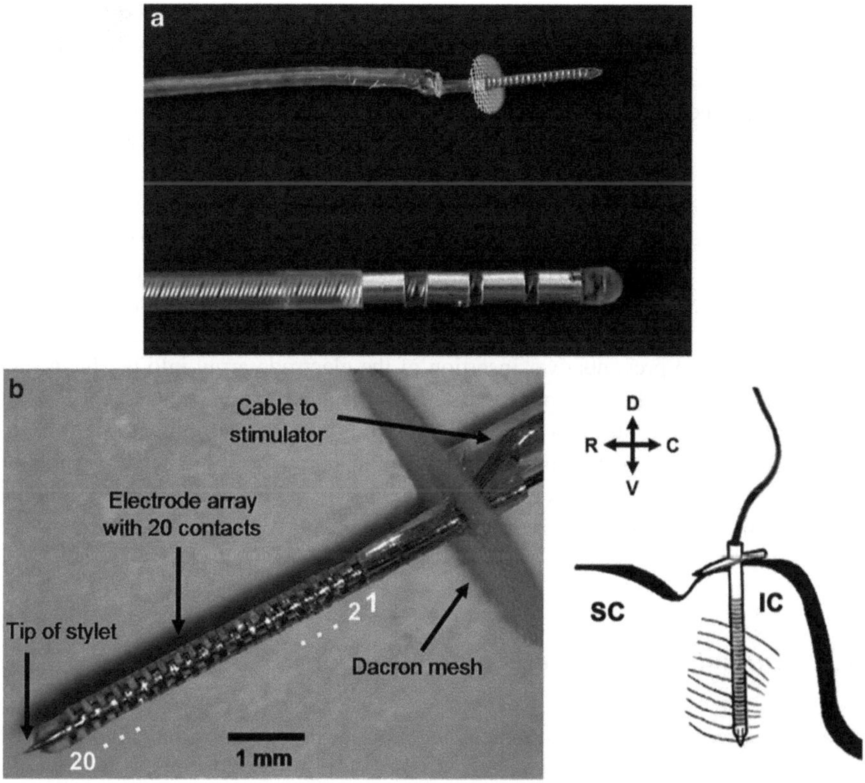

Fig. 9.3 Auditory midbrain implant (AMI) array. (**a**) Image of the AMI array next to a standard deep brain stimulation (DBS) array (Medtronic Inc., Minneapolis, MN, USA). The DBS array consists of 4 platinum-iridium contacts (2 mm center-to-center separation) each with a ring diameter of 1.27 mm, width of 1.5 mm, and surface area of ~6 mm^2. (**b**) Magnified image of the AMI array, which is 6.2 mm long (from Dacron mesh to tip of silicone carrier without stylet). Each of the 20 platinum ring electrodes (0.2 mm center-to-center separation) has a diameter of 0.4 mm, width of 0.1 mm, and surface area of ~0.00126 mm^2. The AMI array is designed to be positioned along the tonotopic gradient of the inferior colliculus (IC), particularly its central nucleus. The array was developed by Cochlear Ltd. SC, superior colliculus. Anatomical directions: *C* caudal, *D* dorsal, *R* rostral, *V* ventral (Taken from Lenarz et al. (2006b) and Samii et al. (2007) and reprinted with permission from Lippincott Williams and Wilkins)

4.1 AMI Concept and Design

The AMI array (Fig. 9.3a) was derived from the CI technology developed by Cochlear Ltd.; thus it consisted of a design and material already shown to be safe for neural stimulation in humans. Although a three-dimensional array would enable more effective activation across the ICC, such a device was not yet approved for human use and would increase the surgical risks for implantation. Instead, a CI array was reduced in dimensions to create an AMI array that was small enough to insert into the ICC with the goal of stimulating the different layers of the ICC (Fig. 9.3b).

The AMI electrode array is 6.4 mm long (from Dacron mesh to tip of stylet) with a diameter of 0.4 mm. It consists of 20 platinum ring electrodes linearly spaced at an interval of 200 μm. Each site has a width of 100 μm (surface area of 126,000 μm^2) and is connected to a parylene-coated 25-μm-thick wire (90% platinum/10% iridium). The body (carrier) of the electrode array is made from silicone rubber and is concentrically hollow. A stiffening element (stylet) made of stainless steel is positioned through the axial center of this silicone carrier to enable insertion of the electrode array into the IC. After the electrode array is in its final position in the midbrain, the stylet can be removed and the softer silicone carrier remains in the tissue (further surgical details presented in Sect. 4.3). The Dacron mesh anchors the electrode array onto the surface of the neural tissue to minimize movement after implantation. This Dacron mesh also prevents over insertion of the electrode array into the IC during implantation. The other components of the AMI system are similar to the latest Nucleus CI system (Patrick et al. 2006) consisting of a behind-the-ear microphone and processor that transmits the electromagnetic signals to the receiver-stimulator implanted under the skin. This receiver-stimulator is implanted in a bony bed on the skull near the craniotomy and is connected with a cable to the electrode array.

4.2 *Animal Feasibility and Safety Studies*

The AMI array was designed to match the tonotopic organization and dimensions of the human IC, and it was hypothesized that AMI stimulation of the ICC would achieve frequency-specific activation (Lenarz et al. 2006b; Lim and Anderson 2006). Furthermore, it was expected that lower thresholds than CI stimulation could be achieved because of the ability to stimulate ICC neurons directly, compared to the distant nature of neural activation (across the bony modiolar wall) for cochlear stimulation. Lenarz et al. (2006a) confirmed these hypotheses performing acute experiments in ketamine-anesthetized guinea pigs in which different regions along the tonotopic axis of the ICC were electrically stimulated and the corresponding neural activity across the tonotopic gradient of the primary auditory cortex (A1) were recorded. AMI stimulation achieved lower thresholds (by ~10 dB) and more frequency-specific cortical activation than what has been reported for CI stimulation (Bierer and Middlebrooks 2002).

In addition to electrophysiological assessment of the AMI array, Lenarz et al. (2007) performed chronic cat studies to ensure that surgical implantation and continuous stimulation of the AMI array within the midbrain tissue would be safe for daily use in humans. The cat was selected as the animal model because its IC is similar in cytoarchitecture and size to the human IC. Eight cats were chronically implanted for 3 months, in which 4 of them were additionally stimulated for 60 days (4 h/day) starting 4 weeks after implantation to assess if clinically relevant stimuli further affected the tissue response. The electrical stimuli were presented using the SPEAK strategy (Cochlear Ltd.) and driven by continuous sound from a radio.

An important component of this study was that the researchers used a surgical approach to expose the IC surface and implant the AMI array, as well as stimulation patterns, that were similar to those used in the human patients. Overall, the histomorphological findings demonstrated that minimal neuronal damage occurs around the electrode array because of chronic implantation and stimulation of the AMI array. Those results are similar to what has been observed with other deep brain neural implants currently used in human patients (Haberler et al. 2000; McCreery, Chap. 8). Furthermore, all 8 animals were healthy throughout the 3 month implant period without any observable complications associated with the surgical approach. In electrically stimulating the IC in cats, as well as in humans, it is possible to stimulate neighboring structures that may elicit non-auditory and even adverse effects (Moore 1987; Kretschmann and Weinrich 1992; Trepel 2004). Activation of the spinothalamic tract (caudal and ventral to the IC) and the trigeminal tract (medial and ventral) can elicit pain, temperature, and pressure sensations in the body and face, respectively. The trochlear nerve (caudal) and the superior colliculus (rostral) are associated with ocular movements. Stimulation of regions more medial and ventral to the IC, such as the periaqueductal gray and cuneiform area, can elicit pain sensations and changes in arterial blood pressure and heart rate. In all 4 of the stimulated animals, none experienced any abnormal eye movements, irregular heart rates, or behavioral responses indicative of painful sensations.

4.3 Surgical Approach

Considering that NF2 patients are the largest initial group of candidates for an AMI, it was essential to develop a combined surgical approach that enables removal of acoustic neuromas and AMI implantation at the same surgical setting. Thomas Lenarz and Minoo Lenarz worked with Madjid Samii, the Director of the International Neuroscience Institute (Hannover, Germany), to develop a successful surgical approach for the AMI (Samii et al. 2007). Figure 9.4a shows the location of the skin incision and craniotomy required for the lateral suboccipital approach used for AMI implantation. Once the neurosurgeon cuts through the dura and folds it over the edges to expose the brain (to later allow the dura to be closed via sutures), the cerebellum and tentorium become visible (Fig. 9.4b). The cerebellum must be retracted medially (to the right in the figure) to expose the auditory nerve and acoustic neuroma. Once the neurosurgeon removes the tumor, the cerebellum can be retracted downwards to expose the surface of the IC. Because of the semi-sitting position and gravity, the cerebellum actually drops downward without any forced retraction as shown in Fig. 9.4c. Both Fig. 9.4c, d show the surface of the IC after the neurosurgeon has carefully cut through the surrounding arachnoid and pushed aside several blood vessels covering the midbrain surface. Once the surface of the IC is exposed, the AMI can be inserted into the IC along the tonotopic gradient of the ICC (Fig. 9.4d). Three-dimensional intraoperative navigation with CT and MRI images

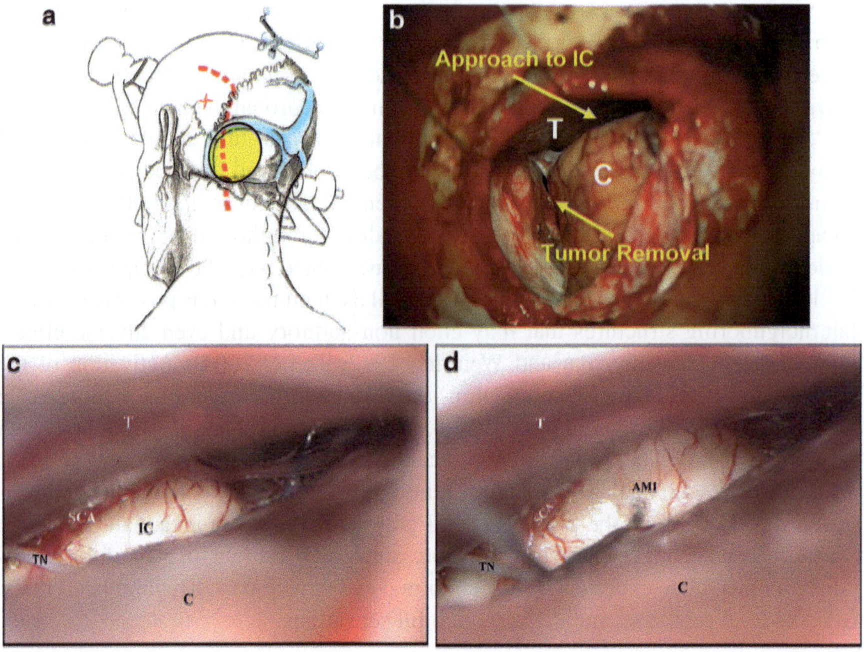

Fig. 9.4 Surgical approach to the inferior colliculus (IC). (**a**) Schematic drawing of the fixed head in a semi-sitting position and showing the skin incision (*red dotted line*), appropriate location for the receiver-stimulator of the auditory midbrain implant (AMI) in the temporoparietal area (*red star*), and the location of the modified lateral suboccipital craniotomy (*yellow circle*) exposing the inferior margin of the transverse sinus and the medial margin of the sigmoid sinus (*blue shaded regions*). The antenna placed at the top of the head is for the three-dimensional intraoperative navigation system. (**b**) After the skull is removed and the dura flaps pulled to the side, the tentorium (T) and cerebellum (C) are visible. The cerebellum is retracted medially (*right*) to expose the auditory nerve and tumor. Because of gravity, the cerebellum drops downward to expose the IC. (**c**) View of the left IC, trochlear nerve (TN), and the caudal branch of the superior cerebellar artery (SCA) through the lateral supracerebellar infratentorial approach after the neurosurgeon has removed the overlying arachnoid and pushed aside several blood vessels. (**d**) The cable extends from the AMI array that has been implanted into the IC. (**a, c, d**) were taken from Samii et al. (2007) and reprinted with permission from Lippincott Williams and Wilkins. (**b**) was taken from Lim et al. (2009b) and reprinted with permission from Springer

based on the bone-anchored registration method can be used to aid in the placement of the array. This surgical approach was first developed and tested in several fixed and fresh human cadavers to obtain approval for clinical trials (Samii et al. 2007).

4.4 Performance in the First Patients

The AMI array has been safely implanted into 5 NF2 patients (Lim et al. 2009a). Performance results for 3 patients have been previously reported (Lim et al. 2007) and are summarized in the following sections.

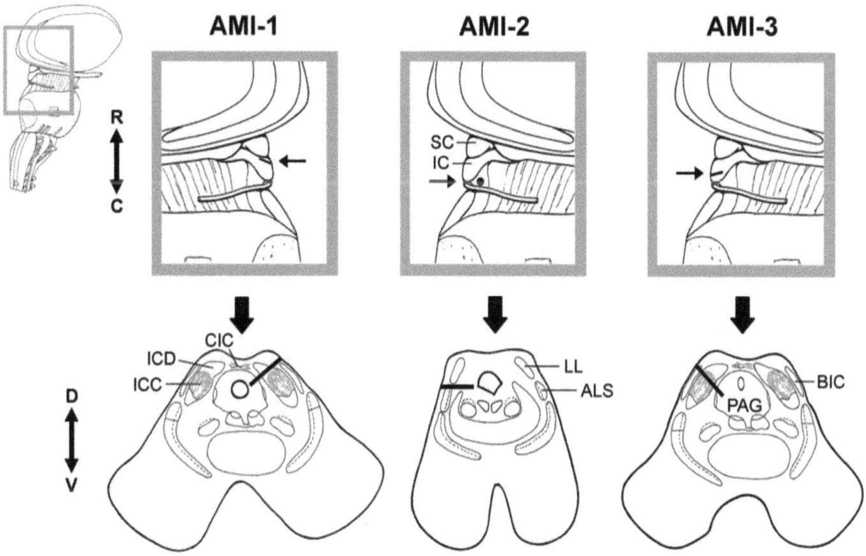

Fig. 9.5 Array placement across patients. Parasagittal (*top*) and axial (*bottom*) sections showing the location and orientation of the array within the midbrain of each patient. *Arrow* in parasagittal section points to the caudorostral location of the array and the corresponding axial section below. The *black line* (or *dot* for AMI-2) representing the array in each section corresponds to the trajectory of the array across several superimposed CT-MRI slices. *ALS* anterolateral system, *BIC* brachium of *IC*, *CIC* commissure of *IC*, *IC* inferior colliculus, *ICC* inferior colliculus central nucleus, *ICD* inferior colliculus dorsal nucleus, *LL* lateral lemniscus, *PAG* periaqueductal gray, *SC* superior colliculus. Anatomical directions: *C* caudal, *D* dorsal, *R* rostral, *V* ventral (Taken from Lim et al. (2007) and reprinted with permission from the Society for Neuroscience)

4.4.1 Array Placement

Figure 9.5 provides a summary of the different locations of the array across the 3 patients. In the first patient (AMI-1), the array was implanted in a rostral and medial location resulting in its placement into the dorsal cortex of the IC. In the second patient (AMI-2), the array was inserted more caudally and laterally resulting in its placement along the surface of the lateral lemniscus. By the third patient (AMI-3), the array was finally positioned into the target region of the ICC. Even with increasing experience with the surgical approach, there were still some difficulties in accurately placing the array into the ICC in the fourth and fifth patients (Lim et al. 2009a).

4.4.2 Patient Fitting and Psychophysical Findings

The patients returned 5 to 7 weeks after AMI implantation for their first fitting session. They did not observe any adverse or painful side effects to electrical stimulation. Non-auditory sensations consisted of paresthesia, mild temperature changes in

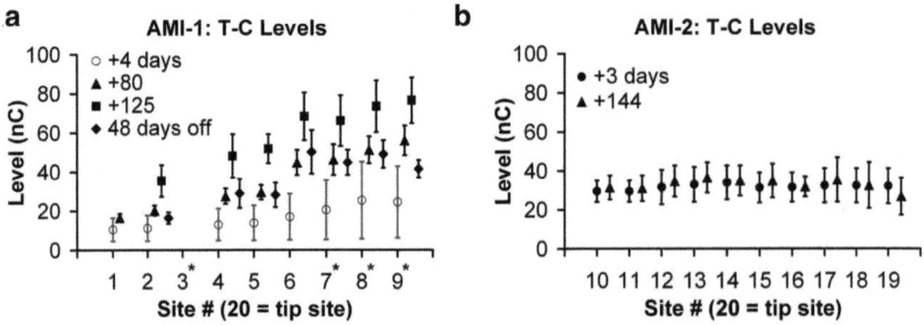

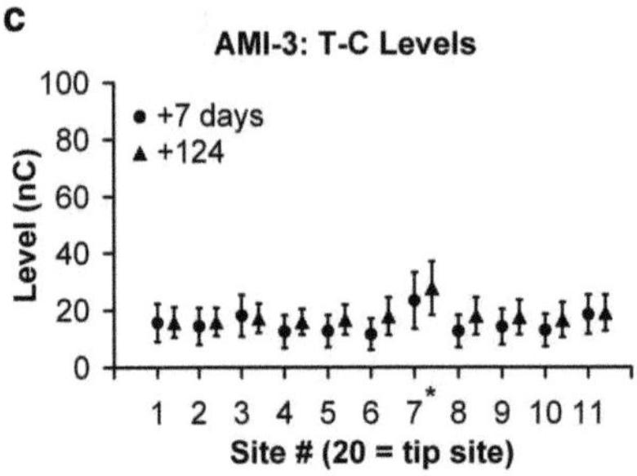

d

T-C Summary

	T nC	T ρ μC/cm²	C nC	C ρ μC/cm²	DR dB
AMI-1	~4-5	~3-4	~16-29	~13-23	~11-15
AMI-2	23-25	18-20	35-43	28-34	3-5
AMI-3	6-12	5-10	17-25	13-20	7-9

ρ, charge density; DR, dynamic range
site area = 126000 μm²

Fig. 9.6 Activation levels for auditory midbrain implant stimulation. Threshold (T) and comfortable (C) levels measured in each patient using 500 ms on-off pulse trains with 250 pps, 100 μs/phase monopolar pulses. (**a**) T-C levels (*endpoints of bar*) for AMI-1 measured for 4 different time points (symbols) from when the implant was initially turned on. Because of rising levels over time, the implant was turned off for 48 days (after the +127 day measurement) and then T-C levels were measured again. At +4 days, only the modified T-C levels used for the daily processor rather than the actual measured values were available. Thus they are labeled with an *open symbol* and *lighter shaded bars*. (**b, c**) T-C levels for AMI-2 and AMI-3 measured at 2 different time points and demonstrating stability over time. (**d**) Summary of values for each patient only for the values from the first testing session shown in (**a–c**). *Asterisks* correspond to sites shorted to other inactive sites, except for site 3 (shorted to active site 9) for AMI-1, that were turned off (Taken from Lim et al. (2007) and reprinted with permission from the Society for Neuroscience)

different parts of the face and body, some dizziness, and mild facial twitches. However, all these side effects were avoided by turning off the corresponding sites for daily stimulation. As for auditory sensations, the patients described the percepts as tonal in nature but that some sites elicited a broad spectral percept with multiple pitches. The patients also described the sounds as having an electronic quality mixed in with the tonal percept. Furthermore, pitch and temporal percepts could be altered by changing the stimulation pulse rate and pattern as well as location of activation. These qualitative results were encouraging for AMI implementation since they suggest that at the level of the midbrain, sound still appears to be somewhat coded into elementary perceptual features of sound (i.e., tonal sounds that can be systematically elicited with varying temporal percepts).

Figure 9.6 presents the threshold (T) and comfortable (C) levels measured for different sites in each patient over the first 3 to 4 months as the patients continued to use their implant on a daily basis. The T and C levels need to be measured and programmed into the patient's processor to deliver the appropriate current levels for daily stimulation. The T and C levels for AMI-1 (Fig. 9.6a) continued to rise reaching the compliance voltage of the stimulator (at +125 days). Because of this rise in levels, the processor was turned off for 48 days to assess if levels would return to usable levels. The activation levels decreased dramatically but not completely to the initial values. The cause of these adaptive effects is not clear. One hypothesis proposed by Lim et al. (2007) is that the stimulation rates and patterns are overdriving the neurons located within the dorsal IC region, which receives a large number of projections from higher auditory and non-auditory centers (Winer 2005) and may be designed for adapting to and modulating various stimuli (Perez-Gonzalez et al. 2005). The other 2 patients exhibited stable activation levels over time (Fig. 9.6b, c) suggesting that location of stimulation, thus the type of neurons activated, is important for AMI implementation. It is interesting that the thresholds, at least for AMI-3 who is implanted into the ICC, still required high activation levels that ranged from 6 to 12 nC (Fig. 9.6d). This is comparable to what has been typically observed for CI patients (5–20 nC) using similar stimuli (Shannon 1985; Pfingst et al. 1997) even though cochlear stimulation requires activation of distant neurons beyond a bony modiolar wall, while ICC stimulation stimulates neurons directly in the vicinity of the sites. Lim et al. (2007) hypothesize that these high thresholds are associated with suboptimal placement within the ICC that may also limit hearing performance, which is further discussed in Sect. 5.1. Although high, the AMI activation levels for ICC stimulation (6–25 nC across T to C levels) are still lower than the general range observed for surface IC stimulation (~10–100 nC) (Colletti et al. 2007), which further suggests that surface stimulation of the IC may not be a favorable approach for effectively activating central auditory neurons for hearing restoration.

Encouragingly, Lim et al. (2007) reported that the auditory sensations and/or side effects associated with each site have generally remained stable, indicating minimal movement of the implant over time. Monotonic loudness growth functions were also observed in all 3 patients in whom higher current levels induced louder percepts. There were initial concerns that loudness percepts would not change systematically with current level, considering that animal studies have revealed a complex pattern of excitatory and inhibitory projections from different brainstem nuclei into the ICC (Loftus et al. 2004; Oliver 2005; Cant and Benson 2006) as well as the existence of both monotonic and non-monotonic rate-level functions for neurons throughout the ICC (Ramachandran et al. 1999; LeBeau et al. 2001).

In addition to monotonic loudness growth functions, programming the processor requires ordering the stimulation sites based on the different pitch percepts to ensure that specific frequency information is transmitted to the appropriate neural regions. Lim et al. (2007) performed various tests (i.e., pitch scaling and pitch ranking) in the 3 patients. Although each patient can detect differences in pitch percepts depending on the stimulated site, a systematic pitch organization was only observed for AMI-3. This is consistent with findings from animal studies in that the array in AMI-1 and AMI-2 is not aligned along any known tonotopic organization, whereas the array in AMI-3 is aligned along the tonotopic gradient of the ICC (i.e., lower pitch percepts are more superficial and higher pitch percepts are in deeper regions). Interestingly, this systematic pitch organization was not observed for the first 6 months of stimulation (Lim et al. 2009a). In fact, stimulation of all sites elicited predominantly low pitch sounds. However, during the 6-month follow-up session, AMI-3 expressed that annoyingly high pitch percepts could be perceived during daily stimulation. After performing extensive pitch tests, a systematic pitch organization within the human ICC consistent with animal findings was revealed. It appears that dramatic plastic effects to midbrain stimulation is possible and may be reversing deficits induced by long periods of deafness (AMI-3 was deaf for 6 years and had only low frequency residual hearing prior to complete deafness).

4.4.3 Speech Results

Overall, it is apparent that location of stimulation can greatly affect hearing performance (Lim et al. 2007, 2009a). AMI-1, who is implanted in the dorsal cortex of the IC, obtains the least benefit from the AMI. This is mainly the result of the adaptive effects experienced during daily stimulation in which the loudness decreases and thresholds increase over time. During the speech tests, in which silent (recovery) periods are followed by speech presentation as well as during daily situations when intermittent sound is presented and perceived at a loud enough level, the patient is able to extract some temporal and pitch information from the sound signal. Generally, improvements in hearing have been limited to lip-reading enhancement and environmental awareness. AMI-2, who is implanted on the surface of the lateral lemniscus, obtains slightly greater improvements in vowel, number, and consonant recognition than AMI-1. Both patients use their implants on a daily basis and have expressed the importance of the implant for enhancing lip-reading and environmental cues. However, the overall speech perception performance has been limited because of the inappropriate placement of the arrays. A more encouraging outcome has been observed in AMI-3, whose array was implanted within the intended target, the ICC. AMI-3 achieved speech scores that exceeded the average score for NF2 ABI patients in the same clinic (Lim et al. 2009a). In particular, AMI-3 obtained a lip-reading enhancement on speech tracking of 26 words per minute compared to about 12 words per minute for ABI patients (average across 14 patients). However, further improvements are still needed considering that normal hearing subjects obtain a speech tracking score of about 85 to 100 words per minute (Strauss-Schier et al. 1995), which is much greater than the 33 words per minute (with lip-reading) obtained by AMI-3.

5 Future Directions

The initial results in the first AMI patients have been encouraging in terms of the ability to implant the array into the auditory midbrain safely and to restore some hearing function that has improved the daily lives of the implanted patients. However,

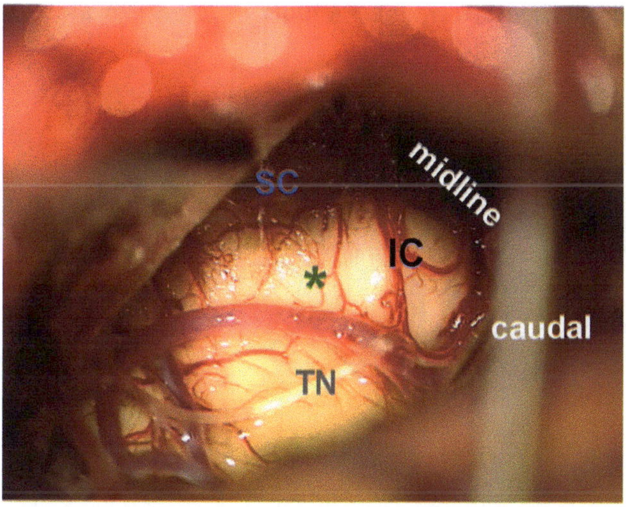

Fig. 9.7 Surgical exposure of the left human midbrain for array implantation. The midline, superior colliculus (SC), trochlear nerve (TN), and inferior colliculus (IC) are visible. However, the caudal edge of the IC and true midline are not clearly visible with the angled view of the midbrain. The *asterisk* corresponds to the hypothesized location of the start of the brachium of the IC (a potential landmark for the lateral of the IC) based on surface IC stimulation results presented in Fig. 9.8 and described in Sect. 5.1 (Taken from Lim et al. 2009a and reprinted with permission from SAGE Publications)

none of the patients has yet achieved open-set speech perception without lip-reading cues. One major limitation with midbrain stimulation has been the difficulty in optimally placing the array into the ICC. As will be discussed in Sect. 5.1, improvements in hearing performance may be achieved by ensuring optimal array placement within specific regions of the ICC. It is also possible that implanting only a single shank array into the ICC may not be sufficient to restore open-set speech perception considering the ICC is a three-dimensional structure. This will be discussed in Sect. 5.2.

5.1 Optimal Array Placement

Upon exposing the surface of the midbrain for AMI implantation (Fig. 9.7), the border between the IC and superior colliculus can be identified by a dip in the midbrain surface between the two structures. However, it is difficult to determine how medial the midline is and how caudal the caudal edge of the IC is because of the limited and distorted surgical view of the midbrain surface. One solution is to enlarge the craniotomy to expose the opposite IC as well as the exit point of the trochlear nerve (located at the caudal edge of the IC). This should provide exposure of the true midline plane and the most caudal portion of the IC, respectively. Midline approaches to the IC with larger craniotomies have been safely performed and reported previously (Stein 1979; Samii et al. 1996; Colletti et al. 2007), so this is a potential option for improving AMI implantation.

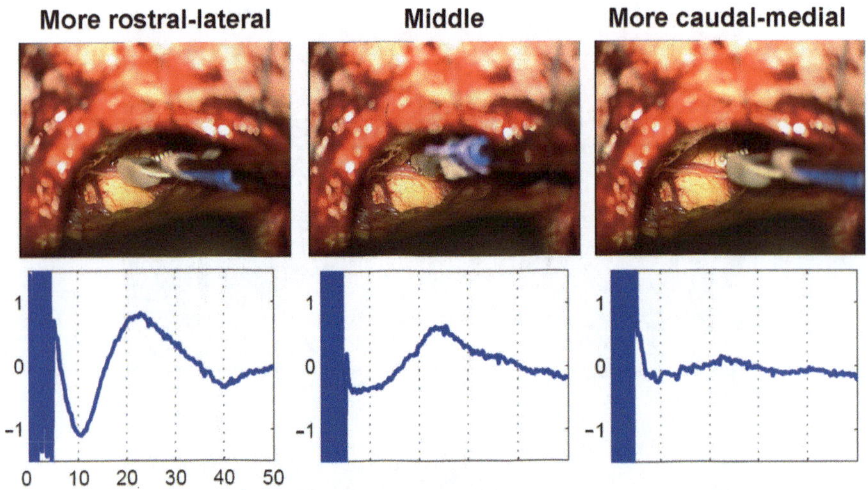

Fig. 9.8 Surface midbrain stimulation method. Electrically evoked middle latency responses were recorded to surface midbrain stimulation during implant surgery under sufentanil anesthesia. A bipolar electrode is positioned in 3 different locations (*top images*) and stimulated to induce evoked potentials (*bottom plots*) recorded with surface needles (signal: high forehead; reference: nape of neck, ground: low forehead). The stimulus consisted of a short burst of 5 pulses (100 μs/phase, 7 μs interphase gap) at 1000 pps, which was repeated at 7.4 Hz. The current level was 220 CU (a level unit used by the Freedom implant system from Cochlear Ltd.), which corresponds to 930 μA. The curves are averages of 500 stimulus repetitions. The abscissa of the curves corresponds to time in milliseconds and the ordinate corresponds to amplitude in microvolts. A millimeter scale was placed on the surface of the midbrain and is visible to the right of the electrode. The evoked response increases in magnitude as the electrode stimulates a more rostral and lateral surface location. If the start of the brachium of the inferior colliculus corresponds to where the response begins to increase in magnitude (i.e., effective activation of output passing fibers to the thalamus), then it should be located somewhere between the "middle" and "more caudal-medial" positions (i.e., the location of the *asterisk* in Fig. 9.7) (Taken from Lim et al. (2009a) and reprinted with permission from SAGE Publications)

The benefit in identifying the caudal edge of the IC is to enable normalization of the distance between that edge and the border between the superior colliculus and IC to locate the center of the IC. This will at least ensure placement of the array into the ICC with respect to the caudal-to-rostral direction. However, another landmark along the medial-to-lateral direction is still needed. By using a larger surgical exposure, the true midline can be identified, but there is still no landmark for the lateral edge of the IC. One possible solution is to stimulate different regions along the surface of the IC and assess the electrically evoked middle latency responses, which represent cortical activity recorded from subcutaneous needle electrodes. Figure 9.8 presents images of three different stimulation locations of a bipolar electrode that was positioned along the IC surface in one patient during AMI surgery (Lim et al. 2009a). The image corresponds to the same exposure shown in Fig. 9.7. The presented results are consistent with the results presented in Sect. 3 in that stimulation of the IC surface requires high current levels to elicit sufficient activation. A current level of 220 CU (~930 μA) was required to observe a response for all three locations. Interestingly, stimulation of more rostral and lateral regions resulted in larger responses with the same stimulus. The magnitude of the response increased dramatically over a 6 mm distance. This differential activation pattern may provide a neurophysiological marker for identifying the lateral edge of the IC to improve AMI implantation. Animal experiments can be performed to confirm these results initially, and then they can be verified across a larger number of patients during future AMI surgeries.

The use of an enlarged surgical exposure combined with surface stimulation will hopefully enable insertion of the AMI array into the ICC and even specific regions of the ICC. AMI-3 is implanted into a caudal-dorsal ICC region (Fig. 9.5). She achieves the best performance across the 3 AMI patients. However, she has not achieved open-set speech perception comparable to CI patients and exhibits high stimulation thresholds. Based on animal studies, there appears to exist two spatially distinct regions perpendicular to the frequency axis (i.e., along the isofrequency domain): a caudal-dorsal and a rostral-ventral region (Fig. 9.5; Lim and Anderson 2007). Stimulation of the caudal-dorsal region exhibits extremely high threshold levels, consistent with the findings for AMI-3, and small evoked activity in A1 (Lim and Anderson 2007; Neuheiser et al. 2010). Stimulation of the rostral-ventral ICC region can elicit activation with current levels ~20 dB lower than those required for effective activation of the caudal-dorsal region. Furthermore, stimulation of the rostral-ventral region elicits smaller discriminable level steps, shorter latencies, greater spiking precision, and possibly less degraded frequency-specific activation with level (Lim and Anderson 2007; Lim et al. 2008). These findings are consistent with other animal studies demonstrating the existence of segregated pathways from the brainstem to the auditory cortex, in which one pathway (i.e., the rostral pathway) may exhibit properties for more robustly transmitting temporal, level, and spectral sound features (Fig. 9.9; Rodrigues-Dagaeff et al. 1989; Cant and Benson 2006, 2007). Based on CI and normal hearing studies, the ability to achieve stronger and more spatially synchronized activation, enhanced temporal precision, better level coding, and more localized frequency-specific activation are all important for speech perception (Pfingst et al. 1983; Shannon et al. 1995; Friesen et al. 2001; Rance et al. 2002). Thus the limited performance in AMI-3 may be the result of implantation of the array into a caudal-dorsal ICC region. Targeting rostral-ventral ICC regions may provide significant improvements in hearing performance in future AMI patients.

5.2 *Three-Dimensional Stimulation*

The ICC is a three-dimensional structure in which one dimension codes for frequency. The other two dimensions corresponding to the isofrequency domain may serve to code for other important sound features spatially, such as temporal cues. There is some convincing evidence from animal studies demonstrating a shift in

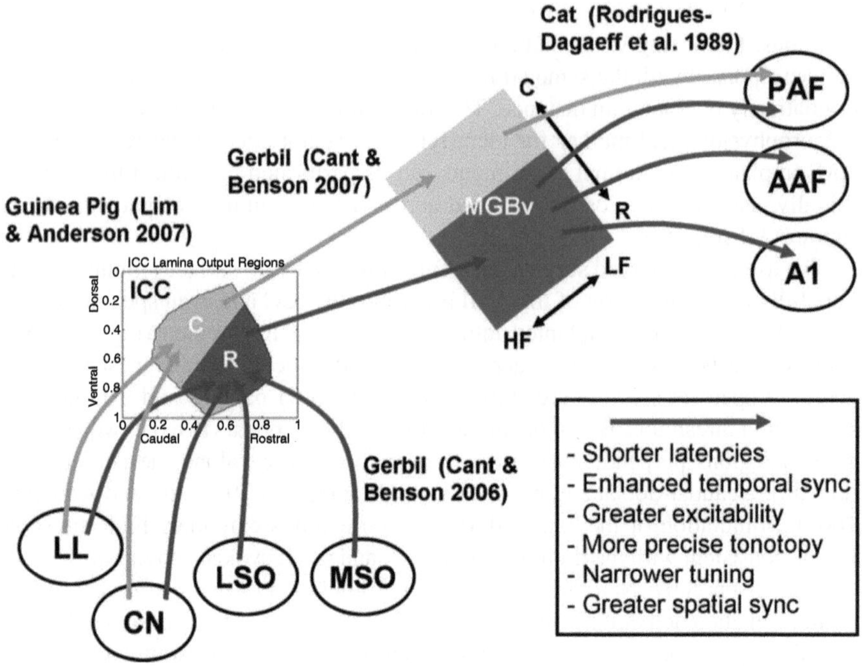

Fig. 9.9 Simplified schematic of segregated functional pathways. There appears to exist some segregation of functional pathways corresponding to different coding properties from the brainstem up through the inferior colliculus central nucleus (ICC), ventral division of the medial geniculate body (MGBv), and several subregions of the auditory cortex as described in previous animal studies (Rodrigues-Dagaeff et al. 1989; Cant and Benson 2006, 2007; Lim and Anderson 2007). There exists at least two segregated functional pathways originating from the ICC and maintained up to the auditory cortex in which one pathway (*dark gray*) corresponds to properties that may be more favorable for an auditory midbrain implant. Note that there are some overlapping projections across regions that are not displayed. *A1* primary auditory cortex, *AAF* anterior auditory field, *CN* cochlear nucleus, *HF* high frequency, *LF* low frequency, *LL* lateral lemniscus, *LSO* lateral superior olive, *MSO* medial superior olive, *PAF* posterior auditory field, sync synchronized activity. Anatomical directions: *C* caudal, *R* rostral (Taken from Lim et al. (2008) and reprinted with permission from the Elsevier)

temporal coding properties from the cochlea up to the auditory cortex (Phillips et al. 1989; Lu and Wang 2000; Snyder et al. 2000). In particular, higher auditory neurons become less capable of synchronizing to high rate stimuli. Cochlear neurons are capable of synchronizing to acoustic and electrical stimuli that repeat at rates exceeding 1000 Hz whereas neurons at the midbrain and cortical levels generally synchronize to rates of a few hundred and tens of hertz, respectively. One ongoing hypothesis is that the representation of high rate changes in the stimulus waveform become coded less by a temporal code and more by a spatial code (possibly through a spike rate and/or interval timing code) in higher auditory centers (Wang et al. 2008). Thus temporal features important for speech understanding may be coded spatially along the isofrequency laminae. Animal studies have already shown that some temporal features, such as periodicity and latencies, are coded systematically along the ICC laminae (Schreiner and Langner 1988; Langner et al. 2002). Based on these findings, improving speech understanding may require sufficient stimulation across both the frequency and isofrequency dimensions of the ICC.

The initial motivation for developing a single shank AMI array was simply that no three-dimensional array technology was available for DBS applications while a single shank array could be directly translated from CI technology already approved for human use (Lenarz et al. 2006b). Furthermore, extensive safety studies would have been required to demonstrate that a three-dimensional array developed from new electrode technologies could be safely implanted and used in humans for daily stimulation. Although it is possible to push such technology forward, it is crucial, first, to demonstrate in animal models that stimulation in multiple regions throughout the three-dimensional structure of the ICC can significantly improve transmission of sound coding features important for speech understanding. This in itself is a difficult task yet must be addressed considering the added surgical risks associated with implanting multiple arrays into a deep brain structure. Currently, Thomas Lenarz at Hannover Medical University, Hubert Lim at the University of Minnesota, and Cochlear Ltd. are investigating new three-dimensional AMI arrays and stimulation strategies in animal models to assess their functional benefits and potential for translation into clinical trials.

5.3 Concluding Remarks

There are already over 1000 deaf patients who have been implanted with an ABI. This number is expected to increase more rapidly in the future as will the number of patients being implanted with a DBS array (already >75,000 patients). Central nervous system stimulation to restore hearing and neurological function, in general, is no longer considered a rare and radical treatment. With this growing acceptance of DBS, more and more deaf patients will seek central auditory prostheses even if the devices cannot provide open-set speech perception. Preliminary results from the first patients implanted with the AMI have demonstrated both its safety and technical feasibility in human. There is an urgent need to continue to push forward such research and technological efforts in keeping with the clinical momentum. The goal is for the research and clinical communities to work together to improve the next generation of central auditory prostheses significantly, providing deaf patients with more than just a lip-reading supplement, but instead with intelligible speech perception.

Acknowledgements The authors thank Gert Joseph, Urte Rost, Joerg Pesch, and Rolf-Dieter Battmer for involvement with AMI patient fitting and testing at Hannover Medical University; Madjid Samii, Amir Samii, and the International Neuroscience Institute (Hannover, Germany) for successful AMI surgery; and the engineers and scientists at Cochlear Ltd. (Lane Cove, Australia), including James F. Patrick, Frank Risi, Godofredo (JR) Timbol, and Peter Gibson, for AMI development and technical assistance. The authors also thank David J. Anderson for providing the scientific pathway for performing the initial AMI feasibility experiments at the University of Michigan; and

Günter Reuter, Uta Reich, Gerrit Paasche, and Alexandru C. Stan for involvement with the animal safety studies at Hannover Medical University. Funding was provided by Cochlear Ltd. with contributions from the German Research Foundation (SFB 599) and NIH through P41 EB2030, P30 DC05188, T32DC00011, and F31DC007009.

References

Bance, M., & Ramsden, R. T. (1999). Management of neurofibromatosis type 2. *Ear Nose & Throat Journal*, 78(2), 91–94, 96.

Benabid, A. L., Chabardes, S., Mitrofanis, J., & Pollak, P. (2009). Deep brain stimulation of the subthalamic nucleus for the treatment of Parkinson's disease. *Lancet Neurology*, 8(1), 67–81.

Benabid, A. L., Pollak, P., Louveau, A., Henry, S., & de Rougemont, J. (1987). Combined (thalamotomy and stimulation) stereotactic surgery of the VIM thalamic nucleus for bilateral Parkinson disease. *Applied Neurophysiology*, 50(1–6), 344–346.

Bierer, J. A., & Middlebrooks, J. C. (2002). Auditory cortical images of cochlear-implant stimuli: dependence on electrode configuration. *Journal of Neurophysiology*, 87(1), 478–492.

Breit, S., Schulz, J. B., & Benabid, A. L. (2004). Deep brain stimulation. *Cell and Tissue Research*, 318(1), 275–288.

Cant, N. B., & Benson, C. G. (2003). Parallel auditory pathways: projection patterns of the different neuronal populations in the dorsal and ventral cochlear nuclei. *Brain Research Bulletin*, 60(5–6), 457–474.

Cant, N. B., & Benson, C. G. (2006). Organization of the inferior colliculus of the gerbil (Meriones unguiculatus): differences in distribution of projections from the cochlear nuclei and the superior olivary complex. *Journal of Comparative Neurology*, 495(5), 511–528.

Cant, N. B., & Benson, C. G. (2007). Multiple topographically organized projections connect the central nucleus of the inferior colliculus to the ventral division of the medial geniculate nucleus in the gerbil, Meriones unguiculatus. *Journal of Comparative Neurology*, 503(3), 432–453.

Casseday, J. H., Fremouw, T., & Covey, E. (2002). The inferior colliculus: A hub for the central auditory system. In D. Oertel, R. R. Fay & A. N. Popper (Eds.), *Springer Handbook of Auditory Research: Integrative functions in the mammalian auditory pathway (Vol. 15)* (pp. 238–318). New York: Springer-Verlag.

Colletti, V., & Shannon, R. V. (2005). Open set speech perception with auditory brainstem implant? *Laryngoscope*, 115(11), 1974–1978.

Colletti, V., Shannon, R., Carner, M., Sacchetto, L., Turazzi, S., Masotto, B., & Colletti, L. (2007). The first successful case of hearing produced by electrical stimulation of the human midbrain. *Otology & Neurotology*, 28(1), 39–43.

Colletti, V., Shannon, R., Carner, M., Veronese, S., & Colletti, L. (2009). Outcomes in nontumor adults fitted with the auditory brainstem implant: 10 years' experience. *Otology & Neurotology*, 30, 614–618.

Crea, K. N., Shivdasani, M. N., Argent, R. E., Mauger, S. J., Rathbone, G. D., O'Leary, S. J., & Paolini, A. G. (2009). Acute cochlear nucleus compression alters tuning properties of inferior colliculus neurons. *Audiology & Neurotology*, 15(1), 18–26.

Dahmen, J. C., & King, A. J. (2007). Learning to hear: plasticity of auditory cortical processing. *Current Opinion in Neurobiology*, 17(4), 456–464.

Dobelle, W. H., Stensaas, S. S., Mladejovsky, M. G., & Smith, J. B. (1973). A prosthesis for the deaf based on cortical stimulation. *Annals of Otology, Rhinology, and Laryngology*, 82(4), 445–463.

Ehret, G. (1997). The auditory midbrain, a "shunting yard" of acoustical information processing. In G. Ehret & R. Romand (Eds.), *The Central Auditory System* (pp. 259–316). New York: Oxford University Press, Inc.

Ehret, G., & Romand, R. (1997). *The Central Auditory System*. New York: Oxford University Press, Inc.

Friesen, L. M., Shannon, R. V., Baskent, D., & Wang, X. (2001). Speech recognition in noise as a function of the number of spectral channels: comparison of acoustic hearing and cochlear implants. *Journal of the Acoustical Society of America*, 110(2), 1150–1163.

Geniec, P., & Morest, D. K. (1971). The neuronal architecture of the human posterior colliculus. A study with the Golgi method. *Acta Oto-Laryngologica Supplementum*, 295, 1–33.

Green, A. L., Wang, S., Owen, S. L., Xie, K., Bittar, R. G., Stein, J. F., Paterson, D. J., & Aziz, T. Z. (2006). Stimulating the human midbrain to reveal the link between pain and blood pressure. *Pain*, 124(3), 349–359.

Haberler, C., Alesch, F., Mazal, P. R., Pilz, P., Jellinger, K., Pinter, M. M., Hainfellner, J. A., & Budka, H. (2000). No tissue damage by chronic deep brain stimulation in Parkinson's disease. *Annals of Neurology*, 48(3), 372–376.

Hage, S. R., & Ehret, G. (2003). Mapping responses to frequency sweeps and tones in the inferior colliculus of house mice. *European Journal of Neuroscience*, 18(8), 2301–2312.

Heath, R. G. (1954). *Studies in Schizophrenia. A Multidisciplinary Approach to Mind-Brain Relationships.* Cambridge, MA: Harvard University Press.

Jen, P. H., Sun, X., & Chen, Q. C. (2001). An electrophysiological study of neural pathways for corticofugally inhibited neurons in the central nucleus of the inferior colliculus of the big brown bat, Eptesicus fuscus. *Experimental Brain Research*, 137(3–4), 292–302.

Kelly, J. B., & Caspary, D. M. (2005). Pharmacology of the inferior colliculus. In J. A. Winer & C. E. Schreiner (Eds.), *The inferior colliculus* (pp. 248–281). New York: Springer Science+Business Media, Inc.

Keuroghlian, A. S., & Knudsen, E. I. (2007). Adaptive auditory plasticity in developing and adult animals. *Progress in Neurobiology*, 82(3), 109–121.

Kretschmann, H. J., & Weinrich, W. (1992). *Cranial neuroimaging and clinical neuroanatomy: magnetic resonance imaging and computed tomography* (2nd ed.). New York: Thieme Medical Publishers, Inc.

Langner, G., Albert, M., & Briede, T. (2002). Temporal and spatial coding of periodicity information in the inferior colliculus of awake chinchilla (Chinchilla laniger). *Hearing Research*, 168(1–2), 110–130.

LeBeau, F. E., Malmierca, M. S., & Rees, A. (2001). Iontophoresis in vivo demonstrates a key role for GABA(A) and glycinergic inhibition in shaping frequency response areas in the inferior colliculus of guinea pig. *Journal of Neuroscience*, 21(18), 7303–7312.

Lenarz, M., Lim, H. H., Patrick, J. F., Anderson, D. J., & Lenarz, T. (2006a). Electrophysiological validation of a human prototype auditory midbrain implant in a guinea pig model. *Journal of the Association for Research in Otolaryngology*, 7, 383–398.

Lenarz, T., Lim, H. H., Reuter, G., Patrick, J. F., & Lenarz, M. (2006b). The auditory midbrain implant: a new auditory prosthesis for neural deafness-concept and device description. *Otology & Neurotology*, 27(6), 838–843.

Lenarz, M., Lim, H. H., Lenarz, T., Reich, U., Marquardt, N., Klingberg, M. N., Paasche, G., Reuter, G., & Stan, A. C. (2007). Auditory midbrain implant: histomorphologic effects of long-term implantation and electric stimulation of a new deep brain stimulation array. *Otology & Neurotology*, 28(8), 1045–1052.

Lenarz, T., Reuter, G., Paasche, G., Lenarz, M., Gibson, P., & Mackiewicz, M. (2003). Auditory Midbrain Implant (AMI) – Ein neues Therapiekonzept fur neurale Taubheit. *HNO*, 28(2), 146.

Lim, H. H., & Anderson, D. J. (2003). Feasibility experiments for the development of a midbrain auditory prosthesis. *Proceedings of the 1st International IEEE EMBS Conference on Neural Engineering*, 1, 193–196.

Lim, H. H., & Anderson, D. J. (2006). Auditory cortical responses to electrical stimulation of the inferior colliculus: implications for an auditory midbrain implant. *Journal of Neurophysiology*, 96(3), 975–988.

Lim, H. H., & Anderson, D. J. (2007). Spatially distinct functional output regions within the central nucleus of the inferior colliculus: Implications for an auditory midbrain implant. *Journal of Neuroscience*, 27(32), 8733–8743.

Lim, H. H., Lenarz, T., Joseph, G., Battmer, R. D., Samii, A., Samii, M., Patrick, J. F., & Lenarz, M. (2007). Electrical stimulation of the midbrain for hearing restoration: insight into the functional organization of the human central auditory system. *Journal of Neuroscience*, 27(49), 13541–13551.

Lim, H. H., Lenarz, T., Anderson, D. J., & Lenarz, M. (2008). The auditory midbrain implant: effects of electrode location. *Hearing Research*, 242(1–2), 74–85.

Lim, H. H., Lenarz, M., & Lenarz, T. (2009a). Auditory midbrain implant: a review. *Trends in Amplification*, 13(3), 149–180.

Lim, H. H., Lenarz, M., & Lenarz, T. (2009b). A new auditory prosthesis using deep brain stimulation: development and implementation. In D. Zhou & E. Greenbaum (Eds.), *Implantable neural prostheses 1: devices and applications* (pp. 117–154). New York: Springer Science+Business Media, LLC.

Lockwood, L. (2003). *Beethoven: the music and the life*. New York: W. W. Norton & Company, Inc.

Loftus, W. C., Bishop, D. C., Saint Marie, R. L., & Oliver, D. L. (2004). Organization of binaural excitatory and inhibitory inputs to the inferior colliculus from the superior olive. *Journal of Comparative Neurology*, 472(3), 330–344.

Lu, T., & Wang, X. (2000). Temporal discharge patterns evoked by rapid sequences of wide- and narrowband clicks in the primary auditory cortex of cat. *Journal of Neurophysiology*, 84(1), 236–246.

Matthies, C., Thomas, S., Moshrefi, M., Lesinski-Schiedat, A., Frohne, C., Battmer, R. D., Lenarz, T., & Samii, M. (2000). Auditory brainstem implants: current neurosurgical experiences and perspective. *Journal of Laryngology and Otology Supplement*, 27, 32–36.

McCreery, D. B. (2008). Cochlear nucleus auditory prostheses. *Hearing Research*, 242(1–2), 64–73.

Moore, J. K. (1987). The human auditory brain stem: a comparative view. *Hearing Research*, 29(1), 1–32.

Moore, J. K., & Osen, K. K. (1979). The cochlear nuclei in man. *American Journal of Anatomy*, 154(3), 393–418.

Nashold, B. S., Jr., & Wilson, W. P. (1966). Central pain. Observations in man with chronic implanted electrodes in the midbrain tegmentum. *Confinia Neurologica*, 27(1), 30–44.

Nashold, B. S., Jr., Wilson, W. P., & Slaughter, D. G. (1969). Sensations evoked by stimulation in the midbrain of man. *Journal of Neurosurgery*, 30(1), 14–24.

Nayagam, D. A., Clarey, J. C., & Paolini, A. G. (2006). Intracellular responses and morphology of rat ventral complex of the lateral lemniscus neurons in vivo. *Journal of Comparative Neurology*, 498(2), 295–315.

Neuheiser, A., Lenarz, M., Reuter, G., Calixto, R., Nolte, I., Lenarz, T., & Lim, H. H. (2010). Effects of pulse phase duration and location of stimulation within the inferior colliculus on auditory cortical evoked potentials in a guinea pig model. *Journal of the Association for Research in Otolaryngology*, 11, 689–708.

Oliver, D. L. (2005). Neuronal organization in the inferior colliculus. In J. A. Winer & C. E. Schreiner (Eds.), *The inferior colliculus* (pp. 69–114). New York: Springer Science+Business Media, Inc.

Owen, S. L., Green, A. L., Nandi, D. D., Bittar, R. G., Wang, S., & Aziz, T. Z. (2007). Deep brain stimulation for neuropathic pain. *Acta Neurochirurgica Supplementum*, 97(2), 111–116.

Patrick, J. F., Busby, P. A., & Gibson, P. J. (2006). The development of the Nucleus Freedom Cochlear implant system. *Trends in Amplification*, 10(4), 175–200.

Penfield, W. (1958). *The excitable cortex in conscious man*. Springfield: Charles C Thomas.

Penfield, W., & Rasmussen, T. (1950). *The cerebral cortex of man*. New York: The Macmillan Company.

Perez-Gonzalez, D., Malmierca, M. S., & Covey, E. (2005). Novelty detector neurons in the mammalian auditory midbrain. *European Journal of Neuroscience*, 22(11), 2879–2885.

Pfingst, B. E., Burnett, P. A., & Sutton, D. (1983). Intensity discrimination with cochlear implants. *Journal of the Acoustical Society of America*, 73(4), 1283–1292.

Pfingst, B. E., Zwolan, T. A., & Holloway, L. A. (1997). Effects of stimulus configuration on psychophysical operating levels and on speech recognition with cochlear implants. *Hearing Research*, 112(1–2), 247–260.

Phillips, D. P., Hall, S. E., & Hollett, J. L. (1989). Repetition rate and signal level effects on neuronal responses to brief tone pulses in cat auditory cortex. *Journal of the Acoustical Society of America*, 85(6), 2537–2549.
Ramachandran, R., Davis, K. A., & May, B. J. (1999). Single-unit responses in the inferior colliculus of decerebrate cats. I. Classification based on frequency response maps. *Journal of Neurophysiology*, 82(1), 152–163.
Rance, G., Cone-Wesson, B., Wunderlich, J., & Dowell, R. (2002). Speech perception and cortical event related potentials in children with auditory neuropathy. *Ear and Hearing*, 23(3), 239–253.
Reuter, G., Reich, U., Marquardt, N., Klingberg, M., Lenarz, M., & Lenarz, T. (2004). Frequency-specific activity of the auditory pathway with auditory midbrain implants. *Biomedizinische Technik*, 49(2), 896–897.
Reuter, G., Stan, A., Reich, U., Marquardt, N., Klingberg, M., Paasche, G., Patrick, J. F., Lenarz, T., & Lenarz, M. (2005). Histologische Untersuchungen des Colliculus inferior nach chronischer elektrischer Stimulation. *Biomaterialien*, 6, 38–39.
Rodrigues-Dagaeff, C., Simm, G., De Ribaupierre, Y., Villa, A., De Ribaupierre, F., & Rouiller, E. M. (1989). Functional organization of the ventral division of the medial geniculate body of the cat: evidence for a rostro-caudal gradient of response properties and cortical projections. *Hearing Research*, 39(1–2), 103–125.
Samii, M., Carvalho, G. A., Tatagiba, M., Matthies, C., & Vorkapic, P. (1996). Meningiomas of the tentorial notch: surgical anatomy and management. *Journal of Neurosurgery*, 84(3), 375–381.
Samii, A., Lenarz, M., Majdani, O., Lim, H. H., Samii, M., & Lenarz, T. (2007). Auditory midbrain implant: a combined approach for vestibular schwannoma surgery and device implantation. *Otology & Neurotology*, 28(1), 31–38.
Schreiner, C. E., & Langner, G. (1988). Periodicity coding in the inferior colliculus of the cat. II. Topographical organization. *Journal of Neurophysiology*, 60(6), 1823–1840.
Schwartz, M. S., Otto, S. R., Shannon, R. V., Hitselberger, W. E., & Brackmann, D. E. (2008). Auditory brainstem implants. *Neurotherapeutics*, 5(1), 128–136.
Shannon, R. V. (1985). Threshold and loudness functions for pulsatile stimulation of cochlear implants. *Hearing Research*, 18(2), 135–143.
Shannon, R. V., Zeng, F. G., Kamath, V., Wygonski, J., & Ekelid, M. (1995). Speech recognition with primarily temporal cues. *Science*, 270(5234), 303–304.
Shannon, R. V., Fu, Q. J., & Galvin, J., 3 rd. (2004). The number of spectral channels required for speech recognition depends on the difficulty of the listening situation. *Acta Oto-Laryngologica Supplementum*, 552, 50–54.
Simmons, F. B., Mongeon, C. J., Lewis, W. R., & Huntington, D. A. (1964). Electrical stimulation of acoustical nerve and inferior colliculus. *Archives of Otolaryngology*, 79, 559–568.
Slattery, W. H., 3 rd, Brackmann, D. E., & Hitselberger, W. (1998). Hearing preservation in neurofibromatosis type 2. *American Journal of Otology*, 19(5), 638–643.
Snyder, R. L., Vollmer, M., Moore, C. M., Rebscher, S. J., Leake, P. A., & Beitel, R. E. (2000). Responses of inferior colliculus neurons to amplitude-modulated intracochlear electrical pulses in deaf cats. *Journal of Neurophysiology*, 84(1), 166–183.
Stein, B. M. (1979). Supracerebellar-infratentorial approach to pineal tumors. *Surgical Neurology*, 11(5), 331–337.
Stiebler, I. (1986). Tone-threshold mapping in the inferior colliculus of the house mouse. *Neuroscience Letters*, 65(3), 336–340.
Strauss-Schier, A., Battmer, R. D., Rost, U., Allum-Mecklenburg, D. J., & Lenarz, T. (1995). Speech-tracking results for adults. *Annals of Otology, Rhinology, and Laryngology Supplement*, 166, 88–91.
Trepel, M. (2004). *Neuroanatomie. Struktur und Funktion* (3 rd ed.). Muenchen: Elsevier GmbH.
Vince, G. H., Herbold, C., Coburger, J., Westermaier, T., Drenckhahn, D., Schuetz, A., Kunze, E., Solymosi, L., Roosen, K., & Matthies, C. (2010). An anatomical assessment of the supracerebellar midline and paramedian approaches to the inferior colliculus for auditory midbrain

implants using a neuronavigation model on cadaveric specimens. *Journal of Clinical Neuroscience*, 17(1), 107–112.

Wang, X., Lu, T., Bendor, D., & Bartlett, E. (2008). Neural coding of temporal information in auditory thalamus and cortex. *Neuroscience*, 157(2), 484–494.

Wichmann, T., & Delong, M. R. (2006). Deep brain stimulation for neurologic and neuropsychiatric disorders. *Neuron*, 52(1), 197–204.

Winer, J. A. (2005). Three systems of descending projections to the inferior colliculus. In J. A. Winer & C. E. Schreiner (Eds.), *The inferior colliculus* (pp. 231–247). New York: Springer Science+Business Media, Inc.

Chapter 10
Central Auditory System Development and Plasticity After Cochlear Implantation

Anu Sharma and Michael Dorman

1 Introduction

Cortical development is dependent on both intrinsic and extrinsic (stimulus-driven) factors. The absence of sensory input from birth, as in congenital deafness, inhibits the normal growth and connectivity needed to form a functional sensory system, resulting in deficits in spoken language. Cochlear implants (CIs) bypass peripheral cochlear damage, directly stimulating the auditory nerve and thus making it possible, in principle, to avoid many of the deleterious effects of stimulus deprivation. From this point of view, children and adults who receive implants provide a platform from which we can examine the characteristics of deprivation-induced and experience-dependent plasticity in the central auditory system. Plasticity, by its very nature, may be both beneficial and harmful. Studies using electrophysiological and brain imaging techniques (such as PET, SPECT, fMRI and MEG) have delineated the time course of, and constraints on, development and plasticity in the central auditory pathways following cochlear implantation. This chapter reviews the literature on sensitive periods for cortical plasticity in congenitally deaf children who receive implants and then examines issues related to cross-modal reorganization in pre- and postlingually deafened adults who receive cochlear implants.

A. Sharma (✉)
Speech, Language and Hearing Sciences, University of Colorado at Boulder,
2501 Kittredge Loop Drive, Boulder, CO 80309-0409, USA
e-mail: anu.sharma@colorado.edu

F.-G. Zeng et al. (eds.), *Auditory Prostheses: New Horizons*, Springer Handbook of Auditory Research 39, DOI 10.1007/978-1-4419-9434-9_10,

2 Development and Plasticity in Congenital Deafness

In one of developmental neurobiology's classic experiments, David Hubel and Torsten Wiesel described a brief period during early infancy when sensory deprivation can profoundly and irrevocably alter central processes in vision (Wiesel and Hubel 1963; Hubel and Wiesel 1965, 1967, 1970). By examining cortical maturation in kittens that were dark-reared until various ages, they demonstrated that there is a limited time window or a critical period in early development during which stimulation must be delivered to a sensory system if that system is to develop normally. Other relevant findings included that the effects of deprivation were more obvious at cortical levels than at lower levels, and that a major factor shaping development was competition for cortical resources, rather than deprivation *per se* (Hubel 1995). The groundbreaking work of Hubel and Wiesel in kittens has since become relevant to the effects of deprivation on plasticity and development of sensory systems in humans.

In order to examine comparable aspects of cortical development and plasticity resulting from auditory deprivation, one would need congenitally deaf children and a method to "re-start hearing" at different ages. With the advent of Food and Drug Administration (FDA) approved pediatric cochlear implantation in 1990, more than 25,000 children in the United States have received a cochlear implant at ages ranging from 6 months to 21 years (National Institute on Deafness and Other Communication Disorders 2009). This wide spread of pediatric implantation has resulted in an elegant natural experiment with congenitally deaf children who experience different durations of sensory deprivation before their hearing is re-initiated with a cochlear implant. Using non-invasive electrophysiological techniques and invasive brain imaging, several laboratories (see Sharma and Dorman 2006; Kral and Eggermont 2007; and Giraud and Lee 2007 for reviews) have systematically examined the effects of deafness on development and plasticity of the central auditory system.

Plasticity can be defined in a number of ways (see Pallas 2001 for a review), but it essentially refers to the ability of neurons and neuronal networks to change their function as a consequence of their previous activity and/or stimulating environment. Plasticity is based, in part, on changes in synaptic activity, on changes in synchronization of neural networks, and on changes in interneuronal connection patterns within neural networks (see Kral 2007 for a review). Plasticity increases as one ascends along the central auditory pathways with the auditory cortex and the thalamus showing a higher degree of plasticity than subcortical structures (e.g., cochlear nucleus and inferior colliculus). Similarly, the higher-order cortex has greater capacity for plastic changes than primary areas (see Kral and Tillein 2006 for a review). Accordingly, this chapter focuses on examining plasticity at the cortical level, including primary and higher-order cortex. Finally, plasticity can be beneficial (adaptive) or harmful (maladaptive). Adaptive plasticity can improve function by compensating for injury or deprivation, e.g., congenitally deaf children learning oral language via a cochlear implant. On the other hand, maladaptive plasticity, for example, may result in cross-modal remapping across brain areas potentially reducing other functions through competition or "crowding." This chapter will examine both adaptive and maladaptive plasticity that is a consequence of auditory deprivation.

2.1 Normal Development of Central Auditory Pathways

Long-term development of the human auditory cortex can be measured by a number of techniques. Huttenlocher and Dabholkar (1997) counted synapses in the post-mortem auditory cortex and found, first, an increase in the number of synapses (synaptogenesis) from birth until early childhood (i.e., age 2 to 4 years) and then a decrease in the number of synapses (pruning) until adult levels were reached during early adolescence (see Fig. 10.1, panel a). Devous et al. (2006) report on regional cerebral blood flow (rCBF) levels in single photon emission computed tomography (SPECT) scans which suggest that neuronal populations in the primary auditory cortex stabilize by age 7 years in contrast to those in higher-order auditory cortex which continue to mature until adolescence. Moore and colleagues examined the laminar development of auditory cortex and found that in early childhood, from age 6 months to 5 years, there is a progressive maturation of the thalamic projections to the cortex (Moore and Guan 2001; Moore and Linthicum 2007). Differential development of axonal densities between the infragranular and supragranular cortical layers and development of the supragranular layers and their intracortical connections do not reach maturity until early adolescence (Moore and Linthicum 2007). Because these developmental changes are reflected in non-invasive scalp-recorded auditory evoked potentials (Moore and Linthicum 2007), long-term changes in cortical development in humans have been charted using cortical auditory evoked potentials or CAEPs (Ponton et al. 1996a; Sharma et al. 1997; Tonnquist-Uhlen et al. 2003).

Auditory evoked potentials reflect synchronized electroencephalographic (EEG) activity in response to sound. In infants and children, the cortical auditory evoked response is dominated by a large positivity, the P1. In infants and children, the P1 is seen at latencies of 100 to 300 ms after the onset of stimulation. As age increases, an invagination appears in the broad P1 waveform in the form of the N1 component of the CAEP. Although the N1 component can be seen in children as young as 3 to 5 years of age at a slow stimulation rate (e.g., 1 stimulus every 2 s), it is consistently seen in children by about age 12 years at standard stimulation rates of about 2 stimuli per second (Gilley et al. 2005). Ponton and Eggermont (2001) suggest that the surface positivity of the P1 response is consistent with "a relatively deep sink ([in cortical] layers IV and lower III) and a superficial current return." P1 generators include primary and secondary auditory areas, while N1 likely reflects activation in higher-order auditory cortex as a result of intra- and interhemispheric activity (Makela and Hari 1992; Makela and McEvoy 1996; Ponton et al. 2000a).

Besle et al. (2008) recorded from depth electrodes in the human auditory cortex and found initial responses to sound 14 to 40 ms after stimulus onset in the medial portion of the transverse gyrus. Activation spread to the lateral transverse gyrus and

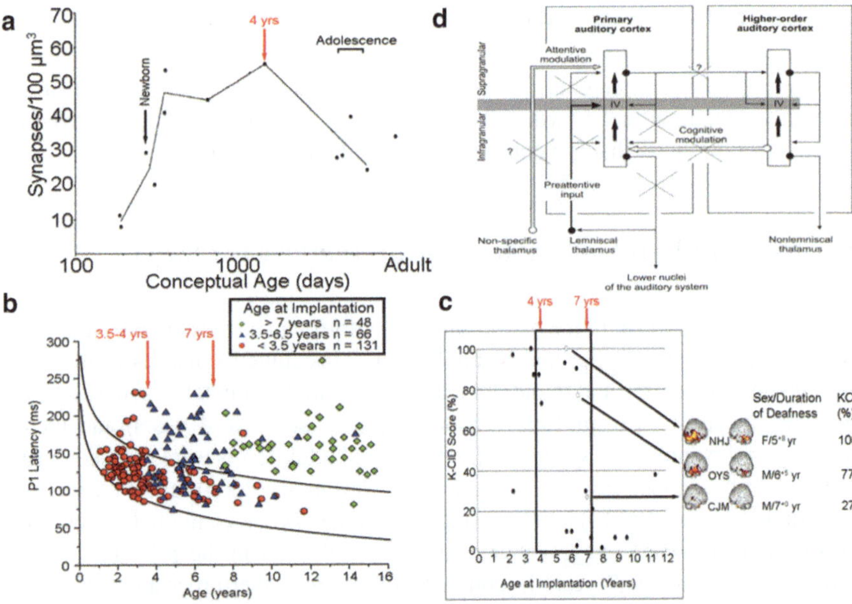

Fig. 10.1 Major determinants of the sensitive period for central auditory development in congenitally deaf children who receive cochlear implants. (**a**) In normal development, the period of intrinsically regulated synaptogenesis or synaptic offshoot in the human auditory cortex increases from the neonatal period until approximately age 4 years, after which synaptic density decreases as a result of synaptic pruning resulting from mainly extrinsic stimulation. Adapted from Huttenlocher and Dabholkar (1997) by permission of John Wiley and Sons. (**b**) Electrophysiologic studies using the latency of the P1 cortical auditory evoked potential as a biomarker for cortical development describe a sensitive period for central auditory development in congenitally deaf children who receive a cochlear implant. Children implanted under age 3.5 years show normal P1 latencies within 3 to 6 months of implant use (*red circles*). This period corresponds closely to the period of synaptic offshoot shown in panel (**a**). There is considerable variability in P1 development for children implanted from 3.5 to 6.5 years (*blue triangles*). However, the sensitive period closes by age 7 years as suggested by data from children who are implanted after age 7 years and who show abnormal P1 latencies and morphology (*green diamonds*) even years after implant use. Adapted from Sharma and Dorman (2006) by permission of Wolters Kluwer Health and S. Karger AG. (**c**) Pre-implantation PET scan studies also document a sensitive period for cortical hypometabolism in congenital deafness which is very similar to that described in panel (**b**). Children implanted under age 4 years show larger areas of pre-implant hypometabolism in auditory cortex (correlated with high speech perception scores after implantation). On the other hand, children implanted after age 7 years show smaller areas of hypometabolism (correlated with poor speech perception scores), suggesting that these areas of auditory cortex have been recruited by other modalities. Similar to the P1 latencies for children implanted between 3.5 and 7 years, the areas of pre-implant hypometabolism for children implanted between 4 and 7 are variable (red areas on MRI); however, they correlate with post-implant speech perception score (percent correct K-CID scores shown on the y-axis). Adapted from Oh et al. (2003) by permission of Informa Medical and Pharmaceutical Science Journals. (**d**) Model of deficits and functional decoupling in the auditory cortex in deafness. Lemniscal input targets A1 mainly in layer IV but also in supragranular and infragranular layers. Neurons in infragranular layers project to layer IV and layer IV projects to supragranular layers. Supragranular layers project back to layer IV and infragranular layers (feedback). Infragranular layers send descending fibers to subcortical nuclei (corticofugal). Feed-forward coupling to the higher-order auditory areas is accomplished via supragranular layers, descending projections from higher-order cortex target the infragranular layers in A1 ("cognitive modulation"). *Dashed crosses* show which connections are supposed not functional in congenitally deaf cats (Based on data from Kral et al. 2000, 2005, 2006; adapted from Kral 2007 by permission of Informa Medical and Pharmaceutical Science Journals)

planum temporale at approximately 65 ms. Further spread of activity to the superior temporal gyrus, the supramarginal gyrus, and the superior temporal sulcus occurred at approximately 120 ms after stimulus onset. Given these data, it is reasonable to assume that the P1 and N1 reflect second-order processing in the auditory cortex, including input from feedback and recurrent loops between primary auditory and association areas (Kral and Eggermont 2007).

Sharma and colleagues have established 95% confidence intervals for the latency of P1 at different ages (Sharma et al. 2002a). A newborn may have a P1 latency of around 300 ms. Rapid neural development during the first 2 to 4 years leads to a rapid decrease in P1 latency: 4-year-olds have a P1 latency of about 125 ms. Adults have a P1 latency of around 60 ms. Because the latency of the CAEP varies with age, latency can be used as a biomarker for maturation of central auditory pathways (Eggermont 1988; Albrecht et al. 2000; Pang and Taylor 2000; Wunderlich et al. 2006). Sharma et al. (1997, 2002a) have described in detail the normal developmental trajectory of the P1 response from birth to adulthood. They examined the latency of P1 as a function of age in 190 normal-hearing subjects ranging in age from 0.1 to 20 years. The results revealed a strong negative correlation between age and latency of P1. The decrease in P1 latency with increasing age suggests more efficient synaptic transmission over time and may reflect a more refined (or pruned) auditory pathway. These normative data provide a standard against which the central development of congenitally deaf children who are fitted with cochlear implants can be compared.

2.2 *Electrophysiologic Evidence for a Brief Sensitive Period for Central Auditory Development*

In a series of studies, Ponton, Eggermont, and colleagues (Eggermont 1988; Ponton et al. 1993, 1996a, b, 1999, 2000a, b, 2002; Ponton and Eggermont 2001; Eggermont and Ponton 2002, 2003) first described cortical development in children and adults fit with implants by comparing the waveform morphologies to those from age-matched normal-hearing persons. Those studies revealed a developmental delay in the morphology and latency of CAEPs in prelingually deaf children after implantation. The authors theorized that the auditory part of the brain is "frozen in time" as a result of sensory deprivation. However, once stimulation is restored via a cochlear implant, maturation proceeds at a normal rate. As a consequence, P1 latencies reflect the "time in sound" experienced by the implanted child. These studies provided the first critical evidence that the potential for normal plasticity of the auditory system is maintained in deaf children during their years of deprivation.

Over the last decade, Sharma and colleagues have conducted several large scale studies examining cortical maturation and functioning in congenitally deaf children fitted with unilateral cochlear implants at varying ages. For example, Sharma et al. (2002b; Sharma and Dorman 2006) examined P1 latencies in 245 congenitally deaf children fit with a cochlear implant and reported that children who received stimulation via and implant early in childhood (<3.5 years) showed normal P1 latencies, while children who received cochlear implant stimulation late in childhood (>7 years) had abnormal cortical response latencies. A group of children receiving CIs between 3.5 and 7 years revealed highly variable response latencies (Fig. 10.1, panel b). In general, for the majority of late implanted children, response latencies did not reach normal limits even after several years of experience with the implant (Sharma et al. 2005a). In another large scale study, Sharma and colleagues examined individual longitudinal trajectories of P1 response development after cochlear implantation in 231 children (Sharma et al. 2007). Children implanted under age 3.5 years showed normal P1 response latencies within 6 to 8 months of implantation. Children implanted after age 7 also showed latency decreases over time, but their developmental trajectories were abnormal and P1 latencies never reached normal limits even after years of implant usage. Waveform morphologies, another measure of central auditory development following the onset of stimulation, showed clear differences between children implanted before age 3.5 years and after age 7 years. In the early-implanted children studied by Sharma and Dorman (2006), waveform morphology was normal and characterized by a broad positivity within a week following the onset of stimulation. For the late-implanted children, waveforms were commonly abnormal and characterized by a polyphasic waveform or a generally low-amplitude waveform. Because P1 cortical evoked potentials are non-invasive and easily obtained in clinical settings, they have been used extensively as a clinical tool to evaluate cortical development and functioning in cochlear implanted children (Sharma et al. 2005b, 2009; Sharma and Dorman 2006; Nash et al. 2008).

Overall, the P1 data described above suggest a sensitive period for central auditory development of about 3.5 years when central auditory pathways are most receptive to stimulation with a cochlear implant. There is some variability in the data from ages 3.5 to 7 years. However, in all likelihood, the sensitive period ends at age 7 years. That is, implantation after age 6 to 7 years occurs into a cortex that shows reduced plasticity to stimulation via a cochlear implant. This finding of a sensitive period for central auditory development in humans is consistent with other studies in animals (Ryugo et al. 1997; Kral et al. 2000, 2001) and in humans (Lee et al. 2001; Eggermont and Ponton 2003; Schorr et al. 2005).

It is interesting to note that there is yet no evidence from studies of congenitally deaf children that suggests the existence of a sensitive period regarding expression of neural plasticity at the level of the auditory brainstem as measured by the auditory brainstem response (ABR). Gordon et al. (2003, 2005) and Sharma and Dorman (2006) reported rapid development of the ABR response in implanted children regardless of the age at which the implant was fitted. These ABR results in humans are not consistent with animal studies of congenitally deaf cats, which suggest rapid alteration of synaptic terminals in brainstem nuclei following early deprivation (Ryugo et al. 1997, 2005). It may be that measures of latency and morphology of the auditory brainstem response are not sensitive measures of the effects of deprivation on lower levels of the auditory pathway. Conversely, it is possible that lower levels of the auditory pathways are not as susceptible to the effects of deprivation and/or sensitive periods in humans as they are in congenitally deaf cats.

2.3 Rates of Post-Implantation P1 Development

Although the studies from the Sharma/Dorman and Ponton/Eggermont groups revealed high levels of plasticity after early implantation, they had different interpretations of the rate of development of P1 latencies following implantation. Using cross-sectional data, Ponton et al. (1996b) interpolated that post-implantation development proceeded at a normal rate, which led them to theorize that cortical development in implanted children would be delayed by roughly the age at which the child received the implant. However, when Sharma and colleagues charted actual longitudinal developmental trajectories from 231 implanted children, they found that in early-implanted children (i.e., children implanted under age 3.5 years), P1 latencies changed very rapidly, reaching normal limits within months (rather than years) following stimulation. On the other hand, in many late-implanted children, although there was clear evidence of a latency decrease in the initial months, the latencies never reached normal limits even after 8 to 10 years of implant usage.

The neurophysiologic mechanisms underpinning the rapid changes in P1 latency are not entirely clear. However, Kral et al. (2000) have shown that congenitally deaf cats show (i) a restricted (atypical) pattern of activation within the layers of the primary auditory cortex compared to normal hearing cats, and (ii) signs of desynchronization of activity among different cortical layers. As a consequence, there are changes in morphology of local field potentials recorded at the cortical surface. If similar processes are at work in young deaf children, then it can be hypothesized that a change in interaction between different cortical layers, driven by the onset of electrical stimulation, results in a re-arrangement in the generators of the P1 response (for cats, see Klinke et al. 1999). Such a process, which would be maximized during the period of synaptic overshoot (which occurs during the first 4 years of life; see Fig. 10.1, panel a), is likely to be the major factor in the rapid changes in waveform morphology and the latency reported in early-implanted children by Sharma et al. (2005a). A corresponding effect of rapid changes after intervention is seen in the visual system. Maurer et al. (1999) assessed visual acuity in human infants who were congenitally deprived of patterned visual input by cataracts. The cataracts were removed at 1 week to 9 months of age. Maurer et al. (1999) found that acuity improved rapidly, with some improvement apparent after as little as 1 h of visual input. Critically, the rate of development of visual acuity after cataract removal was significantly greater than normal (relative to age-matched controls). It is parsimonious that similar evidence of a highly plastic system is found in both congenital deafness and blindness.

2.4 Converging Evidence from Brain Imaging Studies

Positron emission tomography (PET) imaging studies have provided important evidence regarding the age cut-offs for the sensitive period. Measurements of resting cortical metabolic rate and rCBF in PET are believed to correspond to synaptic density of neurons and decreases in brain metabolism which occur in accordance to the synaptic pruning process (Catalan-Ahumada et al. 1993; de Volder et al. 1999; Hirano et al. 2000). Studies such as those of Lee and colleagues (Lee et al. 2001, 2003, 2005a) found a cut-off of about age 4 years and are generally in good agreement with the electrophysiologic studies of Sharma and colleagues (2002a; 2005a; Sharma and Dorman 2006) described in Sect. 2.2. These PET imaging studies made use of recordings of resting glucosemetabolism rates in the auditory cortices of prelingually deafened children and adults before cochlear implantation and related these rates to speech perception scores after implantation. The degree of glucose metabolism pre-implantation was taken to be an indicator of the degree to which cross-modal recruitment of the auditory cortex had occurred. Thus, the auditory cortices should be hypometabolic because of years of auditory deprivation. However, if the cortices had been recruited by other cortical functions, then the cortices would not be hypometabolic. Lee et al. (2001, 2005a) reported that the degree of hypometabolism before implantation (which was greater for younger subjects) was positively correlated with the speech perception scores after implantation. In general, children who were implanted before age 4 years showed the highest degree of hypometabolism in the auditory cortices before implantation and, following implantation, these children had the highest speech perception scores. The age cut-off (4 years) is consistent with the 3.5 years cut-off for maximal plasticity of the central auditory pathways suggested by Sharma et al. (2002b). Lee's data also suggest that following 6.5 to 7.5 years of deprivation significant cross-modal reorganization occurs in the auditory cortices. This finding is concordant with the Sharma et al. (2002b) finding of increased P1 latencies and abnormal P1 morphology following 7 years of auditory deprivation. Finally, Oh et al. (2003) describe that children who received their implant between 5 and 7 years showed highly variable cortical metabolism and speech scores with their implant. This period roughly corresponds to the group of children described by Sharma et al. (2002b), who were implanted between 3.5 and 6.5 years and who showed variable P1 latencies (see Fig. 10.1, panels b and c). The remarkable correspondence between the P1 latency and PET data provides clear converging evidence for a brief sensitive period of central auditory development in congenital deafness.

There is a close correspondence between the age cut-offs for the sensitive period described above and the speech and language performance of congenitally deaf, implanted children. Several investigators have reported that children implanted under age 3 to 4 years show significantly higher speech perception scores and better language skills compared to children implanted after age 6 to 7 years (Geers 2006; Holt and Svirsky 2008; Wang et al. 2008). For a review of sensitive periods as they relate to speech perception and language acquisition in children with cochlear implants, see Harrison et al. (2005) and Holt and Svirsky (2008).

2.5 Mechanisms Underlying the Sensitive Period

There are likely multiple mechanisms at the genetic, molecular, and neural levels which underlie the sensitive period described above. For one, developmental changes as a result of synaptic plasticity are a major contributor to sensitive periods (see Kral and Eggermont 2007 for a review). There are relatively few synapses in the cerebral cortex of newborns. New synapse formation begins in the prenatal period but shows massive increases in the postnatal period and continues for the first 4 years of life, i.e., a period of "synaptic overshoot" (Conel 1939–1967; Huttenlocher and Dabholkar 1997). Kral (2007) suggests that the period of synaptic overshoot is meaningful, in that it allows the brain the flexibility to cope with many different environmental conditions, including sensory deprivation. Critically, this period of synaptogenesis appears to be intrinsically regulated, i.e., it is independent, to a large extent, of the auditory experiences of the child (Huttenlocher and Dabholkar 1997). In that sense, intrinsically regulated synaptic development may have a "protective effect" with regard to the potential for development of the central pathways until about age 4 years. Thus, there is likely a close correspondence between the sensitive period for central auditory development in cochlear implanted children (3.5 to 4 years) and the age of synaptic overshoot prior to the onset of synaptic elimination (approx. 4 years) (see Fig. 10.1, panels a and b). After synaptic density reaches a peak by age 4 years, it subsequently shows a marked decrease or refinement which is experience dependent. Those synapses that are not activated by sensory stimulation are eliminated or "pruned" and those that are repeatedly activated are stabilized and strengthened. Cochlear implantation within this sensitive period would then provide the auditory experience needed for synaptic elimination to refine central auditory pathways. For example, as described earlier, although early-implanted children show clear effects of auditory deprivation (as reflected in P1 waveform morphology and in auditory cortical hypometabolism), they are able to overcome or reverse these deficits rapidly when implanted by age 4 years (Sharma et al. 2005a). In contrast, if appropriate auditory stimulation is not provided within this period, then essential synapses are not established and inappropriate ones are not eliminated, resulting in the naïve cortex not being able to process incoming activity appropriately (Kral et al. 2000, 2006). Taken together, these findings suggest that, in humans, the end of synaptogenesis corresponds to the end of the sensitive period.

2.6 Functional Decoupling of Cortical Areas at the End of the Sensitive Period

The end of the sensitive period has consequences for development and organization of cortical areas and pathways. Congenitally deaf white cats are a useful model system to study cortical organization following deafness and cortical development following the onset of stimulation by a cochlear implant. Kral et al. (2005) have described a sensitive period in congenitally deaf cats of approximately 4 months of age. During this period, congenitally deaf cats show a restricted pattern of activation within the layers of the primary auditory cortex and desynchronized activity among cortical layers. As a consequence, there are changes in morphology of local field potentials recorded at the cortical surface (Kral et al. 2002, 2005; Kral and Tillein 2006). When electrical stimulation is started after 4 months of deafness (i.e., after the end of the sensitive period for central auditory development in cats) there is a delay in the activation of supragranular layers of the cortex and a near absence of activity at longer latencies and in infragranular layers (layers V and VI) (Kral et al. 2005). The near absence of outward currents in layers IV and III of congenitally deaf cats suggests incomplete development of inhibitory synapses and an alteration of information flow from layer IV to supragranular layers. The higher-order auditory cortex projects back to A1 (primary auditory cortex), mainly to the infragranular layers and the infragranular layers (V and VI), sending long range feedback projections to the subcortical auditory areas. The absence of activity in infragranular layers can be interpreted as a functional decoupling of primary cortex from higher-order auditory cortex, also affecting feedback projections to subcortical auditory structures (Kral et al. 2000, 2002, 2005) (see Fig. 10.1, panel d). Kral has speculated that a similar partial or complete decoupling between A1 and higher-order cortex may occur in congenitally deaf children at the end of the sensitive period.

In the absence of auditory stimulation, the naïve primary auditory cortex appears to maintain a rudimentary capacity to process auditory information (Kral et al. 2001, 2005). However, without hearing experience, there are no appropriate higher-order representations established that are associated with incoming auditory stimuli. With longer term deprivation of auditory input, not only does the bottom-up (feed-forward) capacity for information processing decrease, but because of the functional decoupling, there is an inability to integrate the afferent information with cognitive top-down (feed-back) modulatory influences by higher-order cortex. This results in a fundamental decrease in the capacity of the auditory cortex to analyze auditory stimuli efficiently (see Kral 2007 for a review). Because the higher-order auditory cortex is inherently multi-modal, such a decoupling may allow other sensory input to predominate in the higher-order auditory cortex in children deprived of sound for a long period. These mechanisms are cited by Kral as the reasons auditory processing becomes difficult after the sensitive period; specifically, modulation of the primary auditory areas is changed and cortical areas important for auditory and linguistic processing area re-purposed by other systems, making it challenging for auditory processing and learning.

Kral's hypothesis of functional decoupling in deaf children after long durations of deafness is consistent with evidence from imaging and electrophysiologic studies in humans. Kang et al. (2003) examined functional connectivity as consequence of deprivation in prelingually deaf children by examining interregional metabolic correlation with fluorodeoxyglucose (FDG)-PET. The mean activity of FDG uptake in the cytoarchitectonically defined A1 region served as a co-variate for their intra-cortical and interhemispheric analyses. They reported that the functional connectivity of the primary auditory cortex with adjacent regions was greater in younger than in older prelingually deaf children. Interestingly, the functional connectivity of A1 in the left hemisphere was more restricted in deaf children compared to the right hemisphere, suggesting that left hemispheric cortices were less closely coupled with primary auditory cortex in deafness (Kang et al. 2003).

Electrophysiologic studies are also consistent with the notion of a partial or total decoupling of primary auditory cortex from surrounding higher-order cortical areas. Evoked potential studies have documented the absence of the N1 response in children who are implanted after long durations of deafness. As stated earlier, the N1 auditory evoked potential is predominantly generated in higher-order auditory cortex and its generators include cortico-cortical reciprocal loops between primary and secondary auditory cortices (Liegeois-Chauvel et al. 1994). Eggermont and Ponton (2003) found that the N1 component in the CAEP was absent in cochlear implanted subjects who had been deaf for a period of at least 3 years under the age of 6 years. On the basis of this finding, Eggermont and Ponton (2003) suggested that this period reflects a critical period for cortical maturation and for achieving useful speech perception. Sharma and Dorman (2006) also showed that children who are implanted after age 7 years never develop an N1 response. On the other hand, children implanted before the sensitive period of 3.5 years will develop an N1 component that is similar in morphology and latency to that found in normal-hearing children (Sharma and Dorman 2006). Given that N1 is generated in secondary auditory cortex, Kral and Eggermont (2007) suggest that the missing N1 wave in late-implanted children is indicative of improper activation of higher-order areas likely because of partial or total decoupling of higher-order areas from the primary auditory cortex. Finally, consistent with the decoupling hypothesis, there is a large body of evidence for recruitment of the higher-order auditory cortex by other modalities such as vision (Nishimura et al. 1999; Bavelier and Neville 2002; Lee et al. 2003) and somatosensation (Sharma et al. 2007) after the sensitive period in prelingually deaf persons (Fig. 10.2). However, there is virtually no evidence for re-purposing of primary auditory cortex in long-term deafness (see Kral 2007 for a review).

What might result in partial as opposed to complete functional decoupling after the end of the sensitive period? There are anecdotal reports from clinicians that a minority of prelingually deafened patients, implanted as adults, are able to attach appropriate meanings to sounds (especially environmental sounds), implying at least a partial connection between higher-order and primary areas. A common characteristic of these patients is that they received intensive auditory therapy as children. Thus, experience-dependent plasticity resulting from intensive auditory-aural rehabilitation may be one potential mechanism to prevent a complete decoupling of A1 from higher-order cortex. Similarly, as we will see below, it is possible that in these patients, intensive rehabilitation results in the establishment of alternate pathways from A1 to higher-order multisensory areas allowing for the meaningful interpretation of sounds and even possibly speech via an implant. Clearly more research is needed in congenitally deaf, implanted adults to establish these connections.

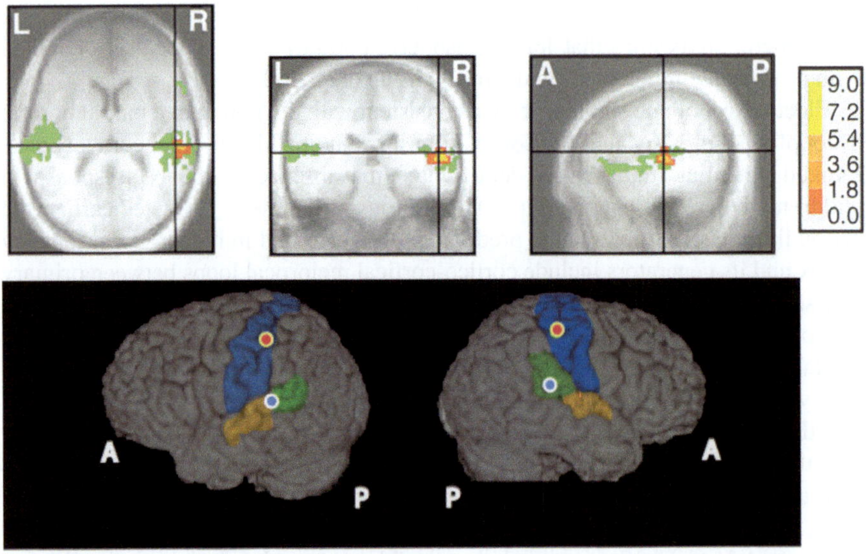

Fig. 10.2 Visual and somatosensory cross-modal plasticity in deafness. *Upper Panel*: Functional magnetic resonance imaging (fMRI) reveals activation in response to visual stimulation in the temporal cortex (including higher-order auditory cortical areas) of deaf adults. Adapted from Finney et al. (2001) by permission of Nature Publishing Group. *Lower panel*: Magnetoencephalograpy (MEG) dipole reconstructions (*circles*) reflect activation of both somatosensory cortex (*blue regions*) and higher-order auditory cortex (*green regions*) (including Wernicke's area, in the left hemisphere) in response to tactile stimulation in deaf adults. *A* Anterior, *P* Posterior, *L* Left, *R* Right (Adapted from Sharma et al. 2007 by permission of Informa Medical and Pharmaceutical Science Journals)

2.7 Cortical Reorganization after the Sensitive Period

In order to examine the manner in which the cortex gets re-organized after congenital deafness, Gilley et al. (2006) recorded 64-channel EEG in normal-hearing, early-implanted, and late-implanted children while they listened passively to a speech sound /ba/. Current density reconstructions using standardized low resolution electro-magnetic brain tomography (sLORETA) and dipole source analyses were performed in the time frame of the P1 & N1 CAEP responses. As expected, auditory stimulation activated the superior temporal sulcus (STS) bilaterally and the right inferior temporal gyrus in normal-hearing children. For children implanted before age 3.5 years (i.e., within the sensitive period), activation was observed along the superior temporal sulcus contralateral to the implanted ear and along the right inferior temporal gyrus activation independent of the ear stimulated. In addition, a minor source of activity was localized to the anterior parietotemporal cortex. Overall, these early-implanted children showed activation that was similar to normal-hearing children. On the other hand, children implanted after the end of the sensitive period (i.e., after age 7 years) showed low amplitude, diffuse activity from the primary generators identified in the normal-hearing and early-implanted children. These late-implanted children primarily showed activation of anterior parietotemporal cortex, insula (which are known multi-modal areas), and areas of visual cortex contralateral to the stimulated ear. The results of the Gilley et al. (2006) study suggest that auditory stimulation in early-implanted children activates a network of auditory areas mostly associated with normal auditory processing. In contrast, primarily multi-modal cortical areas are activated in late-implanted children suggesting an atypically distributed network of brain areas associated with auditory processing underlying their generally poor performance with the implant.

2.8 *Functional Consequences of Cortical Reorganization*

If the cortex is fundamentally reorganized at the end of the sensitive period (Gilley et al. 2006; Kral 2007), then what is the functional consequence of that reorganization? Kral and Eggermont (2007) discussed the mechanisms by which top-down processing modulates the development of the auditory system and suggested that a lack of reciprocating modulation from bottom-up processes may result in a reorganized auditory system, such that the ability to learn in a normal fashion (i.e., across multiple modalities) is compromised following sensory deprivation. If reorganization has already taken place amongst the sensory systems, and a new system (e.g., hearing) is later introduced, what are the functional consequences to those systems currently organized in the auditory areas? One possibility is that introducing new stimulation would impede the functioning of those reorganized sensory processes as well. That is, introducing a new sensory stimulus also introduces a new competition for resources in the cortex. Such activity should then have implications for multisensory processing.

In fact, several studies have implicated a lack of multisensory integration for children who experience long durations of deafness after cochlear implantation. Bergeson et al. (2005) examined the relative roles of auditory and visual input in speech perception performance between children implanted prior to age approximately 4 years to children implanted after that age. They reported that children implanted early in childhood performed better on the auditory-alone (A) and auditory-visual (AV) tasks than children implanted late, while the late-implanted children performed better on the visual-alone (V) task. A comparison of performance in V and AV conditions revealed that the early-implanted children showed greater benefit from the additional auditory input than later implanted children. This auditory gain for early-implanted users continued to improve after several years of implant use, while the auditory gain for late-implanted users remained relatively stable. These findings suggest that there may be a general lack of auditory integration skills in late-implanted CI users.

Schorr and colleagues (Schorr et al. 2005) examined auditory-visual fusion (the McGurk effect) in 36 prelingually deafened children with CIs and 35 normally-hearing children. The McGurk effect occurs when one speech sound is paired with the visual cue from another speech sound differing in place of articulation, resulting in the percept of an altogether different speech sound (McGurk and MacDonald 1976). For example, when visual /ba/ is paired with acoustic /ga/, listeners commonly report /da/. Consistent bimodal fusion was observed in 57% of hearing children and 38% of children who had received an implant before the age of 2.5 years, but in none of the late-implanted children.

Gilley et al. (2010) examined reaction times to the detection of basic auditory (A), visual (V), and combined auditory-visual (AV) stimuli in a group of normally hearing children and in children implanted before age 3.5 years and after age 3.5 years. All children showed the appropriate redundant signal effect (RSE) where AV stimuli were processed faster than A alone or V alone stimuli. However, only normal-hearing and early-implanted children violated Miller's test for inequality, which is a statistical measure of co-activation of multi-modal sensory input. In explaining the lack of co-activation of A and V modalities for late-implanted children, Gilley et al. (2010) suggested that when early input is dominated by one sensory system (in this case vision), such an early dominance leads to a long-lasting bias in sensory processing and organization toward that dominant modality.

2.9 Using Cortical Organization and Plasticity to Predict Cochlear Implant Outcome

While children implanted within the sensitive period generally outperform those implanted after it, there is considerable variability in the performance of even early-implanted children. Giraud and Lee (2007) have described large amounts of performance variability among congenitally deaf children implanted at various ages, including children implanted under 3 years of age (Fig. 10.1; Giraud and Lee 2007). While it is likely that factors such as socio-economic status, mode of communication, and the amount of rehabilitation likely influence outcome in implanted children (Wang et al. 2008), the possibility remains that there is some degree of age-independent plasticity which might influence outcome. That is, certain cognitive circuits activated in deafness may be more beneficial than other brain networks with regard to learning aural language after implantation.

In a series of studies using resting FDG PET images, Lee and colleagues have investigated the relationship between preoperative brain metabolism and CI outcome in deaf children (Lee et al. 2005a; 2007; Giraud and Lee 2007). As expected, the studies showed a clear effect of age at implantation on hypometabolism in the temporal cortex. As a next step, Lee and colleagues factored out the effects of age to examine brain areas associated with superior outcome with the implant. In a group of 22 prelingually deaf children tested between ages of 1 and 11 years, they report that good speech perception outcomes after implantation were correlated with decreased FDG uptake in the medial part of Heschl's gyrus outside the primary auditory cortex and in the depth of the right superior temporal sulcus, as well as increased FDG uptake in the left dorsolateral prefrontal (DLPF). At the same time, poor outcomes were associated with higher metabolic activity in large ventral regions, including fusiform regions and lower occipital regions. The strongest negative correlation with CI outcome was located in the depth of the right STS, which has been shown functionally to contribute to auditory complex pattern recognition, suggesting that higher metabolism in this area (associated with poor speech outcomes) is likely associated with a change in functional specialization. That is, regions normally underlying temporal pattern recognition become involved with other cognitive networks (Lee et al. 2007). Similarly, there was a strong effect of the left DLPF: children showing spontaneous activation of this area were those who later became good CI performers. Lee et al. (2007) suggest that because functionally the left DLPF has been shown to participate in higher cognitive functions such as reasoning, attentional control, or working memory, it is possible that listening to degraded implant speech signals requires an increased engagement of DLPF, conferring an advantage in acquisition of auditory language. The PET results also appear to be consistent with results from studies which suggest a link between behavioral measures of working memory and cochlear-implant performance in deaf children (Fagan et al. 2007).

Consistent with the results of Lee et al. (2007), Giraud and Lee (2007) showed that deaf children who demonstrated a high metabolism in dorsal cortical regions (including prefrontal and parietal) pre-implantation were most likely to learn speech with a cochlear implant successfully. Conversely, ventrotemporal brain regions, in particular, right temporal/auditory and occipital/visual areas correlated negatively with speech outcome. The higher dorsal metabolism in the future good performers was unlikely the result of higher initial language levels, because no children with sign language abilities participated in the study. Giraud and Lee speculate that the dorsal engagement might reflect a potential to recruit areas of the brain generally associated with higher general intelligence, executive control, working memory, and attention. A tendency to involve brain regions at rest that are normally active in higher cognitive functions indicates that the children probably engage these regions in speech perception and language learning after implantation. Conversely, ventral brain regions related to poor outcome may suggest some degree of cross-modal reorganization even in early childhood, which may result in a competition for cortical resources after implantation.

Other studies have found that good and poor implant users can be separated on the bases of their auditory (Gordon et al. 2005; Dinces et al. 2009) and visual (Doucet et al. 2006) evoked potentials. These studies report that good users, in general, displayed more typical waveform morphology and topography compared to poor users who displayed aberrant morphology and scalp topography. Additionally, consistent with the PET data (Lee et al. 2007), the Doucet study found that poor performers showed topographical evidence of visual-to-auditory cross-modal reorganization likely underlying their limited performance with the cochlear implant. Overall, the studies described above suggest that cortical organization in deafness may be predictive of future outcome with a cochlear implant.

3 Plasticity in Adults Who Receive an Implant

Following the above discussion on cortical organization in children fit early *vs.* fit late with an implant, implantation in congenitally deaf adults should present an even greater challenge to cortical plasticity. Based on the work of Kral et al. (2000, 2002, 2005) with congenitally deaf cats, one should expect to find, in the primary auditory cortex of human patients, a restricted pattern of activation within the layers of the cortex and desynchronized activity among cortical layers. One should also expect to find (i) a functional decoupling of the primary cortex (BA 41) from both higher-order auditory cortex and subcortical structures and (ii) cross-modal reorganization of higher-order auditory cortex (but not primary auditory cortex). Together, these factors presage very poor or no speech understanding in congenitally deaf adults fit with a cochlear implant. Indeed this is usually the case (e.g., Zwolan et al. 1996).

There is abundant evidence for cross-modal reorganization of higher-order auditory cortex in congenitally deaf adults. This cortex is activated by sign language input (e.g., Neville et al. 1998), by lip-reading (e.g., Sadato 2005), by moving non-biological dot patterns (e.g., Finney et al. 2001) and by somatosensory signals (e.g., Sharma et al. 2007) (see Fig. 10.2). Lee et al. (2007) argued that latent multi-modal connectivity rather than long-term reorganization is responsible for the response to speech reading in the left posterior superior temporal cortex of congenitally deaf patients.

The PET literature paints a contradictory picture of cortical activity evoked by electrical (auditory) stimulation in the relatively few congenitally deaf patients who choose a cochlear implant. One the one hand, Nishimura et al. (1999) reported that the primary auditory cortex, but not the higher-order auditory cortex, is activated by electrical stimulation. On the other hand, Okazawa et al. (1996) reported no activation of primary auditory cortex but activation of posterior temporal and inferior frontal cortex. Both types of outcome fit with the hypothesis of a decoupling of primary and higher-order cortex in congenitally deaf patients. Hirano et al. (2000), like Nishimura et al. (1999), found only primary auditory cortex activation in 2 prelingually deaf patients who showed no improvement in speech understanding with electrical stimulation. However, both the primary and higher-order auditory cortices were activated in 1 patient who did show an improvement in speech understanding.

For speech to be understood, signals from primary cortex must first make bilateral contact with receptive areas in the superior temporal gyrus that are sensitive to frequency and amplitude modulated signals. Hickok and Poeppel (2004) specifically suggested that this information is directed in a ventral stream to cortex in the superior temporal sulcus and in the posterior inferior temporal lobe that serves as an interface between auditory-based representations and meaning. Simultaneously, signals are sent in a dorsal stream to frontal regions involved in motor planning via an area at the boundary between parietal and temporal lobes. The PET evidence of little or no electrically evoked activity in higher-order auditory cortex in congenitally deaf adults fit with cochlear implants suggests little or no elaboration of signals directed to the primary auditory cortex (see Tucker 1998 for a personal account of learning to hear after 50 years of profound deafness).

How does the cortical function in *postlingually* deaf patients who receive an implant? Because these patients were not deprived of auditory stimulation during the sensitive period for development of central auditory pathways, the underlying cortical organization for speech understanding should be normal. Are there plastic changes in cortical organization brought on by either the abnormal stimulation provided by an implant and/or by the experience of at least partial auditory deprivation for various amounts of time?

Giraud and colleagues have described, using PET, multiple differences in cortical activation for cochlear implant patients and for listeners with normal hearing (Giraud et al. 2000, 2001, 2004). Between-group differences in regional cerebral blood flow in response to speech and non-speech stimuli were noted in the posterior superior temporal cortex, Wernicke's area, Broca's area, the posterior inferior temporal cortex, the left dorsal occipital cortex, the precuneus, the parahippocampal gyrus, and bilaterally at the temporoparietal junction. The left dorsal occipital cortex was the only area that was exclusively recruited by patients and not controls. Giraud et al. (2001) concluded that for cochlear-implant patients the ability to distinguish speech from other kinds of sounds is decreased in the auditory and association cortices. The preliminary processing that takes place at early levels is compensated by top-down processing input from higher cognitive level functions, such as memory, internal visualization ability, and visual attention. This increase in activity required for early processing is accompanied by concomitant decreases in later processing in semantic and speech production areas. Together these outcomes suggest changes in strategies for processing acoustic signals rather than re-purposing of cortex following the onset of deafness.

Similar to their pediatric studies, Lee et al. (2005b) suggested better performance for postlingually deafened adults who experience shorter durations of deafness prior to implantation. Patients with high levels of activity in frontotemporal regions pre-implant (e.g., the right anterior superior temporal gyrus and the left superior frontal gyrus) tend to have high levels of speech understanding post-implant. Conversely, patients with high, resting, pre-implant levels of metabolic activity in visual cortex (e.g., bilateral lingual gyrus/cuneus/superior occipital gyrus, and the left and right fusiform gyrus) tend to have low levels of speech understanding post-implant. Lee et al. (2005b) suggested that a reliance on both visual input and prefrontal top-down modulation is the *cognitive trait* that accounts for some of the variance in performance in postlingually deaf patients fit with an implant.

So far the studies have suggested that variability in performance with the implant is linked to differential activation of neural networks during auditory processing in both children and adults. Investigations are underway to try to enhance plasticity of optimal neural circuits with a view to improving patient performance. One route for increasing plasticity is via direct pharmacological intervention. For example, Tobey et al. (2005) found that administering 10 mg of d-amphetamine to adult cochlear implant patients prior to a 1.5 h intensive aural rehabilitation session resulted in a significant increase in magnitude of activation of primary and association auditory cortex, correlated with a 43% increase in auditory-only speech tracking scores. Another well established option for enhancing experience-dependent neural plasticity and corresponding behavioral performance is by administering intensive training to patients (Kilgard and Merzenich 1998; see also Chap. 11 by Fu and Galvin).

4 Summary

Studies of congenitally deaf children fitted with cochlear implants, utilizing electrophysiologic and brain imaging methodologies, have established the existence and limits of a sensitive period for cochlear implantation. The optimal time for cochlear implantation is within the first 3.5 to 4 years of life when central auditory pathways show the maximum plasticity to sound stimulation. This period in early childhood coincides with the period of maximal synaptogenesis in the auditory cortex. The end of the sensitive period (approximately age 7 years in humans) has consequences for the reorganization of cortical areas and pathways. Animal models suggest that the primary auditory cortex may be completely, or partially, decoupled from higher-order auditory areas, because of restricted development of inter- and intra-cortical connections. This decoupling may result in recruitment of higher-order auditory cortex by other sensory modalities (e.g., vision and somatosensation) and may be responsible for the well documented difficulties in oral speech and language skills seen in late-implanted, congenitally deaf children.

Much more research needs to be done to understand the tremendous individual variability seen in cochlear implanted children, including those who receive their implants within the sensitive period. It is likely that individual variability in performance is affected by both peripheral factors and by cortical factors (e.g., whether cross-modal reorganization has occurred), the extent to which it can be reversed by electrical stimulation, and the specific cognitive circuits that are engaged as a result of rehabilitation. It would be useful to develop new paradigms using *auditory, visual, and somatosensory* potentials to explore (i) cross-modal reorganization and (ii) cognitive circuitry related to memory, executive function, and attention. A more comprehensive understanding of the neural correlates of individual variability will be critical to developing better habilitation options that are aimed at, and customized for, individual patients.

The cortical response to electrical stimulation in *postlingually* deafened adults is different from the cortical response to acoustic stimulation in normal-hearing listeners. This is the case even for patients with high levels of speech understanding. The data from PET imaging studies points to changes in cortical activation driven by (i) the minimal representation of speech features in the electrically transmitted input signal; (ii) increased reliance on visual input from periods of severe and profound hearing loss; and, perhaps, (iii) individual differences in cognitive strategies to manage (i) and (ii). More research should be directed towards examining pharmacological interventions which increase adaptive plasticity in postlingually deafened adults and towards developing sensory- or cognitive- based training programs targeted to optimize patient performance.

Acknowledgements We gratefully acknowledge the assistance of Julia Campbell in preparation of the manuscript. We would like to thank Andrej Kral for many fruitful discussions. Research supported by NIH NIDCD R01 06527 to A.S.

References

Albrecht, R., Suchodoletz, W., & Uwer, R. (2000). The development of auditory evoked dipole source activity from childhood to adulthood. *Clinical Neurophysiology*, 111(12), 2268–2276.

Bavelier, D., & Neville, H. J. (2002). Cross-modal plasticity: where and how? *Nature Reviews Neuroscience*, 3(6), 443–452.

Bergeson, T. R., Pisoni, D. B., & Davis, R. A. (2005). Development of audiovisual comprehension skills in prelingually deaf children with cochlear implants. *Ear and Hearing*, 26(2), 149–164.

Besle, J., Fischer, C., Bidet-Caulet, A., Lecaignard, F., Bertrand, O., & Giard, M. H. (2008). Visual activation and audiovisual interactions in the auditory cortex during speech perception: intracranial recordings in humans. *Journal of Neuroscience*, 28(52), 14301–14310.

Catalan-Ahumada, M., Deggouj, N., De Volder, A., Melin, J., Michel, C., & Veraart, C. (1993). High metabolic activity demonstrated by positron emission tomography in human auditory cortex in case of deafness of early onset. *Brain Research*, 623(2), 287–292.

Conel, J.L. (1939–1967). *The Post-natal development of human cerebral cortex* (Vols. I–VIII). Cambridge, MA: Harvard University Press

de Volder, A. G., Catalan-Ahumada, M., Robert, A., Bol, A., Labar, D., Coppens, A., Michel, C., & Veraart, C. (1999). Changes in occipital cortex activity in early blind humans using a sensory substitution device. *Brain Research*, 826(1), 128–134.

Devous, M. D., Sr., Altuna, D., Furl, N., Cooper, W., Gabbert, G., Ngai, W. T., Chiu, S., Scott, J. M. III, Harris, T. S., Payne, J. K., & Tobey, E. A. (2006). Maturation of speech and language functional neuroanatomy in pediatric normal controls. *Journal of Speech, Language, and Hearing Research*, 49(4), 856–866.

Dinces, E., Chobot-Rhodd, J., & Sussman, E. (2009). Behavioral and electrophysiological measures of auditory change detection in children following late cochlear implantation: a preliminary study. *International Journal of Pediatric Otorhinolaryngology*, 73(6), 843–851.

Doucet, M. E., Bergeron, F., Lassonde, M., Ferron, P., & Lepore, F. (2006). Cross-modal reorganization and speech perception in cochlear implant users. *Brain*, 129(Pt. 12), 3376–3383.

Eggermont, J. J. (1988). On the rate of maturation of sensory evoked potentials. *Electroencephalography and Clinical Neurophysiology*, 70(4), 293–305.

Eggermont, J. J., & Ponton, C. W. (2002). The neurophysiology of auditory perception: from single units to evoked potentials. *Audiology & Neurotology*, 7(2), 71–99.

Eggermont, J. J., & Ponton, C. W. (2003). Auditory-evoked potential studies of cortical maturation in normal hearing and implanted children: correlations with changes in structure and speech perception. *Acta Oto-Laryngologica*, 123(2), 249–252.

Fagan, M. K., Pisoni, D. B., Horn, D. L., & Dillon, C. M. (2007). Neuropsychological correlates of vocabulary, reading, and working memory in deaf children with cochlear implants. *Journal of Deaf Studies and Deaf Education*, 12(4), 461–471.

Finney, E. M., Fine, I., & Dobkins, K. R. (2001). Visual stimuli activate auditory cortex in the deaf. *Nature Neuroscience*, 4(12), 1171–1173.

Geers, A. E. (2006). Factors influencing spoken language outcomes in children following early cochlear implantation. *Advances in Oto-Rhino-Laryngology*, 64, 50–65.

Gilley, P. M., Sharma, A., Dorman, M., & Martin, K. (2005). Developmental changes in refractoriness of the cortical auditory evoked potential. *Clinical Neurophysiology*, 116(3), 648–657.

Gilley, P. M., Sharma, A., Dorman, M., & Martin, K. (2006). Abnormalities in central auditory maturation in children with language-based learning problems. *Clinical Neurophysiology*, 117(9), 1949–1956.

Gilley, P. M., Sharma, A., Mitchell, T. V., & Dorman, M. F. (2010). The influence of a sensitive period for auditory-visual integration in children with cochlear implants. *Restorative Neurology and Neuroscience*, 28(2), 207–218.

Giraud, A. L., & Lee, H. J. (2007). Predicting cochlear implant outcome from brain organisation in the deaf. *Restorative Neurology and Neuroscience*, 25(3–4), 381–390.

Giraud, A. L., Truy, E., Frackowiak, R. S., Gregoire, M. C., Pujol, J. F., & Collet, L. (2000). Differential recruitment of the speech processing system in healthy subjects and rehabilitated cochlear implant patients. *Brain*, 123(Pt. 7), 1391–1402.

Giraud, A. L., Price, C. J., Graham, J. M., & Frackowiak, R. S. (2001). Functional plasticity of language-related brain areas after cochlear implantation. *Brain*, 124(Pt. 7), 1307–1316.

Giraud, A. L., Kell, C., Thierfelder, C., Sterzer, P., Russ, M. O., Preibisch, C., & Kleinschmidt, A. (2004). Contributions of sensory input, auditory search and verbal comprehension to cortical activity during speech processing. *Cerebral Cortex*, 14(3), 247–255.

Gordon, K. A., Papsin, B. C., & Harrison, R. V. (2003). Activity-dependent developmental plasticity of the auditory brain stem in children who use cochlear implants. *Ear and Hearing*, 24(6), 485–500.

Gordon, K. A., Papsin, B. C., & Harrison, R. V. (2005). Effects of cochlear implant use on the electrically evoked middle latency response in children. *Hearing Research*, 204(1–2), 78–79.

Harrison, R. V., Gordon, K. A., & Mount, R. J. (2005). Is there a critical period for cochlear implantation in congenitally deaf children? Analyses of hearing and speech perception performance after implantation. *Developmental Psychobiology* 46(3), 252–261

Hickok, G., & Poeppel, D. (2004). Dorsal and ventral streams: a framework for understanding aspects of the functional anatomy of language. *Cognition*, 92(1–2), 67–99.

Hirano, S., Naito, Y., Kojima, H., Honjo, I., Inoue, M., Shoji, K., Tateya, Il, Fujiki, N., Nishizawa, S., & Konishi, J. (2000). Functional differentiation of the auditory association area in prelingually deaf subjects. *Auris Nasus Larynx*, 27(4), 303–310.

Holt, R. F., & Svirsky, M. A. (2008). An exploratory look at pediatric cochlear implantation: is earliest always best? *Ear and Hearing*, 29(4), 492–511.

Hubel, D. (1995). *Eye, Brain, and Vision* (pp. 191–219). New York: Scientific American Library.

Hubel, D. H., & Wiesel, T. N. (1965). Binocular interaction in striate cortex of kittens reared with artificial squint. *Journal of Neurophysiology*, 28(6), 1041–1059.

Hubel, D. H., & Wiesel, T. N. (1967). Cortical and callosal connections concerned with the vertical meridian of visual fields in the cat. *Journal of Neurophysiology*, 30(6), 1561–1573.

Hubel, D. H., & Wiesel, T. N. (1970). The period of susceptibility to the physiological effects of unilateral eye closure in kittens. *The Journal of Physiology*, 206(2), 419–436.

Huttenlocher, P. R., & Dabholkar, A. S. (1997). Regional differences in synaptogenesis in human cerebral cortex. *Journal of Comparative Neurology*, 387(2), 167–178.

Kang, E., Lee, D. S., Lee, J. S., Kang, H., Hwang, C. H., Oh, S. H., Kim, C. S., Chung, J. K., Lee, M. C., Jang, M. J., Lee, Y. J., Morosan, P., & Zilles, K. (2003). Developmental hemispheric asymmetry of interregional metabolic correlation of the auditory cortex in deaf subjects. *Neuroimage*, 19(3), 777–783.

Kilgard, M. P., & Merzenich, M. M. (1998). Cortical map reorganization enabled by nucleus basalis activity. *Science*, 279(5357), 1714–1718.

Klinke, R., Kral, A., Heid, S., Tillein, J., & Hartmann, R. (1999). Recruitment of the auditory cortex in congenitally deaf cats by long-term cochlear electrostimulation. *Science*, 285(5434), 1729–1733.

Kral, A. (2007). Unimodal and cross-modal plasticity in the "deaf" auditory cortex. *International Journal of Audiology*, 46(9), 479–493.

Kral, A., & Tillein, J. (2006). Brain plasticity under cochlear implant stimulation. *Advances in Oto-rhino-Laryngology*, 64, 89–108.

Kral, A., & Eggermont, J. J. (2007). What's to lose and what's to learn: development under auditory deprivation, cochlear implants and limits of cortical plasticity. *Brain Research Reviews*, 56(1), 259–269.

Kral, A., Hartmann, R., Tillein, J., Heid, S., & Klinke, R. (2000). Congenital auditory deprivation reduces synaptic activity within the auditory cortex in a layer-specific manner. *Cerebral Cortex*, 10(7), 714–726.

Kral, A., Hartmann, R., Tillein, J., Heid, S., & Klinke, R. (2001). Delayed maturation and sensitive periods in the auditory cortex. *Audiology & Neurotology*, 6(6), 346–362.

Kral, A., Hartmann, R., Tillein, J., Heid, S., & Klinke, R. (2002). Hearing after congenital deafness: central auditory plasticity and sensory deprivation. *Cerebral Cortex*, 12(8), 797–807.

Kral, A., Tillein, J., Heid, S., Hartmann, R., & Klinke, R. (2005). Postnatal cortical development in congenital auditory deprivation. *Cerebral Cortex*, 15(5), 552–562.

Kral, A., Tillein, J., Heid, S., Klinke, R., & Hartmann, R. (2006). Cochlear implants: cortical plasticity in congenital deprivation. *Progress in Brain Research*, 157, 283–313.

Lee, D. S., Lee, J. S., Oh, S. H., Kim, S. K., Kim, J. W., Chung, J. K., Lee, M. C., & Kim, C. S. (2001). Cross-modal plasticity and cochlear implants. *Nature*, 409(6817), 149–150.

Lee, H. J., Kang, E., Oh, S. H., Kang, H., Lee, D. S., Lee, M. C., & Kim, C. S. (2005a). Preoperative differences of cerebral metabolism relate to the outcome of cochlear implants in congenitally deaf children. *Hearing Research*, 203(1–2), 2–9.

Lee, H. J., Oh, S. H., Kim, C. S., Giraud, A. & Lee, D. D. (2005b). Pre-operative cerebral glucose metabolism and CI outcome in postlingually deafened adults. Poster session presented at the 25th Politzer Society Meeting, Seoul, Korea.

Lee, H. J., Giraud, A. L., Kang, E., Oh, S. H., Kang, H., Kim, C. S., & Lee, D. S. (2007). Cortical activity at rest predicts cochlear implantation outcome. *Cerebral Cortex*, 17(4), 909–917.

Lee, J. S., Lee, D. S., Oh, S. H., Kim, C. S., Kim, J. W., Hwang, C. H., Koo, J., Kang, E., Chung, J. K., & Lee, M. C. (2003). PET evidence of neuroplasticity in adult auditory cortex of postlingual deafness. *The Journal of Nuclear Medicine*, 44(9), 1435–1439.

Liegeois-Chauvel, C., Musolino, A., Badier, J. M., Marquis, P., & Chauvel, P. (1994). Evoked potentials recorded from the auditory cortex in man: evaluation and topography of the middle latency components. *Electroencephalography and Clinical Neurophysiology*, 92(3), 204–214.

Makela, J. P., & Hari, R. (1992). Neuromagnetic auditory evoked responses after a stroke in the right temporal lobe. *Neuroreport*, 3(1), 94–96.

Makela, J. P., & McEvoy, L. (1996). Auditory evoked fields to illusory sound source movements. *Experimental Brain Research*, 110(3), 446–454.

Maurer, D., Lewis, T. L., Brent, H. P., & Levin, A. V. (1999). Rapid improvement in the acuity of infants after visual input. *Science*, 286(5437), 108–110.

McGurk, H., & MacDonald, J. (1976). Hearing lips and seeing voices. *Nature*, 264(5588), 746–748.

Moore, J. K., & Guan, Y. L. (2001). Cytoarchitectural and axonal maturation in human auditory cortex. *Journal of the Association for Research in Otolaryngology*, 2(4), 297–311.

Moore, J. K., & Linthicum, F. H., Jr. (2007). The human auditory system: a timeline of development. *International Journal of Audiology*, 46(9), 460–478.

Nash, A, Sharma A, Martin K, & Biever A. (2008). Clinical applications of the P1 cortical auditory evoked potential (CAEP) biomarker. A sound foundation through early amplification: proceedings of a Fourth International Conference. Stafa, Switzerland: Phonak, AG.

National Institute on Deafness and Other Communication Disorders (NIDCD). (2009). *Cochlear Implants*. Retrieved December 19, 2008 from http://www.nidcd.nih.gov/health/hearingcoch.asp.

Neville, H. J., Bavelier, D., Corina, D., Rauschecker, J., Karni, A., Lalwani, A., Braun, A., Clark, V., Jezzard, P., & Turner, R. (1998). Cerebral organization for language in deaf and hearing subjects: biological constraints and effects of experience. *Proceedings of the National Academy of Sciences U S A*, 95(3), 922–929.

Nishimura, H., Hashikawa, K., Doi, K., Iwaki, T., Watanabe, Y., Kusuoka, H., Nishimura, T., & Kubo, T. (1999). Sign language "heard" in the auditory cortex. *Nature*, 397(6715), 116.

Oh, S. H., Kim, C. S., Kang, E. J., Lee, D. S., Lee, H. J., Chang, S. O., Ahn, S. H., Hwang, C. H., Park, H. J., & Koo, J. W. (2003). Speech perception after cochlear implantation over a 4-year time period. *Acta Oto-Laryngologica*, 123(2), 148–53.

Okazawa, H., Naito, Y., Yonekura, Y., Sadato, N., Hirano, S., Nishizawa, S., Magata, Y., Ishizu, K., Tamaki, N., Honjo, I., & Konishi, J. (1996). Cochlear implant efficiency in pre- and postlingually deaf subjects. A study with H2(15)O and PET. *Brain*, 119(Pt. 4), 1297–1306.

Pallas, S. L. (2001). Intrinsic and extrinsic factors that shape neocortical specification. *Trends in Neuroscience*, 24(7), 417–423.

Pang, E. W., & Taylor, M. J. (2000). Tracking the development of the N1 from age 3 to adulthood: an examination of speech and non-speech stimuli. *Clinical Neurophysiology*, 111(3), 388–397.

Ponton, C. W., & Eggermont, J. J. (2001). Of kittens and kids: altered cortical maturation following profound deafness and cochlear implant use. *Audiology & Neurotology*, 6(6), 363–380.

Ponton, C. W., Don, M., Waring, M. D., Eggermont, J. J., & Masuda, A. (1993). Spatio-temporal source modeling of evoked potentials to acoustic and cochlear implant stimulation. *Electroencephalography and Clinical Neurophysiology*, 88(6), 478–493.

Ponton, C. W., Don, M., Eggermont, J. J., Waring, M. D., & Masuda, A. (1996a). Maturation of human cortical auditory function: differences between normal hearing children and children with cochlear implants. *Ear and Hearing*, 17(5), 430–437.

Ponton, C. W., Don, M., Eggermont, J. J., Waring, M. D., Kwong, B., & Masuda, A. (1996b). Auditory system plasticity in children after long periods of complete deafness. *Neuroreport*, 8(1), 61–65.

Ponton, C. W., Moore, J. K., & Eggermont, J. J. (1999). Prolonged deafness limits auditory system developmental plasticity: evidence from an evoked potentials study in children with cochlear implants. *Scandinavian Audiology Supplement*, 51, 13–22.

Ponton, C. W., Eggermont, J. J., Kwong, B., & Don, M. (2000a). Maturation of human central auditory system activity: evidence from multi-channel evoked potentials. *Clinical Neurophysiology*, 111(2), 220–236.

Ponton, C. W., Eggermont, J. J., Don, M., Waring, M. D., Kwong, B., Cunningham, J., & Trautwein, P. (2000b). Maturation of the mismatch negativity: effects of profound deafness and cochlear implant use. *Audiology & Neurotology*, 5(3–4), 167–185.

Ponton, C., Eggermont, J. J., Khosla, D., Kwong, B., & Don, M. (2002). Maturation of human central auditory system activity: separating auditory evoked potentials by dipole source modeling. *Clinical Neurophysiology*, 113(3), 407–420.

Ryugo, D. K., Pongstaporn, T., Huchton, D. M., & Niparko, J. K. (1997). Ultrastructural analysis of primary endings in deaf white cats: morphologic alterations in endbulbs of Held. *Journal of Comparative Neurology*, 385(2), 230–244.

Ryugo, D. K., Kretzmer, E. A., & Niparko, J. K. (2005). Restoration of auditory nerve synapses in cats by cochlear implants. *Science*, 310(5753), 1490–1492.

Sadato, N. (2005). How the blind "see" Braille: lessons from functional magnetic resonance imaging. *Neuroscientist*, 11(6), 577–582.

Schorr, E. A., Fox, N. A., van Wassenhove, V., & Knudsen, E. I. (2005). Auditory-visual fusion in speech perception in children with cochlear implants. *Proceedings of the National Academy of Sciences U S A*, 102(51), 18748–18750.

Sharma, A., & Dorman, M. F. (2006). Central auditory development in children with cochlear implants: clinical implications. *Advances in Oto-Rhino-Laryngology*, 64, 66–88.

Sharma, A., Kraus, N., McGee, T. J., & Nicol, T. G. (1997). Developmental changes in P1 and N1 central auditory responses elicited by consonant-vowel syllables. *Electroencephalogry and Clinical Neurophysiology*, 104(6), 540–545.

Sharma, A., Dorman, M., Spahr, A., & Todd, N.W. (2002a). Early cochlear implantation in children allows normal development of the central auditory pathways. *Annals of Otology, Rhinology, and Laryngology Supplement*, 189, 38–41.

Sharma, A., Dorman, M.F., & Spahr, A.J. (2002b). A sensitive period for the development of the central auditory system in children with cochlear implants: implications for age of implantation. *Ear and Hearing*, 23(6), 532–539.

Sharma, A., Dorman, M. F., & Kral, A. (2005a). The influence of a sensitive period on central auditory development in children with unilateral and bilateral cochlear implants. *Hearing Research*, 203(1–2), 134–143.

Sharma, A., Martin, K., Roland, P., Bauer, P., Sweeney, M. H., Gilley, P., & Dorman, M. (2005b). P1 latency as a biomarker for central auditory development in children with hearing impairment. *Journal of the American Academy of Audiology*, 16(8), 564–573.

Sharma, A., Gilley, P. M., Dorman, M. F., & Baldwin, R. (2007). Deprivation-induced cortical reorganization in children with cochlear implants. *International Journal of Audiology*, 46(9), 494–499.

Sharma, A., Nash, A. A., & Dorman, M. (2009). Cortical development, plasticity and reorganization in children with cochlear implants. *Journal of Communication Disorders*, 42(4), 272–279.

Tobey, E. A., Devous, M. D., Sr., Buckley, K., Overson, G., Harris, T., Ringe, W., & Martinez-Verhoff, J. (2005). Pharmacological enhancement of aural habilitation in adult cochlear implant users. *Ear and Hearing*, 26(4, Suppl.), 45 S–56 S.

Tonnquist-Uhlen, I., Ponton, C. W., Eggermont, J. J., Kwong, B., & Don, M. (2003). Maturation of human central auditory system activity: the T-complex. *Clinical Neurophysiology*, 114(4), 685–701.

Tucker, B. P. (1998). Deaf culture, cochlear implants, and elective disability. *The Hastings Center Report*, 28(4), 6–14.

Wang, N. Y., Eisenberg, L. S., Johnson, K. C., Fink, N. E., Tobey, E. A., Quittner, A. L., Niparko, J. K., & DDaCI Investigative Team. (2008). Tracking development of speech recognition: longitudinal data from hierarchical assessments in the Childhood Development after Cochlea Implantation Study. *Otology & Neurotology*, 29(2), 240–245.

Wiesel, T. N., & Hubel, D. H. (1963). Effects of visual deprivation on morphology and physiology of cells in the cats lateral geniculate body. *Journal of Neurophysiology*, 26, 978–993.

Wunderlich, J. L., Cone-Wesson, B. K., & Shepherd, R. (2006). Maturation of the cortical auditory evoked potential in infants and young children. *Hearing Research*, 212(12), 185–202.

Zwolan, T. A., Kileny, P. R., & Telian, S. A. (1996). Self-report of cochlear implant use and satisfaction by prelingually deafened adults. *Ear and Hearing*, 17(3), 198–210.

Chapter 11
Auditory Training for Cochlear Implant Patients

Qian-Jie Fu and John J. Galvin III

1 Introduction

The cochlear implant (CI) is a true medical miracle, providing hearing to the deaf. Certainly, those who developed the implant technology deserve credit for the success of the CI, as do the surgeons who implant the device and the audiologists who program the processors. However, the success of the CI would not be possible without the plasticity of the human brain. In this sense, CI patients have been their own "miracle workers," because they have learned to make sense of the crude electrical signals provided by the implant device. When the CI was first introduced, many thought that the device would provide only limited benefit. Clearly they were wrong, because many CI users are capable of auditory-only speech perception (e.g., telephone conversation), greatly exceeding initial expectations.

As CI technology has improved over the past 30 years, so has CI users' overall speech performance (see Fig. 11.1). Relaxed candidacy criteria and better surgical techniques have surely contributed to this steady improvement in patient outcomes. While mean CI speech performance in quiet is generally good, considerable variability in patient outcomes remains. Even for good performers, speech understanding in noise remains quite difficult. Variability in CI performance is increased for challenging listening tasks (e.g., music perception, talker identification, etc.). Thus, the variability in patient outcomes can be influenced by many factors, including the state of CI technology, the state of the implanted ear, and the demands of the listening task. While it is important to understand the variability in patient outcomes and factors that might limit performance, it is equally if not more important to improve

Q.-J. Fu (✉)
Division of Communication and Auditory Neuroscience, House Ear Institute,
2100 West Third Street, Los Angeles, CA 90057, USA
e-mail: qfu@hei.org

F.-G. Zeng et al. (eds.), *Auditory Prostheses: New Horizons*, Springer Handbook of Auditory Research 39, DOI 10.1007/978-1-4419-9434-9_11,

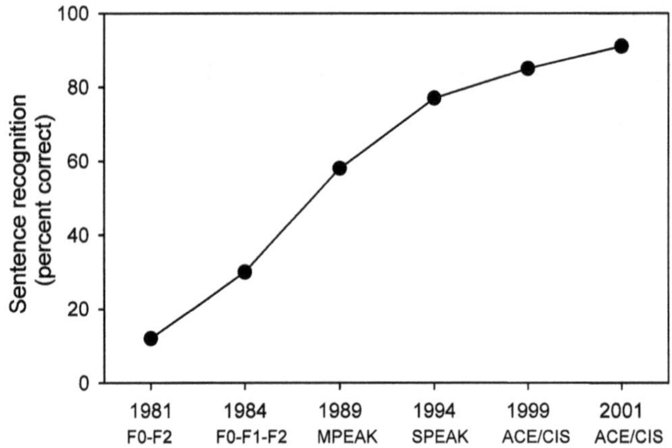

Fig. 11.1 Mean sentence recognition (in quiet) over time for Nucleus CI users. The x-axis shows the year and contemporaneous speech processing strategy. (Modified from Zeng 2004)

poor outcomes. Unfortunately, even state-of-the-art CI technology has shown only limited success in improving performance for poor CI users or for good CI users in difficult listening conditions (e.g., speech in noise, music perception, etc.).

1.1 Variability in CI Speech Performance

While overall performance has improved over the years (Fig. 11.1), considerable variability remains in CI patient outcomes. Some CI patients receive little benefit from the latest implant technology, even after many years of experience with their device. Different factors may contribute to variability in patient outcomes. Patient-related factors (e.g., etiology of deafness, duration of deafness, age at implantation, health and proximity of auditory neurons to implanted electrodes, electrode insertion depth, etc.) have been significantly correlated with speech performance (Eggermont and Ponton 2003; Kelly et al. 2005). Psychophysical measures such as electrode discrimination (Donaldson and Nelson 2000), temporal modulation detection (Cazals et al. 1994; Fu 2002; Luo et al. 2008), and gap detection (Cazals et al. 1991; Muchnik et al. 1994; Busby and Clark 1999) have been correlated with speech performance and provide some indication of individual patients' perceptual limits. Processor-related factors (e.g., signal-processing strategy, microphone type/array) can also contribute to variability. Test-related factors (e.g., availability of contextual cues, previous familiarity with materials, and procedural learning) may also reveal different degrees of patient variability. Figure 11.2 shows acutely measured vowel recognition in quiet for 30 Nucleus CI users (SPEAK or ACE strategies) who have participated in research conducted over the years at House Ear Institute; there is a wide range in performance (approximately, 6% to 85% correct). If only mean recognition of simple sentences in quiet is considered (as in Fig. 11.1), the breadth of

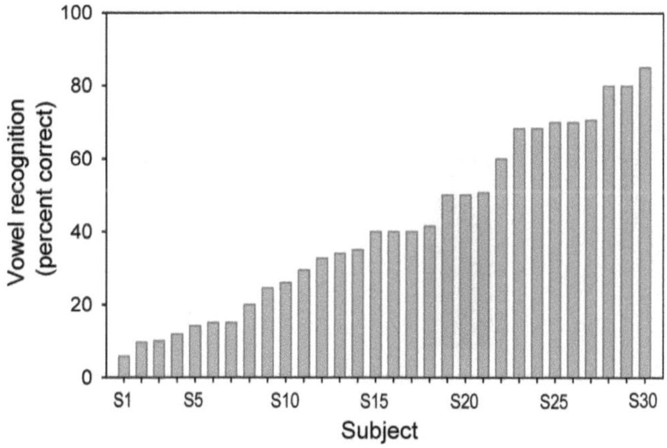

Fig. 11.2 Range of vowel recognition scores for Nucleus CI users who have participated in research at the House Ear Institute

CI performance may not be apparent. Indeed, different listening tasks may better reveal individual differences in patient performance.

1.2 *Susceptibility to Background Noise*

Noise is a near constant presence for all listeners, whether from the environment (e.g., traffic noise, wind, or rain), appliances (e.g., fans or motors), or other people ("cocktail party effect"). For even the best CI users, speech understanding deteriorates rapidly as noise levels increase (Dowell et al. 1987; Skinner et al. 1994; Kiefer et al. 1996). Innovations in electrode design and speech processing have shown some improvements in performance in noise (Kiefer et al. 1996; Skinner et al. 1994). However, CI performance is much poorer than that of normal-hearing (NH) subjects listening to comparable acoustic CI simulations (Friesen et al. 2001). CI users are especially susceptible to competing speech or temporally modulated noise (Nelson et al. 2003; Fu and Nogaki 2005; Zeng et al. 2005). Figure 11.3 shows mean speech recognition thresholds (SRTs) for modulated and steady noise, defined as the signal-to-noise ratio (SNR) required for 50% correct whole sentence recognition (Plomp and Mimpen 1979). With steady noise, the mean SRT is about −5 dB for NH listeners and +10 dB for CI users. With modulated noise (4 Hz), the SRT improves to about −22 dB for NH listeners but worsens to about +12 dB for CI users (Fu and Nogaki 2005). NH subjects seem able to "listen in the dips" of modulated noise while CI subjects require relative high SNRs, whether or not the noise was modulated. Fu and Nogaki (2005) proposed that CI patients' increased susceptibility to noise was partly the result of poor spectral resolution (i.e., the limited number of implanted electrodes and channel interactions between the electrodes).

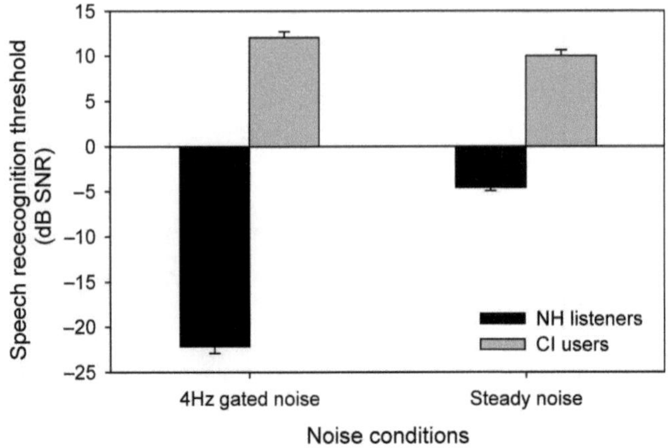

Fig. 11.3 SRTs with modulated and steady noise for NH and CI listeners. The *error bars* show the standard error

Much recent research has been aimed at improving the spectral resolution of the CI. Spatially compact bi- or tripolar stimulation has been proposed to reduce electrode interaction (e.g., Bierer and Middlebrooks 2002; Bierer 2007; Landsberger and Srinivasan 2009). High stimulation rates and/or conditioner pulses have been proposed to improve temporal sampling and improve channel selectivity (e.g., Rubinstein et al. 1999; Litvak et al. 2003). Current steering and virtual channels have been proposed to increase the number of spectral channels beyond the number of physical electrodes (e.g., Wilson et al. 1997; Donaldson et al. 2005). However, these approaches have yet to show clear or consistent improvements in speech recognition in quiet or in noise.

1.3 Music Perception and Appreciation

Music perception and appreciation are challenging for many CI users. The relatively coarse spectral resolution, while good enough for speech understanding in quiet, is not sufficient to support complex musical tasks (McDermott and Mckay 1997; Smith et al. 2002; Shannon et al. 2004). CI users hear musical rhythm components nearly as well as NH listeners (Kong et al. 2004) but seem to receive only limited melodic pitch information. Familiar melody identification (FMI) without rhythm cues has often been used (e.g., Kong et al. 2004; Drennan and Rubinstein 2008) but does not quantify listeners' melodic pitch perception very well. Melodic contour identification (MCI) has been used to quantify CI users' melodic pitch perception (e.g., Galvin et al. 2007, 2008, 2009). In the MCI test, subjects are asked to identify 5-note melodic contours which vary in terms of pitch direction (e.g., rising, falling, etc.) and in terms of spacing between notes (e.g., 1 to 5 semitones). Large inter-subject variability in MCI performance has been observed among CI users, with more than half of the subjects performing at the chance level (see Fig. 11.4).

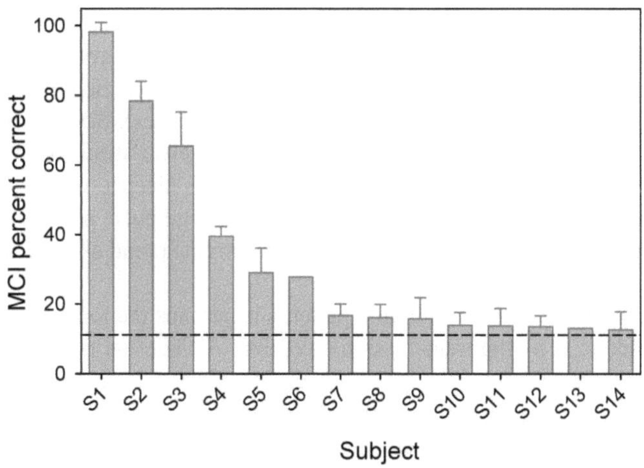

Fig. 11.4 Acutely measured melodic contour identification (MCI, across all semitone spacing conditions) for CI users who participated in music research studies at the House Ear Institute. The *dashed line* shows chance level performance. The *error bars* show one standard deviation of the mean

Top performers are able to identify nearly all contours with 2 semitones between notes, while the poor performers could correctly identify only 30% of the contours even with 5 semitones between notes. Because top performers tended to have more music training than poor performers, the variability in CI users' music perception may be related to musical experience before and after implantation. This suggests that experience or training may help CI users make better use of the limited information provided by their device.

1.4 Limits to Transmission, Reception, and Perception of CI Stimulation Patterns

Much recent research has been directed at improving spectral resolution, with some studies demonstrating existence of virtual channels (e.g., Donaldson et al. 2005; Firszt et al. 2007) or increased channel selectivity (e.g., Bierer and Middlebrooks 2002; Berenstein et al. 2008; Landsberger and Srinivasan 2009), at least for single-channel psychophysics. However, these methods have not resulted in significant gains in multi-channel performance (speech or otherwise). At present, research and development has been aimed at providing more channels at higher rates, despite the limited evidence that CI users will benefit from these costly efforts.

Electric hearing can be broken into three stages: (1) transmission of acoustic information, (2) reception of electric information, and (3) perception of electric stimulation. Transmission of acoustic cues involves the speech processor (i.e., the microphone, "pre-processing" of the acoustic input, the acoustic-to-electric amplitude and frequency mapping, the stimulation rate, etc.). CI research has been largely focused on improving the transmission stage. However, as many studies have shown, transmitting more channels does not mean that listeners necessarily receive more channels. The reception of electric information is largely limited by patient-related factors (e.g., proximity of electrodes to healthy neurons) but also by device-related factors (e.g., the current spread from stimulated electrodes). Some recent research has been directed at improving the reception of electric stimulation by current-focusing. However, current-focusing requires greater electric charge, which may result in tradeoffs with stimulation rate (because of large pulse phase duration) and/or pulse amplitude (which may result in similar current spread as with broader stimulation modes).

Perception of electrical stimulation patterns is perhaps the most important stage, and it is influenced by the transmission/reception of electrical patterns, but in relation to the central patterns. For postlingually deafened CI patients, these patterns were developed during periods of normal hearing; these listeners must adapt the new electric patterns to the previous acoustic patterns. For speech signals, this adaptation can be fairly rapid, and many postlingually deafened CI users perform quite well under optimal listening conditions. However, if there is a shallow electrode insertion depth (Fu and Shannon 1999), or if there is a "dead region" (Shannon et al. 2002; Moore 2004), adaptation may be more difficult. Pediatric CI patients develop central patterns with electric hearing only and seem to approach NH levels of performance in quiet listening conditions. Still, these listeners must sometimes adapt to new electric patterns (e.g., changes in frequency allocation, the addition of a second CI, etc.) in relation to the previous central patterns. Prelingually deafened CI users who experience long duration of deafness most likely have weak auditory pattern recognition. While these listeners may be able to discriminate between electric patterns, they cannot identify these patterns because no central templates exist. These templates can take a very long time to develop in these listeners. While recent research has been directed at improving perception of electrical patterns via auditory training, the perception stage has received less attention than the transmission and reception stages.

These three stages – transmission, reception, and perception – should be considered interactively to maximize the benefit of implantation. CI users might be trained to hear acoustic information transmitted at high rates, or virtual channels in a multichannel context, or differences in spectral envelopes with current focusing. The training should direct attention to cues that are useful in difficult listening tasks (e.g., melodic pitch, talker identification, speech in noise, etc.), because high-context tests and training (e.g., words or sentences in quiet) may not depend on these enhanced cues. Differences in CI performance may well depend on listeners' utilization of available cues. Auditory training may be the key to improving utilization.

2 Passive Learning by Cochlear Implant Users

Postlingually deafened CI users are able to adapt to electric hearing without explicit training ("passive learning"). Longitudinal studies show that the greatest gains in performance occur during the first 3 months after activation (e.g., Waltzman et al. 1986;

Spivak and Waltzman 1990; Loeb and Kessler 1995). Continued improvement has been observed over longer periods (Tyler et al. 1997). Even after this initial adaptation, CI users must sometimes adapt to new stimulation patterns because of changes in hardware or signal processing. This adaptation also largely occurs during the first 3 to 6 months of exposure, beyond which there is little improvement (Dorman and Loizou 1997; Pelizzone et al. 1999).

While these longitudinal studies revealed learning trends, differences in CI patients' initial listening experiences and motivation may influence outcomes in ways that are difficult to know. In some ways, it may be experimentally preferable to measure experienced CI users' adaptation to changes in speech processing. This way, processor-related factors may be controlled. Fu et al. (2002) studied CI users' adaptation to a severe spectral mismatch (about 1 octave) between the acoustic information and place of stimulation in the cochlea over an extended 3-month learning period. Acutely measured vowel and sentence recognition dropped sharply with the spectral shift. However, performance improved over the study period, with the greatest gains occurring during the first 2 weeks of adaptation. At the end of the study, performance had improved greatly but remained significantly poorer than with subjects' clinically assigned processors.

During passive learning, CI listeners most likely adapt to reduced, shifted, and/or distorted spectral patterns. For postlingually deafened CI users, the vowel formant perceptual space approaches that of NH listeners 6 months to a year after activating the implant (Harnsberger et al. 2001; Svirsky et al. 2001, 2004a). For severe spectral shifts or distortions, there may be limits to passive adaptation. While central patterns may be reshaped by a spectral shift, uncertainty regarding the shifted pattern persisted even after 3 months of continuous exposure (Fu et al. 2002; Sagi et al. 2009). In addition, even though there was partial adaptation to the severe shift, performance with the original clinical processor was unchanged after 3 months of continuous exposure to the experimental processor, suggesting that CI users are able to develop additional representations, rather than replacing previous patterns (Fu et al. 2002). Li and Fu (2007) showed that NH subjects were able to adapt to moderately shifted speech "automatically" but not to severely shifted speech. Thus, CI users may "automatically" learn novel electrical stimulation patterns, as long as the patterns are sufficiently close to previous patterns developed with acoustic hearing.

The manner in which a spectral shift is introduced may also influence the degree and or time course of adaptation via passive learning. Svirsky et al. (2004b) suggested that a gradual (rather than abrupt) introduction to CI speech processing may allow patients to first adapt to the reduced spectral resolution and then the spectral shift. While the degree of adaptation may be similar for abrupt or gradual introduction to a moderate mismatch, the stress level may be reduced with gradual introduction. With a severe mismatch, gradual introduction might provide better adaptation. Indeed, one of the subjects who participated in the Fu et al. (2002) study later was gradually exposed to the same severe shift over an 18-month study period (Fu and Galvin 2007). Results showed significantly better adaptation with gradual exposure, especially for sentence recognition in quiet (see Fig. 11.5).

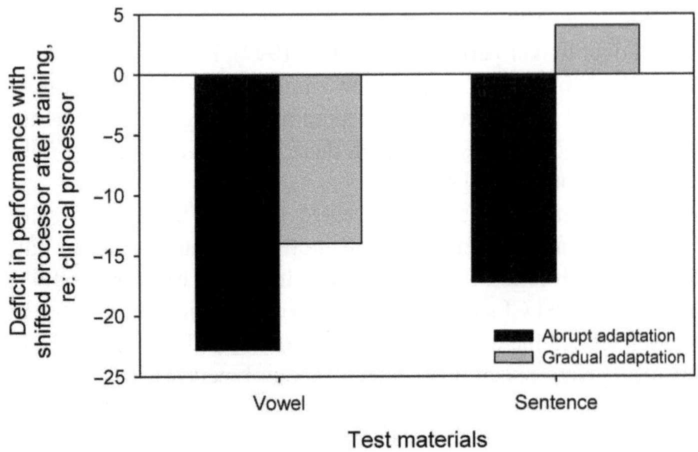

Fig. 11.5 Deficit in performance (percentage points) with the abrupt or gradual adaptation protocol at the end of the adaptation period, re: baseline performance with clinical processors

3 Active Learning by Cochlear Implant Users

There may be limits to passive learning, especially for difficult listening conditions. Active training has been shown to improve CI users' speech performance, even after years of experience with their device and/or signal processing. Some early studies showed limited benefit for auditory training in poor-performing CI patients. Busby et al. (1991) observed only minimal changes in speech performance after 10 1-h training sessions. Dawson and Clark (1997) found small but significant improvements in vowel recognition for 4 out of 5 pediatric CI subjects and that improvements were retained 3 weeks after training was stopped. With improvements in computer technology, computer-based auditory training began to show much greater improvements in performance than observed in these initial studies.

3.1 Training Speech Recognition in Quiet

Computer-based auditory training has been shown to improve English vowel recognition in quiet for CI users (Fu et al. 2005a) and NH subjects listening to CI simulations (e.g., Fu et al. 2005b; Stacey and Summerfield 2005). In the Fu et al. (2005a) study, baseline speech recognition performance (vowel, consonant, and sentence recognition) was first collected for a range of CI performers; all subjects had extensive experience with their devices before participating in the study. Baseline performance was measured for at least 2 weeks or until performance asymptoted (to reduce procedural learning effects). After baseline measures were complete, subjects trained at home on a medial vowel recognition task using custom software installed on their home computers ("Computer-assisted speech training" or CAST).

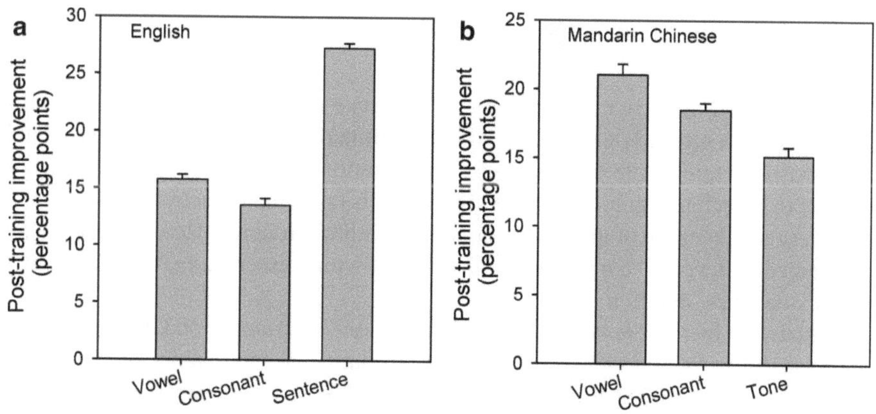

Fig. 11.6 Mean improvement after phonetic contrast training, for postlingually deafened English-speaking CI subjects (**a**) and pediatric Mandarin-speaking CI subjects (**b**). The *error bars* show the standard error

The CAST software automatically adapted the level of difficulty according to individual subjects' performance; the number of response choices were increased and/or the phonemic contrast between choices was reduced. Novel training materials (multi-talker monosyllabic words) were used for training and were not used for testing. Auditory and visual feedback (i.e., audio playback of the correct and incorrect responses, paired with the correct and incorrect response labels) was provided, allowing subjects to compare their answer to the correct response. Subjects trained for approximately 1 h per day, 5 days per week, for 1 month or more. During training, subjects returned to the lab every 2 weeks for re-testing with the baseline speech materials. Subjects also returned to the lab after training was stopped for follow-up measures.

Figure 11.6 shows the mean improvement after training for English (left panel) and Mandarin speaking (right panel) CI subjects. The English training results are from Fu et al. (2005a). The Mandarin training results are from a pediatric Chinese CI study by Wu et al. (2007); subjects were trained using similar methods as in Fu et al. (2005a). Results showed that, for all subjects, vowel, consonant, and tone recognition significantly improved after training, and that improved vowel recognition consonant recognition generalized to improved sentence recognition. While performance improved for all subjects after 4 or more weeks of training, there was significant inter-subject variability in terms of the amount and time course of improvement. For some subjects, performance quickly improved after only a few hours of training (similar to the simulation results observed by Rosen et al. 1999), while others required a much longer time course to show any significant improvements. Follow-up measures remained better than pre-training baseline performance, suggesting that the positive effects of training were retained well after training had been stopped. What is remarkable about these results is that all subjects were very experienced with their devices before training, suggesting that CI listeners could be trained to better use the information provided by their device.

3.2 *Training Speech Recognition in Noise*

While targeted training may improve performance under optimal listening conditions, CI users regularly experience less than optimal conditions. The attention paid to acoustic details while training speech recognition in quiet may or may not contribute to speech recognition in noise. Indeed, a listener cannot recover "lost" acoustic cues (e.g., because of interfering noise) via auditory training. However, they may be trained to develop "coping" strategies to deal with noise, e.g., make better use of contextual cues, duration cues, etc.

Relatively few studies have assessed the benefits of training on CI performance in noise. Fu and Galvin (2007) showed preliminary data for 2 subjects after extensive training with interfering speech babble. Subjects were trained using a phoneme-in-noise or keyword-in-sentence protocol. The phoneme-in-noise protocol was similar to that in Fu et al. (2005a). In the keyword-in-sentence protocol, an IEEE (1969) sentence was presented and subjects selected from a closed set of keywords. In both protocols, the noise level was adapted from trial to trial according to subject response. Similar to Fu et al. (2005a), subjects trained at home using the CAST software. The keyword-in-sentence protocol seemed to provide the greater benefit, suggesting that listeners may develop strategies to deal with noise rather than improve perception of acoustic cues (although the two strategies are related).

More recently Oba et al. (2011) used a digits-in-noise protocol (i.e., recognition of number sequences in competing noise, e.g., "3-5-7") to train CI users' speech recognition in noise. If CI listeners can be trained to improve listening strategies to cope with noise, then a simple digit recognition task might be an effective approach. Similar to previous studies, baseline performance was measured until achieving asymptotic performance and included digit recognition in noise, HINT (Nilsson et al. 1994) SRTs in noise, and IEEE sentence recognition at various SNRs. Subjects trained at home using custom software ("Sound Express," developed by Fu and colleagues and similar to the previous CAST program). Subjects trained for approximately a half-hour per day for 1 month, using a digit-in-noise training protocol; subjects trained with interfering speech babble. The SNR was adapted from trial to trial according to subject response. Auditory and visual feedback was provided. Similar to previous studies, subjects returned to the lab every 2 weeks for repeated testing, and then 1 month after training was stopped for follow-up testing.

Figure 11.7 shows SRTs (the SNR needed for 50% correct) for digits or HINT sentences in the presence of steady noise or speech babble, before and after the training.After training, digit recognition thresholds improved by nearly 3 dB in babble and nearly 4 dB in steady noise (even though steady noise was not used for training). HINT SRTs improved by nearly 2 dB in babble and 1 dB in steady noise, even though sentence recognition was not trained with either noise types. Similarly, mean IEEE sentence recognition in noise improved by 10.5 percentage points in babble and 6 points in steady noise.

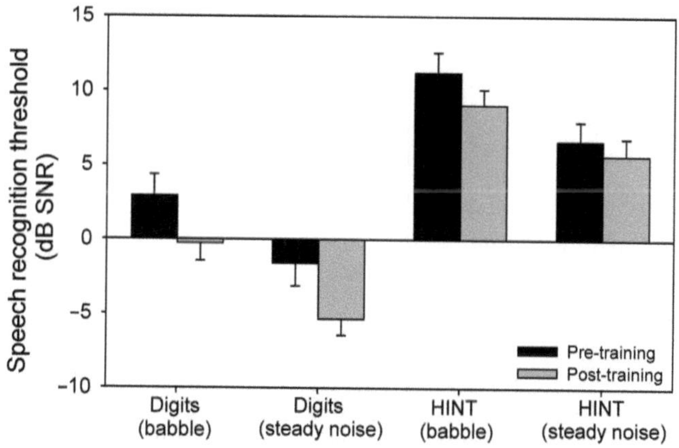

Fig. 11.7 Mean SRTs for digits and sentences, before and after digit-in-noise training. The *error bars* show the standard error

3.3 Training Music Perception

Musical experience seems to affect CI users' perception and appreciation of music. Gfeller et al. (2000) reported that CI users' music appreciation was affected by prior familiarity with a musical piece, i.e., familiar music may sound better (or worse) than similar but unfamiliar music because of the force of memory. Galvin et al. (2008, 2009) found that the top performing subjects in several difficult music perception tasks had extensive music experience, before and after implantation. If music experience benefits CI music perception, auditory training may benefit less experienced CI users.

Structured music training has been shown to improve CI patients' musical timbre perception (Gfeller et al. 2002) and melodic contour identification, or MCI (Galvin et al. 2007, 2009). In Galvin et al. (2007), subjects were trained to identify melodic contours with novel pitch ranges not used for testing; subjects used similar home training software as in Fu et al. (2005a). The level of difficulty was adapted according to subject performance by adjusting the semitone interval between successive notes in the contour. Auditory and visual feedback was provided, and subjects trained at home for approximately1 h a day for 1 month or longer. Results showed that all subjects significantly improved their MCI performance and that these gains were largely retained in follow-up measures more than 1 month later. Even better, subjects' FMI performance improved after the MCI training, even though FMI was not explicitly trained. In the Galvin et al. (2009) study, subjects received MCI training with piano samples (as opposed to the 3-tone complexes used in Galvin et al. 2007) to see whether training with a relatively difficult instrument would improve MCI performance. While the gains were smaller than those reported in Galvin et al. (2007), subjects' MCI performance with the piano improved, and generalized to

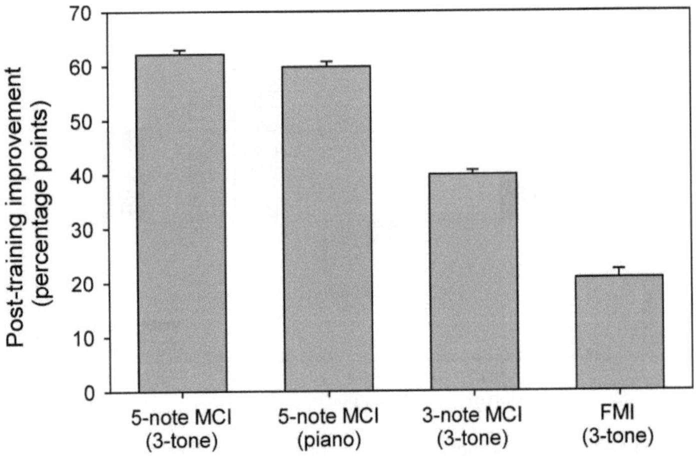

Fig. 11.8 Performance improvement after MCI training for different music perception tasks. The *left three bars* show mean data from Wu et al. (2007) and the *far right bar* shows data from Galvin et al. (2007). The *error bars* show the standard error

other instruments that were not trained. In Wu et al. (2007), Mandarin-speaking pediatric CI users' were trained using a modified version of the MCI training (i.e., 5-tone complexes instead of 3-tone complexes). Similar to the other studies, the MCI training significantly improved MCI performance. Figure 11.8 summarizes performance gains in the Wu et al. (2007) and Galvin et al. (2007) study after MCI training. In these studies, music training seems to benefit CI users' music perception. Indeed, Looi and She (2010) reported that CI users are very interested in receiving some sort of music training to improve their music appreciation.

4 Considerations When Designing a Training Program for CI Users

While the above studies have shown significant benefits for auditory training in CI users, one area of ongoing research is the design of training protocols that maximize benefit while minimizing cost and effort. Some considerations in this regard include the frequency of training, the type of feedback, the stimuli used for training, and the amount of time spent training. Much research has been conducted by Wright and colleagues (e.g., Wright 2001; Wright and Zhang 2006, 2009) and by Moore and colleagues (e.g., Moore et al. 2003) with NH subjects listening to relatively simple stimuli during psychophysical tasks (e.g., frequency discrimination, gap detection, etc.). Wright and Fitzgerald (2001) found rapid procedural learning effects (i.e., learning of test method/stimuli/environment) in a sound localization task. Wright and Fitzgerald (2005) later found that training with one type of stimulus (amplitude modulated noise) at one frequency (150 Hz) in one task (modulation frequency discrimination) did not readily generalize to other stimuli, other frequencies, or other tasks; i.e., training may be specific to the trained task and materials. Wright and Sabin (2007) also found that a critical amount of training needs to occur before learning can be consolidated; training less than this amount was ineffective and training more than this amount did not offer further improvement. Amitay et al. (2006) proposed three processes for perceptual learning – sensitization (exposure), attention, and arousal – and that these processes may depend on the task to be trained.

The above psychophysical studies with NH listeners offer some guidance regarding training stimuli and training method. However, learning novel speech patterns may require different approaches. For example, training might target top-down processing to improve attention to weak signals in noise or to supra-segmental speech cues (e.g., voice pitch). Alternatively, training might target bottom-up processing to improve sensitivity to individual acoustic cues (e.g., modulation, pitch, etc.) that might improve speech perception. While speech training does not offer the experimental control as with simple psychophysical training, the outcome (potentially) has far greater impact. Some recent studies have explored some issues in designing effective and efficient speech training protocols for CI users.

4.1 *Effect of Training Materials*

Different training materials may influence training outcomes. For example, is it better to train with multiple talkers or a single talker? Do digits, phonemes, words, or sentences provide the greatest improvement for sentence recognition? Training with acoustically modified speech has been shown to improve speech understanding for listeners with specific language impairment and dyslexia (e.g., Tallal et al. 1996; Habib et al. 2002). Nagarajan et al. (1998) used enhanced speech signals (stretching duration, envelope expansion) to train language learning-impaired children; during the 4-week training period, recognition of enhanced and unprocessed speech gradually increased until reaching near normal levels.

Some recent studies have looked at the effect of training materials on CI users' training outcomes. Amitay et al. (2005) found that training with wide range of standard frequencies (rather than a small range) provided better frequency discrimination. Amitay et al. (2006) also found significant benefit to training with identical stimuli or even training with an unrelated visual task, suggesting that the benefits to auditory training may have to do with improved attention, rather than auditory perception. The better training outcomes reported by Fu et al. (2005a), compared with previous CI speech training studies, may have been the result of the large database (>1000 monosyllable words) produced by multiple talkers. Stacey and Summerfield (2007) reported that multi-talker stimuli were more effective in training NH subjects listening to CI simulations.

Fu et al. (2005b) reported better improvement in vowel recognition when training with monosyllabic words than with sentences. Fu and Galvin (2007) reported somewhat contradictory pilot data that showed better overall recognition of phonemes and sentences (in steady noise or speech babble) after training with sentences in noise rather than with monosyllable words in noise. For NH subjects listening to acoustic CI simulations, Stacey and Summerfield (2008) reported that training with word and sentences, rather than with phonemes, provided greater improvements in word and sentence recognition. While diversity in training materials seems to provide greater benefit, it is unclear whether training stimuli that targets peripheral processes (acoustic or electric differences) or central patterns (lexicons, words, or sentences) are most effective, especially for everyday, noisy listening environments.

4.2 Effect of Training Methods

Different training methods may also influence training outcomes. For example, what type of feedback is most effective (e.g., lexical or non-lexical, auditory or visual, or both)? Does repeated exposure (without feedback) produce a similar benefit to training with feedback? Li and Fu (2007) compared the effect of lexical and non-lexical response labeling on perceptual adaptation to spectrally shifted vowels for NH subjects listening to acoustic CI simulations, hypothesizing that the degree of spectral shift would limit "automatic" learning and that explicit training with lexically meaningful feedback might be needed to adapt to a severe shift. Recognition of moderately or severely shifted vowels was trained (i.e., 5 min preview of stimuli) and tested using lexical ("heed," "had," "hood," etc.) or non-lexical labels ("a," "b," "c," etc.). While subjects were able automatically adapt to the moderate shift, auditory training provided greater adaptation, even with non-lexical labels. Interestingly, while adaptation to the severe shift was similar for the non-lexical and lexical label training, subjects adapted more quickly with the non-lexical label training. The lexical labels referred to central patterns that strongly "disagreed" with the severely shifted speech; the non-lexical labels may have provided less cognitive "interference," at least initially.

Fu et al. (2005b) compared four different training protocols over a 5-day training period in NH subjects listening to severely shifted CI simulations: (1) test-only without feedback, (2) 5-minute preview of similar stimuli, (3) medial vowel contrast training using monosyllable words, and (4) modified connected discourse. Figure 11.9 shows the mean improvement after training with four different training protocols. Vowel recognition significantly improved with the preview and vowel contrast protocols but not with the test-only or connected discourse protocols; the vowel contrast training also improved medial consonant recognition. These results are somewhat in contrast to those of Stacey and Summerfield (2008), who found that sentence and word training provided the most general benefit.

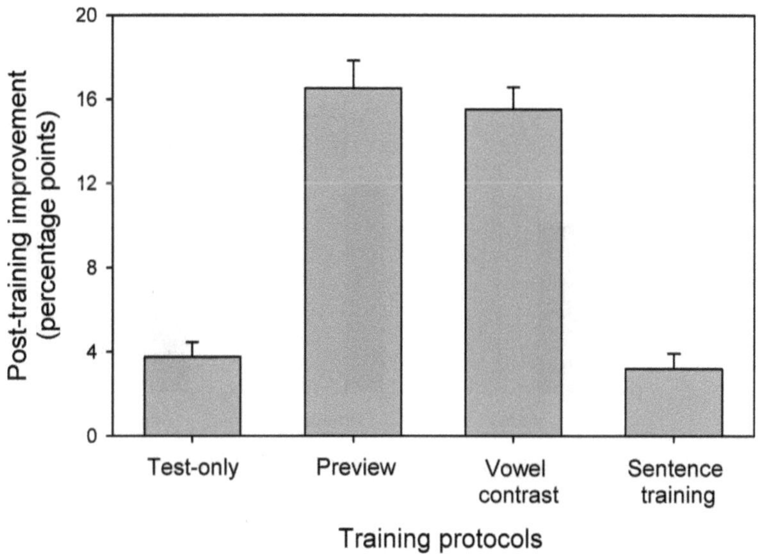

Fig. 11.9 Mean improvement in vowel recognition with different training protocols. The *error bars* show the standard error

4.3 Effect of Training Duration and Frequency

The computer-assisted auditory training used by Fu and colleagues and by Stacey and Summerfield (2008) seem to provide better training outcomes than previous studies (e.g., Busby et al. 1991; Dawson and Clark 1997). While training methods and materials may have differed between studies, another factor that may influence training outcomes is the frequency of training. In the Busby et al. (1991) and Dawson and Clark (1997), subjects trained for approximately 50 min per week, for 10 weeks. In most of the Fu et al. studies (2004, 2005a, 2005b), subjects trained approximately 30 to 60 min per day, 5 days per week, for a period of 1 month or longer.

Nogaki et al. (2007) evaluated the effect of training frequency (5 training sessions over 1, 2, or 5 weeks) on NH listeners' adaptation to 8-channel spectrally shifted speech. Figure 11.10 shows performance gains with for the different training conditions. While more frequent training seemed to provide better adaptation, the frequency of training did not significantly affect training outcomes, suggesting that it may be more important to complete some fixed amount of training over a reasonable time period. If so, trainees may benefit from even occasional training. While increased frequency and amount of training may have contributed to better training outcomes, studies also differed in terms of training methods and materials. Given the relatively minor effect of training frequency, it may be more beneficial to develop better methods and materials than to simply increase trainees' time and effort.

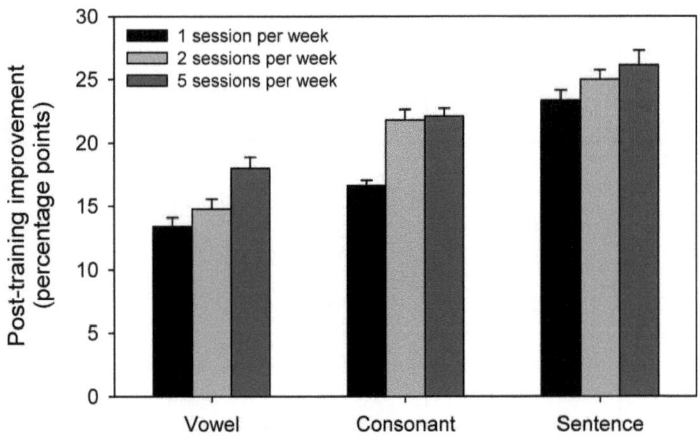

Fig. 11.10 Mean improvement in performance for different training rates. The *error bars* show the standard error

4.4 *Interactions Between Training Methods/Materials and Spectral Distortion to the Acoustic Input*

The results of Li and Fu (2007) and Li et al. (2009) suggest that the type of training and/or training materials may interact with the degree of spectral mismatch being trained. All CI users must adapt to some degree of spectral mismatch between the acoustic signal and the electrode location. For patients with short electrode insertion depths, the mismatch may be severe. In such cases, clinicians must choose between delivering shifted acoustic information (hoping that listeners may adapt) or throwing away acoustic information to minimize the mismatch. Listeners may automatically adapt to small spectral shifts (<3 mm). Larger shifts may require explicit training.

Li and Fu (2007) found that the degree of spectral shift and the type of response label (lexical or non-lexical) interacted with listeners' automatic learning of a spectral shift and that lexical labels may initially interfere with automatic leaning of severely shifted speech. In a related study, Li et al. (2009) measured NH listeners' automatic adaptation to moderately and severely shifted speech, hypothesizing that gradual exposure might influence automatic learning. They found greater adaptation to severely shifted speech when concurrently exposed to moderately shifted speech, suggesting that the moderate shift allowed listeners to bridge the gap between the shifted acoustic input and central pattern. Even with gradual exposure, passive learning did not provide complete adaptation, suggesting that explicit training might be needed to adapt to a severe shift fully. The results are consistent with those of Fu and Galvin (2007), who reported more complete adaptation with "gradual" exposure (over 18 months) than with "abrupt" exposure to a severe shift. These results are also consistent with those of Svirsky et al. (2004b), who found that gradual adaptation was less stressful to CI users.

These studies suggest that CI listeners may automatically learn small spectral shifts. It is unclear whether training might accelerate this learning, or whether the phoneme and sentence recognition data reflect "real-word" hearing function. Even with a small shift, difficult listening tasks (e.g., voice gender recognition, talker identification, speech understanding in noise, etc.) may require training. Auditory training seems necessary to adapt to large spectral shifts, and gradual exposure may help to ease the stress of adaptation and/or make adaptation more complete.

4.5 Other Considerations

There is some doubt whether auditory training really works, or whether trainees are improving their attention rather than auditory perception. One issue is establishing proper control groups for CI training studies. Many early studies had no experimental controls. Many of the studies by Fu and colleagues utilized "within-subject" controls. Because of the variability in CI patient performance, patient etiology, CI device types, duration of deafness, etc., it is difficult to establish experimental control groups. As such, it makes sense to compare baseline and post-training performance within each subject, and compare the relative changes in performance across experimental factors. While some CI simulation studies have used separate control groups (e.g., Stacey and Summerfield 2007), there are presently no CI speech training studies that have implemented across-subject control groups. There is a great need for such controls (both cross-subject and cross-modal) to determine whether trainees' attention or auditory perception is being improved.

Even with within-subject controls, great care should be taken to control for procedural learning effects. In the studies by Fu and colleagues, baseline performance was re-measured over an extended time period until performance asymptoted. Such an extended baseline is necessary to determine whether subjects learned the new signal processing or simply the test environment, stimuli, and/or methods. Another important control is the "test-only" group (Nogaki et al. 2007), to see whether the same amount of exposure to the test stimuli (without preview or feedback) as in training groups also improved performance. Controlling for these procedural learning effects can add great time, cost, and effort to training studies.

Even with proper controls, it may be difficult to ascertain what is being learned. Auditory training may only improve general attention or short-term memory, rather than auditory perception. The willingness to participate in training research may reflect some bias, i.e., the desire to perform better. It is important to understand the source of training benefits when designing training protocols. But whatever the source of improvement, the benefits of auditory training should not be dismissed. In many recent studies (e.g., Fu et al. 2005a, b; Galvin et al. 2007), CI users' performance improved by an average of 15 percentage points, with some subjects improving by 40 points, depending on the task. In these studies, subjects had years of experience with their devices and implant technology but were able to improve performance after moderate auditory training. Training may be the key to maximizing CI patient performance, whether with old or new technology.

Finally, training effects and adaptation should be considered when determining the benefit of CI parameter changes or the efficacy of changes in CI technology. Acute measures may underestimate performance for CI parameters. For example, spectral "holes" in hearing degrade speech understanding (Shannon et al. 2002), but auditory training may help offset this decrement (Smith and Faulkner 2006). Depending on the parameter change, auditory training may compensate for acute deficits or allow the full benefit of technological advances (e.g., bilateral implants, current shaping, high stimulation rates, etc.) to be realized.

5 Summary

Much research and development has been directed at improving CI hardware and software, at great expense. However, relatively little research has been directed toward understanding the auditory plasticity that drives the success of the implant. More regrettably, few resources have been directed toward auditory rehabilitation of CI recipients, even though the benefits of training often exceed performance gains associated with the latest CI technology, at a fraction of the cost. While the CI may provide the sound, it is the brain that hears. By exploiting neural plasticity, auditory training may be the key to maximizing the benefit of auditory prostheses. As more is understood about how training can reshape auditory perception, rehabilitation approaches undoubtedly will improve, along with CI users' speech and music perception.

Acknowledgements The authors would like to thank all of the research participants who graciously gave their time and support toward these studies. The authors also acknowledge NIH funding support. Finally, the authors would like to thank Bob Shannon for many years of guidance, and more importantly, for many years of friendship.

References

Amitay, S., Hawkey, D. J., & Moore, D. R. (2005). Auditory frequency discrimination learning is affected by stimulus variability. *Perception & Psychophysics*, 67(4), 691–698.

Amitay, S., Irwin, A., & Moore, D. R. (2006). Discrimination learning induced by training with identical stimuli. *Nature Neuroscience*, 9(11), 1446–1448.

Berenstein, C. K., Mens, L. H., Mulder, J. J., & Vanpoucke, F. J. (2008). Current steering and current focusing in cochlear implants: comparison of monopolar, tripolar, and virtual channel electrode configurations. *Ear and Hearing*, 29(2), 250–260.

Bierer, J. A. (2007). Threshold and channel interaction in cochlear implant users: evaluation of the tripolar electrode configuration. *Journal of the Acoustical Society of America*, 121(3), 1642–1653.

Bierer, J. A., & Middlebrooks, J. C. (2002). Auditory cortical images of cochlear-implant stimuli: dependence on electrode configuration. *Journal of Neurophysiology*, 87(1), 478–492.

Busby, P. A., & Clark, G. M. (1999). Gap detection by early-deafened cochlear-implant subjects. *Journal of the Acoustical Society of America*, 105(3), 1841–1852.

Busby, P. A., Roberts, S. A., Tong, Y. C., & Clark, G. M. (1991). Results of speech perception and speech production training for three prelingually deaf patients using a multiple-electrode cochlear implant. *British Journal of Audiology*, 25(5), 291–302.

Cazals, Y., Pelizzone, M., Kasper, A., & Montandon, P. (1991). Indication of a relation between speech perception and temporal resolution for cochlear implantees. *Annals of Otology, Rhinology & Laryngology*, 100(11), 893–895.

Cazals, Y., Pelizzone, M., Saudan, O., & Boex, C. (1994). Low-pass filtering in amplitude modulation detection associated with vowel and consonant identification in subjects with cochlear implants. *Journal of the Acoustical Society of America*, 96(4), 2048–2054.

Dawson, P. W., & Clark, G. M. (1997). Changes in synthetic and natural vowel perception after specific training for congenitally deafened patients using a multichannel cochlear implant. *Ear and Hearing*, 18(6), 488–501.

Donaldson, G. S., & Nelson, D. A. (2000). Place-pitch sensitivity and its relation to consonant recognition by cochlear implant listeners using the MPEAK and SPEAK speech processing strategies. *Journal of the Acoustical Society of America*, 107(3), 1645–1658.

Donaldson, G. S., Kreft, H. A., & Litvak, L. (2005). Place-pitch discrimination of single- versus dual-electrode stimuli by cochlear implant users (L). *Journal of the Acoustical Society of America*, 118(2), 623–626.

Dorman, M. F., & Loizou, P. C. (1997). Changes in speech intelligibility as a function of time and signal processing strategy for an Ineraid patient fitted with continuous interleaved sampling (CIS) processors. *Ear and Hearing*, 18(2), 147–155.

Dowell, R. C., Seligman, P. M., Blamey, P. J., & Clark, G. M. (1987). Speech perception using a two-formant 22-electrode cochlear prosthesis in quiet and in noise. *Acta Oto-Laryngologica*, 104(5–6), 439–446.

Drennan, W. R., & Rubinstein, J. T. (2008). Music perception in cochlear implant users and its relationship with psychophysical capabilities. *Journal of Rehabilitation Research & Development*, 45(5), 779–789.

Eggermont, J. J., & Ponton, C. W. (2003). Auditory-evoked potential studies of cortical maturation in normal hearing and implanted children: correlations with changes in structure and speech perception. *Acta Oto-Laryngologica*, 123(2), 249–252.

Firszt, J. B., Koch, D. B., Downing, M., & Litvak, L. (2007). Current steering creates additional pitch percepts in adult cochlear implant recipients. *Otology & Neurotology*, 28(5), 629–636.

Friesen, L. M., Shannon, R. V., Baskent, D., & Wang, X. (2001). Speech recognition in noise as a function of the number of spectral channels: comparison of acoustic hearing and cochlear implants. *Journal of the Acoustical Society of America*, 110(2), 1150–1163.

Fu, Q. J. (2002). Temporal processing and speech recognition in cochlear implant users. *Neuroreport*, 13(13), 1635–1639.

Fu, Q. J., & Galvin, J. J., 3 rd. (2007). Perceptual learning and auditory training in cochlear implant recipients. *Trends in Amplification*, 11(3), 193–205.

Fu, Q. J., & Nogaki, G. (2005). Noise susceptibility of cochlear implant users: the role of spectral resolution and smearing. *Journal of the Association for Research in Otolaryngology*, 6(1), 19–27.

Fu, Q. J., & Shannon, R. V. (1999). Effects of electrode location and spacing on phoneme recognition with the Nucleus-22 cochlear implant. *Ear and Hearing*, 20(4), 321–331.

Fu, Q. J., Shannon, R. V., & Galvin, J. J., 3 rd. (2002). Perceptual learning following changes in the frequency-to-electrode assignment with the Nucleus-22 cochlear implant. *Journal of the Acoustical Society of America*, 112(4), 1664–1674.

Fu, Q. J., Galvin, J. J., 3 rd., Wang, X., & Nogaki, G. (2004). Effects of auditory training on adult cochlear implant patients: a preliminary report. *Cochlear Implants International*, 5(Suppl. 1), 84–90.

Fu, Q. J., Galvin, J. J., 3 rd., Wang, X., & Nogaki, G. (2005a). Moderate auditory training can improve speech performance of adult cochlear implant users. *Journal of the Acoustical Society of America*, 6, 106–111.

Fu, Q. J., Nogaki, G., & Galvin, J. J., 3 rd. (2005b). Auditory training with spectrally shifted speech: implications for cochlear implant patient auditory rehabilitation. *Journal of the Association for Research in Otolaryngology*, 6(2), 180–189.

Galvin, J. J., 3 rd, Fu, Q. J., & Nogaki, G. (2007). Melodic contour identification by cochlear implant listeners. *Ear and Hearing*, 28(3), 302–319.

Galvin, J. J., 3 rd, Fu, Q. J., & Oba, S. (2008). Effect of instrument timbre on melodic contour identification by cochlear implant users. *Journal of the Acoustical Society of America*, 124(4), EL189–195.

Galvin, J. J., 3 rd, Fu, Q. J., & Shannon, R. V. (2009). Melodic contour identification and music perception by cochlear implant users. *Annals of the New York Academy of Sciences*, 1169, 518–533.

Gfeller, K., Christ, A., Knutson, J. F., Witt, S., Murray, K. T., & Tyler, R. S. (2000). Musical backgrounds, listening habits, and aesthetic enjoyment of adult cochlear implant recipients. *Journal of the American Academy of Audiology*, 11(7), 390–406.

Gfeller, K., Witt, S., Adamek, M., Mehr, M., Rogers, J., Stordahl, J., & Ringgenberg, S. (2002). Effects of training on timbre recognition and appraisal by postlingually deafened cochlear implant recipients. *Journal of the American Academy of Audiology*, 13(3), 132–145.

Habib, M., Rey, V., Daffaure, V., Camps, R., Espesser, R., Joly-Pottuz, B., & Démonet, J. (2002). Phonological training in children with dyslexia using temporally modified speech: a three-step pilot investigation. *International Journal of Language & Communication Disorders*, 37(3), 289–308.

Harnsberger, J. D., Svirsky, M. A., Kaiser, A. R., Pisoni, D. B., Wright, R., & Meyer, T. A. (2001). Perceptual "vowel spaces" of cochlear implant users: implications for the study of auditory adaptation to spectral shift. *Journal of the Acoustical Society of America*, 109(5, Pt. 1), 2135–2145.

IEEE Subcommittee (1969). IEEE recommended practice for speech quality measurements. *IEEE Transactions on Audio & Electroacoustics*, AU-17(3), 225–246.

Kelly, A. S., Purdy, S. C., & Thorne, P. R. (2005). Electrophysiological and speech perception measures of auditory processing in experienced adult cochlear implant users. *Clinical Neurophysiology*, 116(6), 1235–1246.

Kiefer, J., Muller, J., Pfennigdorff, T., Schon, F., Helms, J., von Ilberg, C., Baumgartner, W., Gstöttner, W., Ehrenberger, K., Arnold, W., Stephan, K., Thumfart, W., & Baur, S. (1996). Speech understanding in quiet and in noise with the CIS speech coding strategy (MED-EL Combi-40) compared to the multipeak and spectral peak strategies (Nucleus). *Journal for Oto-Rhino-Laryngology and Its Related Specialties*, 58(3), 127–135.

Kong, Y. Y., Cruz, R., Jones, J. A., & Zeng, F. G. (2004). Music perception with temporal cues in acoustic and electric hearing. *Ear and Hearing*, 25(2), 173–185.

Landsberger, D. M., & Srinivasan, A. G. (2009). Virtual channel discrimination is improved by current focusing in cochlear implant recipients. *Hearing Research*, 254(1–2), 34–41.

Li, T., & Fu, Q. J. (2007). Perceptual adaptation to spectrally shifted vowels: training with nonlexical labels. *Journal of the Association for Research in Otolaryngology*, 8(1), 32–41.

Li, T., Galvin, J. J., 3 rd, & Fu, Q. J. (2009). Interactions between unsupervised learning and the degree of spectral mismatch on short-term perceptual adaptation to spectrally shifted speech. *Ear and Hearing*, 30(2), 238–249.

Litvak, L., Delgutte, B., & Eddington, D. (2003). Improved neural representation of vowels in electric stimulation using desynchronizing pulse trains. *Journal of the Acoustical Society of America*, 114(4, Pt. 1), 2099–2111.

Loeb, G. E., & Kessler, D. K. (1995). Speech recognition performance over time with the Clarion cochlear prosthesis. *Annals of Otology, Rhinology & Laryngology*, Supplement 166, 290–292.

Looi, V., & She, J. (2010). Music perception of cochlear implant users: a questionnaire, and its implications for a music training program. *International Journal of Audiology*, 49(2), 116–128.

Luo, X., Fu, Q. J., Wei, C. G., & Cao, K. L. (2008). Speech recognition and temporal amplitude modulation processing by Mandarin-speaking cochlear implant users. *Ear and Hearing*, 29(6), 957–970.

McDermott, H. J., & McKay, C. M. (1997). Musical pitch perception with electrical stimulation of the cochlea. *Journal of the Acoustical Society of America*, 101(3), 1622–1631.

Moore, B. C. (2004). Dead regions in the cochlea: conceptual foundations, diagnosis, and clinical applications. *Ear and Hearing*, 25(2), 98–116.

Moore, D. R., Amitay, S., & Hawkey, D. J. (2003). Auditory perceptual learning. *Learning & Memory*, 10(2), 83–85.

Muchnik, C., Taitelbaum, R., Tene, S., & Hildesheimer, M. (1994). Auditory temporal resolution and open speech recognition in cochlear implant recipients. *Scandanavian Audiology*, 23(2), 105–109.

Nagarajan, S. S., Wang, X., Merzenich, M. M., Schreiner, C. E., Johnston, P., Jenkins, W. M., et al. (1998). Speech modifications algorithms used for training language learning-impaired children. *IEEE Transactions on Rehabilitation Engineering*, 6(3), 257–268.

Nelson, P. B., Jin, S. H., Carney, A. E., & Nelson, D. A. (2003). Understanding speech in modulated interference: cochlear implant users and normal-hearing listeners. *Journal of the Acoustical Society of America*, 113(2), 961–968.

Nilsson, M., Soli, S. D., & Sullivan, J. A. (1994). Development of the Hearing in Noise Test for the measurement of speech reception thresholds in quiet and in noise. *Journal of the Acoustical Society of America*, 95(2), 1085–1099.

Nogaki, G., Fu, Q. J., & Galvin, J. J., 3 rd. (2007). Effect of training rate on recognition of spectrally shifted speech. *Ear and Hearing*, 28(2), 132–140.

Oba, S. I., Fu, Q.-J., & Galvin, J. J., 3 rd. (2011). Digit training in noise can improve cochlear implant users' speech understanding in noise. *Ear and Hearing* (in press).

Pelizzone, M., Cosendai, G., & Tinembart, J. (1999). Within-patient longitudinal speech reception measures with continuous interleaved sampling processors for Ineraid implanted subjects. *Ear and Hearing*, 20(3), 228–237.

Plomp, R., & Mimpen, A. M. (1979). Speech-reception threshold for sentences as a function of age and noise level. *Journal of the Acoustical Society of America*, 66(5), 1333–1342.

Rosen, S., Faulkner, A., & Wilkinson, L. (1999). Adaptation by normal listeners to upward spectral shifts of speech: implications for cochlear implants. *Journal of the Acoustical Society of America*, 106(6), 3629–3636.

Rubinstein, J. T., Wilson, B. S., Finley, C. C., & Abbas, P. J. (1999). Pseudospontaneous activity: stochastic independence of auditory nerve fibers with electrical stimulation. *Hearing Research*, 127(1–2), 108–118.

Sagi, E., Fu, Q. J., Galvin, J. J., 3 rd, & Svirsky, M. A. (2010). A model of incomplete adaptation to a severely shifted frequency-to-electrode mapping by cochlear implant users. *Journal of the Association for Research in Otolaryngology*, 11(1), 69–78.

Shannon, R. V., Galvin, J. J., 3 rd, & Baskent, D. (2002). Holes in hearing. *Journal of the Association for Research in Otolaryngology*, 3(2), 185–199.

Shannon, R. V., Fu, Q. J., & Galvin, J., 3 rd. (2004). The number of spectral channels required for speech recognition depends on the difficulty of the listening situation. *Acta Oto-Laryngologica Supplement*, (552), 50–54.

Skinner, M. W., Clark, G. M., Whitford, L. A., Seligman, P. M., Staller, S. J., Shipp, D. B., Shallop, J. K., Everingham, C., Menapace, C. M., Arndt, P.L., et al. (1994). Evaluation of a new spectral peak coding strategy for the Nucleus 22 Channel Cochlear Implant System. *American Journal of Otology*, 15(Suppl. 2), 15–27.

Smith, M. W., & Faulkner, A. (2006). Perceptual adaptation by normally hearing listeners to a simulated "hole" in hearing. *Journal of the Acoustical Society of America*, 120(6), 4019–4030.

Smith, Z. M., Delgutte, B., & Oxenham, A. J. (2002). Chimaeric sounds reveal dichotomies in auditory perception. *Nature*, 416(6876), 87–90.

Spivak, L. G., & Waltzman, S. B. (1990). Performance of cochlear implant patients as a function of time. *Journal of Speech and Hearing Research*, 33(3), 511–519.

Stacey P. C., & Summerfield, A. Q. (2005). Auditory-perceptual training using a simulation of a cochlear-implant system: a controlled study. *Proceedings from the ISCA Workshop on Plasticity in Speech Perception (PSP2005)* (pp. 143–146).

Stacey P. C., & Summerfield, A. Q. (2007). Effectiveness of computer-based auditory training in improving the perception of noise-vocoded speech. *Journal of the Acoustical Society of America*, 121, 2923–2935.
Stacey, P. C., & Summerfield, A. Q. (2008). Comparison of word-, sentence-, and phoneme-based training strategies in improving the perception of spectrally distorted speech. *Journal of Speech, Language, and Hearing Research*, 51(2), 526–538.
Svirsky, M. A., Silveira, A., Suarez, H., Neuburger, H., Lai, T. T., & Simmons, P. M. (2001). Auditory learning and adaptation after cochlear implantation: a preliminary study of discrimination and labeling of vowel sounds by cochlear implant users. *Acta Oto-Laryngologica*, 121(2), 262–265.
Svirsky, M. A., Silveira, A., Neuburger, H., Teoh, S.W., & Suarez, H. (2004a). Long-term auditory adaptation to a modified peripheral frequency map. *Acta Oto-Laryngologica*, 124, 381–386.
Svirsky, M. A., Talavage, T. M., Sinha, S., & Neuburger, H. (2004b, February). Adaptation to a shifted frequency map: gradual is better. Paper presented at the Annual Meeting of the American Academy for the Advancement of Science, Seattle, WA.
Tallal, P., Miller, S. L., Bedi, G., Byma, G., Wang, X., Nagarajan, S. S., Schreiner, C., Jenkins, W. M., & Merzenich, M. M. (1996). Language comprehension in language-learning impaired children improved with acoustically modified speech. *Science*, 271(5245), 81–84.
Tyler, R. S., Parkinson, A. J., Woodworth, G. G., Lowder, M. W., & Gantz, B. J. (1997). Performance over time of adult patients using the Ineraid or nucleus cochlear implant. *Journal of the Acoustical Society of America*, 102(1), 508–522.
Waltzman, S. B., Cohen, N. L., & Shapiro, W. H. (1986). Long-term effects of multichannel cochlear implant usage. *Laryngoscope*, 96(10), 1083–1087.
Wilson, B., Finley, C., Zerbi, M., Lawson, D., & van den Honert, C. (1997). Speech processors for auditory prostheses. Seventh Quarterly progress report, Neural Prosthesis Program (NIH project N01-DC-5-2103). Bethesda, MD: National Institutes of Health.
Wright, B. A. (2001). Why and how we study human learning on basic auditory tasks. *Audiology & Neurotology*, 6(4), 207–210.
Wright, B. A., & Fitzgerald, M. B. (2001). Different patterns of human discrimination learning for two interaural cues to sound-source location. *Proceedings of the National Academy of Sciences U S A*, 98(21), 12307–12312.
Wright, B. A., & Fitzgerald, M. B. (2005). Learning and generalization of five auditory discrimination tasks as assessed by threshold changes. In D. Pressnitzer, A. de Cheveigne, S. McAdams, & L. Collet (Eds.), *Auditory signal processing: physiology, psychoacoustics & models* (pp. 509–515). New York: Springer.
Wright, B. A., & Sabin, A. T. (2007). Perceptual learning: how much daily training is enough? *Experimental Brain Research*, 180(4), 727–736.
Wright, B. A., & Zhang, Y. (2006). A review of learning with normal and altered sound-localization cues in human adults. *International Journal of Audiology*, 45(Suppl. 1), S92–98.
Wright, B. A., & Zhang, Y. (2009). A review of the generalization of auditory learning. *Philosophical Transactions of the Royal Society B: Biological Sciences*, 364(1515), 301–311.
Wu, J. L., Yang, H. M., Lin, Y. H., & Fu, Q. J. (2007). Effects of computer-assisted speech training on Mandarin-speaking hearing-impaired children. *Audiology & Neurotology*, 12(5), 307–312.
Zeng, F. G. (2004). Trends in cochlear implants. *Trends in Amplification*, 8(1), 1–34.
Zeng, F. G., Nie, K., Stickney, G. S., Kong, Y. Y., Vongphoe, M., Bhargave, A., Wei, C., & Cao, K. (2005). Speech recognition with amplitude and frequency modulations. *Proceedings of the National Academy of Sciences U S A*, 102(7), 2293–2298.

Chapter 12
Spoken and Written Communication Development Following Pediatric Cochlear Implantation

Sophie E. Ambrose, Dianne Hammes-Ganguly, and Laurie S. Eisenberg

1 Introduction

To appreciate fully the significance of language development for deaf children who receive a cochlear implant, one must understand the intricacies involved with language development for those with normal hearing. Language refers to the unique human capacity of using a rule-bound system of sounds, words, and symbols to communicate within a like-speaking community. Whereas language may be expressed in a manual form (e.g., American Sign Language (ASL)), for children with normal hearing, language more commonly is expressed in a spoken or written form. Language can be considered relative to two main aspects, reception and expression. Receptive language refers to an individual's ability to understand what others say to him or her. Expressive language refers to an individual's ability to use words, sentences, and conversation to express one's needs, thoughts, and wishes to others. Barring impeding factors, children with normal hearing develop spoken language seamlessly. The first section of this chapter will provide a general description of the processes involved in normal speech and spoken language development. Literacy skill development will also be discussed as an extension of language.

In the next section of this chapter, the impact of hearing loss on various aspects of language development will be discussed. Following this section, the communication achievements of children with cochlear implants will be highlighted. These advances have changed the trajectory of spoken language development for many children with deafness.

S.E. Ambrose (✉)
Center for Childhood Deafness, Boys Town National Research Hospital, 425 N. 30th Street, Omaha, NE 68131, USA
e-mail: sophie.ambrose@boystown.org

F.-G. Zeng et al. (eds.), *Auditory Prostheses: New Horizons*, Springer Handbook of Auditory Research 39, DOI 10.1007/978-1-4419-9434-9_12,

However, these trajectories vary greatly for individual cochlear implant users: some children develop spoken language skills comparable to their peers with normal hearing, while others receive limited benefit even with the best efforts of the parent, child, and intervention team. In the last section of this chapter, the factors that contribute to the great individual variability in cochlear implant outcomes will be explored.

2 Development of Speech, Language, and Literacy Skills in Children with Normal Hearing

The purpose of this section is to provide a broad overview of development in the areas of speech, language, and literacy for typically developing children who have normal hearing.

2.1 Typical Patterns of Speech Development

Children learn speech through repeatedly hearing the speech patterns of speakers in their environment. Children listen to sound patterns such as rhythm, pitch, loudness, and phoneme combinations (e.g., consonant and vowel patterns, syllables). Over time, these listening experiences, coupled with innate phonological processing abilities, allow children to learn to recognize common sound patterns and to identify where the unique patterns begin and end. Having the ability to recognize the sound patterns and sound boundaries of the language is necessary for recognizing words and sentences, and consequently, for developing spoken language.

In the early weeks of life the vocalizations of infants are believed to result from reflexive vocal productions, that is, as a result of fussiness, digestive, or other random vegetative functions (e.g., breathing or sneezing) (Owens 1996). However, relatively quickly, infants progress from reflexive vocal productions to more speech-like vocalizations. The first sounds typically produced are squeals, shrieks, or cooing, typical of babies 3 to 5 months of age. In the fourth and fifth months of life, vowel-like productions begin to form, followed by true, adult-like vowels and the introduction of consonants by 6 to 7 months of age. With the introduction of consonants, babbling forms. Babbling typically begins in a repetitive way (e.g., "mamamama") and eventually becomes variegated with differing syllable types and lengths (e.g., "uhgoosleebuhdabeenoe"). In this period of time, from 8 to 12 months of age, infants experiment widely with their voice and seem to practice various sound combinations. Whereas initially the patterns may be formed quite randomly, as infants grow nearer to 12 months of age, they become increasingly skilled at engaging in vocal play with adults. The skills include imitating simple sounds in play (e.g., "mamama" used by the adult). By 12 months of age, the first words typically appear. These often contain sounds frequently present in early babble such as "mama" for "mom," "dada" for "dad," or "baba" for "bottle." The act of formulating sounds to produce words is called speech production.

Speech intelligibility refers to the ease in which an individual's speech production is understood. Based on a formula, Flipsen (2006) estimates that at 1 year of age, it can be expected that approximately 25% of a child's utterances are intelligible, and by 4 years of age, a child's speech is generally understood in conversation by both familiar and non-familiar listeners. Spoken communication is fostered by good speech intelligibility.

2.2 *Typical Patterns of Language Development*

The capacity for language development is present at birth (Lenneberg 1967). The outward signs of communication and language development become increasingly evident in the first 12 months of life. Whereas in the first 6 months, infants rely heavily on their caregivers to anticipate their needs, in the second 6 months of life infants begin to respond to the actions and words of others. Initially, this may be more apparent receptively. Infants begin to recognize their name by 5 months of age (Mandel et al. 1995; Owens 1996). Also at this age, infants begin to respond to vocal tones, differentiating between happy and angry voices. As infants become a bit older, instead of needing direct contact for comfort, they can often be calmed by the voice of a familiar person speaking to them from nearby. Recognition of words also develops in this time period. As infants approach 1 year of age, they begin to look at named referents. For example, in response to "Where's mama?" the child will look toward the parent, or if prompted to "find the ball" the child may touch or reach toward the ball.

Expressively, between 6 to 12 months of age, infants may squeal or use their voice to gain attention and communicate through gestures or pointing. For example, to reject an object or interaction, the infant may turn away or use his or her hand to push away the rejected item. In this time frame, the infant may also begin to wave "bye-bye" or "hi." Word learning occurs slowly at first but then greatly accelerates once the child reaches an expressive vocabulary of approximately 50 words (Bloom 2001). This vocabulary explosion occurs at approximately 16 months of age (Owens 1996). At this point, children often begin to combine words. By 24 months of age most children frequently use two-word phrases and are quickly learning to produce longer phrases. By age 3, children have an expressive vocabulary of approximately 900 to 1000 words and communicate in simple sentences, albeit with some errors. By kindergarten, children's use and understanding of sentence patterns and their ability to communicate is remarkably similar to adults. Refining is still necessary in all areas of language development (e.g., use and understanding of syntax or morphology, semantic structures, and social language), but by this age children have a good understanding of the basic form of the language and have a vocabulary of 2100 to 2200 words. Many pragmatic (i.e., social) language abilities have yet to develop. In addition, use of language for understanding and expressing abstract ideas, figurative language, reasoning, logic, negotiating, and for higher-order thought processing also continues to develop. Such language development occurs in tandem with advances in underlying cognitive abilities throughout the school-age years and into adulthood.

2.3 Literacy Skill Development in Individuals with Normal Hearing

Literacy refers to the abilities to read and write. According to the National Early Literacy Panel (NELP) Report (2008), children's literacy outcomes are correlated with a number of preliteracy abilities that can be measured when children are preschool or kindergarten. These include alphabet knowledge, phonological awareness, phonological memory, the ability to write letters or one's name, and the ability to name a series of random letters or numbers quickly (rapid automatic naming). Good oral language and grammatical skills and good visual processing skills were also found to correlate positively with later literacy achievement.

Good literacy skills impact every area of a person's daily life. The National Center for Education Statistics (NCES) undertook a nationwide survey of literacy skills in more than 26,000 adults (Kirsch et al. 2002). It was found that those who obtained the highest literacy levels typically had higher levels of education, were more likely to be employed, earned higher incomes, and had lower incidence of poverty and incarceration.

3 Effects of Hearing Loss on Speech Production, Language, and Literacy Skill Development

Speech and language are mostly auditory events. Literacy skills are developed as an extension of these auditory based events. Owens (1996) outlines the speech perception abilities that are needed for speech and language development. Among the skills are the ability to attend selectively to speech, the ability to discriminate the sounds of one's native language, and the ability to remember sound sequences in the proper order. Individuals must also be able to distinguish between sound sequences, compare those sequences to stored internal models of sounds or sound patterns, and discriminate intonation patterns. Each of these skills serves a distinct purpose in speech, language, and/or literacy skill development. Considering these needed abilities, one begins to understand the serious impact that the presence of severe to profound hearing loss can have on speech, spoken language, and literacy skill development. The sections below outline the characteristic development in these three areas for children with severe to profound hearing loss prior to the availability of cochlear implant technology (for more detailed review, see Kretschmer and Kretschmer 1978).

3.1 Characteristics of Speech Production Prior to Cochlear Implant Technology

Whereas some children with severe to profound hearing loss prior to the availability of cochlear implants did succeed in developing speech skills, most did not. Even for those who did have some level of success in developing speech, the quality of their speech usually differed significantly from that of children with normal hearing. The speech differences would typically be present throughout adulthood.

Vocal development typically reflects the acoustic cues to which the child has access. For example, intonation and pitch patterns are centered in the low frequency speech regions, vowel information is heavily concentrated in the low to mid-pitch range, and the distinguishing information for consonants falls largely in the mid- to high frequency pitch ranges. Depending on the degree of hearing loss and the amount of benefit the child receives from hearing aids, his or her speech may sound more or less similar to the speech of normal-hearing infants and children.

For children with such significant hearing loss that they are candidates for a cochlear implant, whereas their speech may appear similar to that of peers with normal hearing during the early stages of infancy, by the stage of canonical babbling (6 to 10 months of age), speech production skills typically diverge substantially (Oller 2000). Recall that it is at 6 or 7 months of age when children with normal hearing begin to produce consonant sounds and babble. With limited auditory access to the full speech spectrum, children with hearing loss may not reach such a milestone until late in their development, if at all.

Even small amounts of residual hearing may aid in the development of some aspects of speech. With auditory access to low and/or mid-pitch speech frequencies, development of suprasegmental speech features (e.g., pitch, intonation, stress, and syllable number) is often possible. If mid-frequency speech sounds are audible, some vowel development may also be possible. Development of consonants is typically delayed. Comparing the early vocalizations of 94 infants with normal hearing to those of 37 infants with severe to profound hearing loss, Eilers and Oller (1994) found that there was no overlap in the onset of vocal babble between the two groups of children. All of the children with normal hearing began use of canonical babbling by 10 months of age. The earliest babbling onset in the group of deaf children was 11 months but ranging up to 49 months of age. When babble does emerge, it is often more stereotypical (e.g., reduplicative babbling such as "bababa" instead of variegated babbling such as "bagamada) in infants who are deaf than in infants with normal hearing (Mavilya 1972). Relying on lip reading, children with hearing loss are commonly found to use front consonants (e.g., /w, b, m/) more often than back (e.g., /k, g/) consonants (Sykes 1940; Carr 1953; Lach et al. 1970) and notoriously have difficulty making voicing distinctions (e.g., /k/ *vs.* /g/ and /p/ *vs.* /b/) (Carr 1953; Markides 1970).

For those infants with the most limited residual hearing, the development of suprasegmental features may also be delayed or absent. Children who receive very limited benefit from hearing aids may have difficulty hearing syllable information, stress patterns, intonation, and timing patterns (Hudgins and Numbers 1942; Smith 1975; Osberger and McGarr 1982). As a result of difficulty hearing segmental (e.g., consonant or vowel sound differences, sound combinations and patterns, sound boundaries) and suprasegmental patterns of speech (e.g., pitch, timing, etc.), many speech production errors are present. As a result, sound quality is typically distorted and speech intelligibility is often very limited. For example, Smith (1975) reports a mean speech intelligibility rating of 18.7% for a group of 40 children, deaf since birth, ranging in age from 10 to 15 years. Not having been able to learn from and assign meaning to the sound patterns and sound differences of speech, progression to first words and use of spoken language is often delayed or prohibited.

3.2 Characteristics of Spoken Language Prior to Cochlear Implant Technology

Prior to the availability of cochlear implants, the most viable option for developing language for children with limited hearing was through visual communication, such as through American Sign Language, Manually Coded English (e.g., using a Total Communication approach which combines use of spoken English with sign language), or Cued Speech. Some children did develop spoken language even with minimal hearing by relying on residual hearing that was amplified by hearing aids and the use of speech reading, but this was often laborious and frustrating. Not surprisingly, degree of loss and the amount of benefit from hearing aids have a significant relationship with spoken language outcomes including vocabulary levels and syntax (Pressnell 1973; Quigley et al. 1976a, b). As hearing levels become more limited, so do spoken language prospects. Without intervention (therapeutic and with respect to amplification), spoken language development is most limited for children with profound hearing loss.

At the age of 12 months, when children with normal hearing progress to words, children who are deaf may continue to use gestures (Grewel 1963). Words often develop slowly. At 18 months of age, as opposed to a vocabulary of 20 to 50 words for children with normal hearing, a child with hearing loss may have a vocabulary of less than 10 words (Schafer and Lynch 1980). At age 5, instead of a vocabulary of more than 2000 words, a child who is deaf may have a spoken vocabulary of 250 words (Dale 1974).

Although differences do exist, some researchers have observed that the skills of children who are deaf often mirror the skills or error patterns of children who are chronologically younger but who have normal hearing (Pressnell 1973; Kretschmer and Kretschmer 1978). Thus, the 5-year-old child who is deaf and has a vocabulary of 250 words may demonstrate linguistic abilities that are similar to a child of 2 to 2½ years of age (i.e., the age at which children with normal hearing typically demonstrate a vocabulary of that size). In a study of more than 420 10- to 18-year-old children who are deaf, Quigley and colleagues found that at an age when children with normal hearing typically are using adult-like sentence patterns (8 to 10 years of age), the deaf children demonstrated many grammatical errors. Although many errors were consistent with those of much younger children with normal hearing, some error patterns or areas of difficulty appeared unique to children with hearing loss (1976a; 1976b).

In short, spoken language development in the majority of children with severe to profound hearing loss prior to cochlear implant technology was often noted to be delayed or disordered throughout the life span. Aside from the direct impact of hearing loss that prevents deaf children from hearing key aspects of speech, children with severe to profound hearing loss develop language slowly in part because each structure must be taught directly. Recall that when normal hearing is present, children learn by hearing the patterns of the language around them. In part, vocabulary and language are expanded by overhearing parents or other family members in conversation. As children become older, they also learn from hearing peers, teachers, and other audio media (e.g., radio or television). The greater the severity of hearing loss, the less able the child is to learn from these sorts of auditory activities, and the more reliant they become on learning from direct, face-to-face communication. This is an extremely inefficient way of learning spoken language.

Development of connected language (i.e., sentences or conversation) is further complicated for children who are deaf by virtue of their language delays and by speech intelligibility factors. Parents may interact differently with their child who is deaf than with a child who is hearing, as might other people in the child's environment (Kretschmer and Kretschmer 1978). Consequently, language for a deaf child may not be fostered in the same way that it is for children who have normal hearing. For example, even if a child who is deaf establishes good knowledge of the rules of the language, communication interactions may remain challenging because unfamiliar listeners may have difficulty understanding the child's speech. Similarly, the child may have difficulty understanding unfamiliar speakers, especially when communicating with those who are not accustomed to communicating with individuals who are deaf. These two factors result in ineffective communicative interactions and therefore may result in the child or potential communication partners avoiding or overly simplifying communication. This, in turn, results in a less rich language environment for the child and fewer opportunities to practice using spoken language. All of these factors severely impact the language and communication abilities of children who are deaf.

3.3 Characteristics of Literacy Skill Development Prior to Cochlear Implant Technology

Literacy skills develop as an extension of auditory events. In infancy and early childhood, children learn to add meaning to the sound units (consonants, vowels, syllables, phrases, and sentence patterns) of the language. In the preschool years, they begin to relate the sounds to print (e.g., letter-sound associations). In the kindergarten years, children learn to relate the auditory patterns of language to written words, patterns, and sentences. By third grade, for academic success, children must have mastered these skills well enough to prepare them for the transition from learning to read, to reading to learn. As can be seen in this depiction, hearing is an integral link to literacy skill development. Limited hearing often results in poor phonological awareness (i.e., awareness of the sound structure of words, such as knowing that "shoot" is made up of three sounds or that "shoot" without the "t" sound is "shoe"). As pointed out earlier in the chapter, children with poor phonological awareness often have difficulty developing literacy skills. This difficulty is reflected by historic reports that children who are deaf have difficulty achieving literacy skills above those demonstrated by third to fourth grade children with normal hearing (Krose et al. 1986).

Goda (1959) measured the speech reading, oral language, writing, and reading skills of 56 adolescents who are deaf. Those children with better expressive language abilities in general used a greater number of words when speaking and writing. In addition, the children with better expressive language abilities also used comparatively longer and more complex written sentences. Goda noted that children tended to either perform well in all areas, or do poorly across all areas.

4 Effects of Cochlear Implantation on Speech, Language, and Literacy

Studies from the past three decades have documented communication achievements of children with cochlear implants, initially as compared to their deaf peers and more recently as compared to their normally hearing peers. Achievements are noted in a variety of communication areas including speech production, receptive and expressive oral language, and literacy. Factors that are believed to contribute to variability in individual outcomes are outlined in a later section of this chapter.

4.1 *Speech Production and Intelligibility*

Early studies on speech production and intelligibility focused on comparisons between children with implants and control groups of their non-implanted deaf peers who utilized either tactile aids or hearing aids. In the first study published about the speech and language of children with cochlear implants, Kirk and Hill-Brown (1985) compared speech production scores that were collected from children presenting for pre-cochlear implant evaluations to those of age-matched children presenting for 6, 12, or 18-month post-cochlear implant evaluations. The speech production scores achieved by the children with (single channel) cochlear implants were higher than those demonstrated by children without implants in almost every single comparison made. In a 1996 study, Miyamoto et al. compared the speech intelligibility of children with multichannel cochlear implants to their deaf peers who used hearing aids. The average speech intelligibility of the cochlear implant group surpassed that of the hearing aid users with hearing levels of 101 to 110 dB HL sometime around 2 to 2 ½ years after cochlear implant.

Geers and Tobey (1995) compared the speech perception, speech production, and spoken language development of three groups of deaf children over a 3-year period: children who utilized cochlear implants, children who utilized tactile aids, and children who utilized conventional hearing aids. The children were all being educated in the same auditory-oral educational program. The authors found that the tactile aid did not have a significant impact on the acquisition of speech or language. However, for children who did not receive spectral information through hearing aids, the cochlear implant was associated with significantly faster speech and language acquisition than would be anticipated with hearing aid use. Other similar studies also found that children who utilized cochlear implants made greater progress in the area of speech production than did their peers who utilized tactile aids (Kirk et al. 1995; Ertmer et al. 1997).

More recent studies have compared speech production and intelligibility of children with cochlear implants to those of children with normal hearing. Should children with cochlear implants progress at a similar rate to children with normal hearing, they would be expected to master the ability to articulate single words by 7 years, 8 months after cochlear implant.

Children with normal hearing are expected to be fully intelligible by the age 4 (Flipsen 2006). However, studies indicate that children with cochlear implants may never reach full speech intelligibility. Chin et al. (2003) compared speech intelligibility data from 52 children with cochlear implants to data from 47 children with normal hearing. As expected, the children with normal hearing achieved adult-like intelligibility by 4 years of age. However, even when controlling for age and length of auditory experience, the children with cochlear implants did not approach comparable levels to their peers with normal hearing. Peng et al. (2004a) examined the speech intelligibility of children with cochlear implants who had received their implant between 2½ and 11 years of age (mean of 61 months) and who had been using their implants for an average of 84 months. In this investigation, the children attained an intelligibility level of about 72%. Thus, after approximately 7 years of cochlear implant use, these children were not achieving the same levels of speech intelligibility that would be expected of children with normal hearing at 4 years of age.

Tomblin et al. examined the speech production abilities of 27 children who were implanted at an average of 4½ years of age and who had been using their cochlear implant for at least 8 years (2008), the age by which children with normal hearing typically master normal articulation. The children with cochlear implants were shown to demonstrate improvements in articulation skills during the first 5 years of implant use, reaching a plateau after 6 years of experience with the implant. This plateau remained even after removing the 8 children who achieved especially high levels of speech accuracy after 6 or 7 years of implant experience. Similar results have been found for production of speech intonation; that is, children with cochlear implants do not show mastery of intonation in their speech to the same extent as their normally hearing peers (Peng et al. 2008).

It must be noted that the children in these studies were implanted at rather late ages by today's standards. Thus, the full extent of the benefits possible for children implanted at the youngest ages may not be fully evident in these studies. Further, it must be remembered that beyond a certain point, even with errors, children with normal hearing can be understood relatively well. Recall that even for children implanted at slightly older ages than current standards, Peng et al. (2004a) reported speech intelligibility rates of near 72%. This is a drastic improvement from the 18.7% reported by Smith (1975) for children in the era before cochlear implants. Even at this level of intelligibility, spoken communication is greatly enhanced. It remains for future analyses to determine whether children implanted at much younger ages do in fact achieve normal speech production skills.

The previously outlined studies examined the speech production abilities of English-speaking children. Indeed, the majority of research on the speech and language development of children with cochlear implants has been conducted with English-speaking subjects. However, more recently, researchers have begun to examine the speech perception and speech production development of children with cochlear implants who utilize tonal languages, such as Mandarin. This area of study is especially interesting because cochlear implants are known to be poor at encoding the voice pitch information necessary for perceiving intonation and tone. For English-speakers, this may impact the listener's ability to determine whether a speaker is asking a question versus making a statement or their ability to determine the feelings behind a speaker's message (Peng et al. 2004b). However, for a Mandarin-speaker who relies on tonal information for lexical distinctions, the inability of cochlear implants to encode voice pitch information accurately can be even more problematic. Studies of Mandarin-speaking children with cochlear implants show that these children are significantly disadvantaged in comparison to their normal hearing peers on measures of tonal perception and production (Peng et al. 2008).

Despite the fact that children with cochlear implants, as a group, are not performing on par with their normal-hearing peers, the gains made in the areas of speech production and speech intelligibility far exceed that which would be expected with hearing aids for children with this magnitude of hearing loss. Some children with implants do, in fact, perform on par with their normal-hearing peers, and children with implants as a whole often perform within 1 standard deviation of the mean of normal-hearing children (Blamey et al. 2001a; Flipsen and Colvard 2006).

4.2 *Receptive and Expressive Oral Language*

Early studies unequivocally indicated that profoundly deaf children with cochlear implants made faster progress in language development than they would have made with hearing aids (Robbins et al. 1997; Svirsky et al. 2000). This finding has been further supported by research that directly compares children with cochlear implants to their non-implanted peers who were appropriate candidates for a cochlear implant (Truy et al. 1998). One landmark study by Blamey et al. (2001b) indicated that profoundly deaf children with cochlear implants perform similarly to hearing aid users with severe hearing losses. Specifically, on measures of receptive vocabulary and general language, both groups of children demonstrated a rate of language growth that was about one-half to two-thirds the rate expected for children with normal hearing.

Although seminal studies such as those above demonstrate positive findings in relation to deaf peers who use hearing aids, further studies have documented outcomes in relation to children with normal hearing. Geers et al. (2003) directly compared the performance of 8- and 9-year-olds who were implanted under the age of 5 years to the performance of their peers with normal hearing. More than half the children attained receptive and expressive oral language scores that were comparable with those of their hearing peers. Results reported by Schorr et al. (2008) for 5- to 14-year-old children were equally as impressive or even better. The authors reported on the communication abilities of 39 congenitally deaf children with at least 1 year of cochlear implant experience. Between 51 and 66% of children in their study performed on par with their normal hearing peers on measures of receptive and expressive vocabulary and morphology and syntax. In a study of younger-implanted children, Geers et al. (2009) reported results on 153 children with a mean age at testing of 5 years, 10 months who had been using their implants for an average of 3 years, 6 months. Similar to the slightly older children in the Schorr et al. (2008) study, between 39 and 59% of children achieved age-appropriate scores for receptive and expressive vocabulary and general receptive and expressive language.

Niparko et al. (2010) recently reported on 3-year data from a 7-year longitudinal study that is following the childhood development of 188 children who received cochlear implants prior to 5 years of age. For language outcomes, the authors compared post-implant language learning to predicted scores based on the children's pre-implantation language scores. They found that use of a cochlear implant resulted in a more favorable spoken language learning rate than would have been expected based on language growth prior to implantation. Even though language learning rate increased following implantation, children were not found to reach age-appropriate language levels within the first 3 years of cochlear implant use. The fastest learning rates were demonstrated by children implanted under 18 months of age.

Belzner and Seal (2009) recently provided a detailed review of seminal studies of cochlear implant outcomes from the years 2000 through 2007. The trends demonstrated through their literature review are also consistent with the cross section of studies presented here. Namely, children with cochlear implants tend to outperform their profoundly deaf peers who use hearing aids in the area of spoken language development. Moreover, some children perform close to or on par with normal-hearing peers. Not surprisingly, results indicate an advantage for children who receive their cochlear implant at an earlier age. Lastly, the review indicated that speech perception, speech production, and spoken language outcomes are highly variable among children who have cochlear implants.

4.3 Literacy

Prior to the advent of pediatric cochlear implantation, it was well established that children who were deaf often had difficulty surpassing third to fourth grade literacy levels (Furth 1966; Krose et al. 1986). It was hoped that cochlear implants could provide an avenue for children with hearing loss to develop stronger phonological and oral language abilities than the historic expectations, and that this in turn would be an advantage for literacy development. As summarized in the previous section, cochlear implant technology has been instrumental in improving spoken language abilities for children with early-onset severe to profound hearing loss. Additionally, improvement in speech perception have increased children's awareness of the phonemic units that make up speech, translating into improved phonological awareness abilities (James et al. 2005). Both DesJardin et al. (2009) and Ambrose (2009) found that children with cochlear implants performed within 1 standard deviation of the mean of their hearing peers on measures of phonological awareness. However, significant between-group differences indicated that children with CIs still lagged slightly behind their peers in this area.

Similarly, investigations have pointed toward improvement in general literacy scores for children with cochlear implants. In one of the earliest studies on the literacy skills of children with cochlear implants, 54% of 28 children in the fourth to twelfth grades demonstrated reading abilities above the fourth grade level (Spencer et al. 1997). This is in contrast to the studies by Furth (1966) and Krose et al. (1986), which found similarly aged deaf children without cochlear implants to be performing above the fourth grade level in 8% and 14% of cases, respectively. In a more recent study, Spencer et al. (2003) reported on the reading comprehension and writing abilities of children with cochlear implants (mean age of 9 years, 10 months), and their age-matched normal hearing peers. On average, the children with cochlear implants performed within 1 standard deviation of the normal hearing mean for these literacy skill areas. The findings of this study are significant because if children with cochlear implants are able to maintain this level of performance (i.e., literacy abilities within 1 standard deviation of their normal-hearing peers) as they progress through later grades, they will successfully navigate past the fourth grade reading level plateau, a move that has historically proved challenging to deaf high school graduates. Furthermore, these students may have the capability of reaching an eighth grade reading level, which is significant because this is typically the level below which adults are considered illiterate (Office of Educational Research and Improvement 1989).

4.4 *Summary*

The communication gains made by deaf children with cochlear implants are remarkable, particularly when referenced to the outcomes reported in the early literature on the communication abilities of deaf children prior to the widespread availability of cochlear implant technology. Many children with severe to profound hearing loss are now performing on par with their hearing peers in a variety of communication skill areas. As technology continues to improve and professionals update their knowledge and skill in the area of pediatric cochlear implantation, an even greater number of implanted children may be expected to approach these high levels of communication success.

5 Factors Influencing Spoken Language and Literacy Outcomes

Despite impressive reports of high levels of communication outcomes by children with cochlear implants, marked variability is evident in spoken language development following cochlear implantation. This can be seen in Fig. 12.1, where individual

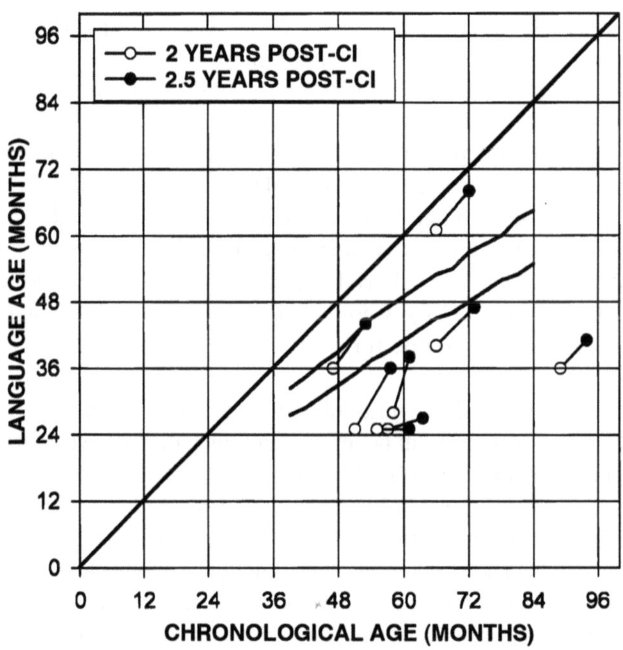

Fig. 12.1 Language results from Svirsky et al. (2000) (Figure 4, p. 157). Reprinted with permission by Sage Publications. Individual data measuring changes in language abilities for cochlear implant (CI) users between 2 and 2.5 years post-implant. The two *black lines* under the diagonal indicate −1 and −2 standard deviations below the mean for the normal-hearing population

data from Svirsky et al. (2000) are depicted. For some children in this study, even after 2 years of cochlear implant experience, their language abilities remained severely delayed. However, after the same length of cochlear implant experience, other children demonstrated expressive language abilities that were similar to those of their hearing peers. A number of factors may explain these individual differences, including factors related to timing, the individual child, the family, and educational services. These factors are not straightforward. They interact in both predictable and unpredictable ways, with the influence of some factors negating or even exacerbating others (Geers et al. 2007).

5.1 Timing Factors

There are a number of timing factors that influence communication outcomes for children who are deaf and hard of hearing. The classic variables are the age at which the child's hearing loss is identified, the age at which the child is fit with hearing aids, and the age at which the child begins early intervention (Moeller 2000). However, for congenitally deaf children, the age at which the child receives his or her cochlear implant may have the largest impact on his or her post-implant success (Nicholas and Geers 2006). Earlier implanted children have two main advantages over their later implanted peers: access to speech information during prime language-learning years and a reduced period of auditory deprivation.

It has long been accepted that humans acquire speech and language with greater facility at younger ages. This knowledge led to the idea of a critical or sensitive period for speech and language learning (Tomblin et al. 2007). Indeed, the literature on pediatric cochlear implantation lends support to this assertion. When cochlear implant clinical trials were first initiated, children who were implanted before 5 years of age were considered to be "early implanted." These children were reported to outperform their later implanted peers (Tye-Murray et al. 1995).

Many studies in the early to mid-2000s found that children who received their implant by 2 years of age perform better on communication measures than their later implanted peers (Svirsky et al. 2004; Miyamoto et al. 2008). A very recent (ongoing) prospective study of 188 children implanted under the age of 5 found that children who were implanted under 18 months of age demonstrated the greatest advantages (Niparko et al. 2010). The children implanted under 18 months of age demonstrated significantly higher rates of receptive and expressive language learning than did children implanted after 18 months of age. They also demonstrated learning trajectories more closely resembling (although not identical to) those of their normally hearing peers than did children implanted after 18 months of age. Although cochlear implants are yet to be approved by the Food and Drug Administration for children under 12 months of age, studies are emerging to suggest that children implanted before their first birthday outperform their peers implanted after 12 months of age (Svirsky et al. 2004; Miyamoto et al. 2005; Dettman et al. 2007).

The advantages of earlier implantation exist even when controlling for the amount of post-implant experience (Nicholas and Geers 2007). That is, early implanted children are not outperforming their later implanted peers simply because they have a longer period of sound exposure. In fact, children implanted at younger ages appear to demonstrate an increased rate of language growth when duration of cochlear implant experience is controlled statistically (Tomblin et al. 2005; Nicholas and Geers 2007).

The ability to better avail patients of a critical or sensitive period for language development is not the only advantage of early implantation. Research by Sharma et al. has shown that the central auditory system also appears to have a sensitive period for development (Sharma et al. 2002b; Bauer et al. 2006; Sharma et al. 2009). Sharma et al. (2009) reported that, as with other sensory systems, if children's auditory systems remain quiescent for an extended period of time, reorganization of the auditory cortex can be expected. Thus, in order for the auditory system to recover from the impact of auditory deprivation, early implantation is critical (Sharma et al. 2002a; Bauer et al. 2006).

5.2 *Child Factors*

There are a number of non-timing related child factors that are also correlated with post-implant communication outcomes. These include variables such as gender, the presence of additional disabilities, non-verbal IQ, and residual hearing (pre-implant residual hearing and post-implant hearing in the non-implanted ear).

5.2.1 Gender

In relation to gender, females generally appear to have an advantage over males in the domain of language development (Maccoby and Jacklin 1974). Research on the relationship between gender and communication outcomes following cochlear implantation have been mixed, but results showing a significant relationship between gender and communication abilities have mirrored the results of findings reported in the literature on children with normal hearing. That is, girls show an advantage over boys (Geers et al. 2003, 2009).

5.2.2 Additional Disabilities

A factor that stands to have a stronger impact on communication outcomes following cochlear implantation is the presence of an additional disability beyond the child's hearing loss. Examples of these disabilities include severe cognitive delays and neurological disorders such as autism. The literature on children with normal hearing suggests these children are clearly at risk for language delays and disorders. The addition of a hearing loss further complicates the language learning process. Children with special needs who receive cochlear implants are likely to demonstrate more limited communication gains than their implanted peers without special needs (Baldassari et al. 2009). This has been shown to be the case even when those special needs are as minimal as mild cognitive delays (Holt and Kirk 2005).

5.2.3 Cognitive Abilities

Beyond cognitive disabilities and neurological disorders, subtle differences exist in children's cognitive abilities. For example, children with cochlear implants vary in their working memory abilities and non-verbal intelligence scores. These differences have the potential to impact children's communication outcomes.

Geers et al. (2009) assessed the language abilities of a group of implanted children whose nonverbal IQ scores ranged from 70 (two standard deviations below the normative mean) to 140. The results indicated that the children's non-verbal intelligence scores predicted more variance in language outcomes than any other factor they considered (e.g., age at implant, parent education). However, not all studies found such a strong relationship between non-verbal IQ and communication outcomes, as indicated in a recent study by Hayes et al. (2009) on the growth of vocabulary skill over time. The difference in findings between studies may be the result of the relatively stronger non-verbal IQ scores reported in the Hayes et al. study; all children had non-verbal IQ scores of 89 or above (i.e., less than 1 standard deviation below the mean or higher).

Another cognitive skill explored as a predictor of children's post-implant performance is phonological processing. A non-word repetition task was used by Dillon and Pisoni (2006) to measure phonological processing in a group of children with cochlear implants. Non-word recognition draws on children's speech perception abilities, working memory abilities, and speech production abilities, as well as other skills. The authors found that children who performed more accurately on the non-word repetition task also demonstrated superior non-word reading abilities, single-word reading abilities, and written sentence comprehension abilities.

Other studies have looked specifically at working memory. Cleary et al. (2002) administered four measures of working memory to 61 children with cochlear implants. Tasks ranged from recalling lists of numbers to remembering a series of sequenced lights. The results indicated that children's receptive vocabulary abilities were related only to the working memory tasks that had an auditory processing component. This result is similar to an earlier finding by Pisoni and Geers (2000), which demonstrated that children's ability to repeat numbers presented without any visual cues is correlated with their communication outcomes (specifically, speech perception, speech production, language, and reading). Thus, as with the previous study, processing of auditory information is significantly related to communication outcomes.

5.2.4 Residual Hearing Prior to Implantation

Only those children with profound hearing loss were originally considered candidates for a cochlear implant. However, criteria have been expanded in recent years to include children with greater residual hearing. Implanted children with aidable residual hearing prior to receipt of a cochlear implant have been shown to outperform both their deaf peers who use hearing aids and their implanted peers who have less residual hearing (Gordon et al. 2001). There are a number of potential explanations for this finding. First, children with more residual hearing may have more intact auditory structures (e.g., absence of cochlear malformation, greater proportion of surviving neural elements, etc.) than those of their peers with more profound hearing loss. Second, pre-implant access to sound, even if minimal, helps lay a rudimentary foundation for listening to sound that may be further enhanced with the cochlear implant. Similarly, the auditory pathways may be better primed to respond to stimuli from the cochlear implant. Lastly, the presence of residual hearing may provide children with some access to speech, enabling early speech development and limited exposure to language concepts prior to cochlear implantation (Nicholas and Geers 2006).

5.2.5 Bilateral Sound Stimulation

The benefits of bilateral implants or bimodal device usage (hearing aid on one ear and a cochlear implant on the other) is a more recent area of study. At this point, the studies have focused almost exclusively on the advantages for speech perception in noise and for localization of sound as opposed to speech, language, or literacy development. Findings have generally been supportive of bilateral or bimodal device usage over unilateral stimulation (Litovsky et al. 2006; Mok et al. 2007; Balkany et al. 2008).

In one of the only studies that has evaluated language as a function of device configuration, Nittrouer and Chapman (2009) compared groups of children who utilized either bilateral cochlear implants, a bimodal configuration, or a unilateral cochlear implant. When grouping children by their device configuration at the time of testing, a linguistic advantage was not found for any of the three groups for the basic language parameters (general language and expressive vocabulary) or generative language parameters (mean length of utterance and number of pronouns) assessed. A generative language advantage was found, however, for children who utilized bimodal stimulation at some point after implantation (e.g., either as their current configuration or prior to receiving a cochlear implant for the second ear). It may be that those who use or have used a bimodal configuration are those children who had more residual hearing and thus demonstrated benefit by virtue of residual hearing. Also, in terms of the overall findings of the study, the children in the study were evaluated at a relatively young age (42 months). The children had not yet reached primary school age and in the scope of development have had very limited listening experience. As stated earlier, the most widely demonstrated benefit of bilateral (and bimodal) stimulation is with respect to listening in noise and sound localization. As children reach school age these binaural listening skills have increasing merit. The children move from mainly listening in one-to-one or small group situations to listening and learning in large groups, which introduces the potential for much more background noise during learning. Thus, the children in this study may not yet have sufficient age or implant experience for the full effects of bilateral stimulation to become evident. Additional research in this area is needed.

5.3 Family Factors

Parental education, income, and interaction styles (e.g., level of participation, talkativeness) are the most commonly reported family factors in the literature on cochlear implants. The first two factors are common measures of socio-economic status (SES). The positive relationship between SES and language development has been well established in the literature on children with normal hearing. Evidence of this relationship similarly exists for children with cochlear implants. Implanted children from higher income homes or homes with more educated parents tend to demonstrate more advanced spoken language and literacy skills than their peers from lower SES backgrounds (Stallings et al. 2000; Geers 2002; Geers et al. 2003; Connor and Zwolan 2004; Niparko et al. 2010).

Parental input is another area that positively influences success with the cochlear implant. Normally hearing children from families with higher SES backgrounds are more likely to be given richer language experiences and be provided with more stimulating language environments (Hoff-Ginsberg 1998; Raviv et al. 2004). DesJardin et al. (2007) investigated maternal perceived self-efficacy and involvement on the language outcomes of children with cochlear implants. The results indicated that mothers of children with cochlear implants who believe in their own ability to influence their child's outcomes, who are highly involved in their child's early intervention and educational programs, and who provide their children with strong linguistic input are more likely to have children who develop strong communicative skills. Research of this type stands to influence early intervention and other educational programming for children with cochlear implants.

5.4 Educational Factors

The aural habilitation and educational services that children with cochlear implants receive varies widely due to parental preference, the child's geographical location (which may impact availability of services), the philosophy of the child's school district, and so forth. However, aside from studies related to communication mode, little research has focused on the varied services offered to children with cochlear implants and the impact of such services on communication outcomes. Geers (2002) investigated key service-related variables for a group of 136 children with cochlear implants including the amount of aural rehabilitation or speech/language therapy hours received, the therapist's experience working with children who are deaf, parent participation in intervention services, school setting (private vs. public), type of class (regular education vs. special education), and the communication mode utilized by the educational program (auditory-oral vs. total communication). The authors examined the influence of these variables on children's speech perception, speech production, spoken language, total language (i.e., speech and sign), and reading abilities. Of these many variables, only type of class and communication mode were correlated with any of the outcomes. Children from programs that emphasized auditory-oral development outperformed their peers in total-communication environments in all areas *except total language*. Children in mainstream classrooms outperformed their peers in self-contained classrooms on measures of spoken language, total language, and reading.

Other studies on the impact of communication mode have not shed much further light on the issue; some studies have found that children in total communication placements outperform their peers in auditory-oral placements (Connor et al. 2000), while others have indicated exactly the opposite, with the advantage being for children in auditory-oral education placements (Somers 1991; Miyamoto et al. 1999; Geers et al. 2000; Tobey et al. 2004), and still other studies have shown no difference in outcomes for children utilizing either of these approaches (Kirk et al. 2002). Drawing conclusions from these studies is difficult because relationships are not unidirectional. Instead, children with better spoken language potential, before or after implantation, may be strategically placed in regular education or auditory-oral classrooms. Similarly, the lack of significant relationships between other variables, such as the number of hours of intervention and spoken language outcomes, could indicate that some children with lower language levels may have been enrolled in more intensive aural habilitation as a direct result of their impoverished skills.

Prospective research on the impact of type, setting, intensity, duration, and quality of services provided to children with cochlear implants is essential for understanding their communication development. Documenting the basis on which parents and educators make decisions about these factors prior to implantation, immediately following a child's receipt of a cochlear implant and during the remainder of the child's education, would be highly beneficial.

6 Summary

Severe to profound hearing loss produces potentially devastating effects on communication development. Cochlear implant technology has greatly expanded communication options for children with this magnitude of hearing loss and improved their potential for speech, spoken language, and literacy skill achievement. Most implanted children achieve spoken communication and literacy skills that far exceed the abilities that were once the norm for deaf children. Additionally, many implanted children develop speech, language, and literacy skills comparable to their peers with normal hearing.

Although such outcomes are encouraging, children with cochlear implants as a group still perform below their normal hearing peers on measures of speech and spoken language abilities. Additionally, marked variability exists in the achievement of individual children with cochlear implants and we cannot fully account for those factors that result in such variability. Further identification and understanding of the variables that contribute to post-implant communication outcomes may allow for more refined cochlear implant candidacy criteria, better counseling of parents regarding realistic expectations for individual children, and more effective habilitation strategies.

References

Ambrose, S. (2009). *Phonological awareness development of preschool children with cochlear implants* (Doctoral dissertation). Retrieved from ProQuest Dissertations and Theses Database. (AAT 3396461).

Baldassari, C. M., Schmidt, C., Schubert, C. M., Srinivasan, P., Dodson, K. M., & Sismanis, A. (2009). Receptive language outcomes in children after cochlear implantation. *Otolaryngology-Head and Neck Surgery*, 140(1), 114–119, doi: 10.1016/j.otohns.2008.09.008.

Balkany, T., Hodges, A., Telischi, F., Hoffman, R., Madell, J., Parisier, S., Gantz, B., Tyler, R., & Peters, R.. (2008). William House Cochlear Implant Study Group: position statement on bilateral cochlear implantation. *Otology & Neurotology*, 29(2), 107–108, doi: 10.1097/mao.0b013e318163d2ea.

Bauer, P. W., Sharma, A., Martin, K., & Dorman, M. (2006). Central auditory development in children with bilateral cochlear implants. *Archives of Otolaryngology-Head & Neck Surgery*, 132(10), 1133–1136, doi: 10.1001/archotol.132.10.1133.

Belzner, K. A., & Seal, B. C. (2009). Children with cochlear implants: a review of demographics and communication outcomes. *American Annals of the Deaf*, 154(3), 311–333.

Blamey, P. J., Barry, J. G., Bow, C. P., Sarant, J. Z., Paatsch, L. E., & Wales, R. J. (2001a). The development of speech production following cochlear implantation. *Clinical Linguistics and Phonetics*, 15(5), 363–382.

Blamey, P. J., Sarant, J. Z., Paatsch, L. E., Barry, J. G., Bow, C. P., Wales, R. J., et al. (2001b). Relationships among speech perception, production, language, hearing loss, and age in children with impaired hearing. *Journal of Speech, Language, and Hearing Research*, 44(2), 264–285.

Bloom, P. (2001). Precis of how children learn the meanings of words. *Behavioral and Brain Sciences*, 24(06), 1095–1103.

Carr, J. (1953). An investigation of the spontaneous speech sounds of five-year-old deafborn children. *Journal of Speech and Hearing Disorders*, 18(1), 22–29.

Chin, S. B., Tsai, P. L., & Gao, S. (2003). Connected speech intelligibility of children with cochlear implants and children with normal hearing. *American Journal of Speech-Language Pathology*, 12(4), 440–451, doi: 10.1044/1058-0360(2003/090).

Cleary, M., Pisoni, D. B., & Kirk, K. I. (2002). Working memory spans as predictors of spoken word recognition and receptive vocabulary in children with cochlear implants. *The Volta Review*, 102(4), 259–280.

Connor, C. M., & Zwolan, T. A. (2004). Examining multiple sources of influence on the reading comprehension skills of children who use cochlear implants. *Journal of Speech, Language, and Hearing Research*, 47(3), 509–526.

Connor, C. M., Hieber, S., Arts, H. A., & Zwolan, T. A. (2000). Speech, vocabulary, and the education of children using cochlear implants: oral or total communication? *Journal of Speech, Language, and Hearing Research*, 43(5), 1185–1204.

Dale, D. M. C. (1974). *Language development in deaf and partially hearing children*. Springfield, IL: Charles C. Thomas.

DesJardin, J. L., & Eisenberg, L. S. (2007). Maternal contributions: supporting language development in young children with cochlear implants. *Ear and Hearing*, 28(4), 456–469. doi: 10.1097/AUD.0b013e31806dc1ab.

DesJardin, J. L., Ambrose, S. E., & Eisenberg, L. S. (2009). Literacy skills in children with cochlear implants: the importance of early oral language and joint storybook reading. *Journal of Deaf Studies and Deaf Education*, 14(1), 22–43, doi: 10.1093/deafed/enn011.

Dettman, S. J., Pinder, D., Briggs, R. J. S., Dowell, R. C., & Leigh, J. R. (2007). Communication development in children who receive cochlear implants younger than 12 months: risks versus benefits. *Ear and Hearing*, 28(2), 11 S–18 S.

Dillon, C. M., & Pisoni, D. B. (2006). Nonword repetition and reading skills in children who are deaf and have cochlear implants. *The Volta Review*, 106, 121–145.

Eilers, R. E., & Oller, D. K. (1994). Infant vocalizations and the early diagnosis of severe hearing impairment. *Journal of Pediatrics*, 124(2), 199–203.

Ertmer, D. J., Kirk, K. I., Sehgal, S. T., Riley, A. I., & Osberger, M. J. (1997). A comparison of vowel production by children with multichannel cochlear implants or tactile aids: Perceptual evidence. *Ear and Hearing*, 18, 307–315.

Flipsen, P. (2006). Measuring the intelligibility of conversational speech in children. *Clinical linguistics & phonetics*, 20(4), 303–312, doi: 10.1080/02699200400024863.

Flipsen, P., & Colvard, L. G. (2006). Intelligibility of conversational speech produced by children with cochlear implants. *Journal of Communication Disorders*, 39(2), 93–108, doi: 10.1016/j.jcomdis.2005.11.001.

Furth, H. G. (1966). A comparison of reading test norms of deaf and hearing children. *American Annals of the Deaf*, 111, 461–462.

Geers, A. E. (2002). Factors affecting the development of speech, language, and literacy in children with early cochlear implantation. *Language, Speech, and Hearing Services in Schools*, 33(3), 172–183.

Geers, A. E., & Tobey, E. A. (1995). Longitudinal comparison of the benefits of cochlear implants and tactile aids in a controlled educational setting. *Annals of Otology, Rhinology and Laryngology Supplement*, 166, 328–329.

Geers, A. E., Nicholas, J., Tye-Murray, N., Uchanski, R., Brenner, C., Davidson, L. S., Toretta, G., & Tobey, E.A. (2000). Effects of communication mode on long term cochlear implant users. *Annals of Otology, Rhinology & Laryngology*, 109(12), 89–92.

Geers, A. E., Nicholas, J. G., & Sedey, A. L. (2003). Language skills of children with early cochlear implantation. *Ear and Hearing*, 24(1, Suppl.), 46S–58S, doi: 10.1097/01.AUD.0000051689.57380.1B.

Geers, A. E., Nicholas, J. G., & Moog, J. S. (2007). Estimating the influence of cochlear implantation on language development. *Audiological Medicine*, 5(4), 262–273.

Geers, A. E., Moog, J. S., Biedenstein, J., Brenner, C., & Hayes, H. (2009). Spoken language scores of children using cochlear implants compared to hearing age-mates at school entry. *Journal of Deaf Studies and Deaf Education*, 14(3), 371–385, doi: 10.1093/deafed/enn046.

Goda, S. (1959). Language skills of profoundly deaf adolescent children. *Journal of Speech and Hearing Research*, 2(4), 369–376.

Goldman, R. (2000). *Goldman-Fristoe Test of Articulation-2*. Circle Pines, MN: American Guidance Service.

Gordon, K. A., Twitchell, K. A., Papsin, B. C., & Harrison, R. V. (2001). Effect of residual hearing prior to cochlear implantation on speech perception in children. *Journal of Otolaryngology*, 30(4), 216–223.

Grewel, F. (1963). Remarks upon the acquisition of language in deaf children. *Language and Speech*, 6(1), 37–45, doi: 10.1177/002383096300600104.

Hayes, H., Geers, A. E., Treiman, R., & Moog, J. S. (2009). Receptive vocabulary development in deaf children with cochlear implants: achievement in an intensive auditory-oral educational setting. *Ear and Hearing*, 30(1), 128–135.

Hoff-Ginsberg, E. (1998). The relation of birth order and socioeconomic status to children's language experience and language development. *Applied Psycholinguistics*, 19(4), 603–629.

Holt, R. F., & Kirk, K. I. (2005). Speech and language development in cognitively delayed children with cochlear implants. *Ear and Hearing*, 26(2), 132–148.

Hudgins, C. V., & Numbers, F. C. (1942). An investigation of the intelligibility of the speech of the deaf. *Genetic Psychology Monographs*, 25, 289–392.

James, D., Rajput, K., Brown, T., Sirimanna, T., Brinton, J., & Goswami, U. (2005). Phonological awareness in deaf children who use cochlear implants. *Journal of Speech, Language, and Hearing Research*, 48(6), 1511–1528, doi: 10.1044/1092-4388(2005/105).

Kirk, K. I., & Hill-Brown, C. (1985). Speech and language results in children with a cochlear implant. *Ear and Hearing*, 6(3, Suppl.), 36S–47S.

Kirk, K. I., Osberger, M. J., Robbins, A. M., Riley, A. I., Todd, S. L., & Miyamoto, R. T. (1995). Performance of children with cochlear implants, tactile aids, and hearing aids. *Seminars in Hearing*, 16(4), 370–380.

Kirk, K. I., Miyamoto, R. T., Lento, C. L., Ying, E. A., O'Neill, T., & Fears, B. (2002). Effects of age at implantation in young children. *Annals of Otology, Rhinology & Laryngology*, 111(5), 69–73.

Kirsch, L. S., Jungeblut, A., Jenkins, L., & Kolstad, A. (2002). *Adult literacy in America: a first look at the findings of the national adult literacy survey*. (NCES 1993–275). Retrieved from http://www.nces.ed.gov/pubs93/93275.pdf.

Kretschmer, R. R., & Kretschmer, L. W. (1978). *Language development and intervention with the hearing impaired*. Baltimore, MD: University Park Press.

Krose, J., Lotz, W., Puffer, C., & Osberger, M. J. (1986). Language and learning skills of hearing impaired children. *ASHA Monographs*, 23, 66–77.

Lach, R., Ling, D., Ling, A. H., & Ship, N. (1970). Early speech development in deaf infants. *American Annals of the Deaf*, 115(5), 522–526.

Lenneberg, E. H. (1967). *Biological foundations of language*. New York, NY: John Wiley & Sons.

Litovsky, R. Y., Johnstone, P. M., & Godar, S. P. (2006). Benefits of bilateral cochlear implants and/or hearing aids in children. *International Journal of Audiology*, 45(Suppl. 1), S78–91, doi: 10.1080/14992020600782956.

Maccoby, E. E., & Jacklin, C. N. (1974). *The psychology of sex differences*. Stanford, CA: Stanford University Press.

Mandel, D. R., Jusczyk, P. W., & Pisoni, D. B. (1995). Infants' recognition of the sound patterns of their own names. *Psychological Science*, 6(5), 314–317, doi: 10.1111/j.1467-9280.1995.tb00517.x.

Markides, A. (1970). The speech of deaf and partially-hearing children with special reference to factors affecting intelligibility. *British Journal of Disorders of Communication*, 5(2), 126–140.

Mavilya, M. (1972). Spontaneous vocalization and babbling in hearing-impaired infants. In G. Fant (Ed.), *International symposium on speech communication ability and profound deafness* (pp. 163–171). Washington, D. C.: Alexander Graham Bell Association for the Deaf.

Miyamoto, R. T., Kirk, K. I., Robbins, A. M., Todd, S., & Riley, A. (1996). Speech perception and speech production skills of children with multichannel cochlear implants. *Acta Oto-Laryngologica*, 116, 240–243.

Miyamoto, R. T., Kirk, K. I., Svirsky, M. A., & Sehgal, S. T. (1999). Communication skills in pediatric cochlear implant recipients. *Acta Oto-Laryngologica, 119*, 219–224.
Miyamoto, R. T., Houston, D. M., & Bergeson, T. (2005). Cochlear implantation in deaf infants. *Laryngoscope*, 115(8), 1376–1380.
Miyamoto, R. T., Hay-McCutcheon, M. J., Kirk, K. I., Houston, D. M., & Bergeson-Dana, T. (2008). Language skills of profoundly deaf children who received cochlear implants under 12 months of age: a preliminary study. *Acta Oto-Laryngologica*, 128(4), 373–377, doi: 10.1080/00016480701785012.
Moeller, M. P. (2000). Early intervention and language development in children who are deaf and hard of hearing. *Pediatrics*, 106(3), E43.
Mok, M., Galvin, K. L., Dowell, R. C., & McKay, C. M. (2007). Spatial unmasking and binaural advantage for children with normal hearing, a cochlear implant and a hearing aid, and bilateral implants. *Audiology & Neurotology*, 12(5), 295–306, doi: 10.1159/000103210.
National Early Literacy Panel. (2008). *Developing early literacy: report of the National Early Literacy Panel. A scientific synthesis of early literacy development and implications for intervention*. Washington, D.C.: National Institute for Literacy.
Nicholas, J. G., & Geers, A. E. (2006). Effects of early auditory experience on the spoken language of deaf children at 3 years of age. *Ear and Hearing*, 27(3), 286–298.
Nicholas, J. G., & Geers, A. E. (2007). Will they catch up? The role of age at cochlear implantation in the spoken language development of children with severe to profound hearing loss. *Journal of Speech, Language, and Hearing Research*, 50(4), 1048–1062.
Niparko, J. K., Tobey, E. A., Thal, D. J., Eisenberg, L. S., Wang, N. Y., Quittner, A. L., Fink, N. E., CDaCI Investigative Team. (2010). Spoken language development in children following cochlear implantation. *Journal of the American Medical Association*, 303(15), 1498–1506, doi: 10.1001/jama.2010.451.
Nittrouer, S., & Chapman, C. (2009). The effects of bilateral electric and bimodal electric-acoustic stimulation on language development. *Trends in amplification*, 13(190), 190–205.
Office of Educational Research and Improvement. (1989). *Library services. LSCA programs: An action report II*. Retrieved from http://www.eric.ed.gov/ERICWebPortal/contentdelivery/servlet/ERICServlet?accno=ED311909.
Oller, D. K. (2000). *The emergence of the speech capacity*. Mahwah, NJ: Lawrence Erlbaum Associates.
Osberger, M. J., & McGarr, N. S. (1982). Speech production characteristics of the hearing impaired. In N. Lass (Ed.), *Speech and language: Advances in basic research and practice* (Vol. 8). New York, NY: Academic Press, 221–283.
Owens, R. E. (1996). *Language development: An introduction* (4th ed.). Needham Heights, MA: Allyn and Bacon.
Peng, S. C., Spencer, L. J., & Tomblin, J. B. (2004a). Speech intelligibility of pediatric cochlear implant recipients with 7 years of device experience. *Journal of Speech, Language, and Hearing Research*, 47(6), 1227–1236.
Peng, S. C., Tomblin, J. B., Cheung, H., Lin, Y.-S., & Wang, L.-S. (2004b). Perception and production of Mandarin tones in prelingually deaf children with cochlear implants. *Ear and Hearing*, 25, 251–264, doi: 10.1097/01.AUD.0000130797.73809.40.
Peng, S. C., Tomblin, J. B., & Turner, C. W. (2008). Production and perception of speech intonation in pediatric cochlear implant recipients and individuals with normal hearing. *Ear and Hearing*, 29(3), 336–351, doi: 10.1097/AUD.0b013e318168d94d.
Pisoni, D. B., & Geers, A. E. (2000). Working memory in deaf children with cochlear implants: correlations between digit span and measures of spoken language processing. *Annals of Otology, Rhinology & Laryngology Supplement*, 185, 92–93.
Pressnell, L. M. (1973). Hearing-impaired children's comprehension and production of syntax in oral language. *Journal of Speech and Hearing Research*, 16(1), 12–21.
Quigley, S. P., Montanelli, D. S., & Wilbur, R. B. (1976a). Some aspects of the verb system in the language of deaf students. *Journal of Speech and Hearing Research*, 19(3), 536–550.

Quigley, S. P., Wilbur, R. B., & Montanelli, D. S. (1976b). Complement structures in the language of deaf students. *Journal of Speech and Hearing Research*, 19(3), 448–457.

Raviv, T., Kessenich, M., & Morrison, F. J. (2004). A mediational model of the association between socioeconomic status and three-year-old language abilities: the role of parenting factors. *Early Childhood Research Quarterly*, 19, 528–547.

Robbins, A. M., Svirsky, M. A., & Kirk, K. I. (1997). Children with implants can speak, but can they communicate? *Otolaryngology-Head and Neck Surgery*, 117, 155–160.

Schafer, D., & Lynch, J. (1980). Emergent language of six prelingually deaf children. *Teachers of the Deaf*, 5, 94–111.

Schorr, E. A., Roth, F. P., & Fox, N. A. (2008). A comparison of the speech and language skills of children with cochlear implants and children with normal hearing. *Communication Disorders Quarterly*, 29(4), 195–210.

Sharma, A., Dorman, M., Spahr, A., & Todd, N. W. (2002a). Early cochlear implantation in children allows normal development of central auditory pathways. *Annals of Otology, Rhinology and Laryngology Supplement*, 189, 38–41.

Sharma, A., Dorman, M. F., & Spahr, A. J. (2002b). A sensitive period for the development of the central auditory system in children with cochlear implants: implications for age of implantation. *Ear and Hearing*, 23(6), 532–539, doi: 10.1097/01.AUD.0000042223.62381.01.

Sharma, A., Nash, A. A., & Dorman, M. (2009). Cortical development, plasticity and re-organization in children with cochlear implants. *Journal of Communication Disorders*, 42(4), 272–279, doi: 10.1016/j.jcomdis.2009.03.003.

Smith, C. R. (1975). Residual hearing and speech production in deaf children. *Journal of Speech and Hearing Research*, 18(4), 795–811.

Somers, M. N. (1991). Speech perception abilities in children with cochlear implants. *American Journal of Otology - Supplement*, 12, 174–178.

Spencer, L. J., Tomblin, J. B., & Gantz, B. J. (1997). Reading skills in children with multichannel cochlear-implant experience. *The Volta Review*, 99(4), 193–202.

Spencer, L. J., Barker, B. A., & Tomblin, J. B. (2003). Exploring the language and literacy outcomes of pediatric cochlear implant users. *Ear and Hearing*, 24(3), 236–247, doi: 10.1097/01.AUD.0000069231.72244.94.

Stallings, L. M., Kirk, K. I., Chin, S. B., & Gao, S. (2000). Parent word familiarity and the language development of pediatric cochlear implant users. *The Volta Review*, 102(4), 237–258.

Svirsky, M. A., Robbins, A. M., Kirk, K. I., Pisoni, D. B., & Miyamoto, R. T. (2000). Language development in profoundly deaf children with cochlear implants. *Psychological Science*, 11(2), 153–158.

Svirsky, M. A., Teoh, S. W., & Neuburger, H. (2004). Development of language and speech perception in congenitally, profoundly deaf children as a function of age at cochlear implantation. *Audiology & Neurotology*, 9, 224–233.

Sykes, J. L. (1940). A study of the spontaneous vocalizations of young deaf children. *Psychological Monograph*, 52, 104–123.

Tobey, E. A., Rekart, D., Buckley, K., & Geers, A. E. (2004). Mode of communication and classroom placement impact on speech intelligibility. *Archives of Otolaryngology-Head & Neck Surgery*, 130(5), 639–643.

Tomblin, J. B., Barker, B. A., Spencer, L. J., Zhang, X., & Gantz, B. J. (2005). The effect of age at cochlear implant initial stimulation on expressive language growth in infants and toddlers. *Journal of Speech, Language, and Hearing Research*, 48(4), 853–867, doi: 10.1044/1092–4388(2005/059).

Tomblin, J. B., Barker, B. A., & Hubbs, S. (2007). Developmental constraints on language development in children with cochlear implants. *International Journal of Audiology*, 46(9), 512–523, doi: 10.1080/14992020701383043.

Tomblin, J. B., Peng, S. C., Spencer, L. J., & Lu, N. (2008). Long-term trajectories of the development of speech sound production in pediatric cochlear implant recipients. *Journal of Speech, Language, and Hearing Research*, 51(5), 1353–1368, doi: 10.1044/1092-4388(2008/07-0083).

Truy, E., Lina-Granade, G., Jonas, A. M., Martinon, G., Maison, S., Girard, J., Porot, M., & Morgan, A. (1998). Comprehension of language in congenitally deaf children with and without cochlear implants. *International Journal of Pediatric Otorhinolaryngology*, 45(1), 83–89.
Tye-Murray, N., Spencer, L. J., & Woodworth, G. G. (1995). Acquisition of speech by children who have prolonged cochlear implant experience. *Journal of Speech and Hearing Research*, 38(2), 327–337.

Chapter 13
Music Perception

Hugh McDermott

1 Introduction

How well do cochlear implant (CI) users perceive music? The short answer is: not very well, at least for the average CI recipient. This chapter reviews much of the published evidence about the perception of musical sounds by CI users, discusses both the design of sound processors and the psychophysical findings that may explain the generally poor perception of music experienced with today's CI systems, and finally presents some suggestions about how music perception may be improved in future.

First, there is the problem of defining precisely what is meant by the word "music." Although dictionary definitions tend to be broad, and individuals may frequently disagree about what sounds musical, there is a widely accepted set of elementary properties that are supposed to characterize music. These include melody, harmony, rhythm, and timbre. In the following it will be assumed that, in general, it is important for information about each of these properties to be presented in auditory form to enable listeners to perceive music satisfactorily (Riess Jones et al. 2010). It is acknowledged, however, that passages of organized sound can easily be identified that are commonly regarded as music but which lack one or more of these fundamental properties. As discussed later, limitations in the perception of some of these properties, especially pitch, underlie the problems experienced by many CI users when listening to music.

H. McDermott (✉)
Bionics Institute, Melbourne, VIC, Australia

Department of Otolaryngology, The University of Melbourne,
384-388 Albert Street, Melbourne, VIC, 3002, Australia
e-mail: hmcdermott@bionicsinstitute.org

F.-G. Zeng et al. (eds.), *Auditory Prostheses: New Horizons*, Springer Handbook of Auditory Research 39, DOI 10.1007/978-1-4419-9434-9_13,

2 Acoustic Elements of Music

The presence of melody is probably one of the most salient subjective features of music. It can be defined in simple terms as a succession of single tones that may differ in pitch. Usually the pitch of each tone in a melody must be restricted to one of a set of predetermined musical notes; for instance, the common chromatic scale is based on the subdivision of each octave into 12 pitch intervals. The ability of a listener to identify a melody in the absence of coincident cues such as a distinctive rhythmic pattern or recognizable lyrics is largely dependent on their ability to perceive the pitch of the constituent notes. Pitch perception and its physical correlates, particularly for CI recipients, are discussed briefly later. Closely related to melody is harmony, which may be thought of as a construct of several different musical notes sounded simultaneously. For example, in many popular songs the melody is accompanied by a succession of chords containing notes that are related to each other in a pleasant-sounding compound. The accurate perception of the notes forming the chords depends on the listener's ability to perceive independently the pitch of each note heard at the same time.

In many types of music, rhythm is contained in a pattern of loudness variations that occur over a relatively short time scale, typically from tens of milliseconds to several seconds. Although other acoustic variations may also be used to create rhythmic patterns, changes in loudness underlie the perception of many common forms of rhythm, including the accentuation of beats within an extended sequence. Sometimes the rhythmic pattern of common tunes is distinctive enough to enable their identification without any melodic pitch information being available acoustically; for instance, "Happy Birthday," which is one of the most familiar common tunes, can be recognized by many listeners when only the rhythm is presented by tapping.

A final elementary property of musical sound is timbre. Sometimes referred to as "tone color," timbre can be defined as that characteristic of sounds which enables them to be distinguished when they have the same pitch and loudness. For example, if a listener can discriminate between identical notes when they are played at the same loudness on two different musical instruments, the distinctiveness can be attributed to a difference in timbre. There are several physical parameters of sounds that may contribute to the perception of timbre, including the change in acoustic intensity over time, the spectral content, and temporal changes in the spectrum. The techniques that convert these acoustic parameters into appropriate patterns of electrical stimulation in CI systems are outlined next.

3 Sound Processing in CI Systems

Cochlear implant systems that are presently available commercially have two main components. One is implanted and includes the electrode array, while the other is a sound processor, which is usually worn on the external ear. These components are coupled by an inductive link which enables power to be transferred from the sound processor into the internal device. The same link also conveys data between the sound processor and the implant so that the required patterns of electric stimulation can be generated.

When sounds are picked up by the microphone, they are delivered to the sound processor for analysis as electric signals. Usually these signals are converted into digital form, and the analysis is based on a periodic estimation of the sound spectrum. The spectral analysis may be carried out using a Fourier transform or a bank of band-pass filters. With either technique the main purpose is to estimate the level in each frequency band every few milliseconds. The bands have partially overlapping frequency responses and the number of bands is chosen to equal the number of active electrode sites in the cochlea. For example, in the Nucleus Freedom devices manufactured by Cochlear Ltd., there are 22 intracochlear electrodes. The sound processor analyses the microphone signal to estimate the levels in each of 22 corresponding frequency bands which are assigned in tonotopic order to the electrodes. The level in each band is converted into an appropriate level of electric current to be delivered to the appropriate electrode. The current is generated in the form of brief, temporally non-overlapping pulses. The pulse rate is generally constant in a given CI user, although the rates used vary widely among CI recipients. The generic sound-processing scheme just described was first employed in a multiple-channel CI system about 30 years ago (MacLeod et al. 1985), and more-recent implementations are referred to as Continuous Interleaved Sampling (CIS) schemes (Wilson et al. 1991; Loizou 1998). An alternative scheme, which is widely used in sound processors made by Cochlear Ltd., differs in that only a subset of the available electrodes is selected for activation at the start of each stimulation cycle. The selected electrodes are those corresponding to frequency bands containing the highest short-term amplitudes; typically 8 to 10 of the 22 electrodes in the device are activated by this process (McKay et al. 1991; McDermott et al. 1992). Comparative studies have suggested that the latter type of scheme (known as Speak, ACE, or "n-of-m") may provide slightly better performance on average to CI users than CIS, at least in terms of speech understanding and subjective preference (Kim et al. 2000; Skinner et al. 2002; Patrick et al. 2006). As discussed later, there are several other types of sound processing that are sometimes used in modern multi-channel CI systems, including experimental schemes that were developed specifically in an attempt to improve music perception, but none are presently in routine use by a large proportion of CI recipients.

When the type of sound processing outlined above is applied to musical sounds, the result is a controlled pattern of current pulses distributed across the intracochlear electrodes. The main parameters of those pulses that are controlled in real time are the current level (or electric charge) and the active electrode. For example, if an acoustic pure tone is picked up by the sound processor, the electrode assigned to the frequency band that is closest to the tone's frequency will be activated and pulses will be delivered to that electrode at a level related to the level of the tone. Because the frequency bands in the sound processor generally overlap partially, some of the electrodes that are spatially close to the first-selected electrode may also be activated, but at lower levels. If the level of the input tone is steady, the levels of the current pulses will also be constant in time.

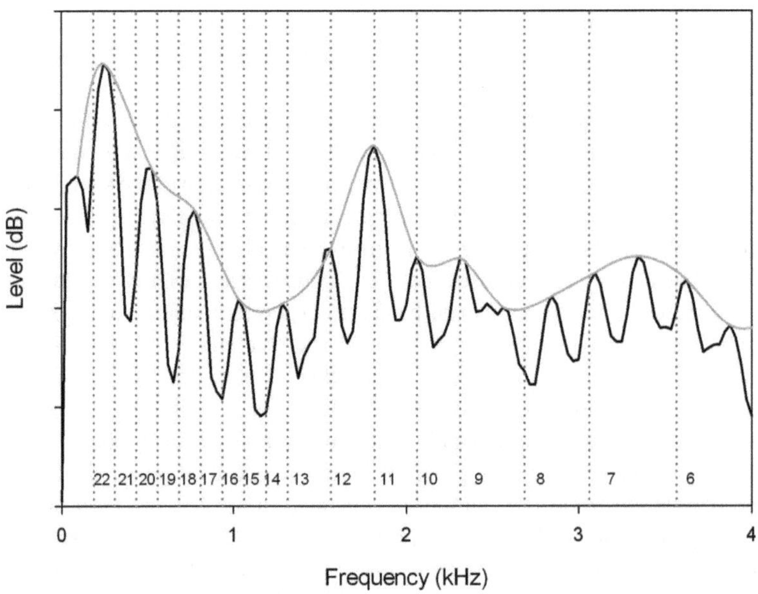

Fig. 13.1 Spectrum of the vowel /i/ sung by a male vocalist at a fundamental frequency of 262 Hz (*black*) and the envelope of that spectrum (*gray*). *Tick marks* on the vertical axis are at 20-dB intervals. The *vertical dotted lines* indicate the approximate frequencies of the boundaries between frequency bands in a widely used sound processor for the Nucleus 22-electrode CI system. The numbers above the horizontal axis show the usual assignment of the frequency bands to the intracochlear electrodes

An example of a more complex sound that has properties of both speech and music is a sung vowel. For instance, Fig. 13.1 shows the spectrum of the vowel /i/ sung by a trained male vocalist at a pitch corresponding to the note C4, which has a fundamental frequency (F0) of 262 Hz. The spectrum is characterized by a number of relatively narrow peaks that repeat at intervals equal to the F0 (i.e., at 262, 524, 786 Hz, and so on). Also shown in Fig. 13.1 is the spectral envelope of this sound (gray curve). The envelope gives an indication of the relative amount of electrical stimulation that would be expected on the CI electrodes when this sound is processed. The vertical dotted lines show the approximate boundaries between adjacent frequency bands, and the numbers near the abscissa between those lines indicate the electrode assignment. As this example is based on the Nucleus 22-electrode CI system, the most apical electrode (i.e., E22) is assigned to the band with the lowest center frequency, which is typically 250 Hz. The bands up to E14 have a constant bandwidth, but at higher frequencies the bands become progressively wider. (Bands assigned to the 5 most basal electrodes cover frequencies above 4 kHz and are not shown in this figure.) The gray curve indicating the envelope of the spectrum shows that most electrical activity would occur on E22, corresponding to the F0 of the sound. The second-highest levels would occur on E12 and E11, in response to a relatively broad peak in the spectrum that is characteristic of a formant (i.e., a resonance in the

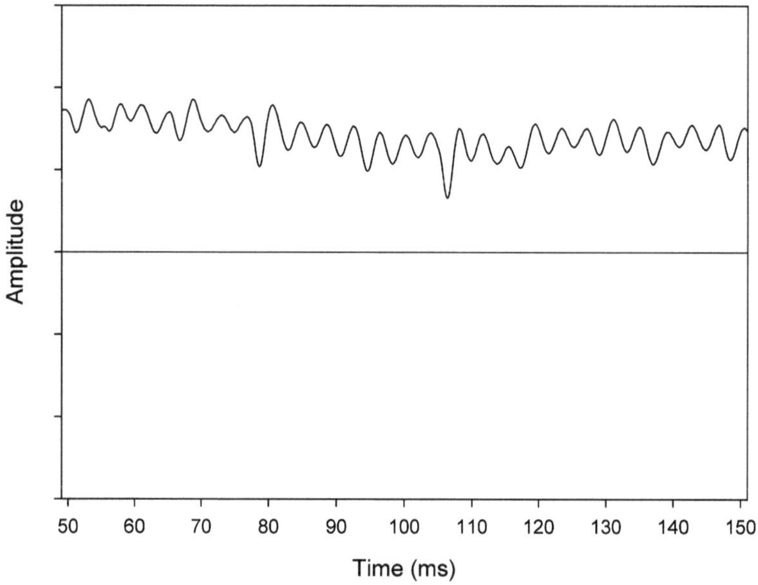

Fig. 13.2 Envelope level at the output of a bandpass filter with a center frequency of 2 kHz when processing the sung vowel whose spectrum is shown in Fig. 13.1. The fluctuation of the signal amplitude (*vertical axis*) over time (*horizontal axis*) has a dominant period of approximately 3.82 ms, which corresponds to the fundamental frequency (262 Hz). The *horizontal line* indicates zero amplitude

vocal tract). This formant appears to have a center frequency close to the boundary between the filters assigned to these two electrodes; i.e., at approximately 1800 Hz.

For a listener with normal hearing (NH), this vowel when sung in isolation would be perceived to have features of loudness, pitch, and timbre. The loudness would be related mainly to the acoustic level, corresponding generally to the current amplitude (or charge per pulse) of electric stimulation for CI users. The pitch could be perceived from either spectral or temporal properties of the signal (or some combination of those properties). As shown in Fig. 13.1, the pitch is directly represented in the F0, which corresponds to the highest peak in the spectrum. The pitch is also represented by the spacing between the spectral harmonics, and consequently could be perceived if at least some of the harmonics are resolved spatially by the tonotopic filtering characteristic of the acoustically stimulated cochlea. In addition, the signal has periodicity in the time domain; for example, the F0 would correspond to a repetition in the temporal waveform having a period of about 3.82 ms. Interestingly, when several adjacent harmonics interact in the same filter of a CI sound processor, a related periodicity becomes apparent. This is illustrated in Fig. 13.2, which shows the envelope output of a bandpass filter having specifications, including bandwidth, similar to those of filters in a typical CI processor, and a center frequency of 2 kHz. The input signal was the same sung vowel. Although the average level of that signal was constant over time, the level in this filter fluctuates with a period of 3.82 ms

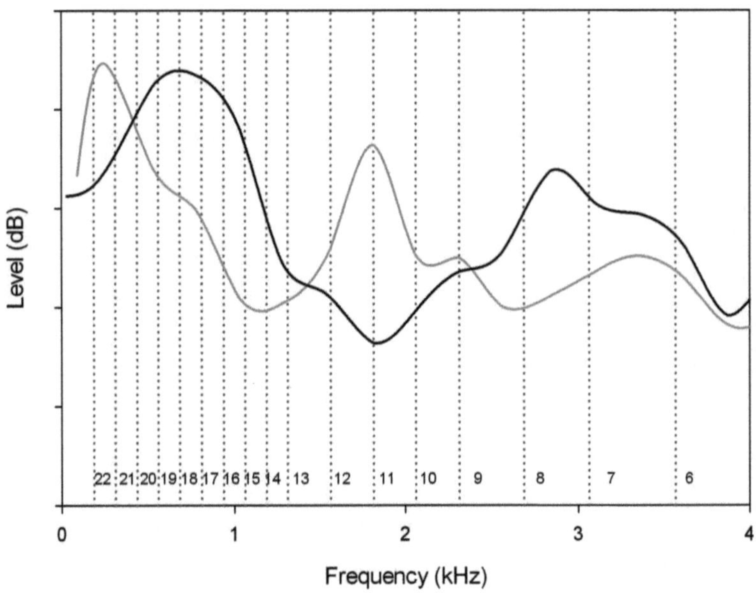

Fig. 13.3 Spectral envelopes of the vowels /i/ (*gray*) and /a/ (*black*) sung by a male vocalist at a fundamental frequency of 262 Hz. Other details as in Fig. 13.1

(i.e., the inverse of the F0). Such amplitude fluctuations are often perceivable by CI users as they are converted to modulations in level of the pulse trains delivered to the active electrodes. Therefore, some temporal information about pitch may be available to CI listeners even though the underlying stimulation pulse rate is constant. Further details about perception of temporal pitch by CI users are discussed later.

Whether CI recipients can extract accurate information about musical pitch from the spatial pattern of electric stimulation is less clear. Many psychophysical studies have reported that changing the position of a single active electrode in a multiple-channel implant generally results in a change in the "pitch" perceived, and that this relationship corresponds to the tonotopic organization of the normal cochlea (Tong et al. 1982; Townshend et al. 1987; McDermott and McKay 1994; Nelson et al. 1995; Zwolan et al. 1997; McKay 2004). However, differences in the quality of a sound can be described loosely as differences in pitch even when such differences could not be used to represent a musical melody. For example, the sounds of two different musical instruments playing the same steady note might be described as being dissimilar in "sharpness" or timbre. This perceptual dissimilarity presumably corresponds principally to a difference in the distribution of acoustic energy across frequencies and therefore can be related to the spectral envelope mentioned earlier. This type of difference is illustrated in Fig. 13.3, which shows the spectral envelopes of the vowels /i/ (gray) and /a/ (black) sung by a male vocalist. As the vowels were sung at the same pitch, the F0 in both cases was 262 Hz. The plots show that the distribution of electrical stimulation along the electrode array would be very different

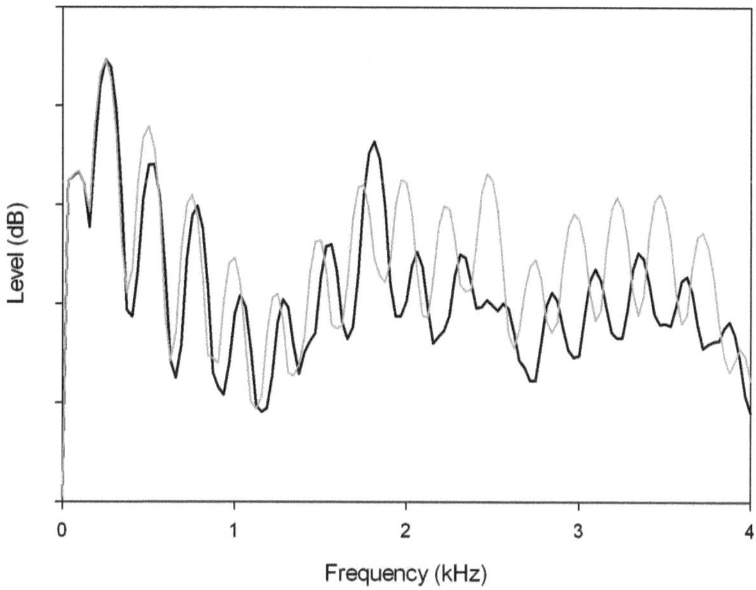

Fig. 13.4 Spectra of the vowel /i/ sung by a male vocalist at fundamental frequencies of 262 Hz (*black*, copied from Fig. 13.1) and 247 Hz, which is one semitone lower (*gray*). *Tick marks* on the vertical axis are at 20-dB intervals

between these vowels and suggest that CI listeners might report that one vowel sounds higher than the other, even though their pitch, according to conventional musical definitions, is identical.

This illustration may be compared with that of Fig. 13.4, which shows two spectra that were obtained when the same vowel /i/ was sung at two F0s that differed by just one semitone. The spectral envelope is similar for the two sounds up to about 2 kHz, and the spectral peaks are so close within that frequency range that little difference would be expected in the resulting patterns of electrical stimulation across the electrode array. This is because most of the peak frequencies below 2 kHz fall within the same frequency bands in the CI sound processor for either signal, as can be inferred by comparing Fig. 13.4 with Fig. 13.1. At higher frequencies, there is more acoustic energy overall in the sound with the lower F0 than the higher F0, and therefore it is possible that CI listeners could report incorrectly that it had the higher pitch, if their perception was dominated by the spatial (rather than temporal) pattern of stimulation. These considerations suggest that, although there appear to be several parameters of electrical stimulation that could, in theory, convey information about musical pitch to CI users, it is also possible that some of those parameters could provide inconsistent or even conflicting perceptual cues. In particular, it seems likely that spectral-envelope information, which is related mainly to perception of timbre by listeners with normal hearing, might interfere perceptually with pitch information for CI users, at least when they are listening to some types of complex musical sound. It is well known that no existing CI sound-processing scheme can reproduce the spatio-temporal patterns of neural excitation that occur when such sounds are heard by people with normal hearing. This is one reason that cues arising from temporal and spatial patterns of electric stimulation may be confused. For example, temporal pitch cues are reported to be absent or unreliable when the frequencies of amplitude modulations in electric stimuli are higher than about 300 Hz, whereas changes in the place of stimulation, which may be associated with components of sounds at much higher frequencies, are often reported by CI listeners as affecting the pitch. The published findings of selected studies investigating the music perception of CI users are reviewed briefly next.

4 Music Perception of CI Listeners

Numerous types of tests of music perception have been developed for listeners who have normal hearing or some kind of perceptual impairment. Many of these tests have been applied to CI users. Most frequently, published studies report on CI users' ability to recognize melodies and to identify musical instruments, and on their subjective appraisal of musical sounds (e.g., their naturalness or pleasantness). As discussed further below, these studies almost always find that CI listeners' perception of musical sounds, on average, is poorer than that of listeners with normal hearing. To explore in some depth the reasons underlying this observation, tests have been developed that investigate specifically CI users' perception of selected elementary features of musical sounds, particularly rhythm, pitch, and timbre.

The wide variety of perceptual tests that have been described in publications can be understood more clearly with the help of the classifications listed in Table 13.1. Taking as an example pitch perception, which has been a focus of many experiments conducted with CI recipients, the table shows six different categories of perception. In general, these categories can be regarded as forming a hierarchy, such that a listener would not be expected to demonstrate perceptual abilities for the categories lower in the table unless they have succeeded (or could succeed) in the tests related to the higher categories. At the top of the table is the category of detection, for which an appropriate test would ask only whether the listener could hear a given sound. Next is discrimination, in which two sounds are presented in succession, and the test determines whether the listener perceives them as the same or different. At the third level is ranking, in which the listener is asked which of two sounds presented in sequence has the higher pitch. The following level is a test of interval estimation, in which two sounds are presented in succession, and the listener is asked to name the musical interval between those sounds. It is important to note that the listener must have been able to provide correct responses in tests related to the first three categories in order to succeed on the test at the fourth level. The fifth category is a test of resolution, which differs from discrimination in that multiple sounds are presented simultaneously rather than sequentially, and the listener is asked about their perception of each of those sounds. In music, an example of this

Table 13.1 Categories of sound perception, including specific examples of how perception of pitch may be investigated experimentally

	Category	Presentation	Example question
1	Detection	Single sound	Can you hear that sound?
2	Discrimination	Two sequential sounds	Are those sounds different?
3	Ranking	Two sequential sounds	Which of those sounds has the higher pitch?
4	Interval estimation	Two sequential sounds	What musical interval do those sounds encompass?
5	Resolution	Two or more simultaneous sounds	What chord is being played?
6	Absolute identification	Single sound	Which musical note is that?

type of test involves playing the listener a chord and asking him or her to describe its note structure. Finally, the category of absolute pitch identification can be addressed using a test in which a single sound is presented, and the listener is asked to identify its note name in terms of the conventional musical scale. This hierarchy of perceptual categories may aid the interpretation of the empirical data which are discussed in the following sections.

4.1 Adult CI Users

4.1.1 Rhythm

The generally high levels of speech understanding in quiet conditions that can be attained by most adult users of modern CI systems (Helms et al. 2004) suggest that sufficient auditory information is available in the pattern of electrical stimulation to enable listeners to perceive basic musical rhythms accurately. This is because the fluctuations of acoustic intensity that occur in speech are broadly similar in both temporal and level dimensions to those that characterize meter and rhythm in many kinds of music. Most studies that have assessed the rhythm perception of adult CI users have confirmed this expectation (McDermott 2004). For example, in one recent study the subjects were asked to discriminate between pairs of short note sequences that were either identical or differed exclusively in the rhythmic pattern (Looi et al. 2008b). This procedure was derived from a standard set of tests known as the Primary Measures of Music Audiation (Gordon 1979) which has been used in several studies involving CI listeners. The subjects included hearing aid (HA) users as well as CI recipients. The scores, averaged across the subjects in each group, were 94% and 93% correct, respectively. As well as being very close to 100%, these scores were not significantly different from each other. In a related study, a different group of HA users whose hearing impairment was so severe that they had elected to receive a CI were tested before and after the activation of the implant (Looi et al. 2008a). Their scores on the above test were close to 95% correct both when listening with their HA before implantation and when listening via their CI afterwards.

Although these reports are representative of most published evaluations of CI users' rhythm perception, there is no doubt that there are individual CI users whose performance on these types of test is poorer than normal. For example, in one study a relatively difficult test requiring identification of complex rhythmic patterns presented at different tempos was used rather than discrimination between pairs of simple rhythmic sequences (Kong et al. 2004). Only 4 listeners with normal hearing and 3 CI recipients participated as subjects. The mean score for the former subject group was close to 100% correct, whereas 2 of the 3 CI listeners had scores that were 10 to 25 percentage points lower. With such a small number of subjects it is difficult to determine whether the poorer scores for those CI users were the result of the complexity of the test, the technical performance of their CI systems, the typical variability among individuals on perceptual tests in general, or other factors.

4.1.2 Pitch

In contrast to rhythm perception, pitch perception by CI listeners is widely acknowledged to be much poorer than normal. This perceptual deficiency has little effect on the semantic comprehension of speech in languages such as English but is presumably important for the understanding of speech in tone languages. As discussed in the following chapter (see Xu and Zhou, Chap. 14), changes in pitch can affect the lexical meaning of utterances in tone languages such as Mandarin and Cantonese. In any language, voice pitch may convey information about the age and sex of the speaker, and pitch variations can sometimes be utilized to apply emphasis or to distinguish questions from statements. Systematic changes of pitch are, of course, also fundamental to almost all forms of music, and the combination of speech with a conventional system of pitch largely defines what is probably the most common type of musical production: singing.

Melodies derived from well-known songs have often been used in tests of music perception for CI listeners. In such a test, a relatively short list of familiar tunes is given to the subject, and, after each of a number of those melodies is played, the subject is asked to name the tune. Outcomes with this type of test depend on several factors, including whether recognizable lyrics are sung with the tunes, how many different tunes are in the test set, whether the tunes have distinctive rhythmic patterns, and how well each subject can remember each tune. Thus, although it may be presumed that accurate pitch information must be perceived for a listener to obtain a high score on a melody identification test, in fact it may be that sufficient cues unrelated to pitch are available to permit good performance. For example, in one study 12 familiar melodies were presented to 6 CI users both with the original rhythmic cues included and with all rhythm information removed (Kong et al. 2004). In the condition with rhythm, the mean score for the CI listeners was 63% correct, which was significantly lower than the score for a group of NH subjects who completed the same test. Those subjects scored nearly 100% whether or not the rhythmic cues were included. The CI listeners' mean score was only about 12% in the condition with rhythmic cues removed, which was close to the score expected with

Table 13.2 Summary of selected results from published tests of melody recognition by cochlear implant users

Publication	Subjects	Number of melodies	Listening condition			Mean score
			Rhythm	Lyrics	Harmony	
Gfeller et al. 2002a	49 adults	12	Y	N	Y and N	19%
Leal et al. 2003	29 adults	7 or 8	Y	Y	Y	77%
			Y	N	Y	21%
			Y	N	N	45%
Kong et al. 2004	6 adults	12	Y	N	N	63%
			N	N	N	12%
Looi et al. 2004	15 adults	10	Y	N	N	51%
Kong et al. 2005	5 adults	12	N	N	N	35%
Olszewski et al. 2005	57 adults	up to 9	Y	N	N	75%
Olszewski et al. 2005	40 children	up to 9	Y	N	N	42%
Vongpaisal et al. 2006	10 children	3 or 5	Y	Y	Y	86%
			Y	N	Y	58%
			Y	N	N	14%
Galvin et al. 2007	11 adults	12	Y	N	N	60%
			N	N	N	28%
Mitani et al. 2007	17 children	3 or 5	Y	Y	Y	35% [a]
			Y	N	Y	Chance
			Y	N	N	Chance
Dorman et al. 2008	15 adults	5	N	N	N	52%
Hsiao 2008	20 children	6	N	N	N	39%
			Y	N	N	59%
			Y	Y	N	96%
Looi et al. 2008a	9 adults	10	Y	N	N	80%
Looi et al. 2008b	15 adults	10	Y	N	N	52%
El Fata et al. 2009	14 adults	15	Y	Y	Y	75%
			Y	N	Y	34%
Kang et al. 2009	42 adults	12	N	N	N	25%
Singh et al. 2009	11 adults	12	N	N	N	44%
Sucher and McDermott 2009	5 adults	7	Y	N	N	46%
Vongpaisal et al. 2009	17 children	4	Y	Y	Y	65%
			Y	N	Y	38%
			Y	N	N	37%

[a] Chance-corrected score

random responses. These findings are broadly similar to those of other researchers who have employed similar measures (McDermott 2004; Olszewski et al. 2005; Kang et al. 2009; Singh et al. 2009); see also Table 13.2.

The effect of the presence of recognizable lyrics on CI listeners' ability to identify tunes has also been investigated. In a study with 29 CI users, either 7 or 8 melodies were presented to each subject, depending on their familiarity with the selected tunes (Leal et al. 2003). When the melodies were presented with lyrics, 28 of the subjects could identify at least half of them, whereas only 1 subject could do so when the

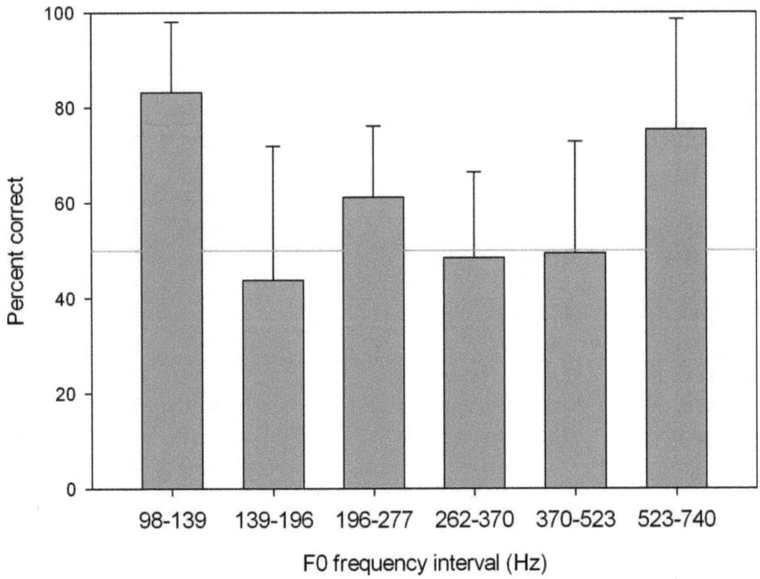

Fig. 13.5 Results of a pitch-ranking experiment with 8 CI users (Sucher and McDermott 2007). The test material comprised pairs of vowels sung at the fundamental frequencies shown along the horizontal axis. In each pair of sounds, the frequencies were separated by an interval of 0.5 octaves, and the subjects were asked to state which of the sounds had the higher pitch. The mean scores (*vertical axis*) are shown for each frequency pair (*horizontal axis*), with *error bars* indicating standard deviations. The *gray line* shows the score expected with uniformly random responses

tunes were presented in an orchestral version which omitted the lyrics. This strong effect on identification scores is not surprising when the generally good performance of CI systems in enabling their users to understand speech is considered.

Because the above observations and other published findings suggest that perception of musical pitch by CI users is generally so poor that even common tunes cannot be recognized reliably when only pitch cues are presented, further investigations have used much simpler patterns of stimuli to obtain more detailed information about pitch perception. In one such test, known as pitch-ranking, listeners hear two sounds in sequence that differ in pitch and are asked to indicate which one is higher. Other parameters of the test stimuli, such as timbre and loudness, are usually held constant within a pitch-ranking experiment. Such an experiment was carried out with 8 adult users of the Speak sound-processing scheme used in some recent CI systems made by Cochlear Ltd. (Sucher and McDermott 2007). The test stimuli were the vowels /a/ and /i/ sung at steady pitches by a male and a female vocalist. The fundamental frequencies of the 2 stimuli presented in each pair always differed by 0.5 octaves (6 semitones). The F0 of the lower-frequency stimulus in each pair was either 98, 139, or 196 Hz for the male singer, and either 262, 370, or 523 Hz for the female singer. Mean results across all subjects for both vowels combined are shown in Fig. 13.5, where the proportion of correct responses is plotted against the F0 of the stimuli in each pair. Note that in any ranking task with only 2 stimuli, the average score expected with completely random responses is 50% correct, as indicated by the horizontal line in the graph. It can be seen that, on average, the scores for all F0 pairs except those having the lowest and highest frequencies were not significantly different from the random-response score.

One plausible explanation for this pattern of experimental data is that the subjects used different perceptual features of the signals, depending on the F0, to provide responses in the tests. For the lowest F0 interval (98 to 139 Hz), pitch information would have been available in the amplitude modulations of the electric stimuli for both sounds in each test pair. As discussed earlier in relation to Fig. 13.2, such modulations are usually present on several of the active electrodes, and psychophysical studies have shown that they can convey a useful pitch sensation for modulation frequencies up to about 300 Hz (McKay et al. 1994, 1995; McKay 2004; Carlyon 2008). On the other hand, the relatively high mean score for the highest F0 interval (523 to 740 Hz) may be the consequence of subjects using cues arising from changes in the spatial pattern of the electric stimuli, as the modulation frequencies would have been too high for perception of temporal pitch cues. Although it is doubtful whether such place of stimulation cues could convey melodic pitch information accurately (McDermott and McKay 1997), they are capable of being ranked systematically in accordance with the tonotopic position of the active electrodes, as mentioned earlier. For the remaining F0 intervals, the subjects' low average scores could be the result of an absence of reliable pitch cues in either the temporal or the spatial domain, or of a perceptual conflict between temporal and spatial cues that were inconsistent. In any case, the observations in Fig. 13.5 suggest that CI users would generally find it difficult to identify or "follow" melodies when they have to rely mainly on pitch changes. In particular, their inability to rank many of the stimulus pairs even though the pitch intervals encompassed 6 semitones is consistent with the poor results reported for tests of melody recognition, as many of the pitch intervals within well-known tunes are relatively small. Several studies have been published with similar findings. For example, in one study 15 CI recipients were assessed in their ability both to rank sounds differing in pitch with several interval sizes and to identify melodies from a set of 10 familiar tunes (Looi et al. 2008b). On average, the score for ranking with an interval size of 0.25 octaves was at chance levels, and the melody identification score was only 52% correct, even though the melodies were presented with intact rhythm cues.

Recently published results from selected tests of melody identification are summarized in Table 13.2. These data show characteristics that are common to those of many other types of test investigating the perceptual abilities of CI recipients. In particular, the mean scores vary over a wide range across studies, and the range among individual listeners (not shown in the table) is even wider. Also, the scores are generally higher, on average, when the test material included a relatively small number of items, or when cues additional to pitch variations, such as lyrics and rhythm, were available. In general, the test conditions in which lyrics were sung in each melody gave the highest scores, but it is questionable whether such conditions are useful tests of music perception if the scores reflected mainly the listeners' ability to recognize some of the words rather than the melodic pitch contour or other musical features. The scores obtained from tests of child CI users appear similar to those of adults, but it is noteworthy that the number of different tunes used in the tests with children was smaller in almost every case than the number used in any melody test with adults. These and other findings related to the music perception of child CI users are discussed later.

Other aspects of pitch perception, such as the perception of harmony, seem to have received relatively little attention from researchers so far. This is presumably because the generally poor results of melody identification and pitch-ranking tests imply that experiments devised specifically to address harmony perception may be too difficult for the majority of CI users. In an experiment that required listeners to indicate how many simultaneously presented pitches were present in a tone complex (i.e., one, two, or three), the average scores of CI users were much lower than those of subjects with normal hearing (Donnelly et al. 2009). In fact, the CI listeners' scores were close to those expected from random responses for stimuli containing two or three simultaneous tones. These results suggest that CI listeners seem to perceive some multi-tone complexes as if they were fused into one sensation with a single pitch (or no identifiable pitch). Accurate perception of musical harmony requires listeners to resolve two or more notes that are sounded simultaneously, and to be able to perceive the pitch of each note independently. A similar perceptual ability would enable listeners to segregate pitch streams and thus be able to "follow" each of several melodies heard concurrently. There appears to be no published evidence that CI recipients are capable of completing these kinds of tests successfully. Moreover, there seem to be no reports of CI users having the ability to identify musical pitch without reference to a standard pitch. Whether there are any CI recipients who had so-called "perfect pitch" (or "absolute pitch") preoperatively while they had adequate acoustic hearing sensitivity, and also report having absolute pitch post-operatively while listening with their CI, is unknown.

4.1.3 Timbre

In most published assessments of CI users' musical timbre perception, subjects were asked to listen to a number of different sounds and to identify the source of each sound by using only auditory cues. In a typical, recent study (Looi et al. 2008b), the sounds were divided into three categories: single (solo) instruments, solo instruments with background accompaniment, and musical ensembles. Four different excerpts of 12 different recordings of music were presented in each category. Fifteen adult CI users listened to each musical excerpt and were asked to name the source of the sound, referring to pictures of the instruments and ensembles to avoid possible ambiguities. The mean scores obtained in each category were 61%, 45%, and 43% correct, respectively. Although these scores were all higher than the score expected from purely random responses (i.e., about 8%), they were much lower than the near-perfect scores obtained from a group of listeners with normal hearing who completed the same tests. For the CI subjects, the score difference between the solo instrument category and the other two categories was statistically significant, suggesting that the relatively simple acoustic properties of sounds from a single instrument are easier to recognize when presented via a CI than the acoustically more complex sounds produced by multiple instruments playing together. These findings are consistent with those reported in other publications (Gfeller et al. 2002b; McDermott 2004; McDermott and Looi 2004). As discussed later, the most likely reason for the poor ability of CI recipients to recognize musical sounds is that existing CI systems cannot enable perceptual resolution equivalent to that of normal hearing, particularly for the spectra of complex sounds.

4.1.4 Appraisal

How much do CI users enjoy listening to music, and how do they rate the quality of musical sounds? These questions have been addressed in a number of published studies, most of which have been based on questionnaires. A commonly reported finding is that most CI users who had usable acoustic hearing before receiving the device consider that listening to music via the CI is less enjoyable than their earlier music listening experiences (Gfeller et al. 2003; Mirza et al. 2003; McDermott 2004). In one study of 35 CI users, each respondent was asked to score their level of enjoyment of music from 0 ("not at all") to 10 ("very much") both before the onset of profound hearing loss and after receiving the implant (Mirza et al. 2003). The mean score across subjects for the former condition was 8.7, whereas in the latter condition it was 2.6. Furthermore, the mean score for a subgroup of 16 subjects who reported that they listened to music at least some of the time after implantation was 5.6; that score was also significantly lower than the corresponding score for the before-deafness condition.

Whereas such studies rely on the recollection of CI users about their listening experiences before the onset of deafness, a more recent publication reported on a comparison between the subjective quality ratings of HA users and CI recipients for musical sounds (Looi et al. 2007). The group of HA users had audiometric characteristics that met the criteria for implantation. Each subject was asked to provide a numerical rating on a scale of 1 to 10 to indicate the perceived pleasantness of sounds from various musical instruments and ensembles. Interestingly, there were no statistically significant differences between the mean scores for each group. Surprisingly, an additional group of 9 subjects who received an implant after a period of relying on HAs provided pleasantness ratings that were, on average, significantly higher for listening to musical sounds with the CI than previously with acoustic HAs. The apparent contrast between these results and those summarized in the preceding paragraph may be explained by the times at which the subjects' reported listening experiences occurred. For example, the relatively high score mentioned above for the before-deafness condition (i.e., 8.7) presumably reflected experiences that, in some instances, may have happened long before implantation, whereas the mean pleasantness rating provided by the HA users immediately prior to implantation was only 4.6 on the 1 to 10 scale (Looi et al. 2007). The mean rating of the latter subjects after implantation increased to 6.6, which is comparable with the post-implantation score (i.e., 5.6) from the earlier publication mentioned above (Mirza et al. 2003).

Factors that have been reported by researchers as beneficial to the subjective musical experiences of CI users include listening in an environment with favorable acoustic characteristics (e.g., with minimal background noise) and listening to music that contains easily recognized cues, such as lyrics or strong rhythmic patterns (Gfeller et al. 2000a). However, there are few reports of large differences in performance among alternative sound-processing schemes for CIs when their users listen to music. In one such study, a total of 87 CI recipients were divided into equal numbers of users of the CIS, Speak, and ACE schemes (Brockmeier et al. 2007). Although several different aspects of music listening were investigated via a questionnaire, the results showed only relatively small differences that could be attributed to the use of each type of sound-processing scheme. In general, some of the ratings provided by the respondents showed a trend favoring the CIS and ACE schemes over Speak, perhaps reflecting certain technical improvements in sound processing, because ACE is based on the same functional principles as Speak with an improved implementation. For instance, a significantly larger proportion of Speak users (16 of 29) rated music as sounding "like an unpleasant noise" than ACE users (7 of 29) or CIS users (5 of 29). Furthermore, significantly more ACE users (68%) reported listening to music "for enjoyment" than Speak (43%) and CIS (46%) users.

Another study based on responses of CI users to a questionnaire investigated whether bilateral fittings of CI devices enabled better perception of music than unilateral fittings (Veekmans et al. 2009). As discussed in Chap. 2 of this volume, research into the use of bilateral CI systems has shown that they can provide several perceptual advantages, particularly for sound localization and listening in noisy conditions. Responses to the questionnaire addressing music perception showed that, on average, bilateral CI users appear to rate music as more enjoyable to listen to than users of unilateral CI devices. This finding is consistent with the positive appraisal in general of binaural compared with monaural hearing that has been found in studies assessing the outcomes of bilateral implantation.

4.2 Child CI Users

Unlike adult CI recipients, most children who receive implants under present candidacy criteria had little or no usable acoustic hearing preoperatively. This might be expected to affect their perception and appraisal of music for several reasons. First, a typical child CI user would have minimal experience or recollection of hearing musical sounds acoustically and therefore may have few strong presumptions about how music should sound through the implant. Secondly, child CI recipients may be generally more adaptable to the unnatural stimuli perceived through the device than adult CI users. Thirdly, young children are more likely to participate in formal auditory training using their implants and consequently may learn to perceive details of the electric stimuli differently from adults whose primary educational experiences occurred during a period in which they had to rely on their acoustic hearing, even if it was seriously impaired. As discussed later, these factors may underlie some of the specific differences in music perception that have been reported between adult and child CI users. The perception of musical rhythm, however, is not discussed further, as the evidence suggests that most CI users, regardless of age, can receive rhythmic information satisfactorily via existing devices.

4.2.1 Pitch

As shown by the summary in Table 13.2, average results for child CI listeners from published assessments of melody identification tend to be comparable to or somewhat lower than the corresponding scores for adult CI users. In many tests, the number of different melodies used was relatively small (e.g., 3 to 6), and the frequent use of forced-choice procedures implies that the scores expected from entirely random responses might be as high as 33%. Not surprisingly, child CI users tended to obtain higher scores on melody recognition tests when coincident auditory cues, such as sung lyrics or distinctive rhythmic patterns, were available (Hsiao 2008). For example, in one recently published study, 17 CI recipients aged 5 to 12 years were asked to identify 4 songs with which they were familiar because of their use as theme tunes in popular television programs (Vongpaisal et al. 2009). The songs were presented in versions that were either reduced to the melody alone (played as a flute solo), or to the melody with instrumental accompaniment, or were played in their entire original form. All versions contained intact rhythmic cues. Across the subject group, the mean scores were significantly higher than chance in each condition. However, the scores were 35 to 45 percentage points lower than the corresponding scores for a group of children with normal hearing who completed the same test. As is usual in such tests, the scores of individual CI subjects varied widely. Only for the condition in which the tunes were presented in their original form did the majority of CI subjects obtain a score greater than 50% correct; by comparison, the normally hearing subjects obtained near-perfect scores for the same condition. In the other two conditions, which did not include sung lyrics, the child CI listeners' mean score was almost 30 percentage points lower than their mean score for the condition in which the original tunes were presented.

A study that involved production as well as perception of pitch also provides evidence that the pitch information conveyed to listeners by cochlear implants is impoverished (Nakata et al. 2006). Twelve children aged 5 to 10 years with CIs and a group of 6 normally hearing children of similar age were asked to sing songs with which they were familiar. Recordings of the songs were analyzed mainly to measure acoustic features related to pitch and rhythm. These measurements were compared with values expected from a corresponding analysis of each song when produced in its accepted standard form. The results showed that, in general, the temporal patterns of the songs produced by the children were close to those of the standard forms for both the CI users and the NH children. In contrast, the pitch patterns in the songs of the CI users were significantly different from those of the NH children. In particular, the pitch range produced by the CI users was much smaller than that of the NH children. Furthermore, the direction of pitch change between adjacent notes was essentially random for the CI users whereas it was almost always correct for the NH children. Very similar results were reported from a more recent study with different subjects (Xu et al. 2009). These findings are consistent with the observations made in many experimental studies with both children and adults that involved only perception of pitch. Although the process of producing pitch may be largely independent from that of perception, the fact that the NH children in the above studies were able to sing the songs with approximately correct pitch suggests that the relatively poor performance of the child CI users was mainly a consequence of their inability to perceive accurately the pitch of sung notes in melodies.

4.2.2 Timbre

There appear to be few published reports of child CI users' ability to perceive musical timbre, at least as assessed in tests of sound-source recognition, such as identification of musical instruments. In one study (Sucher 2007), 11 CI users aged 9 to 16 years were asked to identify each of a set of 12 musical sounds. The test material and procedures were broadly similar to those used in one of the studies with adult CI listeners mentioned above (Looi et al. 2008b). The same test was carried out with a group of 11 NH children of similar age, who obtained an average score of 97% correct. The mean score of the CI users was significantly lower than that score at 79% correct. An analysis of the errors in the subjects' responses found that the only confusion that occurred on more than approximately 5% of trials with the NH subjects was between the flute and the recorder. In contrast, the responses of the CI subjects were scattered widely among the 12 possible alternatives; errors occurred both between different instrumental "families" (e.g., flute vs. violin) and within families (e.g., trumpet vs. trombone). The child CI users' overall mean score of 79% is comparable with the score of 61% obtained in the above study with adult CI recipients (Looi et al. 2008b). Although a formal statistical comparison between these data would not be appropriate because the test materials were not identical, it is plausible that the higher score for the child CI listeners may have resulted in part from the exclusion of musical instruments that were considered to be less familiar or less distinctive for the younger subjects. For example, the test material for the adult subjects included the violoncello and clarinet, but those instruments were replaced with the tambourine and recorder for the tests with children. Taking into account these and other procedural differences, it seems likely that the ability of child CI recipients to recognize musical sounds is generally similar to that of adult CI users.

4.2.3 Appraisal

In contrast with the published reports of adult CI recipients' appraisal of musical sounds, the findings of related studies with child CI users seem to be more positive. In one study involving 15 children aged 8 to 14 years with cochlear implants and 32 NH children of similar age, the participants were asked to provide ratings of liking for each of 30 musical excerpts encompassing a wide range of styles (Stordahl 2002). On average, the ratings were very close for the two subject groups.

Furthermore, the participants' responses to a questionnaire that sought information about their musical interests and preferences showed generally similar patterns. For example, all of the NH children and 93% of the children with CIs reported that they participated in musical activities (such as dancing, listening to CDs or the radio, and going to concerts). However, a smaller proportion of CI users than NH respondents reported participation in a choir, whereas a slightly larger proportion reported active involvement in learning to play a musical instrument. These findings are consistent with the experimental data from other studies that were summarized above indicating that child CI users had much poorer than normal ability to perceive the pitch of musical sounds and to sing with accurate intonation. Other publications have also reported that the appraisal of music by children with CIs is relatively favorable (Gfeller et al. 1998; Vongpaisal et al. 2006; Mitani et al. 2007; Sucher 2007). Most probably this reflects the fact that the majority of such children were implanted at an early age. Because they generally lack the experience of ever having listened to music acoustically, these children with CIs do not have the expectations of musical quality that underlie the less favorable attitudes of many adult CI users who recall previously enjoying music with their natural acoustic hearing.

5 Improving Music Perception

The published research reviewed above suggests that today's commercially available CI systems can provide their users with adequate auditory information about rhythmic features of music but perform less satisfactorily in providing information about pitch and timbre. These findings seem to apply to both adult and child CI recipients. It is likely that the perceptual deficiencies of CIs arise from limitations in their technical function and/or the ability of the impaired auditory system to extract relevant information from the electric stimulation patterns. At present, three different potential solutions are showing promise for addressing these problems. First, new or modified sound-processing schemes are being developed that aim to present CI users with more information about musical sounds than existing processing schemes. Secondly, research with CI recipients who have usable acoustic hearing in the non-implanted ear (or both ears) indicates that perception of pitch and timbre can be improved by presenting sounds acoustically as well as via the CI. Thirdly, specific auditory training programs are being applied to help CI users extract as much information as possible when listening to musical sounds with their devices. These topics are discussed briefly next.

5.1 Alternative Sound-Processing Schemes

Researchers and manufacturers of CI systems are continually seeking to improve the performance of CI devices, especially for listening to music and other complex

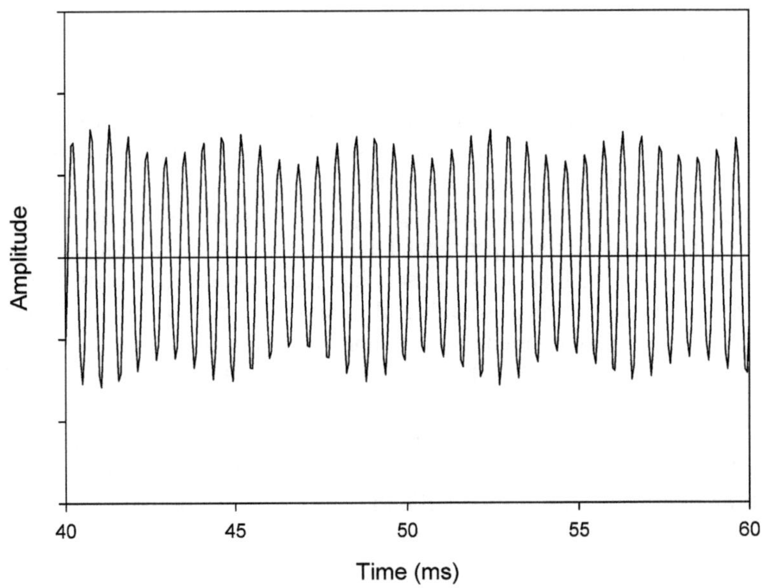

Fig. 13.6 Example of part of the waveform within a bandpass filter centered on 2 kHz when processing the vowel /i/ sung by a male vocalist at a fundamental frequency of 262 Hz. The *horizontal line* indicates zero amplitude. For comparison, the envelope output of the same filter for an extended portion of the same signal is shown in Fig. 13.2

sounds such as speech in noise. As it is well known that existing CIs do not enable their users to perceive the pitch of musical instruments or voices satisfactorily, many recent technical developments have focused on the need to increase or enhance information about pitch that is available in electric stimulation patterns. As outlined in the overview presented earlier in this chapter, psychophysical experiments have confirmed that pitch information can be obtained by CI listeners from temporal and spatial parameters of electric stimuli (McDermott 2004; McKay 2004; Moore and Carlyon 2005). Experimental sound-processing schemes have been developed that aim to improve the delivery of pitch information via each of these parameters.

Temporal information in complex acoustic signals can be divided into two distinct categories. As discussed earlier in relation to Fig. 13.2, the amplitude envelope of the signal at the output of a relatively narrow bandpass filter used in a CI system to analyze incoming sound typically fluctuates at a rate equal to the fundamental frequency of that sound. Such fluctuations are represented by corresponding modulations of the current level of the electric stimuli delivered by the electrode assigned to that filter. Provided that these amplitude modulations are restricted to a range of frequencies no higher than about 300 Hz, CI users are likely to perceive a pitch related to the modulation frequency. The other category of temporal information is sometimes referred to as "fine structure," and includes additional details of the waveform within each frequency band. This is illustrated in Fig. 13.6, which shows part of the waveform within a bandpass filter centered on 2 kHz when processing the vowel /i/ sung by a male vocalist at a fundamental frequency of 262 Hz. For comparison, Fig. 13.2 shows the envelope output of the same filter for an extended portion of the same signal. It can be seen in Fig. 13.6 that the waveform within the filter is dominated by almost periodic fluctuations at a frequency close to the filter's center frequency. The fact that most existing CI sound-processing schemes present information about the envelope and not the fine structure of signals within each frequency band has been hypothesized to underlie the poor pitch perception of CI users in general. Consequently, several innovative schemes have been developed in an attempt to deliver temporal fine-structure information in the electric stimulation patterns produced by the CI electrodes. Some of these schemes are discussed briefly later.

One example of a technique that enhances temporal envelope fluctuations is the "F0mod" scheme developed at the Universiteit Leuven in Belgium (Laneau et al. 2006). In that scheme, the fundamental frequency (F0) of incoming sound signals is estimated and used to modulate the envelope signals in each frequency band by means of a synthesized sinusoidal signal. As a result, the modulation of the envelopes at the estimated F0 is not only relatively large, having a depth of 100%, but also is synchronized across frequency bands. The latter aspect of the processing is based on the supposition discussed further below that maintaining in-phase amplitude modulations across electrodes may improve the ability of CI users to extract pitch information from temporal stimulation patterns (McDermott 2004). When the F0mod scheme was tested with 6 users of the Nucleus CI24 implant system, the results showed that, on average, the subjects obtained significantly better scores for pitch-ranking of complex sounds in comparison with the standard ACE scheme for the two lower F0 values used in the experiments (131 and 185 Hz). Scores did not differ significantly for the highest F0 used (370 Hz). Four of the subjects also completed a melody identification test in which 10 familiar tunes were presented without lyrics or rhythmic cues. The results showed a significant improvement in the proportion of melodies correctly recognized by each subject, especially for a lower range of F0 values (around 185 Hz). However, the mean score improvement was only approximately 10 percentage points (Laneau et al. 2006).

These findings are consistent with those of other studies in which envelope-enhancing schemes have been evaluated. For example, a comparison of various experimental and commercially available schemes used with the Nucleus CI device showed that small but significant benefits could be obtained in pitch-ranking tests when temporal envelope modulation was enhanced (Vandali et al. 2005). Furthermore, these improvements were obtained without any detrimental effect on scores obtained from tests of speech recognition.

Attempts to increase the availability of temporal fine-structure information have generally involved modifying the times at which pulses are delivered to each electrode. In most existing sound-processing schemes, the overall rate of pulsatile stimulation is constant and unrelated to any properties of incoming sounds. A number of alternative schemes deliver pulses to the active electrodes at times related to the waveforms in the frequency bands assigned to those electrodes (Swanson 2008). In one such scheme, "HiResolution," which was developed by Advanced Bionics

Corp., the signals in each bandpass filter are half-wave rectified and then sampled with a relatively high stimulation pulse rate. These techniques preserve some of the temporal fine structure that is typically restricted or discarded by the use of conventional envelope extraction in schemes such as CIS and ACE. However, a study that included 18 adult CI recipients who used HiResolution showed that it provided no benefits relative to other sound-processing schemes on a test of pitch ranking (Gfeller et al. 2007). A similar technique is implemented in the "Fine Structure Processing" (FSP) available in recent CI devices manufactured by MED-EL Corp. (Hochmair et al. 2006). When evaluated with a group of 14 adult CI recipients, FSP was found to give significantly poorer performance than CIS on a test of melody discrimination (Arnoldner et al. 2007). Although 12 of the 14 subjects reported a preference for the FSP scheme over CIS, it is unclear to what extent other effects may have influenced their judgment. For instance, FSP was introduced at the same time as several other changes in the sound-processing algorithms and devices, making it difficult to isolate the specific effects of the altered temporal patterns of pulsatile stimulation.

The often disappointing perceptual outcomes of schemes that augment or enhance fine temporal information in electric stimulation patterns may be explained by two main factors. First, there appears to be an unavoidable limitation on the highest rate at which temporal information can be extracted or utilized by the electrically stimulated auditory nervous system. This limitation is related to the observation that most CI users cannot perceive changes in the stimulation rate when the rate is above approximately 300 Hz (Pijl and Schwarz 1995). As mentioned previously, a similar limitation affects the perception of changes in the modulation frequency when a pulsatile carrier having a constant and relatively high rate is modulated periodically in amplitude. These phenomena imply that, even if improvements can be made to the delivery of temporal information, they may not translate into perceptual improvements in pitch perception when relatively high acoustic frequencies are processed by a CI system. This is a serious problem, because the range of fundamental frequencies relevant to musical pitch extends well beyond 300 Hz. For example, the F0 of middle C on the standard piano keyboard is approximately 262 Hz, and the F0 of some musical instruments may occasionally be above 4 kHz. Furthermore, the frequencies of harmonics that contribute to pitch perception for listeners with normal hearing can be even higher.

The second factor that may underlie the poor performance of temporal fine-structure processing is the spatial distribution of electric stimulation patterns produced by electrodes that are closely spaced in the cochlea. There is some evidence from psychophysical experiments that amplitude modulations in electric pulse trains can be perceived independently only if the active electrodes are relatively far apart (McKay and McDermott 1996). If the active electrodes are close together, the listener is more likely to perceive a combination of the modulations. This finding is consistent with the assumption that the populations of neurons stimulated by each electrode would be at least partially overlapping. Although there appear to be no published studies that have investigated whether or to what extent temporal patterns are combined perceptually when more than 2 intracochlear electrodes are active concurrently, it is plausible that different modulations on multiple electrodes may not be perceived independently particularly in conditions where the electrodes are closely spaced. This means, for example, that amplitude modulations having the same frequency but differing in relative phase across 2 or more active electrodes may not convey pitch information as reliably as a similar modulation pattern presented on a single electrode. Unfortunately, the former condition does sometimes occur with complex periodic sounds in which the F0 is represented by the frequency of amplitude modulations. Phase shifts across frequency in the amplitude modulations can be caused by certain acoustic characteristics of the environment in which the sounds were created or recorded. In such conditions the amplitude modulations are not aligned in time across all active electrodes, and consequently it might be difficult for CI users to perceive the pitch related to the F0 (McDermott 2004). A similar problem most likely affects those sound-processing schemes in which modulations with different frequencies or pulse trains with different rates are intentionally delivered to multiple electrodes concurrently. Although the aim of such schemes is to convey more-detailed temporal information that corresponds specifically to the frequency band assigned to each electrode, the perceptual interference arising from spatial overlaps among the stimulated neural populations may impede the delivery of that information to CI recipients.

An alternative and potentially complementary approach to solving the problem of poor pitch perception by CI users is to emphasize or augment spatial cues in electric stimulation patterns that relate to important frequency components of sounds. Because all sound-processing schemes for existing multiple-channel CI systems contain a bank of bandpass filters or a similar means for analyzing the spectrum of incoming sounds, the frequencies of sound components are represented coarsely by the intracochlear location of the active electrodes. Even when a sound processor analyzes a pure tone, there are often 2 or more electrodes activated, although usually the filter with center frequency closest to the tone frequency causes most activity to occur on the particular electrode assigned to that filter. However, because the filters' frequency responses generally overlap to some extent, some nearby filters and electrodes are also activated, but at lower levels. This effect is illustrated in Fig. 13.7 for the case of a pure tone at 3 different frequencies in the overlap region between 2 hypothetical bandpass filters. The figure shows that relatively small changes in the frequency of a tone can result in changes of the stimulation levels on the active electrodes, leading to a shift in the effective position of the spatial centroid of the stimulated neural population. This can be perceived by CI users as a change in pitch (or timbre) related to the frequency of the tone. As can be deduced by comparing Figs. 13.4 and 13.1, similar changes in the location of stimulation could in principle provide pitch cues for complex sounds as well as for pure tones, provided that at least some of the harmonics are resolved by the spectrum analyzer in the CI sound processor.

Several sound-processing schemes have been developed in attempts to take advantage of these effects. In "HiRes 120," which is an extension of the HiResolution scheme outlined previously, the frequency of a dominant sound component within each frequency band is estimated and then converted into a combination of 2 currents

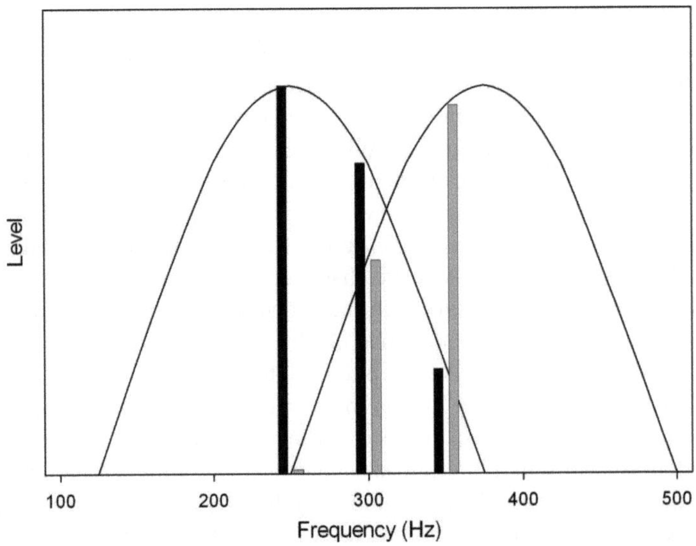

Fig. 13.7 Illustration of the effects on stimulation levels of a pure tone at different frequencies within the overlap region between 2 bandpass filters in a hypothetical CI sound processor. The *black curves* show the frequency responses of filters with center frequencies at 250 and 375 Hz. The *three pairs of columns* indicate the current levels on 2 adjacent electrodes for each of 3 tone frequencies (250, 300, and 350 Hz). The level on the electrode assigned to the 250-Hz filter is shown in *black* while the level on the electrode assigned to the 375-Hz filter is shown in *gray*

delivered simultaneously to 2 adjacent electrodes. As the component frequency increases, the 2 currents are adjusted so that the spatial centroid of stimulation moves in a basal direction. An adjustment in the opposite direction occurs when the component frequency decreases within the band. This process is applied to each of the 15 available adjacent pairs of electrodes in the 16-electrode device, and the combinations of currents are selected from among 8 predetermined ratios. Consequently, it is claimed by the device manufacturer (Advanced Bionics) that 120 separate spectral bands have been created using only 16 intracochlear electrodes by means of this technique, and that spectral resolution is thereby improved. However, the usual definition of resolution (see Table 13.1) refers to the ability of a listener to perceive distinctly each of several simultaneously presented sounds. The HiRes 120 scheme cannot improve resolution in this sense, at least for conditions when multiple components are present within a particular frequency band, because the processing estimates only one frequency per band and then delivers one predetermined combination of currents to the appropriate pair of electrodes. Thus the limitation on spectral resolution imposed by the primary analysis of sound signals in the CI processor is not affected by the use of simultaneous currents in different ratios to represent the frequency within each band. There appear to be no publications that report on the effectiveness specifically of HiRes 120 using objective tests of music perception, although 1 subject who used both HiResolution and HiRes 120 in a study that evaluated a set of music perception tests for clinical assessment obtained comparable results with the two schemes (Nimmons et al. 2008). In another publication, small increases in subjective ratings of the pleasantness of music were found for HiRes 120 compared with HiResolution, but no objective measures of music perception were reported (Firszt et al. 2009).

However, there is evidence that varying the ratios of currents delivered simultaneously by 2 adjacent or nearby electrodes can affect the tonotopic pitch perceived by CI users (Townshend et al. 1987; Donaldson et al. 2005). Although it may seem that the "current steering" technique implemented in HiRes 120 would effectively exploit this perceptual phenomenon, there is also evidence that stimulation by means of current pulses delivered sequentially to nearby electrodes can create pitch percepts intermediate to those associated with each electrode when activated separately (McDermott and McKay 1994; Kwon and van den Honert 2006). Most likely the spatial centroid of the excited neural population can be shifted by changing the relative currents on 2 active electrodes either when those currents are simultaneous or when they are delivered as rapidly interleaved pulse trains.

One implication of the last point is that all CI sound-processing schemes in which pulses are distributed cyclically among multiple active electrodes would inherently produce such intermediate pitches because the frequency responses of the bandpass filters assigned to the electrodes always overlap partially. However, the relative current levels would depend on both the frequency of the signal component within the overlap region between 2 adjacent filters and on the shape of the frequency response of those filters; for an example, see Fig. 13.7. Some researchers have designed filter responses specifically to enhance the resulting pitch cues, particularly in an attempt to improve perception of the F0 of speech signals (Geurts and Wouters 2004). The effects of using a bank of bandpass filters with an approximately triangular shape rather than the conventional frequency response (which is generally bell-shaped with a relatively flat central portion) were to reduce the size of the F0 difference that could be detected, on average, by a small group of CI listeners. Whether the use of a filter bank of this type could improve perception of musical sounds has not been reported.

The experimental findings reviewed above are somewhat disappointing in that no convincing technical solution has been demonstrated so far for the basic deficiencies of CIs when their users listen to music, and particularly their poor performance for pitch perception. Although it is possible that alternative sound-processing schemes will be developed in future that may help to solve these problems, currently the most promising way to improve music perception for CI recipients is to make use of any available acoustic hearing.

5.2 Combined Acoustic and Electric Stimulation

As discussed in Chap. 3 of this volume (Turner and Gantz), at present the number of CI recipients who have usable hearing in at least one ear postoperatively is increasing rapidly. The simultaneous use of acoustic and electric hearing by these

CI users generally provides perceptual benefits, including better understanding of speech in noise, in comparison with the use of the CI by itself (Ching et al. 2007; Schafer et al. 2007; Dorman et al. 2008; Firszt et al. 2008; Olson and Shinn 2008). Moreover, performance for listening to music is also generally improved when acoustic stimulation is used at the same time as a CI (Kong et al. 2005; El Fata et al. 2009; McDermott 2009; Sucher and McDermott 2009). Acoustic and electric stimulation may be combined by means of several possible configurations of hearing devices, depending on the acoustic hearing characteristics of the CI recipient, although the most common arrangement is for a CI user to have an acoustic hearing aid in the ear opposite the implanted ear. Regardless of the physical configuration, however, the availability of an acoustic signal seems to assist most CI listeners particularly with pitch perception. This is assumed to result from better processing of certain pitch-related features of sound by the impaired peripheral auditory system when it is stimulated acoustically rather than electrically.

As outlined briefly earlier, pitch information for complex sounds in normal hearing is available both in the periodicity of sound signals that is related to the F0 and in the distribution of harmonics, which are spaced at F0 intervals. These features are illustrated in Figs. 13.2 and 13.1, respectively. Many psychophysical experiments and modeling studies have investigated the underlying mechanisms by which pitch information is extracted by the acoustically stimulated ear (Moore and Carlyon 2005; Plack and Oxenham 2005). For complex sound signals with relatively low fundamental frequencies, pitch information is more likely to be extracted by the auditory system in conditions enabling resolution of at least the lower harmonics. Because spectral resolution is known to be a principal limiting factor in existing CI devices, it is of interest to examine the potential for harmonics in acoustic signals to be resolved when heard by a CI recipient with some acoustic hearing. This can be investigated using a computational model of impaired acoustic hearing (Moore and Glasberg 1996, 1997). Figure 13.8 shows the output of that model when a complex sound with a fundamental frequency of 262 Hz was presented hypothetically to an impaired ear. Three frequency components (i.e., the fundamental and the next two higher harmonics) at a level of 105 dB SPL were used as input to the model. The overall signal level was selected to be appropriate for the degree of simulated hearing loss. In the model, that hearing loss was specified using the average threshold levels of the CI recipients who participated in a published study on the benefits of combined acoustic and electric stimulation for music perception (Sucher and McDermott 2009). The excitation pattern calculated by the model and plotted in Fig. 13.8 shows distinct peaks corresponding to each of the frequency components, indicating that the lower harmonics of this signal would be at least partly resolved for these listeners. As it can be assumed that the same components would not be resolved perceptually when processed by these listeners' CI devices, the implication is that additional pitch information should be perceivable when acoustic as well as electric stimulation is made available. This prediction is consistent with the experimental results, which showed that the addition of acoustic stimulation added approximately 28 percentage points to the mean scores of the CI users in a test of melody identification (Sucher and McDermott 2009).

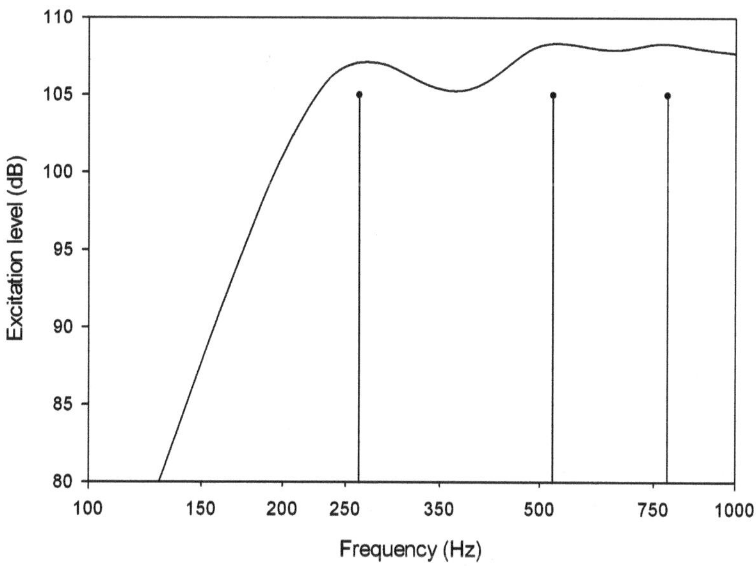

Fig. 13.8 Results of a numerical model (Moore and Glasberg 1997) that estimated the cochlear excitation level (*vertical axis*) versus frequency (*horizontal axis*) for a stimulus comprising the first three harmonics of a complex tone with a fundamental frequency of 262 Hz when delivered to an impaired ear. Each component had a level of 105 dB SPL, as indicated by the *vertical lines* on the graph. The hearing threshold levels used in the model were equal to the mean thresholds of the subjects who participated in a published study of musical sound perception (Sucher and McDermott 2009). These hearing thresholds are therefore typical of those of existing CI users who may obtain benefit from their acoustic hearing

Several other studies have shown similar improvements, and benefits have also been demonstrated for other aspects of auditory perception, including the ability to identify the sounds of musical instruments (Gfeller et al. 2007; Sucher and McDermott 2009). Conversely, there appear to be few reported cases of perceptual disadvantages resulting from the use of acoustic hearing in combination with electric stimulation. Because of these encouraging findings, new electrodes and improved surgical techniques are being developed in an attempt to enable CI devices to be implanted more safely into ears having some usable acoustic hearing sensitivity. In most such ears, there is better sensitivity to sounds at low frequencies than at higher frequencies. This is an advantageous coincidence, in that damage to cochlear structures as a consequence of implantation is more likely to occur near the base where the electrode array is inserted, whereas most functional hair cells are expected to be located near the apex where low frequency sounds are detected. Consequently, several types of electrode arrays that are shorter than those conventionally used in totally deaf ears have been developed and evaluated (Gantz and Turner 2004; Briggs et al. 2006; Gstoettner et al. 2006; Lenarz et al. 2006). In general, these devices seem to be less likely to damage residual hearing, although there are cases both of total hearing loss in the implanted ear with short arrays, and considerable preservation of hearing with conventional arrays (Kiefer et al. 2005; Gantz et al. 2006; Simpson et al. 2009). It is also noteworthy that unilateral cochlear implantation does not affect the ear opposite to the one implanted, and therefore the benefits of acoustic hearing are available to many unilateral CI users even if the surgery does lead to some loss of hearing sensitivity.

5.3 Auditory Training

In addition to the development of innovative CI systems, some researchers have devised and evaluated auditory training programs that attempt to improve the perceptual skills of people with CIs (see also Fu and Galvin, Chap. 11). A few such programs are founded on the assumption that it is worthwhile to direct listeners' attention to specific features of musical sounds and to help them recognize and discriminate among those sounds. Although it is plausible that training would be beneficial to many CI users, especially when listening to music, it is also important to acknowledge certain unavoidable limitations. In particular, it is not feasible to train an observer to perceive aspects of a sensory signal that they cannot resolve.

It may be instructive to consider a visual analogy of this problem. In principle, it would be easy to train an observer to report that the white areas on a television or computer screen are actually made up of red, green, and blue components. The success of the training could then be demonstrated by asking the observer what they see when they look at a white screen, to which they would be expected to reply that there are three distinct colors present. However, although it is true that white areas on a computer screen are created by means of discrete red, green, and blue components, it is not possible for any unaided observer to resolve the individual pixels. In this example, the observer has been trained to label the visual stimulus "correctly," but in fact their underlying perceptual ability has not been affected. To confirm that the training has not changed their ability to resolve the individually colored pixels, the observers could be asked to describe the composition of a white piece of paper. If they have generalized their training as a way of labeling rather than a way of perceiving, then their description of the paper will be incorrect. In relation to perception of musical sounds by CI users, an example of an analogous type of training would be to encourage listeners to label melodies correctly when they are presented without any auditory information other than pitch cues. To determine whether such training has helped the listeners to perceive pitch more accurately in general, it is essential to conduct identification tests using melodies different from those used in the training program. Otherwise it is plausible that the listeners may have learned to assign labels to specific features of the melodies used for training (such as patterns of timbre or loudness variations that may have occurred as unintended effects of the CI sound processing) that cannot be generalized to musical sounds in other listening situations.

Taking into account these qualifications, there is some published evidence that training programs aimed at improving the music perception of CI users can be successful, at least to a limited extent. For example, in one study 11 adult CI recipients participated in a structured 3-month music training program and completed tests of simple melody identification and complex song recognition both before and after the training (Gfeller et al. 2000b). A control group of 9 CI recipients completed the same tests but did not undertake the training. The results showed that the trained subjects obtained a higher mean score on the simple melody identification test after training (23% versus 12% correct), but the score increase was not statistically significant. For the complex song recognition test, the score increase was statistically significant (36% versus 4% correct; $p<0.0001$). The average scores on the same two tests for the control group did not differ significantly. Interestingly, only 5 of the 11 subjects who participated in the training were able to identify any of the newly composed melodies that had been included in the training program. Moreover, each of the items in the complex song recognition test had been included in an extended form in the training program. This probably explains the large and significant score increase associated with the training in the latter test, but it is noteworthy that the absolute mean score was only 36% correct even after training. Generally comparable outcomes were reported for a different procedure in which a set of "melodic contours" was used in training, and both those contours and a set of familiar melodies were used in testing (Galvin et al. 2007). These results suggest that specific training programs may help CI users identify musical excerpts to some extent, but the evidence that such training can be generalized by listeners so that they improve their ability to identify less-familiar musical material seems to be relatively weak.

At the same time, it has been reported that experience listening to music and involvement in formal musical instruction, particularly in the time before onset of severe hearing impairment, are associated with better perceptual ability with a CI. For example, in a study of pitch-ranking using sung vowels as experimental stimuli, the participants included two groups of adult subjects with normal hearing and a group of CI users (Sucher and McDermott 2007). One group of NH subjects had more musical experience and knowledge than the other. On average, the more-experienced NH subjects obtained a much better score on the pitch-ranking test than the musically inexperienced NH group, although even the latter subjects had significantly higher scores than the CI users. These findings imply that part of the reason for the poor ability of CI listeners to rank the pitch of sounds could be related to their typically low levels of musical training relative to that of many adults with normal hearing.

In conclusion, the application of auditory training specifically to improve CI listeners' perception of music shows some promise. However, it is important to verify that the acoustic features of musical sounds that are used in the training are not only audible to the subjects, but also perceivable consistently in a realistic variety of conditions. The latter requirement is necessary to enable the subjects to generalize their learning. Evaluation of the benefit of training programs needs to take account of listeners' normal tendency to improve their test scores over time as a result of gaining experience with their CI through everyday use. A meaningful evaluation can be ensured by including, for example, an appropriately matched control group of subjects in the study who complete the testing without participating in the training. Equally important is the need for the effects of training to be evaluated with formal objective tests when possible, rather than relying exclusively on the participants' subjective appraisal of their own progress.

6 Summary

Over recent years, considerable research has been carried out investigating the perception of music by both adults and children who use cochlear implants. In general, the published findings concur on many significant observations. In particular, CI listeners mostly obtain satisfactory scores on tests of rhythm perception, but their scores on tests that require accurate perception of pitch are much lower than those of typical listeners with normal hearing. The poor pitch perception of most CI users results in very low scores on tests of melody identification unless coincident cues, such as distinctive rhythmic patterns or recognizable sung lyrics, are present in the test material. The ability of children with CIs to sing with approximately correct intonation is much worse than that of NH children of similar age. A related finding is that CI listeners are usually unable to separate perceptually multiple sounds with different pitches presented simultaneously and are therefore probably unable to perceive harmony adequately. Recognition of sound sources, such as musical instruments, is also generally poor if the only information available is from the auditory sensations created by a CI. Despite these deficiencies, some adult CI recipients and most child CI users seem to have a favorable attitude towards the experience of listening to music.

Much research has been directed towards increasing the amount and precision of information about pitch that is available in CI stimulation patterns. Although several innovative techniques have been devised that modify the details of the temporal or spatial stimulation patterns (or both), their effects on the perception of musical sounds have been disappointing. In contrast, the use of acoustic hearing in combination with a CI, where possible, is associated with improved music perception. Even in cases where the acoustic hearing sensitivity of a CI user is relatively poor, the use of that hearing to provide additional auditory information about pitch and other aspects of musical sounds is usually beneficial. However, further research is required to investigate whether or to what extent such benefits are available to children with CIs when listening to music.

For CI recipients who have no usable acoustic hearing postoperatively, improvements in the perception of frequency components of sounds, including better pitch perception, almost certainly require changes to the design of electrode arrays. The lack of substantial benefit found so far with the use of stimulation patterns that are claimed to contain more temporal fine structure than the patterns produced by conventional CI sound-processing schemes (such as Speak, ACE, and CIS) suggests that an improved representation of the spectral fine structure of complex sounds might be more effective. This seems plausible considering that existing electrode arrays provide a maximum of 22 discrete sites of stimulation in the cochlea, whereas the mechanical activity distributed along a normal cochlea is converted into neural excitation by many thousands of hair cells. To approach the spatial resolution enabled by the hair cells in normal hearing, the number of electric stimulation sites would probably need to be increased by at least 1 order of magnitude, and the stimulation distributions may also need to be more focused (van den Honert and Kelsall 2007; O'Leary et al. 2009).

Presuming that CI devices can be developed eventually that deliver electric stimulation selectively from hundreds or thousands of independent sources, the problem remains: how to process sound signals to take advantage of those improved technical capabilities. As it is possible to deliver some information about the temporal fine structure of signals with existing devices, it can be assumed that providing adequate temporal resolution will be practical with future devices as well. However, exploiting the improved spatial resolution will require development of sound-processing schemes that are functionally different from existing schemes. For example, some published theories and models of pitch perception imply that detailed information about acoustic frequency components may be encoded in normal hearing by patterns of synchronized responses in neurons that are excited by activity occurring at different locations in the cochlea (Loeb et al. 1983; Shamma and Klein 2000). If these spatio-temporal patterns of activity underlie the precise perception of frequency components, including components that are important for pitch perception, then future CI systems may need to evoke similar patterns of neural activity electrically. With more-precise control of both the position and the timing of the stimuli, better pitch perception and better perception of musical sounds in general may be enabled for future CI recipients.

Acknowledgments Many colleagues have contributed to the preparation of this chapter, including Brett Swanson, Cathy Sucher, Valerie Looi, Colette McKay, and David MacFarlane. Financial support for some of the reported research was provided by the Garnett Passe and Rodney Williams Memorial Foundation. The Bionics Institute acknowledges the support it receives from the Victorian Government through its Operational Infrastructure Support Program.

References

Arnoldner, C., Riss, D., Brunner, M., Durisin, M., Baumgartner, W.-D., & Hamzavi, J.-S. (2007). Speech and music perception with the new fine structure speech coding strategy: preliminary results. *Acta Oto-Laryngologica*, 127(12), 1298–1303.

Briggs, R. J. S., Tykocinski, M., Xu, J., Risi, F., Svehla, M., Cowan, R. S. C., Stover, T., Erfurt, P., & Lenarz, T. (2006). Comparison of round window and cochleostomy approaches with a prototype hearing preservation electrode. *Audiology & Neurotology*, 11(S1), 42–48.

Brockmeier, S. J., Grasmeder, M., Passow, S., Mawmann, D., Vischer, M., Jappel, A., Baumgartner, W. D., Stark, T., Muller, J. M., Brill, S., Steffens, T., Strutz, J., Keifer, J., Baumann, U., & Arnold, W. (2007). Comparison of musical activities of cochlear implant users with different speech-coding strategies. *Ear and Hearing*, 28(S2), 49S–51S.

Carlyon, R. P. (2008). Temporal pitch processing by cochlear implant users. *Journal of the Acoustical Society of America*, 123(5), 3054.

Ching, T. Y., van Wanrooy, E., & Dillon, H. (2007). Binaural-bimodal fitting or bilateral implantation for managing severe to profound deafness: A review. *Trends in Amplification*, 11(3), 161–192.

Donaldson, G. S., Kreft, H. A., & Litvak, L. (2005). Place-pitch discrimination of single- versus dual-electrode stimuli by cochlear implant users. *Journal of the Acoustical Society of America*, 118(2), 623–626.

Donnelly, P. J., Guo, B. Z., & Limb, C. J. (2009). Perceptual fusion of polyphonic pitch in cochlear implant users. *The Journal of the Acoustical Society of America*, 126(5), EL128–EL133.

Dorman, M. F., Gifford, R. H., Spahr, A. J., & McKarns, S. A. (2008). The benefits of combining acoustic and electric stimulation for the recognition of speech, voice and melodies. *Audiology & Neurotology*, 13(2), 105.

El Fata, F., James, C. J., Laborde, M.-L., & Fraysse, B. (2009). How much residual hearing is 'useful' for music perception with cochlear implants? *Audiology & Neurotology*, 14(Suppl. 1), 14–21.

Firszt, J. B., Reeder, R. M., & Skinner, M. W. (2008). Restoring hearing symmetry with two cochlear implants or one cochlear implant and a contralateral hearing aid. *Journal of Rehabilitation Research and Development*, 45(5), 749–767.

Firszt, J. B., Holden, L. K., Reeder, R. M., & Skinner, M. W. (2009). Speech recognition in cochlear implant recipients: comparison of standard HiRes and HiRes 120 sound processing. *Otology & Neurotology*, 30, 146–152.

Galvin, J. J., 3 rd, Fu, Q. J., & Nogaki, G. (2007). Melodic contour identification by cochlear implant listeners. *Ear and Hearing*, 28(3), 302–318.

Gantz, B. J., & Turner, C. (2004). Combining acoustic and electrical speech processing: Iowa/Nucleus hybrid implant. *Acta Oto-Laryngologica*, 124(4), 344–347.

Gantz, B. J., Turner, C. W., & Gfeller, K. E. (2006). Acoustic plus electric speech processing: preliminary results of a multicenter clinical trial of the Iowa/Nucleus Hybrid implant. *Audiology & Neurotology*, 11(S1), 63–68.

Geurts, L., & Wouters, J. (2004). Better place-coding of the fundamental frequency in cochlear implants. *Journal of the Acoustical Society of America*, 115(2), 844–852.

Gfeller, K. E., Witt, S. A., Spencer, L. J., Stordahl, J., & Tomblin, B. (1998). Musical involvement and enjoyment of children who use cochlear implants. *The Volta Review*, 100(4), 213–233.

Gfeller, K. E., Christ, A., Knutson, J. F., Witt, S. A., Murray, K. T., & Tyler, R. S. (2000a). Musical backgrounds, listening habits, and aesthetic enjoyment of adult cochlear implant recipients. *Journal of the American Academy of Audiology*, 11(7), 390–406.

Gfeller, K. E., Witt, S. A., Stordahl, J., Mehr, M. A., & Woodworth, G. G. (2000b). The effects of training on melody recognition and appraisal by adult cochlear implant recipients. *Journal of the Academy of Rehabilitative Audiology*, 33, 115–138.

Gfeller, K. E., Turner, C. W., Mehr, M. A., Woodworth, G. G., Fearn, R., Knutson, J. F., Witt, S. A., & Stordahl, J. (2002a). Recognition of familiar melodies by adult cochlear implant recipients and normal-hearing adults. *Cochlear Implants International*, 3(1), 29–53.

Gfeller, K. E., Witt, S. A., Woodworth, G. G., Mehr, M. A., & Knutson, J. F. (2002b). Effects of frequency, instrumental family, and cochlear implant type on timbre recognition and appraisal. *Annals of Otology, Rhinology & Laryngology*, 111(4), 349–356.

Gfeller, K. E., Christ, A., Knutson, J. F., Witt, S. A., & Mehr, M. A. (2003). The effects of familiarity and complexity on appraisal of complex songs by cochlear implant recipients and normal hearing adults. *Journal of Music Therapy*, 40(2), 78–112.

Gfeller, K. E., Turner, C. W., Oleson, J., Zhang, X., Gantz, B. J., Froman, R., & Olszewski, C. (2007). Accuracy of cochlear implant recipients on pitch perception, melody recognition, and speech reception in noise. *Ear and Hearing*, 28(3), 412–423.

Gordon, E. E. (1979). *Manual for the primary measures of music audiation and the intermediate measures of music audiation. Music aptitude tests for kindergarten, first, second, third and fourth grade children.* Chicago, IL: GIA Publications, Inc.

Gstoettner, W. K., Helbig, S., Maier, N., Kiefer, J., Radeloff, A., & Adunka, O. F. (2006). Ipsilateral electric acoustic stimulation of the auditory system: results of long-term hearing preservation. *Audiology & Neurotology*, 11(Suppl. 1), 49–56.

Helms, J., Weichbold, V., Baumann, U., von Specht, H., Schon, F., Muller, J., Esser, B., Zieze, M., Anderson, I., & D"Haese, P. (2004). Analysis of ceiling effects occurring with speech recognition tests in adult cochlear-implanted patients. *Journal for Oto-Rhino-Laryngology*, 66(3), 130–135.

Hochmair, I., Nopp, P., Jolly, C., Schmidt, M., Schosser, H., Garnham, C., & Anderson, I. (2006). MED-EL cochlear implants: state of the art and a glimpse into the future. *Trends in Amplification*, 10(4), 201–219.

Hsiao, F. (2008). Mandarin melody recognition by pediatric cochlear implant recipients. *Journal of Music Therapy*, 45(4), 390–404.
Kang, R., Nimmons, G. L., Drennan, W., Longnion, J., Ruffin, C., Nie, K., Won, J. H., Worman, T., Yueh, B., & Rubinstein, J. (2009). Development and validation of the University of Washington Clinical Assessment of Music Perception test. *Ear and Hearing*, 30(4), 411–418.
Kiefer, J., Pok, S. M., Adunka, O. F., Sturzebecher, E., Baumgartner, W. D., Schmidt, M., Tillein, J., Ye, Q., & Gstoettner, W. K. (2005). Combined electric and acoustic stimulation of the auditory system: Results of a clinical study. *Audiology & Neurotology*, 10, 134–144.
Kim, H. N., Shim, Y. J., Chung, M. H., & Lee, Y. H. (2000). Benefit of ACE compared to CIS and SPEAK coding strategies. *Advances in Oto-Rhino-Laryngology*, 57, 408–411.
Kong, Y. Y., Cruz, R. J., Jones, J. A., & Zeng, F. G. (2004). Music perception with temporal cues in acoustic and electric hearing. *Ear and Hearing*, 25(2), 173–185.
Kong, Y. Y., Stickney, G. S., & Zeng, F. G. (2005). Speech and melody recognition in binaurally combined acoustic and electric hearing. *Journal of the Acoustical Society of America*, 117(3, Pt. 1), 1351–1361.
Kwon, B. J., & van den Honert, C. (2006). Dual-electrode pitch discrimination with sequential interleaved stimulation by cochlear implant users. *Journal of the Acoustical Society of America*, 120(1), EL1–EL6.
Laneau, J., Wouters, J., & Moonen, M. (2006). Improved music perception with explicit pitch coding in cochlear implants. *Audiology & Neurotology*, 11, 38–52.
Leal, M. C., Shin, Y. J., Laborde, M. L., Calmels, M. N., Verges, S., Lugardon, S., Andrieu, S., Deguine, O., & Fraysse, B. (2003). Music perception in adult cochlear implant recipients. *Acta Oto-Laryngologica*, 123(7), 826–835.
Lenarz, T., Stover, T., Buechner, A., Paasche, G., Briggs, R. J. S., Risi, F., Pesch, J., & Battmer, R. D. (2006). Temporal bone results and hearing preservation with a new straight electrode. *Audiology & Neurotology*, 11(S1), 34–41.
Loeb, G. E., White, M. W., & Merzenich, M. M. (1983). Spatial cross-correlation. *Biological Cybernetics*, 47(3), 149–163.
Loizou, P. C. (1998). Mimicking the human ear: an overview of signal-processing strategies for converting sound into electrical signals in cochlear implants. *IEEE Signal Processing Magazine*, September, 101–130.
Looi, V., McDermott, H. J., McKay, C. M., & Hickson, L. (2004). Pitch discrimination and melody recognition by cochlear implant users. In *VIII International Cochlear Implant Conference* (Vol. 1273C, pp. 197–200). Indianapolis, IN: Elsevier.
Looi, V., McDermott, H. J., McKay, C. M., & Hickson, L. (2007). Comparisons of quality ratings for music by cochlear implant and hearing aid users. *Ear and Hearing*, 28(S2), 59S–61S.
Looi, V., McDermott, H., McKay, C., & Hickson, L. (2008a). The effect of cochlear implantation on music perception by adults with usable pre-operative acoustic hearing. *International Journal of Audiology*, 47(5), 257–268.
Looi, V., McDermott, H., McKay, C., & Hickson, L. (2008b). Music perception of cochlear implant users compared with that of hearing aid users. *Ear and Hearing*, 29(3), 421–434.
MacLeod, P., Chouard, C.-H., & Weber, J. P. (1985). French device. In R. A. Schindler & M. M. Merzenich (Eds.), *Cochlear implants* (pp. 111–120). New York: Raven Press.
McDermott, H. J. (2004). Music perception with cochlear implants: a review. *Trends in Amplification*, 8(2), 49–82.
McDermott, H. J. (2009). Cochlear implants and music. In M. Chasin (Ed.), *Hearing loss in musicians, prevention and management* (pp. 117–127). San Diego, CA: Plural Publishing.
McDermott, H. J., & McKay, C. M. (1994). Pitch ranking with nonsimultaneous dual-electrode electrical stimulation of the cochlea. *Journal of the Acoustical Society of America*, 96(1), 155–162.
McDermott, H. J., & McKay, C. M. (1997). Musical pitch perception with electrical stimulation of the cochlea. *Journal of the Acoustical Society of America*, 101(3), 1622–1631.
McDermott, H. J., & Looi, V. (2004). Perception of complex signals, including musical sounds, with cochlear implants. In *VIII International Cochlear Implant Conference* (Vol. 1273C, pp. 201–204). Indianapolis, Indiana: Elsevier.

McDermott, H. J., McKay, C. M., & Vandali, A. E. (1992). A new portable sound processor for the University of Melbourne/Nucleus Limited multielectrode cochlear implant. *Journal of the Acoustical Society of America*, 91(6), 3367–3371.

McKay, C. M. (2004). Psychophysics and electrical stimulation. In F. G. Zeng, A. N. Popper, & R. R. Fay (Eds.), *Springer Handbook of Auditory Research: auditory prostheses* (pp. 286–333). New York: Springer-Verlag.

McKay, C. M., & McDermott, H. J. (1996). The perception of temporal patterns for electrical stimulation presented at one or two intracochlear sites. *Journal of the Acoustical Society of America*, 100(2, Pt. 1), 1081–1092.

McKay, C. M., McDermott, H. J., Vandali, A. E., & Clark, G. M. (1991). Preliminary results with a six spectral maxima sound processor for the University of Melbourne/Nucleus multiple-electrode cochlear implant. *Australian Journal of Oto-Laryngology*, 6(5), 354–359.

McKay, C. M., McDermott, H. J., & Clark, G. M. (1994). Pitch percepts associated with amplitude-modulated current pulse trains in cochlear implantees. *Journal of the Acoustical Society of America*, 96(5, Pt. 1), 2664–2673.

McKay, C. M., McDermott, H. J., & Clark, G. M. (1995). Pitch matching of amplitude-modulated current pulse trains by cochlear implantees: The effect of modulation depth. *Journal of the Acoustical Society of America*, 97(3), 1777–1785.

Mirza, S., Douglas, S. A., Lindsey, P., Hildreth, T., & Hawthorne, M. (2003). Appreciation of music in adult patients with cochlear implants: a patient questionnaire. *Cochlear Implants International*, 4(2), 85–95.

Mitani, C., Nakata, T., Trehub, S. E., Kanda, Y., Kumagami, H., Takasaki, K., Miyamoto, I., & Takahashi, H. (2007). Music recognition, music listening, and word recognition by deaf children with cochlear implants. *Ear and Hearing*, 28(S2), 29 S–33 S.

Moore, B. C. J., & Glasberg, B. R. (1996). A revision of Zwicker's loudness model. *Acustica-Acta Acustica*, 82(2), 335–345.

Moore, B. C. J., & Glasberg, B. R. (1997). A model of loudness perception applied to cochlear hearing loss. *Auditory Neuroscience*, 3, 289–311.

Moore, B. C. J., & Carlyon, R. P. (2005). Perception of pitch by people with cochlear hearing loss and by cochlear implant users. In C. J. Plack, R. R. Fay, A. J. Oxenham, & A. N. Popper (Eds.), *Springer Handbook of Auditory Research: pitch perception* (Vol. 24, pp. 234–277). New York: Springer-Verlag.

Nakata, T., Trehub, S. E., Mitani, C., & Kanda, Y. (2006). Pitch and timing in the songs of deaf children with cochlear implants. *Music Perception*, 24(2), 147–154.

Nelson, D. A., van Tasell, D. J., Schroder, A. C., Soli, S. D., & Levine, S. (1995). Electrode ranking of "place pitch" and speech recognition in electrical hearing. *Journal of the Acoustical Society of America*, 98(4), 1987–1999.

Nimmons, G. L., Kang, R. S., Drennan, W. R., Longnion, J., Ruffin, C., Worman, T., Yueh, B., & Rubinstein, J. T. (2008). Clinical assessment of music perception in cochlear implant listeners. *Otology & Neurotology*, *29*(2), 149–155.

O'Leary, S. J., Richardson, R. R., & McDermott, H. J. (2009). Principles of design and biological approaches for improving the selectivity of cochlear implant electrodes. *Journal of Neural Engineering*, 6(5), 055002.

Olson, A. D., & Shinn, J. B. (2008). A systematic review to determine the effectiveness of using amplification in conjunction with cochlear implantation. *Journal of the American Academy of Audiology*, 19(9), 657–671.

Olszewski, C., Gfeller, K. E., Froman, R., Stordahl, J., & Tomblin, B. (2005). Familiar melody recognition by children and adults using cochlear implants and normal hearing children. *Cochlear Implants International*, 6(3), 123–140.

Patrick, J. F., Busby, P. A., & Gibson, P. J. (2006). The development of the Nucleus® Freedom™ cochlear implant system. *Trends in Amplification*, 10(4), 175–200.

Pijl, S., & Schwarz, D. W. (1995). Melody recognition and musical interval perception by deaf subjects stimulated with electrical pulse trains through single cochlear implant electrodes. *Journal of the Acoustical Society of America*, 98(2, Pt. 1), 886–895.

Plack, C. J., & Oxenham, A. J. (2005). The psychophysics of pitch. In R. R. Fay & A. N. Popper (Eds.), *Springer Handbook of Auditory Research: pitch* (Vol. 24, pp. 7–55). New York: Springer.
Riess Jones, M., Fay, R. R., & Popper, A. N. (Eds.). (2010). *Music perception* (Vol. 36). New York: Springer.
Schafer, E. C., Amlani, A. M., Seibold, A., & Shattuck, P. L. (2007). A meta-analytic comparison of binaural benefits between bilateral cochlear implants and bimodal stimulation. *Journal of the American Academy of Audiology*, 18(9), 760–776.
Shamma, S., & Klein, D. (2000). The case of the missing pitch templates: how harmonic templates emerge in the early auditory system. *Journal of the Acoustical Society of America*, 107(5, Pt. 1), 2631–2644.
Simpson, A., McDermott, H. J., Dowell, R. C., Sucher, C., & Briggs, R. J. S. (2009). Comparison of two frequency-to-electrode maps for acoustic-electric stimulation. *International Journal of Audiology*, 48(2), 63–73.
Singh, S., Kong, Y., & Zeng, F. (2009). Cochlear implant melody recognition as a function of melody frequency range, harmonicity, and number of electrodes. *Ear and Hearing*, 30(2), 160–168.
Skinner, M. W., Holden, L. K., Whitford, L. A., Plant, K. L., Psarros, C., & Holden, T. A. (2002). Speech recognition with the Nucleus 24 SPEAK, ACE, and CIS speechcoding strategies in newly implanted adults. *Ear and Hearing*, 23(3), 207–223.
Stordahl, J. (2002). Song recognition and appraisal: a comparison of children who use cochlear implants and normally hearing children. *Journal of Music Therapy*, 39(1), 2–19.
Sucher, C. M. (2007). *Music perception of children who use cochlear implants* (Unpublished Master's thesis). The University of Melbourne, Australia.
Sucher, C. M., & McDermott, H. J. (2007). Pitch ranking of complex tones by normally hearing subjects and cochlear implant users. *Hearing Research*, 230, 80–87.
Sucher, C. M., & McDermott, H. J. (2009). Bimodal stimulation: benefits for music perception and sound quality. *Cochlear Implants International*, 10(S1), 96–99.
Swanson, B. (2008). *Pitch perception with cochlear implants* (Unpublished doctoral thesis). The University of Melbourne, Australia.
Tong, Y. C., Clark, G. M., Blamey, P. J., Busby, P. A., & Dowell, R. C. (1982). Psychophysical studies for two multiple-channel cochlear implant patients. *Journal of the Acoustical Society of America*, 71(1), 153–160.
Townshend, B., Cotter, N., Compernolle, D. V., & White, R. L. (1987). Pitch perception by cochlear implant subjects. *Journal of the Acoustical Society of America*, 82(1), 106–115.
van den Honert, C., & Kelsall, D. C. (2007). Focused intracochlear electric stimulation with phased array channels. *Journal of the Acoustical Society of America*, 121(6), 3703–3716.
Vandali, A. E., Sucher, C. M., Tsang, D. J., McKay, C. M., Chew, J. W., & McDermott, H. J. (2005). Pitch ranking ability of cochlear implant recipients: A comparison of sound-processing strategies. *Journal of the Acoustical Society of America*, 117(5), 3126–3138.
Veekmans, K., Ressel, L., Mueller, J., Vischer, M., & Brockmeier, S. J. (2009). Comparison of music perception in bilateral and unilateral cochlear implant users and normal-hearing subjects. *Audiology & Neurotology*, 14(5), 315–326.
Vongpaisal, T., Trehub, S. E., & Schellenberg, E. G. (2006). Song recognition by children and adolescents with cochlear implants. *Journal of Speech, Language, and Hearing Research*, 49, 1091–1103.
Vongpaisal, T., Trehub, S. E., & Schellenberg, E. G. (2009). Identification of TV tunes by children with cochlear implants. *Music Perception*, 27(1), 17–24.
Wilson, B. S., Finley, C. C., Lawson, D. T., Wolford, R. D., Eddington, D. K., & Rabinowitz, W. M. (1991). Better speech recognition with cochlear implants. *Nature*, 352(6332), 236–238.
Xu, L., Zhou, N., Chen, X., Li, Y., Schultz, H., Zhao, X., & Han, D. (2009). Vocal singing by prelingually-deafened children with cochlear implants. *Hearing Research*, 255(1–2), 129–134.
Zwolan, T. A., Collins, L. M., & Wakefield, G. H. (1997). Electrode discrimination and speech recognition in postlingually deafened adult cochlear implant subjects. *Journal of the Acoustical Society of America*, 102(6), 3673–3685.

Chapter 14
Tonal Languages and Cochlear Implants

Li Xu and Ning Zhou

1 Introduction

As a major part of world languages, tonal languages are spoken in every continent except for Australia. In a tonal language, voice pitch variation (i.e., tone) at the monosyllabic level is a segmental structure that conveys lexical meaning of a word (Duanmu 2000). Mandarin Chinese, a tonal language, is spoken by more people than any other single language, including non-tonal languages. While some dialects in southern Mexico may distinguish as many as 14 tones, Chinese dialects typically have 4–6 contrastive tones.

Multi-channel cochlear implants (CIs) have been a great success in providing profoundly deafened individuals with satisfactory speech perception in quiet. The contemporary speech-processing strategies deliver primarily temporal-envelope information of speech to the auditory nerve of the implantees (see Loizou 2006 for a review). These strategies do not explicitly code pitch information, because they have been mainly designed to accommodate Western languages that use pitch variation only for suprasegmental structures, such as intonation difference between a statement and a question. Because of the lack of pitch coding, tonal-language understanding remains challenging for implant users. The challenge can be tone recognition as well as tone production in language development. Both temporal and spectral approaches have been taken to improve CI pitch perception.

This chapter will describe the acoustical cues for recognition of lexical tones, primarily the Mandarin Chinese tones, and the relative contributions of these cues. This chapter will also discuss results in tone recognition in implant recipients in relation to their differences in demographics, devices, strategies, and psychophysics.

L. Xu (✉)
School of Rehabilitation and Communication Sciences,
Ohio University, Athens, OH 45701, USA
e-mail: xul@ohio.edu

F.-G. Zeng et al. (eds.), *Auditory Prostheses: New Horizons*, Springer Handbook of Auditory Research 39, DOI 10.1007/978-1-4419-9434-9_14,

The effects of frequency-place mismatch on tone recognition, which is a unique problem in CI users, will also be discussed. This chapter will then explore the relationship between music pitch perception and lexical tone recognition. Lastly, this chapter will evaluate results on tone production and vocal singing in prelingually deafened, native tonal-language speaking children with CIs.

2 Acoustical Cues for Tone Recognition

Mandarin Chinese has 4 lexical tones that are commonly known as tone 1, tone 2, tone 3, and tone 4. The Mandarin tones are classified based on both the fundamental frequency (F0) variation patterns and the absolute frequency heights (Howie 1976). Tone 1 has a "high flat" F0 pattern, and tone 2 has a "middle low rising" pattern. Tone 3 has a "dipping and rising" contour, with a possible break on the turning point from dipping to rising. The break reflects the loss of voicing that corresponds to a glottal stop. Sometimes tone 3 can also lose its final rising, resulting in a F0 contour that falls from a moderate level to a low level without rising. Tone 4 has a "high falling" contour. Figure 14.1 shows the F0 contours of the 4 Mandarin tones spoken by multiple talkers.

2.1 Primary and Secondary Cues

The F0 height and contours are the primary intrinsic cues for tone recognition. Speech materials contain redundant F0 information. Liang (1963) demonstrated that high-level tone recognition was preserved even when the Mandarin Chinese speech signals

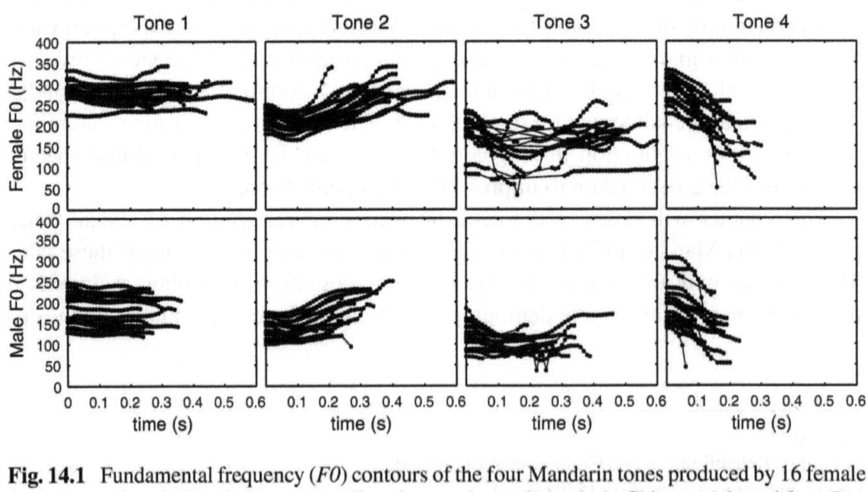

Fig. 14.1 Fundamental frequency (*F0*) contours of the four Mandarin tones produced by 16 female (*upper panel*) and 16 male (*lower panel*) native speakers of Mandarin Chinese. Adapted from Lee and Hung (2008) with permission from the Acoustical Society of America

were highpass filtered at 2.4 kHz, via presumably the unresolved harmonics inducing the F0 residue pitch (Schouten et al. 1962). The F0 contour itself is considered redundant for tone recognition; that is, not all parts of the contour are necessary for tone recognition. Liu and Samuel (2004) found that perception of tone 3 of Mandarin Chinese was not affected when the rising part of the F0 contour was neutralized. Gottfried and Suiter (1997) and Lee (2009) also showed the redundancy of F0 contour by producing better than chance performance with shortened tone stimuli including only the preceding consonant and the following 6 glottal periods of the vowel.

F0 constitutes the most important acoustic characteristic of tones, but secondary acoustic cues are also useful for tone recognition, particularly when F0 is compromised. These secondary cues include duration, amplitude contour, and spectral envelope of the speech signal.

Mandarin tones differ in duration, with tone 3 having the longest duration. Duration differences of the other 3 tones, however, are less consistent (e.g., Howie 1976; Luo and Wang 1981). Fu and Zeng (2000) recorded tones from 6 syllables spoken by 10 talkers and found that the average duration for tone 3 was the longest (463.3 ms), followed by tone 2 (374.7 ms) and tone 1 (339.5 ms), with tone 4 being the shortest (334.4 ms). Xu et al. (2002) and Lee and Hung (2008) found a similar distribution of tone durations (see Fig. 14.1).

The reliability of the duration cue for tone recognition is still in debate. Based on a maximum likelihood model (Green and Swets 1966), Xu et al. (2002) found that tone recognition can be as high as 56.5% correct with duration cues alone. Xu et al. (2002) reported that tone recognition reduced by approximately 15 percentage points for all 4 tones of equal durations as compared to that using the tokens with preserved durations in a vocoder study (more details below).

The other secondary cue is the overall amplitude contour or temporal envelope. It has been shown that tone recognition, using the signal-correlated noise with controlled duration, still remained above chance (Whalen and Xu 1992). Whalen and Xu attributed the performance to the energy distribution in the amplitude contour of the tones. Fu and Zeng (2000) further used signal-correlated noises to explore the roles of amplitude contour, duration, and periodicity in tone recognition. Tone recognition was around 70% correct, when all 3 temporal cues were available. Either the amplitude contour or the periodicity cue alone resulted in approximately 55% correct recognition, and the duration cue alone provided the lowest recognition score of about 35% correct. These results were confirmed in a more recent acoustical and perceptual study by Kuo et al. (2008).

A third secondary cue is related to the speech spectral envelope. In whispered speech, where the spectral envelope is preserved but the voice source is absent, Mandarin Chinese tone recognition is fairly good (i.e., 60–70% correct) (Liang 1963). Kong and Zeng (2006) pointed out that the duration and overall amplitude cues alone cannot account for the significant tone-recognition performance of whispered speech. They reasoned that the formant frequencies represented in the spectral envelope of whispered speech can be used by the listeners to match the voice pitch of the speaker. However, tone recognition of whispered speech in other tonal languages was typically found to be much lower than in Mandarin Chinese: ~40%

correct for Vietnamese (Miller 1961) and 20–45% correct for Thai (Abramson 1978). Since the duration and overall amplitude cues were poorly controlled in the whispered-speech studies, the degree to which the spectral-envelope cue contributed to tone recognition remains unclear.

2.2 *Interaction Between the Temporal and Spectral Cues*

The contributions of spectral and temporal information to tone recognition have been examined in several studies that used a vocoder technique simulating multi-channel cochlear implants. Cochlear implant simulation, using a noise-excited vocoder, typically involves dividing speech signals into spectral bands, extracting temporal envelopes from each of the bands as modulators, modulating wide-band noise spectrally limited by the same bandpass filters, and summing all the amplitude-modulated narrow-band noise to form a reconstructed signal (see Xu and Pfingst 2008). The temporal fine structure in the speech single is therefore replaced by noise in the bands. However, the band-specific temporal-envelope cue is well preserved. It is possible to control the spectral resolution of the output signal by varying the number of spectral bands and the amount of temporal details is typically controlled by varying the cut-off frequency of the lowpass filters that are used to extract the envelopes.

Fu et al. (1998) first reported that an increase from 50 to 500 Hz in the lowpass cut-off frequencies of the envelope extractors greatly improved tone recognition, which indicates that when spectral information is limited, tone recognition can be improved by an increase in temporal details. On the contrary, an increase of spectral bands from 1 to 4 did not seem to improve tone recognition. Xu et al. (2002), however, reported a significant effect of spectral channels when the number was varied from 1 to 12. In particular, Xu et al. (2002) found a trade-off between spectral and temporal cues: tone-recognition performance with higher spectral resolution but less detailed temporal envelope was equivalent to that with low spectral resolution but more detailed temporal envelope (Fig. 14.2).

Besides the trade-off relationship between temporal and spectral cues, Kong and Zeng (2006) observed complementary contribution between temporal periodicity cues and spectral cues for Mandarin tone recognition in quiet and noise conditions. They found that in quiet, tone-recognition performance of 8-band 50-Hz lowpass cutoff condition was worse than that of 1-band 500-Hz lowpass cutoff condition, but this pattern was reversed in noise conditions. This indicates that the coarse spectral information may not be very useful for tone recognition in quiet but is of great importance for perception in noise, because the temporal envelope cues might be more susceptible to noise compared to the spectral cues. A more recent study has suggested that temporal envelope and periodicity information (i.e., ≤500 Hz) within different frequency bands may have differential contribution to tone recognition. Yuen et al. (2007) showed that Cantonese tone recognition was significantly better when the listeners were provided with temporal-envelope information from the 2 higher frequency bands (1–2 kHz and 2–4 kHz) rather than with that of the two lower frequency bands (60–500 Hz and 500–1000 Hz).

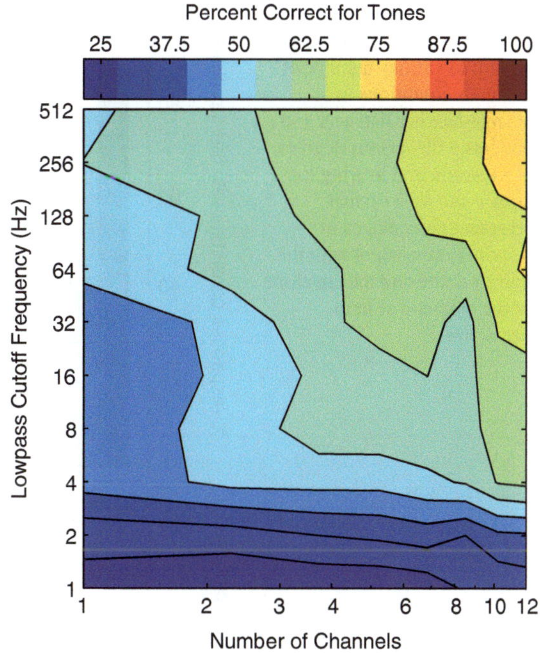

Fig. 14.2 Mean tone-recognition performance in the number-of-channels-versus-low-pass-cutoffs matrix. The data are plotted in a contour format in which the percent correct scores are represented by colors, as indicated by the color bar on the top. Adapted from Xu et al. (2002) with permission from the Acoustical Society of America

2.3 Relative Contributions of Temporal Envelope and Fine Structure

As early as 1988, Lin demonstrated that when the temporal fine structure of a broadband signal exists, the temporal-envelope information has no influence on the recognition of Mandarin tone contrasts. The examination of the relative importance of temporal envelope and fine structure in multiple bands was made possible by using an "auditory chimera" signal-processing technique (Smith et al. 2002). This technique creates chimeric signals that have the temporal envelope of one tone and the fine structure of another tone (Xu and Pfingst 2003). Since the chimeric stimuli contain conflicting tonal information, the listeners' response reveals whether it is the temporal fine structure or the envelopes that they depend on when making a tone judgment. Results of Xu and Pfingst (2003) showed that approximately 90% of the time the responses were consistent with the identity of fine structure of the chimeric stimuli.

The above findings have heralded new development of speech-processing strategies that have aimed at providing fine structure in the electrical stimulation in cochlear implants. Specifically, these are the HiResolution or HiResolution 120 strategies from the Advanced Bionics (Koch et al. 2004; Firszt et al. 2009) and the FSP strategy from Med El (Arnoldner et al. 2007). As reviewed below in Sect. 3, the emerging clinical data indicate little, if any, improvement in lexical-tone recognition using these new strategies. The reasons might be that the CI users are unable to use the fine structure information delivered in the electrical form or that perhaps the auditory system of a deaf individual has a reduced ability to process temporal

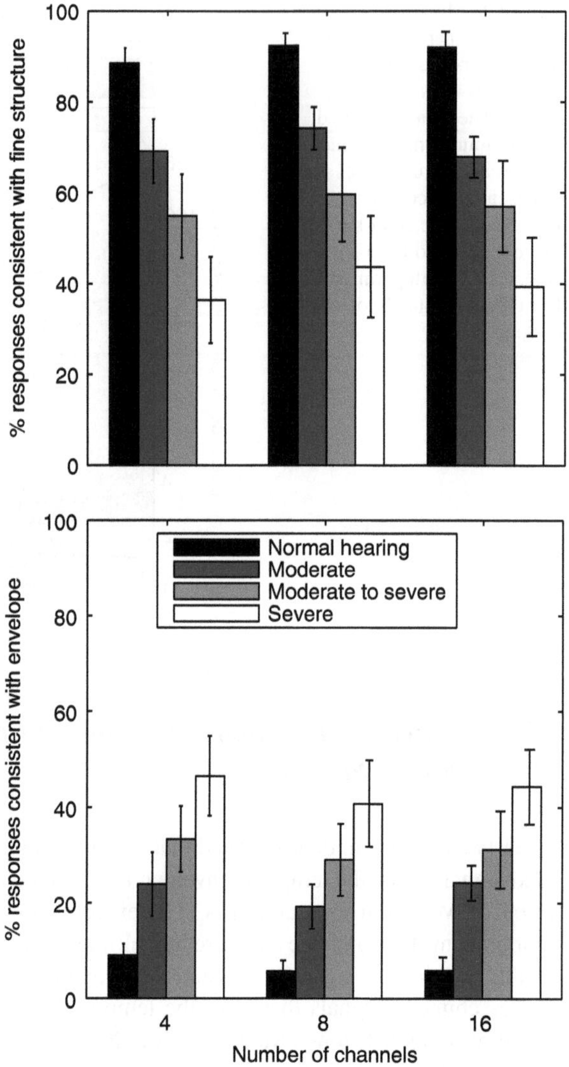

Fig. 14.3 Tone-recognition performance using chimeric stimuli in various number-of-channels conditions in normal-hearing listeners and patients with various degrees of sensorineural hearing loss. *Upper and lower panel* represent percentages of responses consistent with the fine structure and the envelope of the chimeric stimuli, respectively

fine-structure information (e.g., Lorenzi et al. 2006). Recently, Wang et al. (2010) used the auditory chimera technique and tested tone recognition in a group of patients with moderate to severe sensorineural hearing loss. Results clearly indicate that while normal-hearing listeners rely on fine structure for tone recognition, the hearing-impaired patients rely more on the temporal envelope for tone recognition as the hearing loss becomes more severe (Fig. 14.3).

Collectively, the literature suggests that F0 and its harmonic structures of the signal are the most dominant cues for tone recognition. In the absence of explicit F0, such as in CI stimulation or its vocoder simulation, temporal information, particularly the temporal envelopes presented in multiple channels that are equivalent to what are available in current CI technology, contributes to a moderate level (70 to 80% correct) of tone recognition.

3 Tone Recognition

3.1 *Tone-Recognition Performance in CI Users*

Tone-recognition performance in children with CIs who speak tonal languages is highly variable. Data from Mandarin-speaking as well as Cantonese-speaking children with CIs generally indicate that the prelingually deafened children with CIs have deficits in perceiving lexical tones. Wei et al. (2000) measured tone-recognition accuracy in 28 CI children (2–12 years of age) who speak Cantonese. Tone recognition improved significantly from pre-implantation, but performance plateaued at 65% correct after 2 years of training. Ciocca et al. (2002) tested tone recognition in 17 native Cantonese-speaking children aged between 4 to 9 years old. The subjects identified the tone in a monosyllabic target /ji/ in a two-alternative forced-choice test. The tone target was presented in a sentence median position. The subjects responded by pointing to the pictures that represented the auditory stimuli. Performance ranged from chance (i.e., 50% correct) to 61% correct for the 8 tone contrasts. Similar results have been reported by Lee et al. (2002) and Wong and Wong (2004) using either tone discrimination or tone identification paradigms.

Using a tone contrast test for Mandarin tone recognition (chance = 50% correct), Peng et al. (2004) reported a relatively higher average score of approximately 73% correct in a group of 30 pediatric Mandarin-speaking implantees aged from 6 to 12.6 years old. Among the 6 Mandarin Chinese tone contrasts, Peng et al. reported that the pediatric implantees could identify the tone contrasts that involve tone 4 better than other tone contrasts. Note that the researchers used a live voice presentation and did not control for the durations of the tone tokens.

A large-scale study was conducted recently on tone development in children with CIs who speak Mandarin Chinese (Xu et al. 2009b). The study tested tone recognition in 107 children aging from 2.4 to 16.2 years old. The study used a computerized two-alternative forced-choice paradigm in which the durations of the tone tokens were equalized. Performance of individuals varied considerably from chance to as high as a nearly perfect score (Fig. 14.4). Identification of the 6 tone contrasts did not show significant differences. The averaged tone-recognition accuracy of the group was 67% correct, significantly lower than the nearly perfect performance of the typically-developing, normal-hearing control group (N = 112). The children with CIs could perform relatively well whenever tone 1 was contrasted with other tones. Confusion matrix analysis also showed that tone 1 overall was the best recognized tone.

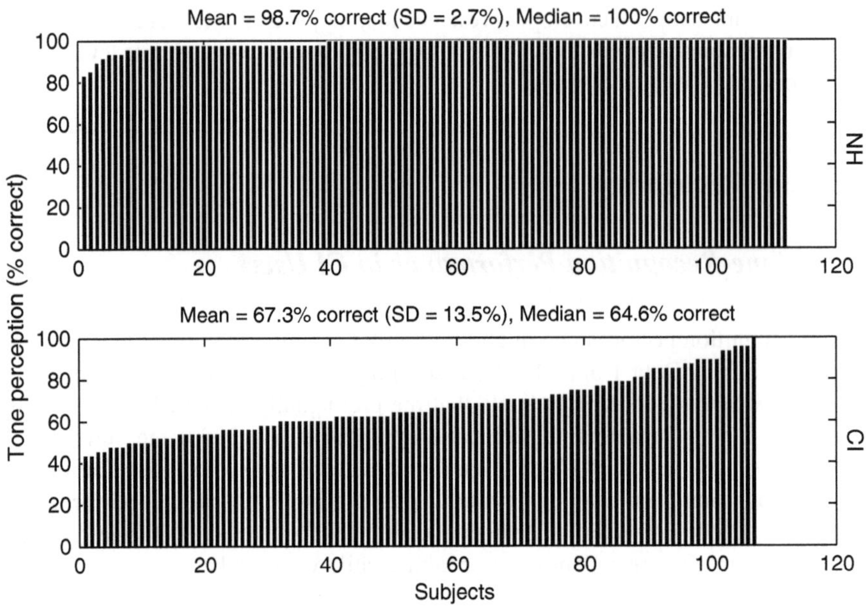

Fig. 14.4 Rank-ordered tone-recognition scores (% correct) by the 112 normal-hearing children (*upper panel*) and 107 children with CIs (*lower panel*)

3.2 Psychophysics and Tone Recognition

Lexical tone-recognition performance in CI users has been shown to correlate with a number of psychophysical measures including electrode discrimination, rate discrimination, gap detection, frequency discrimination, amplitude modulation detection, and amplitude modulation frequency discrimination thresholds. Wei et al. (2004) for example, measured pulse rate discrimination in individual electrodes in 5 CI subjects who speak Mandarin Chinese. Two standard rates (i.e., 100 and 200 pps) were chosen because they were in the range of the voice pitch. The average rate discrimination thresholds varied significantly in individuals ranging from 0.2 to 1.2 measured in Weber's fraction. The 5 CI users were also tested for tone recognition using various numbers of active electrodes. Despite the fact that the results varied largely between electrode conditions and implant subjects, Wei et al. (2004) found that tone-recognition scores using a full-electrode map (i.e., 20 electrodes) were highly correlated ($r = -0.97$) with the subjects' averaged rate discrimination thresholds across the electrode array. Wei et al. (2007) also compared tone recognition in 17 Mandarin-speaking implant users with their gap detection and frequency discrimination thresholds. They found that tone recognition in noise conditions showed stronger strength in correlation with the psychophysical measures. Luo et al. (2008) used a research interface to bypass the clinical speech processor to measure psychophysical performance. Amplitude modulation detection thresholds (AMDT) and amplitude modulation frequency discrimination thresholds (AMFDT) were measured in 10 Mandarin-speaking implant users at their middle electrodes with various stimulation levels. Results showed that the mean AMDTs (averaged for 20- or 100-Hz AM across different levels) and mean AMFDTs (averaged for the 50-Hz standard AM frequency across different levels) were significantly correlated with Mandarin Chinese tone, consonant, and sentence recognition scores, but not with vowel recognition scores. Their results further confirmed the importance of temporal-envelope cues for Chinese speech recognition in CI users.

3.3 The Effects of Frequency-Place Mismatch on Tone Recognition

In the normal cochlea, acoustic signals of different frequencies stimulate corresponding places on the basilar membrane in a tonotopic fashion. In CI systems, a number of forms of frequency-place mismatch may occur as a result of the pathology of hearing loss, shallow insertion of the electrode, or frequency mapping of the device. Localized losses of auditory neurons can result in "holes" in hearing and elevate electrical thresholds of the corresponding electrodes. The increased signal level will likely result in spread of electric current to neural fibers that are not intended to be activated, producing frequency warping around the "holes" in the cochlea (Shannon et al. 2002). Frequency-place mismatch can also take place in shallow insertion conditions that result in an overall shift of the spectrum. Consider the case where the implant electrode array is not fully inserted into the cochlea so that the location of the electrode array does not match the analysis bands. Typically, the output of a low frequency analysis band is delivered to the electrode at a higher frequency place, resulting in a basal shift of the spectrum. Matching the analysis bands to the location of the electrode array nonetheless eliminates frequency coverage especially in the low frequency region. Additionally, because of the limited length of the cochlear implant electrode array, the frequency range stimulated by a cochlear implant does not necessarily cover the entire speech spectrum. As a consequence, frequency compression is another commonly encountered case of frequency-place mismatch. Clinically used maps usually compressively allocate a wider frequency range, sufficient for speech understanding to electrodes that cover a narrower tonotopic length, regardless of the position of the electrode array.

There is a consensus in the literature that suggests a detrimental effect of frequency-place mismatch on consonant, vowel, and sentence recognition in English (see Dorman et al. 1997; Pfingst et al. 2001; Baskent and Shannon 2006). Basal shift also shows a systematic effect on English consonant confusion (Zhou et al. 2010). The effects of basal spectral shift and frequency compression on lexical-tone recognition were examined by Zhou and Xu (2008b). In the study, a noise-excited vocoder was used to simulate a cochlear implant with varying insertion depths. Speech envelopes were delivered to carriers of higher frequencies to simulate basal spectral shift of 1–7 mm. Zhou and Xu (2008b) found that tone recognition was much more resistant to the basal spectral shift compared to English phoneme and sentence recognition. The detrimental effects of basal shift did not show until the carriers were shifted to

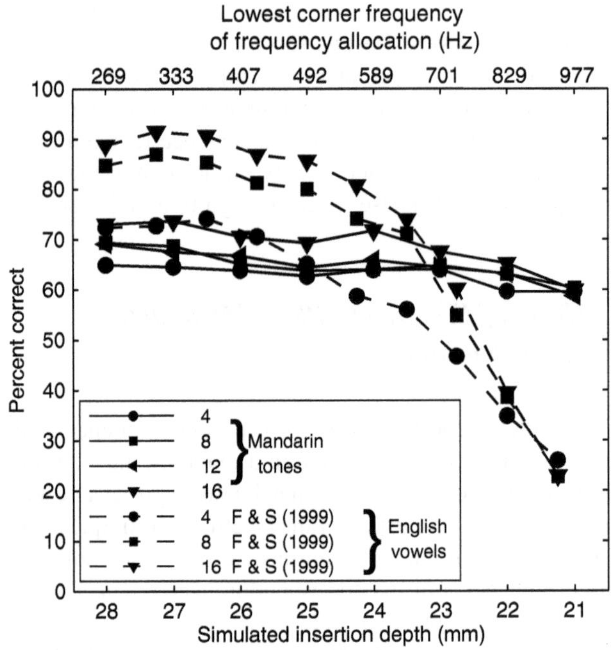

Fig. 14.5 Tone-recognition performance as a function of simulated insertion depth of cochlear implants (*CIs*). Tone-recognition performance is plotted for 4, 8, 12, and 16 channel conditions in *solid lines* with different symbols. The lowest corner frequency of frequency allocations for the carriers is noted for each simulated insertion depth. Data of vowel recognition from Fu and Shannon (abbreviated as F & S in the legend) (1999) are replotted. Simulated insertion depth of 28 mm corresponds to a full insertion or tonotopically matched condition. Adapted from Zhou and Xu (2008b) with permission from Elsevier

almost 2 octaves higher. A 7-mm basal shift of the spectrum only caused tone-recognition performance to decrease from the unshifted condition by approximately 10 percentage points. Compared to the vowel recognition scores measured from similar experimental conditions (Fu and Shannon 1999), the effects of spectral shift on tone recognition appeared to be much smaller (Fig. 14.5).

Zhou and Xu (2008b) also reported the effect of frequency compression on Mandarin tone recognition. Compression of 3 or 5 mm at both frequency ends on the basilar membrane of the cochlea (Greenwood 1990) produced better tone-recognition performance than that without compression. This is consistent with the findings by Baskent and Shannon (2003, 2005) that for English phoneme recognition, in shallow insertion conditions, a moderate amount of compression is better than tonotopic matching with low frequency truncation. The Zhou and Xu study (2008b) suggests that wider frequency allocation that includes low frequency information critical for pitch perception may benefit tone recognition. However, the degree of compression should be controlled so that the effects of frequency mismatch will not cancel out the benefit of frequency coverage.

Research data from a limited number of CI users are consistent with the vocoder study discussed above in that frequency-place mismatch affects Mandarin tone recognition much less than word recognition. In such a study, Liu et al. (2004) tested the effects of frequency shift and compression on Mandarin tone and word recognition in 6 prelingually deafened children fit with Nucleus CI24. The frequency range was kept constant, while electrodes were selectively turned off, to create frequency shift and compression combined conditions. Their results suggested that as long as a sufficient number of electrodes were activated, selection of the stimulation sites did not seem to affect tone recognition. It was not possible however to separate the effects of frequency shift and frequency compression in Liu et al. (2004), because the same frequency range was used in the experiments.

3.4 Demographic Factors Contributing to Tone Recognition

Many studies have tried to explain the variable performance in the CI children in relation to their demographic variables such as age at implantation and the experience with the devices. For example, Lee et al. (2002) reported that tone-recognition performance was related to the duration of CI use and age at implantation. Han et al. (2009) found a consistent relationship between tone recognition performance and age at implantation. Nonetheless, Peng et al. (2004) and Wong and Wong (2004) did not find significant correlation between tone-recognition performance and any of the potential predictive variables. The limited sample sizes in these studies may explain the discrepancies.

Xu et al. (2009b) collected a much larger sample size and were able to study a number of other predictor variables as potential contributors to tone recognition in addition to the demographic variables. These predictors included family variables (such as family size and household income), cochlear implant variables (such as implant type and speech processing strategy), and educational variables (such as communication mode and duration of speech therapy). All predictor variables were entered step-wise into a linear regression model for analysis. The regression analysis, however, showed that only the demographic variables were the significant predictors for tone recognition. The study showed that jointly age at implantation and duration of CI use were significant predictors to tone recognition. Xu et al. (2009b) reported that duration of CI use was a stronger predictor, because it had a significant marginal relationship with tone recognition. Age at implantation alone, however, could not explain a significant amount of the total variance in tone recognition. Therefore, the marginal relationship was not significant. When both variables were entered into the regression model, in the presence of duration of CI use, age at implantation explained a significant amount of unique variance in tone recognition. Jointly, they explained approximately 50% of the total variance for tone-recognition outcome. The results indicate that although the performance of perception is bound to improve as the experience of using the device increases, early implantation may facilitate this improvement.

3.5 CI Stimulation Features and Tone Recognition

Although Xu et al. (2009b) did not find predictor variables associated with tone-recognition performance other than age at implantation and duration of CI use, several studies have indicated that CI tone recognition is related to speech-processing strategies. Fu et al. (2004) showed that ACE and CIS were better than SPEAK for Mandarin tone recognition. They speculated that the stimulation rate in the ACE and CIS strategies are higher than the SPEAK strategy (typically at 250 pps), providing higher temporal resolution to better encode the voice pitch. On the other hand, Barry et al. (2002) found similar Cantonese tone recognition between children who used ACE and SPEAK strategies. These contradictory findings might be the result of differences in stimulus, subject, and experience. For example, in Fu et al. study, all subjects had over 3 years of experience with the high-rate ACE strategy as opposed to limited experience with the SPEAK strategy.

Vandali et al. (2005) tested a novel speech-processing strategy, namely MEM (i.e., multi-channel envelope modulation) that was designed to enhance coding of F0 periodicity cues in the speech signal. In this strategy, the low frequency (80–400 Hz) envelope of the broadband signal, which contains F0 periodicity information for voiced/periodic signals, was used to modulate the envelope of each analysis band of the ACE strategy. CI subjects performed significantly better in pitch-ranking tests using the novel strategy than with ACE and CIS. Based on the encouraging results, Wong et al. (2008) evaluated the MEM strategy in a group of Cantonese-speaking adult CI users. Although tone recognition was not measured in the study, no difference was found in sentence recognition using the Cantonese version of HINT (Wong and Soli 2005) between ACE and MEM strategies.

Based on the psychophysical evidence that simultaneous stimulation of 2 adjacent electrodes results in a pitch percept that is between those elicited by stimulation of the 2 electrodes individually (see Bonham and Litvak 2008 for a review), Advanced Bionics introduced HiResolution with Fidelity 120 (HiRes 120) in 2006. Han et al. (2009) studied whether HiRes 120, which presumably provides much finer spectral resolution than the traditional strategies, would benefit lexical tone recognition. Twenty Mandarin-speaking children who were originally fitted with HiRes were experimentally switched to the new HiRes120 for a period of 6 months. Tone-recognition performance with the HiRes120 was comparable to the baseline performance with the HiRes strategy as a group. Some benefits were observed for approximately one third of the individuals after a period of either 3 or 6 months after the strategy conversion. Similar results were obtained in a separate study on HiRes 120 by Chang et al. (2009).

In an effort to enhance pitch coding, Med El recently launched a Fine Structure Coding Strategy (FSP) in which "packets" of pulses that represent zero-crossings of the acoustical signals in the low frequency bands are delivered to the apical electrodes (Arnoldner et al. 2007; Riss et al. 2009). Schatzer et al. (2010) compared tone-identification performance with the FSP strategy and the traditional CIS strategy in 12 Cantonese-speaking adult CI users. Their preliminary results showed no significant differences between the two strategies in an acute experiment.

There are a number of other speech-processing strategies that were proposed to enhance tonal-language recognition with CIs and were tested in acoustic simulations. Lan et al. (2004) proposed to use F0 as a carrier to replace the fixed-rate carrier in the standard CIS strategy. Nie et al. (2005) extracted slowly varying frequency modulation to encode the temporal fine structure. Luo and Fu (2004) modified the temporal envelope as well as the modulation depth of the periodicity fluctuation in local channels to better resemble the F0 contour. Alternatively, the overall amplitude contour was adjusted based on the F0 contour before the vocoder processing. Yuan et al. (2009) attempted to replace the temporal envelopes in the high frequency bands by a sinusoid with frequency equal to the F0 and found that Cantonese tone-recognition performance significantly improved. All these manipulations in signal processing have not been implemented in CI processors.

3.6 The Relation Between Musical and Lexical Pitch Perception

Because of inadequate representation of pitch information in the cochlear implant systems, implant users also have difficulties in perceiving musical pitch (see McDermott, Chap. 13). Studies indicate that postlingually deafened implant users consistently show impairment in identifying familiar songs without rhythmic cues (e.g., Gfeller et al. 2002, 2007; Leal et al. 2003). Many of them reported that the enjoyment of listening to music declines substantially after implantation (Lassaletta et al. 2007). A handful of studies specifically examined pitch-discrimination ability in CI users that is directly linked to their ability to perceive music. Fujita and Ito (1999) reported that the pitch-ranking thresholds measured from 8 CI users fell in a wide range of 4 semitones to 2 octaves. Looi et al. (2008) reported that CI subjects were unable to rank pitches that were a quarter-octave (i.e., 3 semitones) apart. They were only able to rank pitches that were a half-octave and one-octave apart 64% and 68% of the time correctly, respectively. Similar results of CI users' performance in pitch ranking were reported by Sucher and McDermott (2007). Other studies used adaptive procedures and reported pitch-discrimination thresholds that were in the range of 1–12 semitones (Gfeller et al. 2002; Nimmons et al. 2008; Kang et al. 2009).

Based on the notion that music appreciation and tone recognition both involve pitch perception, intuitively these two aspects of pitch perception should correlate with each other. Such a relationship was reported by Wang et al. (2010) who examined the mechanisms of musical and lexical tone recognition using CIs. Using a novel method to measure music perception that had several advantages over the traditional pitch ranking or familiar melody tests, they found that the discrimination thresholds of the CI subjects were highly variable, ranging from 0.65 to 19 semitones and were significantly worse than those of the normal-hearing controls (Fig. 14.6, left panel). Tone-recognition scores from the CI subjects ranged from 12.5% to 86.8% correct (Fig. 14.6, middle panel). More importantly, a highly significant negative

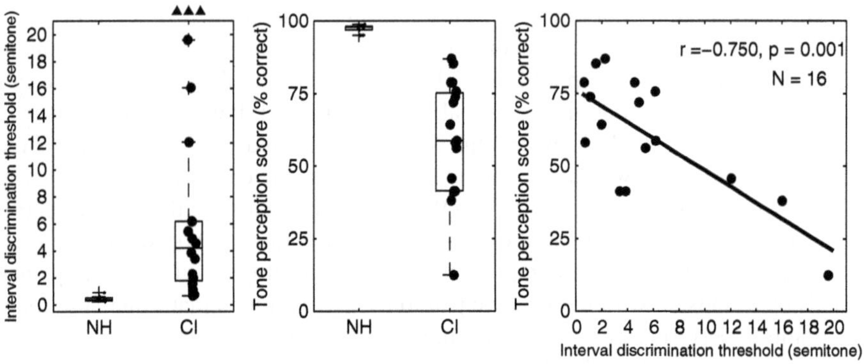

Fig. 14.6 Musical and lexical tone recognition performance. *Left panel*: Box plot of the averaged pitch interval discrimination thresholds in 10 normal-hearing (NH) and 19 CI subjects. The 3 triangles plotted at the top represent the 3 CI subjects who could not perform the interval discrimination test even at a ΔF0 of 2 octaves. *Middle panel*: Box plot of the Mandarin-Chinese tone-recognition scores for normal-hearing (N=10) and CI (N=19) subjects. *Right panel*: Correlation between tone-recognition scores and averaged pitch interval discrimination thresholds in CI subjects. Each symbol represents one subject with a CI. The *solid line* represents the linear fit of the data

correlation was shown between the pitch interval discrimination thresholds and the tone-recognition performance in the CI subjects (r=−0.75, N=16, p<0.01) (Fig. 14.6, right panel).

Wang et al. (2010) suggested that the strong correlation between the musical interval discrimination threshold and the tone-recognition performance indicated a shared mechanism in electric stimulation for musical and voice pitch perception. Musical pitch can be perceived via temporal patterns in amplitude modulation over restricted and relatively low modulation frequencies (e.g., McKay et al. 1994). Likewise, lexical tone information is supported by the periodicity in the temporal envelopes. Although they speculated that the temporal pitch coding underlies the common mechanisms for musical pitch and lexical tone recognition, it should not be ruled out that pitch perception can be realized through different excitation patterns of the electrical stimulation via a more place coding mechanism (e.g., Pretorius and Hanekom 2008).

4 Tone Production

4.1 Tone Development in Normal-Hearing Children

There is a substantial body of literature related to normal-hearing children's phonological development, but only a few on the tone acquisition of children who speak tonal languages. The earliest account of tone acquisition in normal-hearing children was from Chao (1951), who reported that his granddaughter acquired tones very early, and her isolated tones of stressed syllables were practically the same as in standard Mandarin (cited by Li and Thompson 1977 and Tse 1978). Li and Thompson (1977) studied tone acquisition in 17 Mandarin-speaking children ranging in age from 1.5 to 3.0 years old. Although no exact time frame for tone development was provided and the finding was descriptive in nature, Li and Thompson (1977) concluded that tone acquisition is accomplished very early in age. Tone acquisition takes place within a relatively short period of time and is well in advance of the mastery of segmentals. This notion was first suggested in a longitudinal case study (Tse 1978). Tse reported that his son's tone acquisition peaked between 14 and 22 months of age. More recently, Zhu and Dodd (2000) studied phonological acquisition in a large group of Mandarin-speaking children. Tone production in 21 children of their youngest subgroup (1.5–2.0) was reported fairly accurate. Both Li and Thompson (1977) and Zhu and Dodd (2000) used a picture-naming procedure to elicit production. One or two experienced Mandarin-speaking phoneticians transcribed the tone production as a measure for production performance. Wong et al. (2005) used 10 Mandarin-speaking adults to judge the tone production of 13 3-year-old Mandarin-speaking children residing in the United States. They found that the children had not fully mastered the production of the 4 Mandarin Chinese tones in monosyllabic words. From the limited data on tone acquisition, there is no clear agreement on the exact time frame that tone acquisition is complete.

4.2 *Tone-Production Performance in CI Children*

Accompanying the difficulties in perceiving tones, prelingually deafened implant users have also shown challenges in tone production, probably as a result of distorted auditory input of the tone targets (Xu et al. 2004, 2010). A converging finding of tone-production performance in those children is that, like tone recognition, there is a great individual difference among the users (Peng et al. 2004; Xu et al. 2004, 2009b; Han et al. 2007; Lee et al. 2010). Peng et al. (2004) reported tone-production performance in a group of 30 prelingually deafened, Mandarin-speaking children with CIs. The average percent-correct score for the group was 62%. The lowest score was about 20% correct, while 2 of the children scored nearly perfect. Peng et al. found that the children's performance was correlated with their age at implantation but was not related to the experience of the device use. More recently, Lee et al. (2010) evaluated Cantonese tone production in a longitudinal study. Their results indicated that children with earlier implantation (<4 years old) achieved more effective acquisition of lexical tones than those who received CIs after 4 years of age.

Tone production of Mandarin-speaking prelingually deafened children with CIs was further explored by Han et al. (2007). Native adult listeners were not asked to give any subjective judgments, but rather, they were asked to identify the tones they heard out of 4 possible choices (chance = 25% correct). The averaged production accuracy of the CI group was 48% correct, significantly lower than that of the

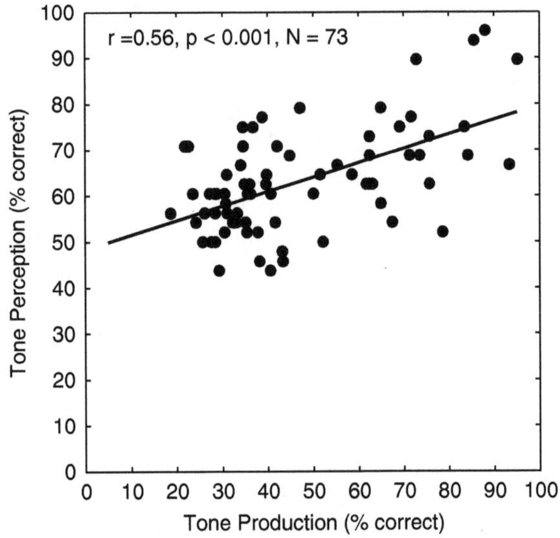

Fig. 14.7 Correlation between tone recognition and production performance in CI children. Tone-production scores were those evaluated by the normal-hearing adult listeners. Each symbol represents 1 subject with a CI. The solid line represents the linear fit of the data

age-matching normal-hearing control group (78% correct). The CI subjects in the study had particular difficulties in producing tone 2 followed by tones 3 and 4. The normal-hearing, native Mandarin-speaking adult listeners often perceived the CI children's intended contour tones as a flat tone (i.e., tone 1). In a follow-up study with more subjects, Xu et al. (2009b) also observed that the implanted children seemed to have particular difficulties with producing tone 2. This is consistent with the report by Peng et al. (2007), which revealed deficits in the production of rising intonation in native English-speaking pediatric CI users. Nonetheless, the difficulty of producing the rising tone 2 cannot seem to be attributed to the ability of perceiving this tone. It is more likely caused by the efforts demanded in producing the rising pitch. The ability of perception, similarly, cannot fully explain the difficulties of the children with CIs had in producing tone 3, which had the most sophisticated pitch contour. Although perception did not seem to predict the children's production performance for each tone, the perception accuracy predicted their overall production performance, and vice versa (Fig. 14.7). This correlation suggested that the development in tone recognition and production in children with CIs in general are two related aspects of language that facilitate each other (Xu et al. 2010).

Studies that examined the acoustic properties of the tone production from children with CIs revealed interesting findings. One of the approaches for the acoustic analysis is based on the distributions of the onsets and offsets of the F0 contours (Barry and Blamey 2004). Such distributions form the tonal ellipses, the overlap of which can be quantified based on Signal Detection Theory (Green and Swets 1966). Figure 14.8 shows an example of the tonal ellipses from 4 normal-hearing, native Mandarin-speaking children and 4 prelingually deafened children with CIs. The tonal ellipses reflect two components of the acoustic properties, which are the variability in the F0 use for one tone (i.e., ellipse size) and the overall F0 span that measures the tonal area (i.e., ellipse overlap).

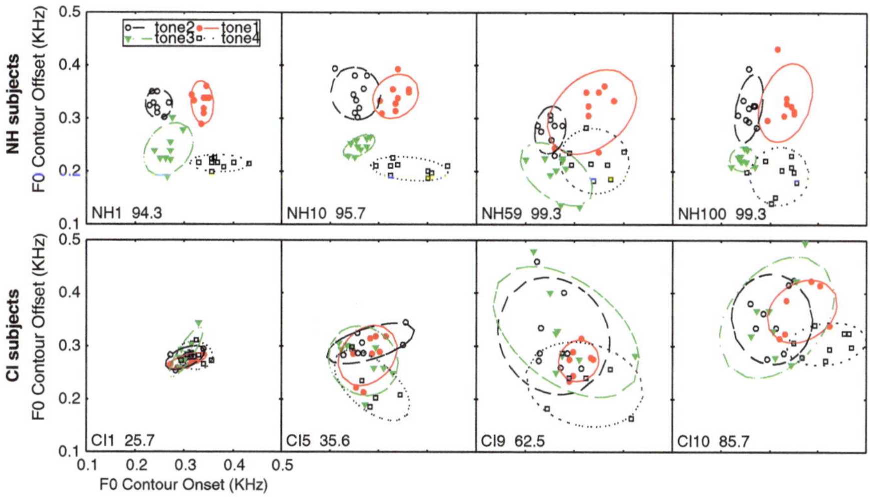

Fig. 14.8 Tonal ellipses based on distributions of F0 onsets and offsets of F0 contours produced by children. The 4 subjects from normal-hearing group (*upper panels*) and 4 from prelingually-deafened CI group (*lower panels*) were randomly selected from those with tone-production scores (judged by the native adult listeners) in the ranges of 0–25, 25–50, 50–75, and 75–100 percentiles of each respective group. Each data point represents a pair of F0 onset-offset value of a monosyllabic word. Symbols in different colors and styles represent different tones, as indicated by the legend in the *top left-hand panel*. Each *ellipse* encompasses 95% of the data points for each tone

The F0 variability, as assessed by the tonal ellipse analysis mentioned above, shows an age-related function in the typically-developing normal-hearing children. Zhou and Xu (2008a) showed that the F0 use of individual tones by the normal-hearing group presents a more confined pattern with age (i.e., ≥6 years of age). Such a development, however, was not found in the CI group with their duration of device use. That means that the F0 variability does not improve even with the accumulated experience with the device (Zhou and Xu 2008a). Xu et al. (2009b) confirmed that a relatively older normal-hearing group (≥6 years of age) have a significantly smaller variability of F0 use for a particular tone than that of the CI group. The F0 span or tonal area that measures the overall F0 range has always been reported to be much larger in the normal-hearing group than the CI children group. Zhou and Xu (2008a) indicated that even though the younger normal-hearing children have not learned to use a certain F0 range consistently for producing individual tones, their large F0 span or tonal area compensates for such variability; thus their tone production remains more differentiable than the CI children.

In addition to the acoustic analysis, artificial neural networks have been used to evaluate tone production in tonal-language speakers (e.g., Xu et al. 2006, 2007; Zhou et al. 2008). A feed-forward multilayer perceptron has been used to recognize the tones produced by a group of Mandarin-speaking children with normal hearing (Xu et al. 2007). The neural network provides direct classification results, from which the recognition percent-correct scores as well as tone confusion matrices can

Fig. 14.9 Correlation of tone-production performance by CI subjects evaluated by a neural network and by normal-hearing adult listeners. Each symbol represents 1 subject with a CI. The *solid line* represents the linear fit of the data

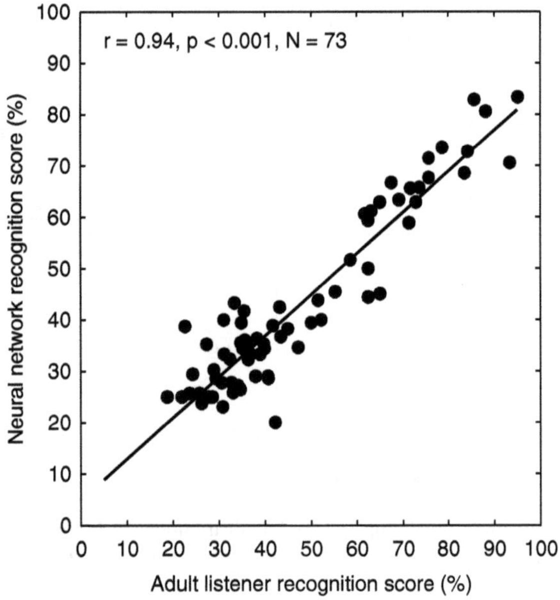

be generated. The error patterns of the neural network were remarkably similar to that of the human listeners (Xu et al. 2007; Zhou et al. 2008). Xu et al. (2009b) tested tone production in a group of 73 prelingually deaf children with CIs and found that the neural-network data correlated strongly with the perceptual judgments by adult native listeners ($r=0.94$, $N=73$, $p<0.001$) (Fig. 14.9).

5 Vocal Singing with CIs

As discussed in 3.6, musical pitch perception and lexical tone recognition may share a similar mechanism in electric hearing. As a consequence of poor pitch perception, prelingually deafened children with CIs have demonstrated poor development in vocal singing similar to that in tone production. Nakata et al. (2005) studied vocal singing in 12 congenitally deafened children (4.9–10.3 years of age) who had received CIs. Children with CIs could sing familiar songs from their memory, although the pitch patterns were largely unrelated to the direction of pitch patterns in the target songs (Nakata et al. 2005). Xu et al. (2009a) further explored vocal singing in 7 prelingually deafened children with CIs (age: 5.4–12.3 years old). The control group consisted of 14 normal-hearing children (age: 4.1–8.0 years old). The production of music pitch was evaluated acoustically. In the study, five metrics were developed based on the acoustic analysis of the F0 contours of the sung notes.

The five metrics included (1) F0 contour direction of the adjacent notes, (2) F0 compression ratio of the entire song, (3) mean deviation of the normalized F0 across the notes, (4) mean deviation of the pitch intervals, and (5) standard deviation of the note duration differences. Compared to the normal-hearing children, the CI group performed significantly poorer in the first four metrics that assessed the pitch-based performance of vocal singing. Similar to tone production, a large individual difference was observed. Singing of the CI children tended to be monotonic, compressed in the F0 range, and largely unrelated to the pitch contour of the target songs. No significant differences were seen between the two groups in the rhythm-based measure (i.e., the fifth metric). Current CI systems can faithfully deliver rhythmic information. Patients with CIs have been reported to perform at a level similar to normal-hearing subjects in rhythmic perception tasks (e.g., Gfeller et al. 1997; Kong et al. 2004; see also McDermott, Chap. 13). Thus, as a result of the preserved rhythmic information, the rhythmic aspect of singing in implanted children might not be significantly affected.

6 Summary

The primary acoustic cue for tone recognition is temporal or spectral fine structure of the signal. When the primary cue is not available, as in electric hearing, the temporal cues such as the amplitude contour, periodicity in the amplitude modulation patterns, or duration serve as secondary cues for tone recognition. Tone recognition is less susceptible to the spectral distortions because of the contribution of the secondary temporal cues.

Tone development in children with CIs demonstrates different patterns than the normal-hearing children, as revealed by the acoustic properties of their tone production and the error patterns of tone recognition. Although the overall performance of tone recognition is correlated with that of tone production in children with CIs, their error patterns of perception and production are not associated. Earlier implantation and more experience with the implant device predict better tone recognition, but only younger age at implantation predicts better tone production in children with CIs. Because of poor pitch encoding in the contemporary CI systems, prelingually deafened CI users who speak tonal languages generally have poor tone production and vocal singing. Some tonal-language users can probably maintain a fairly high level of speech perception using contextual cues when tone information is sparse. The long-term impact of poor tone recognition on tonal-language development remains to be explored in pediatric CI users.

Acknowledgements We thank Heather Schultz and Marisol Gliatas for the technical support during the preparation of the manuscript. The work was supported in part by NIH NIDCD Grants R03-DC006161, R15-DC009504, and F31-DC009919.

References

Abramson, A. S. (1978). Static and dynamic acoustic cues in distinctive tone. Language Speech, 21, 319–325.

Arnoldner, C., Riss, D., Brunner, M., Durisin, M., Baumgartner, W.-D., & Hamzavi, J.-S. (2007). Speech and music perception with the new fine structure speech coding strategy: preliminary results. *Acta Oto-Laryngologica*, 127, 1298–1303.

Barry, J. G., & Blamey, P. J. (2004). The acoustic analysis of tone differentiation as a means for assessing tone production in speakers of Cantonese. *Journal of the Acoustical Society of America*, 116, 1739–1748.

Barry, J. G., Blamey, P. J., Martin, L. F. A., Lee, K. Y. S., Tang, T., Ming, Y. Y., & Van Hasselt, C. A. (2002). Tone discrimination in Cantonese-speaking children using a cochlear implant. *Clinical Linguistics & Phonetics*, 16, 79–99.

Baskent, D., & Shannon, R. V. (2003). Speech recognition under conditions of frequency-place compression and expansion. *Journal of the Acoustical Society of America*, 113, 2064–2076.

Baskent, D., & Shannon, R. V. (2005). Interactions between cochlear implant electrode insertion depth and frequency-place mapping. *Journal of the Acoustical Society of America*, 117, 1405–1416.

Baskent, D., & Shannon, R. V. (2006). Frequency transposition around dead regions simulated with a noiseband vocoder. *Journal of the Acoustical Society of America*, 119, 1156–1163.

Bonham, B. H., & Litvak, L. M. (2008). Current focusing and steering: modeling, physiology, and psychophysics. *Hearing Research*, 242, 141–153.

Chang, Y. T., Yang, H. M., Lin, Y. H., Liu S. H., & Wu, J. L. (2009). Tone discrimination and speech perception benefit in Mandarin speaking children fit with HiRes fidelity 120 sound processing. *Otology & Neurotology*, 30, 750–757.

Chao, Y. R. (1951). The Cantian idiolect: an analysis of the Chinese spoken by a twenty-eight-month-old child. In C. A. Ferguson & D. I. Slobin (Eds.), *Studies of child language development*. New York: Holt, Rinehart and Winston, Inc.

Ciocca, V., Francis, A. L., Aisha, R., & Wong, L. (2002). The perception of Cantonese lexical tones by early-deafened cochlear implantees. *Journal of the Acoustical Society of America*, 111, 2250–2256.

Dorman, M. F., Loizou, P. C., & Rainey D. (1997). Simulating the effect of cochlear implant electrode insertion depth on speech understating. *Journal of the Acoustical Society of America*, 102, 2993–2996.

Duanmu, S. (2000). *The phonology of standard Chinese*. Oxford: Oxford University Press.

Firszt, J. B., Holden, L. K., Reeder, R. M., & Skinner, M. W. (2009). Speech recognition in cochlear implant recipients: comparisons of standard HiRes and HiRes120 sound processing. *Otology & Neurotology*, 30, 146–152

Fu, Q.-J., & Shannon, R. V. (1999). Recognition of spectrally degraded and frequency-shifted vowels in acoustic and electric hearing. *Journal of the Acoustical Society of America*, 105, 1889–1990.

Fu, Q.-J., & Zeng, F.-G. (2000). Identification of temporal envelope cues in Chinese tone recognition. *Asia Pacific Journal of Speech, Language and Hearing*, 5, 45–57.

Fu, Q.-J., Zeng, F.-G., Shannon, R. V., & Soli, S. D. (1998). Importance of tonal envelope cues in Chinese speech recognition. *Journal of the Acoustical Society of America*, 104, 505–510.

Fu, Q.-J., Hsu, C. J., & Horng, M. J. (2004). Effects of speech processing strategy on Chinese tone recognition by nucleus-24 cochlear implant users. *Ear and Hearing*, 25, 501–508.

Fujita, S., & Ito, J. (1999). Ability of nucleus cochlear implantees to recognize music. *Annals of Otology, Rhinology & Laryngology*, 108, 634–640.

Gfeller, K., Woodworth, G., Robin, D. A., Witt, S., & Knutson, J. F. (1997). Perception of rhythmic and sequential pitch patterns by normally hearing adults and adult cochlear implant users. *Ear and Hearing*, 18, 252–260.

Gfeller, K., Turner, C., Mehr, M., Woodworth, G., Fearn, R., Knutson, J. F., Witt, S., & Stordahl, J. (2002). Recognition of familiar melodies by adult cochlear implant recipients and normal-hearing adults. *Cochlear Implant International*, 3, 29–53.

Gfeller, K., Turner, C., Oleson, J., Zhang, X. Y., Gantz, B., Froman, R., & Olszewski, C. (2007). Accuracy of cochlear implant recipients on pitch perception, melody recognition, and speech reception in noise. *Ear and Hearing*, 28, 412–23.

Gottfried, T. L., & Suiter, T. L. (1997). Effect of linguistic experience on the identification of Mandarin Chinese vowels and tones. *Journal of Phonetics*, 25, 207–231.

Green, D. M., & Swets, J. A. (1966). *Signal detection theory and psychophysics*. New York: Wiley.

Greenwood, D. D. (1990). A cochlear frequency-position function for several species-29 years later. *Journal of the Acoustical Society of America*, 87, 2952–2605.

Han, D., Zhou, N., Li, Y., Chen, X., Zhao, X., & Xu, L. (2007). Tone production of Mandarin Chinese speaking children with cochlear implants. *International Journal of Pediatric Otorhinolaryngology*, 71, 875–880.

Han, D., Liu, B., Zhou, N., Chen, X., Kong, Y., Liu, H., Zheng, Y., & Xu, L. (2009). Lexical Tone recognition with HiResolution® 120 Speech-Processing Strategy in Mandarin-Speaking Children. *Ear and Hearing*, 30, 169–177.

Howie, J. (1976). *An acoustic study of Mandarin tones and vowels*. London: Cambridge University Press.

Kang, R., Nimmons, G. L., Drennan, W., Longnion, J., Ruffin, C., Nie, K., Won, J. H., Worman, T., Yueh, B., & Rubinstein, J. (2009). Development and validation of the University of Washington Clinical Assessment of Music Perception test. *Ear and Hearing*, 30, 411–418.

Koch, D. B., Osberger, M. J., Segel, P., & Kessler, D. (2004). HiResolution™ and conventional sound processing in the HiResolution™ bionic ear: using appropriate outcome measures to assess speech recognition ability. *Audiology & Neurotology*, 9, 214–223.

Kong, Y. Y., & Zeng, F. G. (2006). Temporal and spectral cues in Mandarin tone recognition. *Journal of the Acoustical Society of America*, 120(5), 2830–2840.

Kong, Y. Y. Cruz, R., Jones, J. A., & Zeng, F. G. (2004). Music perception with temporal cues in acoustic and electric hearing. *Ear and Hearing*, 25, 173–185.

Kuo, Y.-C., Rosen, S., & Faulkner, A. (2008). Acoustic cues to tonal contrasts in Mandarin: Implications for cochlear implants. *Journal of the Acoustical Society of America*, 123(5), 2815–2824.

Lan, N., Nie, K., Gao, S., & Zeng, F. G. (2004). A novel speech-processing strategy incorporating tonal information for cochlear implants. *IEEE Transactions on Biomedical Engineering*, 51, 752–60.

Lassaletta, L., Castro, A., Bastarrica, M., Perez-Mora, R., Madero, R., de Sarria, J., & Gavilan, J. (2007). Does music perception have an impact on quality of life following cochlear implantation? *Acta Oto-Laryngologica*, 127, 682–686.

Leal, M. C., Shin, Y. J., Laborde, M. L., Calmels, M. N., Verges, S., Lugardon, S., Andrieu, S., Deguine, O., & Fraysse, B. (2003). Music perception in adult cochlear implant recipients. *Acta Oto-Laryngologica*, 123, 826–835.

Lee, C.-Y. (2009). Identifying isolated, multispeaker Mandarin tones from brief acoustic input: A perceptual and acoustic study. *Journal of the Acoustical Society of America*, 125, 1125–1137.

Lee, C.-Y., & Hung, T.-H. (2008). Identification of Mandarin tones by English-speaking musicians and nonmusicians. *Journal of the Acoustical Society of America*, 5, 3235–3248.

Lee, K. Y. S., van Hasselt, C. A., Chiu, S. N., & Cheung, D. M. C. (2002). Cantonese tone recognition ability of cochlear implant children in comparison with normal-hearing children. *International Journal of Pediatric Otorhinolaryngology*, 63, 137–147.

Lee, K. Y. S., van Hasselt, C. A., & Tong, M. C. F. (2010). Age sensitivity in the acquisition of lexical tone production: evidence from children with profound congenital hearing impairment after cochlear implantation. *Annals of Otology, Rhinology & Laryngology*, 119, 258–265.

Li, C. N., & Thompson, S. A. (1977). The acquisition of tone in Mandarin-speaking children. *Journal of Child Language*, 4, 185–199.

Liang, Z. A. (1963). Tonal discrimination of Mandarin Chinese. *Acta Physiologica Sinica* 26, 85–91.

Lin, M.-C. (1988). The acoustic characteristics and perceptual cues of tones in standard Chinese. *Zhongguo Yuwen*, 204, 182–193.

Liu, S., & Samuel, A. G. (2004). Perception of Mandarin lexical tones when f0 information is neutralized. *Language and Speech*, 47, 109–138.

Liu, T.-C., Chen, H.-P., & Lin, H.-C. (2004) Effects of limiting the number of active electrodes on mandarin tone perception in young children using cochlear implants. *Acta Oto-Laryngologica*, 124, 1149–1154.

Loizou P.C. (2006). Speech processing in vocoder-centric cochlear implants. *Advances in Otorhinolaryngology*, 64, 109–143.

Looi, V., McDermott, H., McKay, C., & Hickson, L. (2008). Music perception of cochlear implant users compared with that of hearing aid users. *Ear and Hearing*, 29, 421–434.

Lorenzi, C., Gilbert, G., Carn, H., Garnier, S., & Moore, B. C. J. (2006) Speech perception problems of the hearing impaired reflect inability to use temporal fine structure. *Proceedings of the National Academy of Science*, 103, 18866–18869.

Luo, C., & Wang, J. (1981). *Putong yuyinxue gangyao* [*Outline of general phonetics*], new ed. Beijing: Shangwu Yinshuguan.

Luo, X., & Fu, Q. (2004). Enhancing Chinese tone recognition by manipulating amplitude envelope: Implications for cochlear implant. *Journal of the Acoustical Society of America*, 116, 3659–3667.

Luo, X., Fu, Q.-J., Wei, C.-G., & Cao, K.-L. (2008). Speech recognition and temporal amplitude modulation processing by Mandarin-speaking cochlear implant users. *Ear and Hearing*, 29, 957–970.

McKay, C. M., McDermott, H. J., & Clark, G. M. (1994). Pitch percepts associated with amplitude-modulated current pulse trains in cochlear implantees. *Journal of the Acoustical Society of America*, 96, 2664–2673.

Miller, J.D. (1961). Word tone recognition in Vietnamese whispered speech. *Word*, 17, 11–15.

Nakata, T., Trehub, S. E., Mitani, C., Kanda, Y., Shibasaki, A., & Schellenberg, E. G. (2005). Music recognition by Japanese children with cochlear implants. *Journal of Physiological Anthropology & Applied Human Sciences*, 24, 29–32.

Nie, K. B., Stickney, G. S., & Zeng, F.-G. (2005). Encoding frequency modulation to improve cochlear implant performance in noise. *IEEE Transactions on Biomedical Engineering*, 52, 64–73.

Nimmons, G. L., Kang, R. S., Drennan, W. R., Longnion, J., Ruffin, C., Worman, T., Yueh, B., & Rubinstein, J. T. (2008). Clinical assessment of music perception in cochlear implant listeners. *Otology & Neurotology*, 29, 149–155.

Peng, S. C., Tomblin, J. B., Cheung, C., Lin, Y.-S., & Wang, L. (2004). Perception and production of Mandarin tones in prelingually deaf children with cochlear implants. *Ear and Hearing*, 25, 251–264.

Peng, S. C., Tomblin, J. B., Spencer, L. J., & Hurtig, R. R. (2007). Imitative production of rising speech intonation in pediatric cochlear implant recipients. *Journal of Speech, Language, and Hearing Research*, 50, 1210–1227.

Pfingst, B. E., Franck, K. H., Xu, L., Bauer, E. M., & Zwolan, T. A. (2001). Effects of electrode configuration and place of stimulation on speech perception with cochlear prostheses. *Journal of the Association for Research in Otolaryngology*, 2, 87–103.

Pretorius, L. L., & Hanekom, J. J. (2008). Free field frequency discrimination abilities of cochlear implant users. *Hearing Research*, 244, 77–84.

Riss, D., Arnoldner, C., Reiss, S., Baumgartner, W. D., & Hamzavi, J. S. (2009). 1-year results using the Opus speech processor with the fine structure speech coding strategy. *Acta Oto-Laryngologica*, 129, 988–991.

Schatzer, R., Krenmayr, A., Au, D. K. K., Kals, M., & Zierhofer, C. (2010). Temporal fine structure in cochlear implants: preliminary speech perception results in Cantonese-speaking implant users. *Acta Oto-Laryngologica*, 130, 1031–1039.

Schouten, J. F., Ritsma, R. J., & Cardoz, B. L. (1962). Pitch of the residue. *Journal of the Acoustical Society of America*, 34, 1418–1424.

Shannon, R. V., Galvin, J. J., 3 rd, & Baskent, D. (2002). Holes in hearing. *Journal of the Association for Research in Otolaryngology*, 3, 185–199.

Smith Z. M., Delgutte B., & Oxenham A. J. (2002). Chimaeric sounds reveal dichotomies in auditory perception. *Nature*, 416, 87–90.

Sucher, C. M. & McDermott, H. J. (2007). Pitch ranking of complex tones by normally hearing subjects and cochlear implant users. *Hearing Research*, 230, 80–87.

Tse, J. K., (1978). Tone acquisition in Cantonese: a longitudinal case study. *Journal of Child Language*, 5, 191–204.

Vandali, A. E., Sucher, C., Tsang, D. J., McKay C. M., Chew J. W. D., & McDermott, H. J. (2005). Pitch ranking ability of cochlear implant recipients: a comparison of sound-processing strategies. *Journal of the Acoustical Society of America*, 117, 3126–3138.

Wang, S., Xu, L., & Mannell, R. (2010). Lexical tone recognition in sensorineurally hearing impaired listeners using temporal cues. Paper presented at the American Auditory Society Annual Meeting, Scottsdale, AZ.

Wang, W., Zhou, N., & Xu, L. (2010). Musical pitch and lexical tone recognition with cochlear implants. *International Journal of Audiology*, 50, 270–278.

Wei, C., Cao, K., & Zeng, F. G. (2004). Mandarin tone recognition in cochlear-implant subjects. *Hearing Research*, 197, 87–95.

Wei, W. I., Wong, R., Hui, Y., Au, D. K. K., Wong, B. Y. K., Ho, W. K., Tsang, A., Kung P., & Chung, E. (2000). Chinese tonal language rehabilitation following cochlear implantation in children. *Acta Oto-Laryngologica*, 120, 218–221.

Wei, C., Cao, K., Jin, X., Chen, X., & Zeng, F. G. (2007). Psychophysical performance and Mandarin tone recognition in noise by cochlear implant users. *Ear and Hearing*, 28(2), 62 S–65 S.

Whalen, D. H., & Xu, Y. (1992). Information for Mandarin tones in the amplitude contour and in brief segments. *Phonetica*, 49, 25–47.

Wong, A. O. C., & Wong, L. L. N. (2004). Tone recognition of Cantonese-speaking prelingually hearing-impaired children with cochlear implants. *Otolaryngology-Head and Neck Surgery*, 130, 751–758.

Wong, L. L. N., & Soli, S. D. (2005). Development of the Cantonese HINT. *Ear and Hearing*, 26, 276–289.

Wong, L. L. N., Vandali, A. E., Ciocca, V., Luk, B., Ip, V. W. K., Murray, B., Yu, H. C., & Chung, I. (2008). New cochlear implant coding strategy for tonal language speakers. *International Journal of Audiology*, 47, 337–347.

Wong, P., Schwartz, R. G., & Jenkins, J. J (2005). Perception and production of lexical tones by 3-year-old, Mandarin-speaking children. *Journal of Speech, Language, and Hearing Research*, 48, 1065–1079.

Xu, L., & Pfingst, B. E., (2003). Relative importance of temporal envelope and fine structure in lexical-tone recognition. *Journal of the Acoustical Society of America*, 114, 3024–3027.

Xu, L. & Pfingst, B. E. (2008). Spectral and temporal cues for speech recognition: Implications for auditory prostheses. *Hearing Research*, 242, 132–140.

Xu, L., Tsai, Y., & Pfingst. B. E. (2002). Features of stimulation affecting tonal-speech perception: implications for cochlear prostheses. *Journal of the Acoustical Society of America*, 112, 247–258.

Xu, L., Li, Y., Hao, J. P., Chen, X. W., Xue, S. A., et al. (2004). Tone production in Mandarin-speaking children with cochlear implants: a preliminary study. *Acta Oto-Laryngologica*, 124, 363–367.

Xu, L., Zhang, W., Zhou, N., Lee, C.-Y., Li, Y., et al. (2006). Mandarin Chinese tone recognition with an artificial neural network. *Journal of Otology*, 1, 30–34.

Xu, L., Chen, X., Zhou, N., Li, Y., Zhao, X., et al. (2007) Recognition of lexical tone production of children with an artificial neural network. *Acta Oto-Laryngologica*, 127, 365–369.

Xu, L., Zhou, N., Chen, X., Li, Y., Schultz, H. M., et al. (2009a). Vocal singing by prelingually-deafened children with cochlear implants. *Hearing Research*, 255, 129–134.

Xu, L., Zhou, N., Huang, J., Chen, X., Li, Y., et al. (2009b). Lexical tone development in prelingually-deafened children with cochlear implants. Paper presented at The 12th International Symposium on Cochlear Implants in Children, Seattle, WA.

Xu, L., Chen, X., Lu, H., Zhou, N., Wang S., et al. (2010) Tone recognition and production in pediatric cochlear implants users. *Acta Oto-Laryngologica*, 131, 395–398.

Yuan, M., Lee, T., Yuen, K. C. P., Soli, S. D., van Hasselt, C. A., et al. (2009). Cantonese tone recognition with enhanced temporal periodicity cues. *Journal of the Acoustical Society of America*, 126, 327–337.

Yuen, K. C. P., Yuan, M., Lee, T., Soli, S., Tong, M. C. F., & van Hasselt, C. A. (2007) Frequency-specific temporal envelope and periodicity components for lexical tone identification in Cantonese. *Ear and Hearing*, 28, 107 S–113S.

Zhou, N., & Xu, L. (2008a). Development and evaluation of methods for assessing tone production skills in Mandarin-speaking children with cochlear implants. *Journal of the Acoustical Society of America*, 123, 1653–1664.

Zhou, N., & Xu, L. (2008b). Lexical tone recognition with spectrally mismatched envelopes. *Hearing Research*, 46, 36–43

Zhou, N., Zhang, W., Lee, C.-Y., & Xu, L. (2008). Lexical tone recognition by an artificial neural network. *Ear and Hearing*, 29, 326–335.

Zhou, N., Xu, L. & Lee, C.-Y. (2010). The effects of frequency-place mismatch on consonant confusion. *Journal of the Acoustical Society of America*, 128, 401–409.

Zhu, H., & Dodd, B. (2000). The phonological acquisition of Putonghua (Modern Standard Chinese). *Journal of Child Language*, 27, 3–42.

Chapter 15
Multisensory Processing in Cochlear Implant Listeners

Pascal Barone and Olivier Deguine

1 Introduction

With advances in cochlear implant (CI) technology, CI users' speech recognition in quiet can be very good; many are able to communicate on the telephone. However, other aspects of speech perception remain difficult, including speech understanding in noise, voice identification, and prosody recognition. These difficulties can be related to the limitations of current cochlear implants, which provide much less effective frequency processing than normal hearing. To recover the distorted speech signal, the limited CI processing can be complemented by the other modalities such as visual and somatosensory inputs. The present chapter describes how CI users take significant advantage of multisensory integration, mainly the combination of auditory and visual information, as well as the physiological mechanisms underlying this integration.

2 Multisensory Perception and Cross-Modal Compensation

In a natural environment, humans are exposed simultaneously to many kinds of sensory stimulation in visual, auditory, proprioceptive, somatosensory, or even chemical domains. A large body of psychophysical studies has demonstrated that simultaneous polysensory stimulation results in percepts distinct from those derived

P. Barone (✉)
Université Toulouse, CerCo, Université Paul Sabatier 3, Toulouse, France

Centre de Recherche Cerveau et Cognition UMR 5549, Pavillon Baudot, CHU Purpan, 31052, Toulouse, Cedex, France
e-mail: pascal.barone@cerco.ups-tlse.fr

F.-G. Zeng et al. (eds.), *Auditory Prostheses: New Horizons*, Springer Handbook of Auditory Research 39, DOI 10.1007/978-1-4419-9434-9_15,

from unisensory stimulation (Welch and Warren 1986). In these cases, multisensory integration improves perception by reducing ambiguity in simple detection, complex discrimination, memory, or learning tasks (Lehmann and Murray 2005; Lovelace et al. 2003; Seitz et al. 2006). Multisensory integration allows humans to link the unimodal sensory analysis with the polysensory context in which they appear, producing cross-modal fusion (Stein and Meredith 1993).

In many behavioral experiments, polysensory interactions have been measured by subject reaction time and performance (Welch and Warren 1986). For example, the detection of a visual target in a complex scene is facilitated when it is associated with an irrelevant auditory stimulus (McDonald et al. 2000; Vroomen and de Gelder 2000). The additional auditory stimulus not only reduces reaction time but also increases performances. However, the improved auditory-visual performance can only be achieved when the auditory and visual stimuli are congruent (Stein and Meredith 1993). If the stimuli are incongruent, then they may alter perception leading to illusory phenomena. The most famous examples are the ventriloquism illusion in which a sound source is mislocated towards a synchronous but spatially incongruent visual stimulus (Slutsky and Recanzone 2001; Spence and Driver 2000) and the "McGurk effect," in which conflicting audio and visual stimulation results in a bias in speech perception (McGurk and MacDonald 1976).

Despite vast literature on multi-modal interactions, their neuronal substrate is still poorly understood. Several brain regions have been described as "heteromodal" because they respond to more than one modality and consequently have been identified as sites of multi-modal integration (Calvert et al. 1998, 2000; Downar et al. 2000; Bremmer et al. 2001). Understanding the neuronal processes of multisensory interactions is also of primary importance to the mechanisms of cortical plasticity that occurs after sensory loss such as deafness (Bavelier and Neville 2002; Merabet and Pascual-Leone 2010). Because multisensory integration and cross-modal compensation are probably supported by a common cortical network (Cappe et al. 2009), the loss of one sensory modality may leads to an increase in performance by the remaining modalities (Rauschecker 1991; Roder et al. 1999; Bavelier and Neville 2002; Bavelier et al. 2006; Putzar et al. 2007). Crossmodal compensation of perception is accompanied by functional reorganizations (Kujala et al. 2000; Röder and Rosler 2004) such as colonization of the deprived cortical areas by the remaining modalities (Sadato et al. 1996; Cohen et al. 1997; Weeks et al. 2000; Röder et al. 2002). The degrees of functional reorganization and cross-modal compensation are highly dependent on the age at which the sensory deprivation occurs, as a result of the decreasing capacities of adaptive plasticity of the brain from birth to adulthood (Knudsen 2004).

The CI is an effective neuroprosthesis that allows postlingually deafened subjects to recover partial auditory functions, especially in speech intelligibility. At present, such a functional recovery is not available in other sensory modalities (Hallum et al. 2007). Thus, the access to CI patients provides a unique opportunity to analyze the perceptual strategies and cortical reorganization induced by a long period of deafness, as well as the reverse process, when the auditory function is recovered by the CI.

3 Speech Is Multisensory

While speech is primarily perceived through the auditory channel, numerous studies show that visual information improves speech intelligibility, particularly in a noisy background or with a degraded acoustic signal (Sumby and Pollack 1954; Summerfield 1979; MacLeod and Summerfield 1987; Ross et al. 2006). Visual information can also enhance speech detection and intelligibility, even when acoustic signals are perfectly clear. For example, visual information decreases the intensity threshold of speech detection (Grant and Seitz 2000) and improves speech perception under difficult situations, such as understanding foreign language or speech materials of high semantic complexity (Kim and Davis 2003).

Visual information is particularly important for speech perception in developing children. For instance, infants are highly sensitive to spatial (Aronson and Rosenbloom 1971; Spelke and Owsley 1979), temporal (Dodd 1979), and phonetic congruency between auditory and visual components of speech. Consequently, this audiovisual congruency has been suggested to facilitate speech acquisition in normal-hearing infants (Kuhl and Meltzoff 1982; MacKain et al. 1983), in hearing-impaired infants (Arnold and Kopsel 1996), as well as in cochlear-implanted deaf children (Staller et al. 1991; Tyler et al. 1997a, b; Geers et al. 2003). In contrast, blind infants have more difficulties learning peculiar phonemic contrasts, which are not easily distinguishable without the help of visual information (Mulford 1988).

Recent research has investigated the role of multisensory information in speech prosody perception. Prosody is used to communicate suprasegmental information, such as whether someone asks a question or makes a statement, which words in a phrase are emphasized, and what a speaker's emotional state is. It mainly consists of acoustic changes in the voice pitch, amplitude, and duration patterns (Morton and Jassem 1965). Yet it has been shown that some visual features, such as head and eyebrow movements, correlate with sound changes (Hadar et al. 1984; Munhall et al. 2004), while others such as chin movement conveys speech stress (Scarborough et al. 2009). Cross-modal prosody processing not only enhances the comprehension of speech in general (Munhall et al. 2004), but also compensates for the reduced pitch perception in cochlear implanted deaf patients in particular (Chatterjee and Peng 2008; Donnelly et al. 2009; Foxton et al. 2010).

4 Multisensory Processing in Cochlear-Implanted Deaf Listeners

4.1 *Role of Vision in Speech Comprehension*

In the last three decades, the CI has become the only efficient method to help restore partial hearing in individuals with profound bilateral sensorineural hearing loss (Copeland and Pillsbury 2004; Deggouj et al. 2007). The CI has led to considerable success in the functional rehabilitation of deafness (Moller 2006) in terms of its user's ability to understand spoken speech, identify environmental sounds, and in some cases to listen to music (Pressnitzer et al. 2005; Drennan and Rubinstein 2008; see also McDermott, Chap. 13).

However, the auditory information sent to the brain from the CI is not only spectrally degraded (Shannon et al. 1995), but also lacks the fine temporal structure (Friesen et al. 2001; Zeng et al. 2005; Lorenzi et al. 2006). These limitations likely contribute to an initial period of adaptation, typically several months, required for the CI recipients to reach an asymptotic level of speech performance (Tyler et al. 1997a, b). During this initial period of adaptation, CI recipients rely strongly on visual cues to complement the crude auditory signal delivered by the implant (Summerfield 1992; Tyler et al. 1997a, b; Grant et al. 1998; Kaiser et al. 2003). In fact there has been evidence that the CI recipients have higher levels of speech-reading performance than normal-hearing subjects (Bernstein et al. 2000, 2001; Rouger et al. 2007). There has been debate about whether better CI speech-reading performance is a result of learning processes before or after cochlear implantation (Gray et al. 1995; Bergeson et al. 2005; Giraud et al. 2001a, c). Figure 15.1a shows results from a longitudinal study of speech-reading performance in a population of about 100 postlingually deafened adult CI recipients with post-implantation follow-up up to 8 years (Rouger et al. 2007). Similar to the previously reported results (Gray et al. 1995), this large sample sized study found that, at the time when CI was first activated, CI speech-reading performance was already significantly higher than normal-hearing (NH) participants (35% vs. 9% word recognition). Interestingly, these initial visual speech perception scores did not correlate with their corresponding CI speech perception scores, suggesting different, rather than shared, speech-processing pathways between auditory and visual modalities. This different-pathway hypothesis was further supported by the maintenance of the high-level speech-reading scores over the entire post-implantation period tested. One reason for preserving the high speech-reading ability is that it contributes to bi-sensory integration, which improves speech recognition in noisy environments, a challenge for the majority of CI recipients (Fu et al. 1998; Munson and Nelson 2005).

In spite of the facilitative role of vision in visual-auditory speech, it has been claimed that cross-modal compensations linked to the acquisition of speech-reading skills or sign language in deafness (Nishimura et al. 1999; Petitto et al. 2000) could be detrimental to auditory recovery in CI users. The colonization of the auditory areas by visual speech processing in CI users could interfere with auditory treatment (Champoux et al. 2009). Indeed there is a much larger recruitment of the cortex to process visual information in CI users, particularly in poorly performing CI users (Doucet et al. 2006). This could imply that the extent of cross-modal reorganization directed toward visual processing is directly linked to the level of auditory recovery (Nishimura et al. 2000). In contrast, good CI users have an enhanced response to visual stimuli in the visual areas, which reflect more efficient visual processing to compensate for the crude information delivered by the implant. Furthermore, in CI recipients, the degree of temporal hypometabolism that reflects the level of functional cross-modal reorganization (Wanet-Defalque et al. 1988;

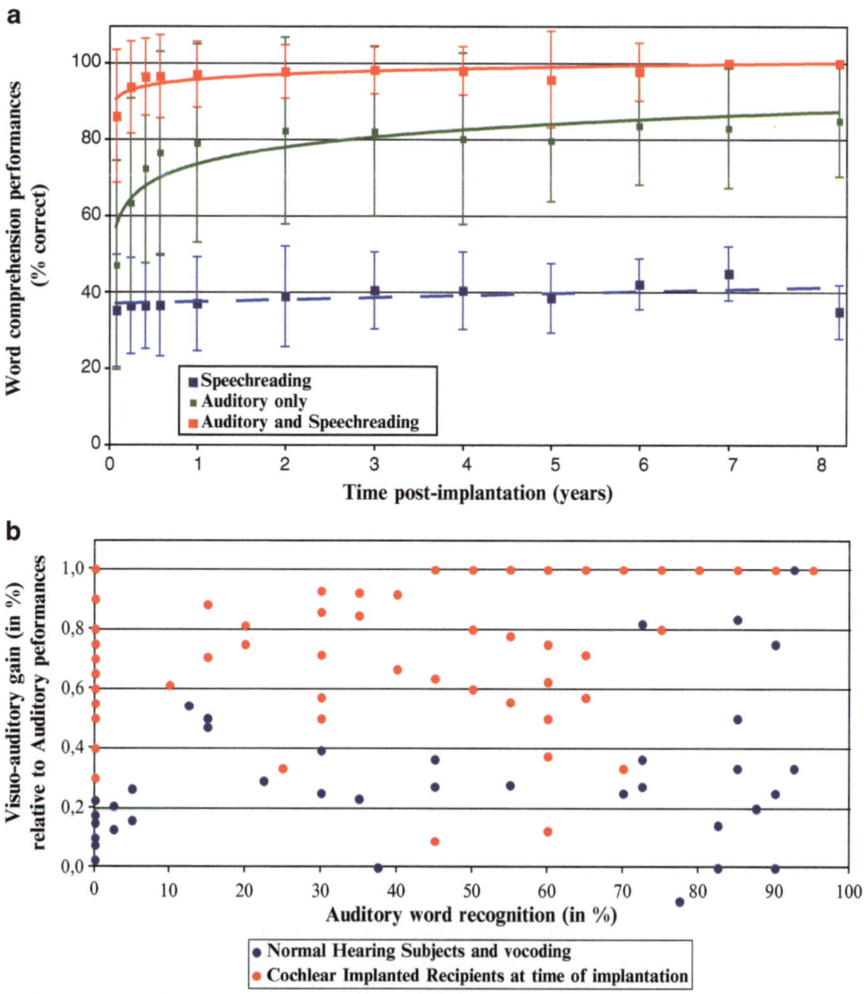

Fig. 15.1 Role of audiovisual integration in speech comprehension in cochlear implant recipients (adapted from Rouger et al. 2007). (**a**) Percent of correct responses for CI users at different periods after cochlear implant activation for speech-reading of words and auditory or audiovisual performance. Note that CI users maintain a high level of speech-reading performance even after complete auditory recovery. (**b**) Comparison of the audiovisual benefits between CI recipients and normal-hearing (NH) subjects with vocoder. For each group the audiovisual gain [(AV-A)/A] is plotted as function of the auditory performance. The CI distribution of multisensory gain is largely above that observed in NH subjects with similar auditory performance

Veraart et al. 1990; De Volder et al. 1997) can be predictive of speech performance after cochlear implantation (Lee et al. 2001, 2007; Giraud and Lee 2007). At present, it is crucial to define role of visual processing in CI users and how it impacts speech processing and auditory recovery, especially in children.

4.2 Mechanisms of CI Audiovisual Integration

To address directly the role of pre-implant deaf experience in audiovisual integration, audiovisual integration was obtained in naïve NH subjects exposed to noise-vocoded speech or CI simulations (Shannon et al. 1995). Figure 15.1b shows that the audiovisual gain in actual CI recipients doubled that in simulated CI subjects (Rouger et al. 2007), especially in conditions of low auditory performance (<30%). Application of an optimal model of multisensory integration to the observed data found that the improved multisensory performance with practice could be entirely explained by the increased auditory performance. This result suggests that, as a result of re-organization in response to the degraded auditory input, the audiovisual integration has been optimally integrated from the very beginning in CI recipients, and more importantly, the improved auditory performance post-implantation does not compromise this nearly optimal integration.

Audiovisual integration is also related to age and experience. In deaf children, Shorr and collaborators (2005) found that the bimodal fusion of the McGurk stimuli depended on the age of implantation, suggesting a sensitive period for the development of audiovisual integration. In postlingually deafened CI recipients, audiovisual integration efficiency and strategy are found to be dependent upon the duration of CI usage (Desai et al. 2008; Rouger et al. 2008). For example, when an auditory dental (/ata/) is dubbed with a visual bilabial (/apa/), most NH subjects show a response of (/aka/) that indicates some form of integrated articulatory components, with only 7% of them showing a response that is purely visual (i.e.,/apa/). In contrast, 98% of CI users reported purely visual percepts. These results suggest that, in perceiving the place of articulation for incongruent audiovisual stimuli, CI users place more weight on visual cues than auditory cues, whereas NH subjects balance the weight based on cues from both modalities.

The analysis based on probability summation of the facilitation in reaction response times (RTs) during a multisensory task, allows us to distinguish between a race model and a co-activation model (Miller 1982). In the "*race model*" (Raab 1962), neural convergence and interactions are not required; rather, stimuli independently compete for response initiation, and the faster of the two stimuli mediates the behavioral response on any given trial ("*the faster the winner*"). In contrast, the co-activation model hypothesizes that multisensory representations converge and are pooled prior to the initiation of the behavioral response (e.g., Miller et al. 1982, 2001). These models have been applied to analyze the response times to simple stimuli in two groups of CI children, early and late implanted (Gilley et al. 2010). First, the RTs obtained in the early and late implanted children tend to be shorter during AV condition compared to than that obtained during a unisensory condition (A or V). This reduction in both groups is in agreement with the behavioral benefits induced by the processes of multisensory interaction (Stein and Meredith 1993). When the "*race model*" is applied, a converging hypothesis (violation of the model) can be inferred from the RTs obtained in the early implanted group. In contrast, in spite of a reduction in RTs, in the late CI children the co-activation model cannot be applied, suggesting that auditory and visual stimuli are processed independently. Again, these results provide evidence for a sensitive period in multisensory integration during development (Bergeson et al. 2005; Schorr et al. 2005), suggesting that following a long period of deafness in early life, the neuronal network processing visual-auditory information is not fully restored by the CI.

4.3 Synergy Between Vision and Audition in Cochlear Implant Listeners

It has been shown that behavioral response to multisensory stimuli is quicker and more accurate in comparison with their unisensory counterparts (Stein 1998; Welch and Warren 1986). Recently multisensory integration benefit has been extended to perceptual learning (Shams and Kim 2010). Audiovisual training not only increases the rate of learning but also improves perceptual performance (Frassinetti et al. 2005; Seitz et al. 2006; Lippert et al. 2007). Visual information may provide strong positive feedback that facilitates the "decoding" of auditory cues because the primary auditory cortex can retain long-term memory traces about the behavioral significance of sounds (Weinberger 2004). A possible underpinning of this neural feedback may lie in Hebbian mechanisms (Rauschecker 1991; Seitz and Dinse 2007), mediating direct heteromodal connections between sensory areas (Falchier et al. 2002, 2010; Cappe et al. 2009).

A close analysis of the strategies and performances of CI users revealed not only strong synergy between vision and audition, but also a gender, as well as experience effect (Strelnikov et al. 2009). In the beginning of CI experience, female CI users significantly outperformed their male counterparts in speech-reading, but gradually over a 2-year period, the male CI users caught up. The initial female superiority in speech-reading could originate from higher sensitivity to visual speech influence (Aloufy et al. 1996; Irwin et al. 2006) or from a different strategy in processing global speech structures (Kansaku and Kitazawa 2001; Ruytjens et al. 2006). The improved CI male performance might derive from the role of multisensory integration in perceptual learning (Shams and Kim 2010). The progressive recovery of auditory and audiovisual speech comprehension after cochlear implantation constitutes a positive feedback that reinforces visual performances.

The strong synergy between audio and visual modalities also has clinical implications. Because audiovisual training facilitates perceptual learning in a single modality, an audiovisually based rehabilitation strategy during the initial months after implantation would significantly improve and hasten the functional recovery of speech intelligibility (Kawase et al. 2009). Similar multisensory integration training might also be used to improve recovery of other auditory functions such as sound localization (Strelnikov et al. 2011).

4.4 Role of Cued Speech for Speech Comprehension

The benefit of visual information for speech perception is well established, especially in a noisy environment (Sumby and Pollack 1954). However, even for high lip-reading performers, speech can only be perceived at 40 to 60% level without knowledge of the semantic context (Montgomery and Jackson 1983). In 1965, Cornett developed the Cued Speech system (CS) as a complement to lip information. CS is a visual speech communication system that makes use of hand-shapes placed in different positions near the face in combination with the natural speech lip-reading to enhance overall speech perception. Combining lip-reading, CS can produce nearly perfect spoken language understanding (Nicholls and Ling 1982; Uchanski et al. 1994). Moreover, CS offers deaf individuals a thorough representation of the phonological system, making a positive impact on their language development (Leybaert 2000; Hage and Leybaert 2006). CS has been used to facilitate oral communication in deaf and even late implanted children (Torres et al. 2006; Kos et al. 2009).

Tactile devices have also been developed to provide complementary information to lip-reading in deaf patients (reviewed in Galvin et al. 1993). More recently, based on the perceptual benefits of auditory-tactile integration (Wilson et al. 2010a, b), it has been shown that tactile aids combined with cochlear implants can improve speech and music perception in totally deafened individuals (Huang et al. 2009).

5 Functional Cross-Modal Reorganization in Cochlear Implant Users

Psychophysical and neuroimaging studies in both animal and human subjects have demonstrated that sensory deprivation from early developmental stages leads to the functional reorganization of the brain that favors the spared modalities (Rauschecker 1995). Such cross-modal compensatory mechanisms have been described in both congenitally blind (Sadato et al. 1996; Weeks et al. 2000; Röder et al. 2002) and deaf subjects (Nishimura et al. 1999; Petitto et al. 2000; Finney et al. 2003). In congenital deaf subjects, the auditory areas within the temporal cortex can be activated by visual sign language (Nishimura et al. 1999; Petitto et al. 2000) or even simple non-biological visual moving stimuli (Finney et al. 2003).

In CI recipients, PET scan studies have shown a reactivation of the neuronal network involved in language processing (Naito et al. 1995; Giraud et al. 2000; Giraud et al. 2001a, b, c; Green et al. 2005; Mortensen et al. 2006). The language network in CI recipients is complex as several studies have revealed strong involvement in cross-modal compensatory mechanisms. For example, the cochlear implant performance is inversely related to the degree of access in the auditory system to process non-auditory information (Lee et al. 2001). Similarly, speech-reading in deaf individuals involves, to some extent, a similar network of cortical areas in auditory speech processing (Nishitani and Hari 2002; Campbell 2008). However, in normal hearing subjects, the involvement of the primary auditory cortex in lip-reading is still controversial (Calvert et al. 1997; MacSweeney et al. 2000; Bernstein et al. 2002; Pekkola et al. 2005), but these discrepancies among studies probably depend on the experimental protocols involving different levels of phonological or lexical processing. In congenitally deaf subjects, the pattern of brain activation during lip-reading largely overlaps with that described in NH subjects, but it presents a much higher level of activity in the auditory areas (Capek et al. 2008). However, the cortical network that supports visual speech perception in CI users probably differs from that in congenitally deaf and in NH subjects. Rouger et al. (2011) analyzed the pattern of brain activity in CI users performing a speech-reading task at two key time intervals: at the time of cochlear implant activation and at the time of plateau speech performance (between 3 and 11 months post-implantation). In addition to cross-modal activation of the auditory temporal areas (BA 22) and the posterior part of the left superior temporal sulcus (parieto-temporo-occipital junction, or PTO), speech-reading surprisingly activated the voice-sensitive region located in the anterior part of the right superior temporal sulcus (STS, Belin et al. 2000). Interestingly, these cross-modal activations declined during the first year after implantation when the CI users continue to improve their CI speech comprehension. PTO and right STS might be the neural correlates of the highly synergic audiovisual speech integration observed in CI users. During a simple auditory-only word perception, CI users displayed cross-modal activations in low-level visual areas, namely, BA 17/18 (Giraud et al. 2001b). Figure 15.2a shows this activation area, which increased progressively during the initial years after implantation.

There are also significant changes in intrinsic brain activity at rest as a result of cortical re-organization in response to long-term adaptive processes for speech processing (Strelnikov et al. 2010). Without any visual or auditory stimulation, CI subjects showed abnormal changes of cerebral blood flow in the visual and auditory cortex, Broca area, and the posterior temporal cortex. Figure 15.2b shows the increment of activity in these areas from the time of activation of the implant to less than a year after the implantation. The increased activity at rest observed in the left posterior and middle temporal region may be related to the previously reported higher activation of this region in CI users during phonological processing (Giraud et al. 2000; Ito et al. 2004). The left posterior and middle temporal region may merge perceptual and semantic information, and its higher activity in CI users at rest likely reflects adaptive and predictive coding needed to map the impoverished information provided by the implant onto semantic representations. The higher level of activity at rest observed in the visual cortex (inferior occipital gyrus) is probably related to the cross-modal compensation induced by the deafness. The higher activity in visual areas at rest may correspond to the higher capacities for visual and audiovisual integration of speech.

As speech-reading is no longer the unique reliable sensory channel, reorganization of the resting state activity may occur as a result of the progressive reactivation of the auditory speech network after cochlear implantation and the development of new audiovisual speech comprehension strategies by CI users. The presence of a

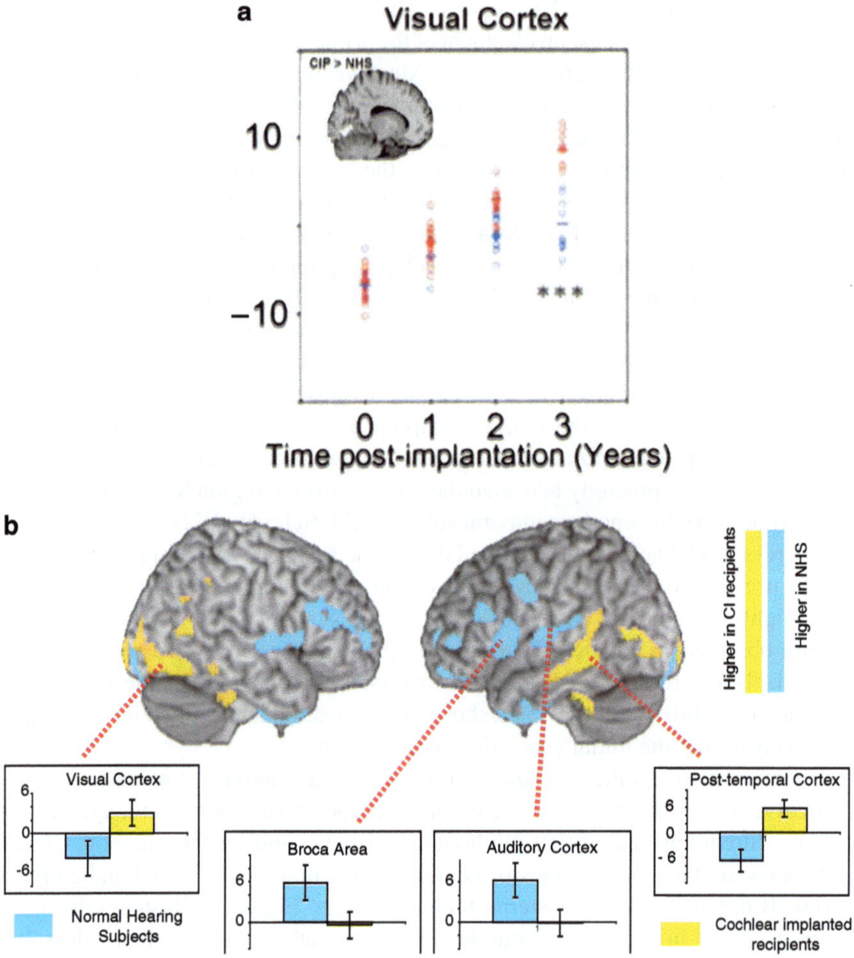

Fig. 15.2 Cross-modal interaction in cochlear implant recipients. (**a**) Visual activation during listening auditory speech. The visual cortex of CI users present a higher level of activity compared to normal listener (inset), and this activity level continues to increase as long as the CI users are using the device. Adapted from Giraud et al. (2001b). (**b**) Differences in regional activity at rest between experienced CI recipients and normal-hearing subjects (for better illustration, an uncorrected $p < 0.00001$ voxel level threshold was used for the figure). Yellow reflects increased activity in CI users, and blue reflects decreased activity in CI users, compared with normal-hearing subjects. In the corresponding activity level plots, yellow bars are for CI recipients and blue bars for normal-hearing subjects (Adapted from Strelnikov et al. 2010)

differential level of activity in CI users at rest may reflect a facilitating mechanism, whereby crucial cortical areas for speech processing are maintained efficiently active. As a result, the crude stimulation by CI can trigger speech-related activity in the brain.

6 Conclusions

There is ample evidence for multisensory interactions in cochlear implant users at both the behavioral and physiological levels. Higher than normal auditory-to-visual and visual-to-auditory cross-modal activations occur after cochlear implantation. This tight cooperation between auditory and visual networks is especially important for the recovery of auditory speech comprehension. The multisensory data are of importance to the development of appropriate therapeutic strategies that maximize speech rehabilitation based on visual and auditory interactions. It is likely that rehabilitation strategies built on visual and audiovisual training will not only improve but also fasten the recovery of auditory speech comprehension in CI users.

References

Aloufy, S., Lapidot, M., & Myslobodskym, M. (1996). Differences in susceptibility to the blending illusion among Native Hebrew and English speakers. *Brain and Language*, 53(1), 51–57.

Arnold, P., & Kopsel, A. (1996). Lipreading, reading and memory of hearing and hearing-impaired children. *Scandinavian Audiology*, 25, 13–20.

Aronson, E., & Rosenbloom, S. (1971). Space perception in early infancy: perception within a common auditory-visual space. *Science*, 172, 1161–1163.

Bavelier, D., & Neville, H. J. (2002). Cross-modal plasticity: where and how? *Nature Review Neuroscience*, 3, 443–452.

Bavelier, D., Dye, M. W., & Hauser, P. C. (2006). Do deaf individuals see better? *Trends in Cognitive Science*, 10, 512–518.

Belin, P., Zatorre, R. J., Lafaille, P., Ahad, P., & Pike, B. (2000). Voice-selective areas in human auditory cortex. *Nature*, 403, 309–312.

Bergeson, T. R., Pisoni, D. B., & Davis, R. A. (2005). Development of audiovisual comprehension skills in prelingually deaf children with cochlear implants. *Ear and Hearing*, 26, 149–164.

Bernstein, L. E., Auer, E. T., Jr., Moore, J. K., Ponton, C. W., Don, M., & Singh, M. (2002). Visual speech perception without primary auditory cortex activation. *Neuroreport*, 13, 311–315.

Bernstein, L. E., Auer, E. T., Jr., & Tucker, P. E. (2001). Enhanced speechreading in deaf adults: can short-term training/practice close the gap for hearing adults? *Journal of Speech, Language, and Hearing Research*, 44, 5–18.

Bernstein, L. E., Demorest, M. E., & Tucker, P. E. (2000). Speech perception without hearing. *Perception & Psychophysics*, 62, 233–252.

Bremmer, F., Schlack, A., Shah, N. J., Zafiris, O., Kubischik, M., Hoffmann, K., et al. (2001). Polymodal motion processing in posterior parietal and premotor cortex: a human fMRI study strongly implies equivalencies between humans and monkeys. *Neuron*, 29, 287–296.

Calvert, G. A., Bullmore, E. T., Brammer, M. J., Campbell, R., Williams, S. C., McGuire, P. K., et al. (1997). Activation of auditory cortex during silent lipreading. *Science*, 276, 593–596.

Calvert, C. A., Brammer, M. J., & Iversen, S. D. (1998). Crossmodal identification. *Trends in Cognitive Science*, 2, 247–253.

Calvert, G. A., Campbell, R., & Brammer, M. J. (2000). Evidence from functional magnetic resonance imaging of crossmodal binding in the human heteromodal cortex. *Current Biology*, 10, 649–657.

Campbell, R. (2008). The processing of audio-visual speech: empirical and neural bases. *Philosophical Transactions of the Royal Society B: Biological Sciences*, 363, 1001–1010.

Capek, C. M., Macsweeney, M., Woll, B., Waters, D., McGuire, P. K., David, A. S., et al. (2008). Cortical circuits for silent speechreading in deaf and hearing people. *Neuropsychologia*, 46, 1233–1241.

Cappe, C., Rouiller, E. M., & Barone, P. (2009). Multisensory anatomical pathways. *Hearing Research*, 258, 28–36.

Champoux, F., Lepore, F., Gagne, J. P., & Theoret, H. (2009). Visual stimuli can impair auditory processing in cochlear implant users. *Neuropsychologia*, 47, 17–22.

Chatterjee, M., & Peng, S. C. (2008). Processing F0 with cochlear implants: Modulation frequency discrimination and speech intonation recognition. *Hearing Research*, 235, 143–156.

Cohen, L. G., Celnik, P., Pascual-Leone, A., Corwell, B., Falz, L., Dambrosia, J., et al. (1997). Functional relevance of cross-modal plasticity in blind humans. *Nature*, 389, 180–183.

Copeland, B. J., & Pillsbury, H. C., 3 rd. (2004). Cochlear implantation for the treatment of deafness. *Annual Review of Medicine*, 55, 157–167.

De Volder, A. G., Bol, A., Blin, J., Robert, A., Arno, P., Grandin, C., et al. (1997). Brain energy metabolism in early blind subjects: neural activity in the visual cortex. *Brain Research*, 750, 235–244.

Deggouj, N., Gersdorff, M., Garin, P., Castelein, S., & Gerard, J. M. (2007). Today's indications for cochlear implantation. *B-Ent*, 3, 9–14.

Desai, S., Stickney, G., & Zeng, F. G. (2008). Auditory-visual speech perception in normal-hearing and cochlear-implant listeners. *Journal of the Acoustical Society of America*, 123, 428–440.

Dodd, B. (1979). Lip reading in infants: attention to speech presented in- and out-of-synchrony. *Cognitive Psychology*, 11, 478–484.

Donnelly, P. J., Guo, B. Z., & Limb, C. J. (2009). Perceptual fusion of polyphonic pitch in cochlear implant users. *Journal of the Acoustical Society of America*, 126, EL128–133.

Doucet, M. E., Bergeron, F., Lassonde, M., Ferron, P., & Lepore, F. (2006). Cross-modal reorganization and speech perception in cochlear implant users. *Brain*, 129, 3376–3383.

Downar, J., Crawley, A. P., Mikulis, D. J., & Davis, K. D. (2000). A multimodal cortical network for the detection of changes in the sensory environment. *Nature Neuroscience*, 3, 277–283.

Drennan, W. R., & Rubinstein, J. T. (2008). Music perception in cochlear implant users and its relationship with psychophysical capabilities. *Journal of Rehabilitation Research and Development*, 45, 779–789.

Falchier, A., Clavagnier, S., Barone, P., & Kennedy, H. (2002). Anatomical evidence of multimodal integration in primate striate cortex. *Journal of Neuroscience*, 22, 5749–5759.

Falchier, A., Schroeder, C. E., Hackett, T. A., Lakatos, P., Nascimento-Silva, S., Ulbert, I., Karmos, G., & Smiley, J. F. (2010). Projection from Visual Areas V2 and Prostriata to Caudal Auditory Cortex in the Monkey. *Cerebral Cortex* 20, 1529–1538.

Finney, E. M., Clementz, B. A., Hickok, G., & Dobkins, K. R. (2003). Visual stimuli activate auditory cortex in deaf subjects: evidence from MEG. *Neuroreport*, 14, 1425–1427.

Foxton, J. M., Riviere, L. D., & Barone, P. (2010). Cross-modal facilitation in speech prosody. *Cognition*, 115, 71–78.

Frassinetti, F., Bolognini, N., Bottari, D., Bonora, A., & Ladavas, E. (2005). Audiovisual integration in patients with visual deficit. *Journal of Cognitive Neuroscience*, 17, 1442–1452.

Friesen, L. M., Shannon, R. V., Baskent, D., & Wang, X. (2001). Speech recognition in noise as a function of the number of spectral channels: comparison of acoustic hearing and cochlear implants. *Journal of the Acoustical Society of America*, 110, 1150–1163.

Fu, Q. J., Shannon, R. V., & Wang, X. (1998). Effects of noise and spectral resolution on vowel and consonant recognition: acoustic and electric hearing. *Journal of the Acoustical Society of America*, 104, 3586–3596.

Galvin, K. L., Cowan, R. S. C., Sarant, J. Z., Blamey, P. J., & Clark, G. M. (1993). Factors in the development of a training-program for use with tactile devices. *Ear and Hearing*, 14, 118–127.

Geers, A., Brenner, C., & Davidson, L. (2003). Factors associated with development of speech perception skills in children implanted by age five. *Ear and Hearing*, 24, 24 S–35 S.

Gilley, P. M., Sharma, A., Mitchell, T. V., & Dorman, M. F. (2010). The influence of a sensitive period for auditory-visual integration in children with cochlear implants. *Restorative Neurology and Neuroscience*, 28, 207–218.

Giraud, A. L., & Lee, H. J. (2007). Predicting cochlear implant outcome from brain organisation in the deaf. *Restorative Neurology and Neuroscience*, 25, 381–390.

Giraud, A. L., Truy, E., Frackowiak, R. S., Gregoire, M. C., Pujol, J. F., & Collet, L. (2000). Differential recruitment of the speech processing system in healthy subjects and rehabilitated cochlear implant patients. *Brain*, 123, 1391–1402.

Giraud, A. L., Price, C. J., Graham, J. M., & Frackowiak, R. S. (2001a). Functional plasticity of language-related brain areas after cochlear implantation. *Brain*, 124, 1307–1316.

Giraud, A. L., Price, C. J., Graham, J. M., Truy, E., & Frackowiak, R. S. (2001b). Cross-modal plasticity underpins language recovery after cochlear implantation. *Neuron*, 30, 657–663.

Giraud, A. L., Truy, E., & Frackowiak, R. (2001c). Imaging plasticity in cochlear implant patients. *Audiology & Neurotology*, 6, 381–393.

Grant, K. W., & Seitz, P. F. (2000). The use of visible speech cues for improving auditory detection of spoken sentences. *Journal of the Acoustical Society of America*, 108, 1197–1208.

Grant, K. W., Walden, B. E., & Seitz, P. F. (1998). Auditory-visual speech recognition by hearing-impaired subjects: consonant recognition, sentence recognition, and auditory-visual integration. *Journal of the Acoustical Society of America*, 103, 2677–2690.

Gray, R. F., Quinn, S. J., Court, I., Vanat, Z., & Baguley, D. M. (1995). Patient performance over eighteen months with the Ineraid intracochlear implant. *Annals of Otology, Rhinology & Laryngology Supplement*, 166, 275–277.

Green, K. M., Julyan, P. J., Hastings, D. L., & Ramsden, R. T. (2005). Auditory cortical activation and speech perception in cochlear implant users: effects of implant experience and duration of deafness. *Hearing Research*, 205, 184–192.

Hadar, U., Steiner, T. J., & Rose, F. C. (1984). Involvement of head movement in speech production and its implications for language pathology. *Advances in Neurology*, 42, 247–261.

Hage, C., & Leybaert, J. (2006). The development of oral language through Cued Speech. In P. Spencer & M. Marschark (Eds.), *The development of spoken language in deaf children* (pp. 193–211). Psychology Press.

Hallum, L. E., Dagnelie, G., Suaning, G. J., & Lovell, N. H. (2007). Simulating auditory and visual sensorineural prostheses: a comparative review. *Journal of Neural Engineering*, 4, S58–71.

Huang, J., Sheffield, B., & Zeng, F. G. (2009). Vibrotactile stimulation enhances cochlear-implant music perception but not speech perception. Paper presented at the 32nd MidWinter Meeting of the Association for Research in Otolaryngology, Baltimore, MD.

Irwin, J. R., Whalen, D. H., & Fowler, C. A. (2006). A sex difference in visual influence on heard speech. *Perception & Psychophysics*, 68(4), 582–592.

Ito, K., Momose, T., Oku, S., Ishimoto, S., Yamasoba, T., Sugasawa, M., & Kaga, K. (2004). Cortical activation shortly after cochlear implantation. *Audiology & Neurotology*, 9, 282–293.

Kaiser, A. R., Kirk, K. I., Lachs, L., & Pisoni, D. B. (2003). Talker and lexical effects on audiovisual word recognition by adults with cochlear implants. *Journal of Speech, Language, and Hearing Research*, 46, 390–404.

Kansaku, K., & Kitazawa, S. (2001). Imaging studies on sex differences in the lateralization of language. *Neuroscience Research*, 41, 333–337.

Kawase, T., Sakamoto, S., Hori, Y., Maki, A., Suzuki, Y., & Kobayashi, T. (2009). Bimodal audio-visual training enhances auditory adaptation process. *Neuroreport*, 20, 1231–1234.

Kim, J., & Davis, C. (2003). Hearing foreign voices: does knowing what is said affect visual-masked-speech detection? *Perception*, 32, 111–120.

Knudsen, E. I. (2004). Sensitive periods in the development of the brain and behavior. *Journal of Cognitive Neuroscience*, 16, 1412–1425.

Kos, M. I., Deriaz, M., Guyot, J. P., & Pelizzone, M. (2009). What can be expected from a late cochlear implantation? *International Journal of Pediatric Otorhinolaryngology*, 73, 189–193.

Kuhl, P. K., & Meltzoff, A. N. (1982). The bimodal perception of speech in infancy. *Science*, 218, 1138–1141.

Kujala, T., Alho, K., & Naatanen, R. (2000). Cross-modal reorganization of human cortical functions. *Trends in Neuroscience*, 23, 115–120.

Lee, D. S., Lee, J. S., Oh, S. H., Kim, S. K., Kim, J. W., Chung, J. K., Lee, M. C., & Kim, C. S. (2001). Cross-modal plasticity and cochlear implants. *Nature*, 409, 149–150.

Lee, H. J., Giraud, A. L., Kang, E., Oh, S. H., Kang, H., Kim, C. S., & Lee, D. S. (2007). Cortical Activity at Rest Predicts Cochlear Implantation Outcome. *Cerebral Cortex*, 17, 909–917.

Lehmann, S., & Murray, M. M. (2005). The role of multisensory memories in unisensory object discrimination. *Cognitive Brain Research*, 24, 326–334.

Leybaert, J. (2000). Phonology acquired through the eyes and spelling in deaf children. *Journal of Experimental Child Psychology*, 75, 291–318.

Lippert, M., Logothetis, N. K., & Kayser, C. (2007). Improvement of visual contrast detection by a simultaneous sound. *Brain Research*, 1173, 102–109.

Lorenzi, C., Gilbert, G., Carn, H., Garnier, S., & Moore, B. C. (2006). Speech perception problems of the hearing impaired reflect inability to use temporal fine structure. *Proceedings of the National Academy of Sciences U S A*, 103, 18866–18869.

Lovelace, C. T., Stein, B. E., & Wallace, M. T. (2003). An irrelevant light enhances auditory detection in humans: a psychophysical analysis of multisensory integration in stimulus detection. *Cognitive Brain Research*, 17, 447–453.

MacKain, K., Studdert-Kennedy, M., Spieker, S., & Stern, D. (1983). Infant intermodal speech perception is a left-hemisphere function. *Science*, 219, 1347–1349.

MacLeod, A., & Summerfield, Q. (1987). Quantifying the contribution of vision to speech perception in noise. *British Journal of Audiology*, 21, 131–141.

MacSweeney, M., Amaro, E., Calvert, G. A., Campbell, R., David, A. S., McGuire, P., Williams, S. C., Woll, B., & Brammer, M. J. (2000). Silent speechreading in the absence of scanner noise: an event-related fMRI study. *Neuroreport*, 11, 1729–1733.

McDonald, J. J., Teder-Salejarvi, W. A., & Hillyard, S. A. (2000). Involuntary orienting to sound improves visual perception. *Nature*, 407, 906–908.

McGurk, H., & MacDonald, J. (1976). Hearing lips and seeing voices. *Nature*, 264, 746–748.

Merabet, L. B., & Pascual-Leone, A. (2010). Neural reorganization following sensory loss: the opportunity of change. *Nature Review Neuroscience*, 11, 44–52.

Miller, J. (1982). Divided attention: evidence for coactivation with redundant signals. *Cognitive Psychology*, 14, 247–279.

Miller, J., Ulrich, R., & Lamarre, Y. (2001). Locus of the redundant-signals effect in bimodal divided attention: a neurophysiological analysis. *Perception & Psychophysics*, 63, 555–562.

Moller, A. R. (2006). Physiological basis for cochlear and auditory brainstem implants. *Advances in Otorhinolaryngology*, 64, 206–223.

Montgomery, A. A., & Jackson, P. L. (1983). Physical characteristics of the lips underlying vowel lipreading performance. *Journal of the Acoustical Society of America*, 73, 2134–2144.

Mortensen, M. V., Mirz, F., & Gjedde, A. (2006). Restored speech comprehension linked to activity in left inferior prefrontal and right temporal cortices in postlingual deafness. *Neuroimage*, 31, 842–852.

Morton, J., & Jassem, W. (1965). Acoustic correlates of stress. *Language and Speech*, 8, 159–181.

Mulford, R. (1988). First words of the blind child. In M. Smith & J. Locke (Eds.), *The emergent lexicon: the child's development of a linguistic vocabulary* (pp. 293–338). New York: Academic Press.

Munhall, K. G., Jones, J. A., Callan, D. E., Kuratate, T., & Vatikiotis-Bateson, E. (2004). Visual prosody and speech intelligibility: head movement improves auditory speech perception. *Psychological Sciences*, 15, 133–137.

Munson, B., & Nelson, P. B. (2005). Phonetic identification in quiet and in noise by listeners with cochlear implants. *Journal of the Acoustical Society of America*, 118, 2607–2617.

Naito, Y., Okazawa, H., Honjo, I., Hirano, S., Takahashi, H., Shiomi, Y., et al. (1995). Cortical activation with sound stimulation in cochlear implant users demonstrated by positron emission tomography. *Cognitive Brain Research*, 2, 207–214.

Nicholls, G. H., & Ling, D. (1982). Cued Speech and the reception of spoken language. *Journal of Speech and Hearing Research*, 25, 262–269.

Nishimura, H., Hashikawa, K., Doi, K., Iwaki, T., Watanabe, Y., Kusuoka, H., Nishimura, T., and Kubo, T. (1999). Sign language 'heard' in the auditory cortex. *Nature*, 397, 116.

Nishimura, H., Doi, K., Iwaki, T., Hashikawa, K., Oku, N., Teratani, T., Hasegawa, T., Watanabe, A., Nishimura, T., & Kubo, T. (2000). Neural plasticity detected in short- and long-term cochlear implant users using PET. *Neuroreport*, 11, 811–815.

Nishitani, N., & Hari, R. (2002). Viewing lip forms: cortical dynamics. *Neuron*, 36, 1211–1220.

Pekkola, J., Ojanen, V., Autti, T., Jaaskelainen, I. P., Mottonen, R., Tarkiainen, A., & Sams, M. (2005). Primary auditory cortex activation by visual speech: an fMRI study at 3 T. *Neuroreport*, 16, 125–128.

Petitto, L. A., Zatorre, R. J., Gauna, K., Nikelski, E. J., Dostie, D., & Evans, A. C. (2000). Speech-like cerebral activity in profoundly deaf people processing signed languages: implications for the neural basis of human language. *Proceedings of the National Academy of Sciences U S A*, 97, 13961–13966.

Pressnitzer, D., Bestel, J., & Fraysse, B. (2005). Music to electric ears: pitch and timbre perception by cochlear implant patients. *Annals of the New York Academy of Sciences*, 1060, 343–345.

Putzar, L., Goerendt, I., Lange, K., Rosler, F., & Roder, B. (2007). Early visual deprivation impairs multisensory interactions in humans. *Nature Neuroscience*, 10, 1243–1245.

Raab, D. H. (1962). Statistical facilitation of simple reaction times. *Transactions of the New York Academy of Sciences*, 24, 574–590.

Rauschecker, J. P. (1991). Mechanisms of visual plasticity: Hebb synapses, NMDA receptors, and beyond. *Physiological Review*, 71, 587–615.

Rauschecker, J. P. (1995). Compensatory plasticity and sensory substitution in the cerebral cortex. *Trends in Neuroscience*, 18, 36–43.

Röder, B., & Rosler, F. (2004). Compensatory plasticity as a consequence of sensory loss. In G. Calvert, C. Spence, & B. E. Stein (Eds.), *The handbook of multisensory processes* (pp. 719–747). Cambridge, MA: MIT Press.

Röder, B., Teder-Salejarvi, W., Sterr, A., Rosler, F., Hillyard, S. A., & Neville, H. J. (1999). Improved auditory spatial tuning in blind humans. *Nature*, 400, 162–166.

Röder, B., Stock, O., Bien, S., Neville, H., & Rosler, F. (2002). Speech processing activates visual cortex in congenitally blind humans. *European Journal of Neuroscience*, 16, 930–936.

Ross, L. A., Saint-Amour, D., Leavitt, V. M., Javitt, D. C., & Foxe, J. J. (2006). Do you see what I am saying? Exploring visual enhancement of speech comprehension in noisy environments. *Cerebral Cortex*, 17, 1147–1153.

Rouger, J., Lagleyre, S., Fraysse, B., Deneve, S., Deguine, O., & Barone, P. (2007). Evidence thant cochlear implanted deaf patients are better multisensory integrators. *Proceedings of the National Academy of Sciences U S A*, 104, 7295–7300.

Rouger, J., Fraysse, B., Deguine, O., & Barone, P. (2008). McGurk effects in cochlear-implanted deaf subjects. *Brain Research*, 1188, 87–99.

Rouger, J., Lagleyre, S., Demonet, J. F., Fraysse, B., Deguine, O., & Barone, P. (2011). Evolution of crossmodal reorganization of the voice area in cochlear-implanted deaf patients. *Human Brain Mapping*, doi:10.1002/hbm.21331.

Ruytjens, L., Albers, F., van Dijk, P., Wit, H., & Willemsen, A. (2006). Neural responses to silent lipreading in normal hearing male and female subjects. *European Journal of Neuroscience*, 24, 1835–1844.

Sadato, N., Pascual-Leone, A., Grafman, J., Ibanez, V., Deiber, M. P., Dold, G., et al. (1996). Activation of the primary visual cortex by Braille reading in blind subjects. *Nature*, 380, 526–528.

Scarborough, R., Keating, P., Mattys, S. L., Cho, T., & Alwan, A. (2009). Optical phonetics and visual perception of lexical and phrasal stress in English. *Language and Speech*, 52, 135–175.

Schorr, E. A., Fox, N. A., van Wassenhove, V., & Knudsen, E. I. (2005). Auditory-visual fusion in speech perception in children with cochlear implants. *Proceedings of the National Academy of Sciences U S A*, 102, 18748–18750.

Seitz, A. R., & Dinse, H. R. (2007). A common framework for perceptual learning. *Current Opinion in Neurobiology*, 17, 148–153.

Seitz, A. R., Kim, R., & Shams, L. (2006). Sound facilitates visual learning. *Current Biology*, 16, 1422–1427.

Shams, L., & Kim, R. (2010) Crossmodal influences on visual perception. *Physics of Life Reviews*, 7, 269–284

Shannon, R. V., Zeng, F. G., Kamath, V., Wygonski, J., & Ekelid, M. (1995). Speech recognition with primarily temporal cues. *Science*, 270, 303–304.

Slutsky, D. A., & Recanzone, G. H. (2001). Temporal and spatial dependency of the ventriloquism effect. *Neuroreport*, 12, 7–10.

Spelke, E. S., & Owsley, C. J. (1979). Intermodal exploration and knowledge in infancy. *Infant Behavior and Development*, 2, 13–27.

Spence, C., & Driver, J. (2000). Attracting attention to the illusory location of a sound: reflexive crossmodal orienting and ventriloquism. *Neuroreport*, 11, 2057–2061.

Staller, S. J., Dowell, R. C., Beiter, A. L., & Brimacombe, J. A. (1991). Perceptual abilities of children with the Nucleus 22-channel cochlear implant. *Ear and Hearing*, 12, 34 S–47 S.

Stein, B. E. (1998). Neural mechanisms for synthesizing sensory information and producing adaptive behaviors. *Experimental Brain Research*, 123, 124–135.

Stein, B. E., & Meredith, M. A. (1993). *The merging of the senses*. Cambridge, MA: MIT Press.

Strelnikov, K., Rouger, J., Lagleyre, S., Fraysse, B., Deguine, O., & Barone, P. (2009). Improvement in speech-reading ability by auditory training: evidence from gender differences in normally hearing, deaf and cochlear implanted subjects. *Neuropsychologia*, 47, 972–979.

Strelnikov, K., Rouger, J., Demonet, J. F., Lagleyre, S., Fraysse, B., Deguine, O., & Barone, P. (2010). Does brain activity at rest reflect adaptive strategies? Evidence from speech processing after cochlear implantation. *Cerebral Cortex*, 20, 1217–1222.

Strelnikov, K., Rosito, M., & Barone, P. (2011). Effect of audiovisual training on monaural spatial hearing in horizontal plane. *PLoS ONE*, 6(3), e18344.

Sumby, W. H., & Pollack, I. (1954). Visual contribution to speech intelligibility in noise. *Journal of the Acoustical Society of America*, 26, 212–215.

Summerfield, Q. (1979). Use of visual information for phonetic perception. *Phonetica*, 36, 314–331.

Summerfield, Q. (1992). Lipreading and audio-visual speech perception. *Philosophical Transactions of the Royal Society of London. Series B, Biological Sciences*, 335, 71–78.

Torres, S., Moreno-Torres, I., & Santana, R. (2006). Quantitative and qualitative evaluation of linguistic input support to a prelingually deaf child with cued speech: a case study. *Journal of Deaf Studies and Deaf Education*, 11, 438–448.

Tyler, R. S., Fryauf-Bertschy, H., Kelsay, D. M., Gantz, B. J., Woodworth, G. P., & Parkinson, A. (1997). Speech perception by prelingually deaf children using cochlear implants. *Otolaryngology-Head and Neck Surgery*, 117, 180–187.

Tyler, R. S., Parkinson, A. J., Woodworth, G. G., Lowder, M. W., & Gantz, B. J. (1997). Performance over time of adult patients using the Ineraid or nucleus cochlear implant. *Journal of the Acoustical Society of America*, 102, 508–522.

Uchanski, R. M., Delhorne, L. A., Dix, A. K., Braida, L. D., Reed, C. M., & Durlach, N. I. (1994). Automatic speech recognition to aid the hearing impaired: prospects for the automatic generation of cued speech. *Journal of Rehabilitation Research and Development*, 31, 20–41.

Veraart, C., De Volder, A. G., Wanet-Defalque, M. C., Bol, A., Michel, C., & Goffinet, A. M. (1990). Glucose utilization in human visual cortex is abnormally elevated in blindness of early onset but decreased in blindness of late onset. *Brain Research*, 510, 115–121.

Vroomen, J., & de Gelder, B. (2000). Sound enhances visual perception: cross-modal effects of auditory organization on vision. *Journal of Experimental Psychology. Human Perception and Performance*, 26, 1583–1590.

Wanet-Defalque, M. C., Veraart, C., De Volder, A., Metz, R., Michel, C., Dooms, G., et al. (1988). High metabolic activity in the visual cortex of early blind human subjects. *Brain Research*, 446, 369–373.

Weeks, R., Horwitz, B., Aziz-Sultan, A., Tian, B., Wessinger, C. M., Cohen, L. G., Hallett, M., & Rauschecker, J. P. (2000). A positron emission tomographic study of auditory localization in the congenitally blind. *Journal of Neuroscience*, 20, 2664–2672.

Weinberger, N. M. (2004). Specific long-term memory traces in primary auditory cortex. *Nature Review Neuroscience*, 5, 279–290.

Welch, R. B., & Warren, D. H. (1986). Intersensory interactions. In K. R. Boff, L. Kaufman, & J. P. Thomas (Eds.), *Handbook of perception and human performance* (Vol. 1, pp. 1–36). New York: Wiley.

Wilson, E. C., Braida, L. D., & Reed, C. M. (2010a). Perceptual interactions in the loudness of combined auditory and vibrotactile stimuli. *Journal of the Acoustical Society of America*, 127, 3038–3043.

Wilson, E. C., Reed, C. M., & Braida, L. D. (2010b). Integration of auditory and vibrotactile stimuli: Effects of frequency. *Journal of the Acoustical Society of America*, 127, 3044–3059.

Zeng, F. G., Nie, K., Stickney, G. S., Kong, Y. Y., Vongphoe, M., Bhargave, A., Wei, C., & Cao, K. (2005). Speech recognition with amplitude and frequency modulations. *Proceedings of the National Academy of Sciences U S A*, 102, 2293–2298.

Index

F.-G. Zeng et al. (eds.), *Auditory Prostheses: New Horizons*, Springer Handbook of Auditory Research 39, DOI 10.1007/978-1-4419-9434-9,

MIX
Papier aus verantwortungsvollen Quellen
Paper from responsible sources
FSC® C105338

If you have any concerns about our products,
you can contact us on
ProductSafety@springernature.com

In case Publisher is established outside the EU,
the EU authorized representative is:
Springer Nature Customer Service Center GmbH
Europaplatz 3, 69115 Heidelberg, Germany

Printed by Libri Plureos GmbH
in Hamburg, Germany